AF248658

A MASTER OF SCIENCE HISTORY

Archimedes

NEW STUDIES IN THE HISTORY AND PHILOSOPHY OF SCIENCE AND TECHNOLOGY

VOLUME 30

Archimedes has three fundamental goals; to further the integration of the histories of science and technology with one another: to investigate the technical, social and practical histories of specific developments in science and technology; and finally, where possible and desirable, to bring the histories of science and technology into closer contact with the philosophy of science. To these ends, each volume will have its own theme and title and will be planned by one or more members of the Advisory Board in consultation with the editor. Although the volumes have specific themes, the series itself will not be limited to one or even to a few particular areas. Its subjects include any of the sciences, ranging from biology through physics, all aspects of technology, broadly construed, as well as historically-engaged philosophy of science or technology. Taken as a whole, *Archimedes* will be of interest to historians, philosophers, and scientists, as well as to those in business and industry who seek to understand how science and industry have come to be so strongly linked.

For further volumes:
http://www.springer.com/series/5644

Jed Z. Buchwald

Editor

A Master of Science History

Essays in Honor of Charles Coulston Gillispie

 Springer

Editor
Prof. Dr. Jed Z. Buchwald
California Institute of Technology
Division of Humanities & Social Sciences
E. California Blvd. 1200
91125 Pasadena California
USA
buchwald@its.caltech.edu

ISSN 1385-0180
ISBN 978-94-007-2626-0   e-ISBN 978-94-007-2627-7
DOI 10.1007/978-94-007-2627-7
Springer Dordrecht Heidelberg London New York

Library of Congress Control Number: 2011944095

Springer is part of Springer Science+Business Media (www.springer.com)

Charles and Emily Gillispie. Photo by Ivor Grattan-Guinness

Born in 1918, Charles Gillispie graduated from Wesleyan University in 1940 and, following service during the Second World War, received his PhD from Harvard in 1949. He thereafter joined the Department of History at Princeton University, where he remained until his retirement as Dayton-Stockton Professor of History, Emeritus in 1987. He has been awarded many honors, including Princeton's Behrman Prize in the Humanities (1981), the Pfizer Prize (1981) and Sarton Medal (1984) of the History of Science Society, la Médaille Alexandre Koyré of the Académie Internationale d'Histoire des Sciences (1985), the Distinguished Alumnus Award of Wesleyan University (1990), the Dibner Award for Distinction in History of Science and Technology (1994), and the Balzan Prize for History and Philosophy of Science (1997). In 2011 he received the D.H.S., Honoris Causa, from Princeton University. Gillispie's work has ranged widely over science and technology history. He is among the founding members of the discipline whose publications and teaching having influenced generations of historians and readers. We are honored to present these essays by his students and colleagues which were completed on August 6, 2011, the day of his 93rd birthday.

Contents

Contributors

Marco Beretta University of Bologna, Bologna, Italy, marco.beretta@unibo.it

Patrice Bret Centre Alexandre Koyré – Centre de recherches en histoire des sciences et des techniques, Paris, France, patrice.bret@yahoo.fr

Jed Z. Buchwald Division of Humanities and Social Sciences, California Institute of Technology, Pasadena, CA, USA, buchwald@its.caltech.edu

Robert Fox Museum of the History of Science, University of Oxford, Oxford, UK, robert.fox@history.ox.ac.uk

Tore Frängsmyr Uppsala University, Uppsala, Sweden, Tore.Frangsmyr@idehist.uu.se

Charles Coulston Gillispie Department of History, Princeton University, Princeton, NJ, USA, gillispie@Princeton.edu

Michael D. Gordin Department of History, Princeton University, Princeton, NJ, USA, mgordin@princeton.edu

Ivor Grattan-Guinness Middlesex University Business School, London, UK, ivor2@MDX.AC.UK

Yung Sik Kim Department of Asian History and Program in History and Philosophy of Science, Seoul National University, Seoul, South Korea, kysik@snu.ac.kr

Knut Kleve University of Oslo, Oslo, Norway, knut.kleve@ifikk.uio.no

Eberhard Knobloch Technische Universität, Berlin, Germany, eberhard.knobloch@tu-berlin.de

Manfred D. Laubichler Center for Biology and Society, Arizona State University, Tempe, AZ 85287-4501, USA; Marine Biological Laboratory, Woods Hole, MA 02543, USA, Manfred.Laubichler@asu.edu

John E. Lesch University of California, Berkeley, CA, USA, jlesch@calmail.berkeley.edu

Jane Maienschein Center for Biology and Society, Arizona State University, Tempe, AZ 85287-4501, USA; Marine Biological Laboratory, Woods Hole, MA 02543, USA, Maienschein@asu.edu

Seymour H. Mauskopf Professor Emeritus, Department of History, Duke University, Box 90719, Durham, North Carolina 27708-0719, USA, shmaus@duke.edu

Evan M. Melhado Department of History and Medical Humanities and Social Sciences Program, University of Illinois at Urbana-Champaign, Urbana, IL 61801, USA, melhado@uiuc.edu

David Oldroyd Department of History and Philosophy of Science, University of New South Wales, Sydney, NSW 2052, Australia, doldroyd@bigpond.com

Theodore M. Porter Department of History, University of California, Los Angeles, CA, USA, tporter@history.ucla.edu

Roshdi Rashed Univ Paris Diderot, Sorbonne Paris Cité, Laboratoire SPHERE, UMR 7219, CNRS, F-75013 Paris, France, rashed@paris7.jussieu.fr

Nicolaas A. Rupke Georg-August University Göttingen, Göttingen, Germany, nrupke@gwdg.de

Nathan Sivin History and Sociology of Science, University of Pennsylvania, Philadelphia, PA 19104-6304, USA, nsivin@sas.upenn.edu

N.M. Swerdlow Division of Humanities and Social Sciences, California Institute of Technology, Pasadena, CA, USA, swerdlow@caltech.edu

David B. Wilson Department of History, Iowa State University, Ames, IA 50011, USA, davidw@iastate.edu

About the Editor

Jed Z. Buchwald is Doris and Henry Dreyfuss Professor of History at Caltech, where he teaches courses in ancient civilization, religion and in the history of physics. His most recent books are *The Zodiac of Paris* (Princeton, 2010), with Diane Greco Josefowicz and *Isaac Newton and the Origin of Civilization*, with Mordechai Feingold (Princeton, 2012). He is co-editor with Jeremy Gray of *Archive for History of Exact Sciences* (Springer), of *Sources and Studies in the History of Mathematics and the Physical Sciences* (Springer), and editor of *Archimedes* (Kluwer) and of *Transformations* (MIT).

About the Authors

Marco Beretta is Professor of history of science at the University of Bologna and Vice. director of the Museo Galileo in Florence. Since 2004 he is editor of *Nuncius, Journal of the Material and Visual History of Science*. He has published extensively on the history of chemistry and on Lavoisier. His latest book is *The Alchemy of Glass. Counterfeit, Imitation and Transmutation in Ancient Glassmaking* (Science History Publications/USA, 2009).

Patrice Bret is Chargé de recherche at the Centre Alexandre Koyré - Centre d'Histoire des Sciences et des Techniques (CAK-CRHST), and responsible for the section Histoire de l'armement at the Institut de Recherche Stratégique de l'École militaire (IRSEM), Paris. Among other works, he has published *L'État, l'armée, la science. L'invention de la recherche publique en France, 1763–1830* (2002, Rennes, Presses universitaires de Rennes), and *L'Égypte au temps de l'expédition de Bonaparte (1798–1801)* (1998, Paris, Hachette Littératures).

Robert Fox is Emeritus Professor of the History of Science at the University of Oxford. His main research interests are in the history of physical sciences since the eighteenth century, with special reference to France. His book *The savant and the state. Science and cultural politics in nineteenth-century France* is currently in press with the Johns Hopkins University Press.

Tore Frängsmyr is Hans Rausing Professor Emeritus in History of Science at Uppsala University. He has published several books on the history of geology during the 18th and 19th century, and also on the Enlightenment. In 2000 he published a two volume work on *Swedish History of Ideas* during a thousand years. For twenty years he was editor of *Les Prix Nobel*, the yearbook of the Nobel Foundation.

Michael D. Gordin is Professor of History at Princeton University, where he teaches the history of modern science. He is the author of *A Well-Ordered Thing: Dmitrii Mendeleev and the Shadow of the Periodic Table* (Basic Books, 2004), *Five Days in August: How World War II Became a Nuclear War* (Princeton University Press, 2007), and *Red Cloud at Dawn: Truman, Stalin,* *and the End of the Atomic Monopoly* (Farrar, Straus & Giroux, 2009). He is also the co-editor of six volumes of essays on history and the history of science.

Ivor Grattan-Guinness is Emeritus Professor of the History of Mathematics and Logic at Middlesex University, England. Editor of the history of science journal *Annals of science* from 1974 to 1981, in 1979 he also founded the journal *History and philosophy of logic*, editing it until 1992. He edited a *Companion Encyclopedia of the History and Philosophy of the Mathematical Sciences* (two volumes, 1994), and Landmark writings in Western mathematics, 1640–1940 (2005). His individual books include *Convolutions in French Mathematics, 1800–1840* (three volumes, 1990); *The Norton History of the Mathematical Sciences* (1997); *The Search for Mathematical Routes, 1870–1940* (2000); and collections of articles in *Routes of Learning* (2009) and *Corroborations and Criticisms* (2010). In July 2009 the International Commission for the History of Mathematics awarded him the Kenneth O. May Medal and Prize in the History of Mathematics for his contributions to the field.

Yung Sik Kim is Professor in the Department of Asian History and the Program in History and Philosophy of Science, Seoul National University. He received his PhD from Princeton University in 1980 with Charles Gillispie as supervisor. His research interests are science and Confucianism, scholars and specialized knowledge in East Asia, and comparative history of science.

Knut Kleve born 1926 in Oslo (Norway), professor of Classics in Bergen 1963–1973 and in Oslo 1974–1996, leader of the Herculaneum papyri opening team in Naples (Italy) 1984–2003, knighted by Berlusconi 2005.

Eberhard Knobloch born in 1943, studied mathematics, classical philology, and history of science and technology; in 1972 PhD in history of science and technology, in 1976 habilitation; since 2002 university professor of history of science and technology at the Technical University of Berlin and Academy Professor at the Berlin-Brandenburg Academy of Sciences and Humanities (former Prussian Academy of Sciences), since 2009 Professor Emeritus. Member of several national and international Academies of Sciences, President of the Inter-national Academy of the History of Science, former president of the European Society for the History of Science; about 350 publications, especially on the history and philosophy of mathematical sciences and Renaissance technology.

Manfred D. Laubichler is Professor of Theoretical Biology and History of Biology at Arizona State University, where he directs the Center for Social Dynamics and Complexity. He is an Adjunct Professor at the Marine Biology Laboratory in Woods Hole, Massachusetts, an External Faculty Member at the Konrad Lorenz Institute for Evolution and Cognition Research in Altenberg, Austria and a Visiting Scholar at the Max Planck Institute for the History of Science (Department 1) in Berlin. His research is on developmental evolution, the evolution of social complexity and the history and philosophy of evolutionary and developmental biology. Most recently, he and Jane Maienschein co-edited *Form and Function in Developmental Evolution*, Cambridge University Press, 2009.

John E. Lesch is Professor Emeritus of History at the University of California, Berkeley, and Distinguished Scholar in Residence at Rutgers University. His publications include *Science and Medicine in France: The Emergence of Experimental Physiology 1790–1855*; *The First Miracle Drugs: How the Sulfa Drugs Transformed Medicine*; and (as editor), *The German Chemical Industry in the Twentieth Century*.

Jane Maienschein is Regents' Professor, President's Professor, and Parents Association Professor at Arizona State University, where she also directs the Center for Biology and Society. At the Marine Biological Laboratory in Woods Hole, Massachusetts, she is Adjunct Professor and Director of the HPS Program. Her scholarship focuses on the history and philosophy of developmental biology, in particular on epistemological issues, as demonstrated in her 3 books, dozen (co)edited books, and many articles. Most recently, she and Manfred Laubichler co-edited *Form and Function in Developmental Evolution*, Cambridge University Press, 2009.

Seymour H. Mauskopf is Professor Emeritus of History at Duke University. His fields of research interest are the history of chemistry (*Crystals and Compounds*, 1976; *Chemical Sciences in the Modern World*, 1993) and the history of marginal science (parapsychology) (*The Elusive Science*, with Michael R. McVaugh, 1980). Currently, he is working on a book on Alfred Nobel's interactions with British munitions scientists in the late nineteenth century. He is co-editor of *Integrating History and Philosophy of Science* (with Tad Schmaltz), Boston Studies in the Philosophy of Science (2012) and *Chemical Knowledge in the Early Modern World* (with William Newman and Matthew Eddy), Osiris (2014). In 1998, he received the Dexter Award for Outstanding Contributions to the History of Chemistry from the American Chemical Society. He taught history of science at Duke University from 1964 through 2010.

Evan M. Melhado holds an appointment in the Department of History at the University of Illinois at Urbana-Champaign (UIUC) and serves as Head of the Medical Humanities and Social Sciences Program at the University of Illinois College of Medicine at Urbana-Champaign (UICOM-UC). Charles Gillispie was among his professors in the late 1960s and early 1970s at Princeton University, in what was then called the Program in History and Philosophy of Science. For some years, Melhado specialized in the history of the physical sciences, especially chemistry, in the eighteenth and early nineteenth centuries; later he turned to recent American health-care policy. In History, he has taught courses in the history of science, of medicine, and of health policy. With colleagues in the College of Medicine, he helped create and sustain a pioneering clerkship in "Medicine and Society," which emphasizes the economic, political and social environment for health care, as well as the social and cultural dimensions of clinical practice. He has worked to foster interest on his campus in teaching and research in the medical humanities and social sciences, particularly in connection with the University of Illinois Medical Scholars Program, an MD/PhD program conducted collaboratively by the university (UIUC) and the medical school (UICOM-UC).

David Oldroyd is an Honorary Professor in the School of History and Philosophy of Science at the University of New South Wales, from which institution he retired in 1996, having been departmental head there. He started his career as a chemistry teacher in London, and took the evening-class MSc program in HPS at University College. He was later employed in two schools in New Zealand before gaining a lectureship at UNSW in 1969. His early studies in history of science were at the interface of mineralogy and chemistry and most of his more important publications have been in history of geology. Since 1996 he has devoted himself exclusively to work in the history of geology and was Secretary General of the International Commission for the History of Geological Sciences for eight years. Presently he is editor of *Earth Sciences History*. His best-known works are *Darwinian Impacts* (1980), *The Arch of Knowledge* (1986), *The Highlands Controversy* (1990) and *Thinking about the Earth* (1996). There is a Variorum edition of some of his earlier papers (1998) and he has edited or co-edited several collections of articles. Translations of his books have appeared variously in Spanish, Italian, German, Turkish, and Chinese.

Theodore M. Porter is Professor of History at UCLA. His most recent books are *The Cambridge History of Science, volume 7: Modern Social Sciences*, coedited with Dorothy Ross (Cambridge University Press, 2003), and *Karl Pearson: The Scientific Life in a Statistical Age* (Princeton University Press, 2004). In an earlier era he wrote *The Rise of Statistical Thinking* (1986) and *Trust in Numbers* (1995). Just now he is exploring the uses of and statistical recording practices and field work in Europe and North America from about 1820 to 1920 to investigate heredity at insane asylums and schools for the "feeble-minded." On the side, he has some recent papers on issues of science and public reason, such as "How Science Became Technical" in ISIS (2009).

Roshdi Rashed is Emeritus Research Director at the NATIONAL CENTER FOR SCIENTIFIC RESEARCH (CNRS – France), and Honorary Professor at the University of Tokyo. His recent publications include *Les Mathématiques infinitésimales du IXe au XIe* siècle, 5 vols (London, al-Furqan Islamic Heritage Foundation, 1993–2006); *Geometry and Dioptrics in Classical Islam* (London, al-Furqan, 2005); *Al-Khwārizmī: The Beginnings of Algebra*, "History of Science and Philosophy in Classical Islam" (London, Saqi, 2009); *Apollonius: Les Coniques* (the seven books), 5 vols (Berlin/New York, Walter de Gruyter,

2008–2010); *D'al-Khwārizmī à Descartes. Études sur l'histoire des mathématiques classiques* (Paris, Hermann, 2011).

Nicolaas A. Rupke is Lower Saxony Research Professor of the History of Science, Georg-August University Göttingen, Germany. Educated at Groningen (B.Sc., 1968) and Princeton (Ph.D., 1972), he held research fellowships at the Smithsonian Institution, Oxford University, Tübingen University, the Netherlands Institute for Advanced Studies, the Wellcome Institute for the History of Medicine, the National Humanities Center, and the Institute of Advanced Studies in Canberra. Rupke was the inaugural holder of the Nelson O. Tyrone Jr. Chair of the History of Medicine at Vanderbilt. He is the author of several scientific biographies, including a study of the Oxford geologist William Buckland, the London biologist Richard Owen and the German scientist and explorer Alexander von Humboldt. Rupke currently works on leading figures in the structuralist tradition of evolutionary biology, starting with Johann Friedrich Blumenbach. He is a fellow of the German Academy of Sciences Leopoldina and the Göttingen Academy of Sciences.

Nathan Sivin is Professor of Chinese Culture and of the History of Science, Emeritus, at the University of Pennsylvania. He is also a fellow of the American Academy of Arts and Sciences, and an honorary professor in the Chinese Academy of Sciences. His most recent book, *Granting the Seasons,* is a study of the high point of Chinese mathematical astronomy in the late 13th century. He is now working on a study of the varieties of health care in China in the 11th century.

N.M. Swerdlow is Professor Emeritus in the Department of Astronomy and Astrophysics and the Department of History at The University of Chicago. He is currently Visiting Associate in the Division of Humanities and Social Sciences at the California Institute of Technology. His research is in the history of the exact sciences, principally mathematical astronomy, from antiquity through the seventeenth century. Among his publications are, with O. Neugebauer, *Mathematical Astronomy in Copernicus's De Revolutionibus* (1984), and *The Babylonian Theory of the Planets* (1998). He is completing a study of Renaissance astronomy, on the work of Regiomontanus, Copernicus, Tycho, Kepler, and Galileo.

David B. Wilson is a member of the History Department at Iowa State University. His research has concentrated on 18th- and 19th-century British physics, including its philosophical, theological, and institutional contexts. His *Seeking Nature's Logic: Natural Philosophy in the Scottish Enlightenment* was published in 2009, and in 2010 he was the Eleanor Searle Visiting Professor of the History of Science at Caltech and the Huntington Library. He is currently doing research on William Whewell (1794–1866), historian and philosopher of science and inventor of the word *scientist*.

Part I
Charles Gillispie

Chapter 1
Publications by Charles Coulston Gillispie

1.1 Books and Separate Publications

1. *Genesis and Geology: A Study in the Relations of Scientific Thought, Natural Theology, and Social Opinion in Great Britain, 1790–1850.* Harvard Historical Studies, Vol. LVIII. Harvard University Press, 1951, xv + 315 pp., 2nd printing, 1969; Harper Torch Book Edition, 1959. New edition, with forward by Nicolaas Rupke and new preface, Harvard University Press, 1996. Chinese translation, 1999.
2. *A Diderot Pictorial Encyclopedia of Trades and Industry: Manufacturing and the Technical Arts in Plates Selected from L'Encyclopèdie of Denis Diderot.* Dover Publications, New York, 1959, 2 volumes; 485 plates; xxx + 920 pp.
3. *The Edge of Objectivity: An Essay in the History of Scientific Ideas.* Princeton University Press, 1960, 562 pp. Oxford University Press edition, 1960. Translations: Japanese, 1966; Greek, 1975; Korean, 1981; Italian, II Mulino, Bologna, 1981; Rumanian, 2010. Reissued with a new preface, Princeton University Press, 1990.
4. *Les Fondements intellectuels de l'introduction des probabilites en physique.* Palais de la découverte, Paris, 1963, 27 pp.
5. Editor, *Dictionary of Scientific Biography*, Scribners, New York, 1970–1980, 16 vols.
6. *Lazare Carnot, Savant.* With an essay by A.P. Youschkevitch. Princeton University Press, 1971, xiii + 359 pp. French translation, Vrin, Paris, 1976.
7. *Science and Polity in France at the End of the Old Regime*, Princeton University Press, xii + 602 pp. 1980. Italian translation, II Mulino, Bologna, 1983. 2nd paperback printing as *Science and Polity in France, the End of the Old Regime*. 2004. To accompany 13, below.
8. *The Montgolfier Brothers and the Invention of Aviation, 1783–84, with a word on the importance of ballooning for the science of heat and the art of building railroads.* Princeton University Press, xiv + 212 pp., 11 plates and 70 illustrations (June 1983). French translation, Actes Sud, 1989. France (1770–1830).
9. *The Professionalization of Science: Kyoto, 1983. 40 pp. compared to the United States (1910–1970)*, Doshisha University Press. Reprinted in A14.

J.Z. Buchwald (ed.), *A Master of Science History*, Archimedes 30,
DOI 10.1007/978-94-007-2627-7_1, © Springer Science+Business Media B.V. 2012

10. *The Princeton Mathematics Community in the 1930s. An Oral History Project.* Administrator. Interviews with Albert W. Tucker et al., Interviewers, A.W. Tucker, William Aspray. Ed. Frederic Nebeker. Princeton, 1985. Trustees of Princeton University.

11. *Monuments of Egypt, the Napoleonic Edition; The Complete Archaeological Plates from "La Description de l'Égypte.* Edited with Introduction and Notes by Charles Coulston Gillispie and Michel Dewachter. New York: Princeton Architectural Press, 1987. xxx+47 pp. (426 plates. Map). 2nd ed., 2 vols. boxed (1988), French translation, *Monuments d'Égypte*, 2 vols. boxed (Paris Editions, Hazan: 1988). Italian translation, 1990. 3rd Printing, 1991; 4th Printing, 1994.

12. *Pierre-Simon Laplace, 1749–1827, A Life in Exact Science.* Princeton University Press, 1997. xii+323 pp. Contributions by Ivor Grattan-Guinness and Robert Fox. Revision of C-7. Paperback ed., 2000.

13. *Science and Polity in France, the Revolutionary and Napoleonic Years.* Princeton University Press. 2004. lx + 751 pp. Illustrated.

14. Essays and Reviews in History and History of Science. *Transactions of the American Philosophical Society*, Vol. 96, Part 5. Philadelphia, 2007.

15. *Lazare and Sadi Carnot, A Scientific and Filial Relationship.* Charles C. Gillispie and Raffaele Pisano. In Preparation for Springer, scheduled for 2011.

1.2 Contributions to Collective Volumes

1. "English Ideas of the University in the Nineteenth Century," Chapter II in *The Modern University*, Margaret Clapp, ed. Cornell University Press, Ithaca, NY, 1950, pp. 25–55. Reprinted in A14.

2. "Fontenelle and Newton," in *Isaac Newton's Papers and Letters on Natural Philosophy*, I. Bernard Cohen, ed. Harvard University Press, Cambridge, 1958, pp. 425–474. 2nd ed. 1982.

3. "Lamarck and Darwin in the History of Science," Chapter X in *The Forerunners of Darwin*, Bentley Glass, ed. The Johns Hopkins Press, Baltimore, 1959, pp. 265–291.

4. "The *Encyclopédie* and the Jacobin Philosophy of Science," Chapter IX in *Critical Problems in the History of Science*, Marshall Clagett, ed., University of Wisconsin Press, Madison, 1959, pp. 255–290. Reprinted in A14.

5. "Intellectual Factors in the Background of Analysis by Probabilities," in *Scientific Change*, A.C. Crombie, ed. Heinemann, London, 1963, pp. 431–453, 499–502. Reprinted in A14.

6. "Commentary on Social or Behavioral Sciences," in *Science in the College Curriculum*, Robert Hoopes, ed. Oakland University, Rochester, Michigan, 1963, pp. 93–96.

7. "Elements of Physical Idealism," *Aventures de l'esprit*, Vol. I of *Mélanges Alexandre Koyré*. Editions Hermann, Cambridge University Press, Paris, 1964, pp. 206–224.

8. "Science and Technology," Chapter V of *War and Peace in an Age of Upheaval, 1793–1830; New Cambridge Modern History*, Indiana University Press, Cambridge, 1965, Volume IX (1793–1830), pp. 118–145.

9. "Charles Darwin," *International Encyclopedia of Social Sciences* (1968), IV, pp. 7–14.

10. "Science and the Literary Imagination: Voltaire and Goethe," in *The Literature of the Western World*, David Daiches and Anthony Thorlby, ed. Aldus, London, 1975. Vol. IV, Chapter 6, pp. 167–194. Reprinted in A14.

11. "The Scientific Work of Lazare Carnot and its' Influence on that of his Son," in *Sadi Carnot et l'essor de la Thermodynamique,* École Polytechnique, Editions du CNRS, Paris, 1976, pp. 23–34.

12. "The liberating influence of science in history," in *Aspects of American Liberty, Philosophical, Historical and Political*. Memoirs of the American Philosophical Society, Vol. 118, Philadelphia, 1977, pp. 37–46.

13. "Scientific Theories and Social Values," in *Science, An American Bicentennial View*. National Academy of Sciences, Washington, 1977, pp. 13–19.

14. "L'enseignement de l'histoire des sciences aux États-Unis," in *Actes du Collogue sur l'enseignement de l'histoire des sciences aux scientifiques,* Jean Dhombres, ed. Université de Nantes, Nantes, 1981, pp. 20–24. First delivered as a lecture, October 8, 1980.

15. "The Invention of Aviation," in *The Balloon: A Bicentennial Exhibition*, University Art Museum, University of Minnesota, Minneapolis, 1983, pp. 20–33.

16. "Science et Société: Le Cas de Laplace et des Probabilités," in *Sciences en Revolution, 1770–1830*. Centre interdisciplinaire d'étude de l'évolution des ideas, des sciences, et des techniques. Université de Paris-Sud. Centre Scientifique d'Orsay, 1983.

17. "Préface," pp. vii–xiv in Goulven Laurent, Paléontologie et évolution en France, 1800–1860: Une histoire des idées de Cuvier et Lamarck à Darwin. Paris, Editions du Comité des Travaux Historiques et Scientifiques, Mémoires de la Section d'Histoire des Sciences et Techniques, No. 4, 1987.

18. "Postface," pp. 281–284, in Maurice Daumas, *Arago, 1786–1853, la jeunesse de la science*. Nouvelle édition, Collection "Un Savant, Une Époque," dirigé par Jean Dhombres (Belin), Paris, 1987.

19. Henry Laurens, Charles C. Gillispie, Jean-Claude Golvin, Claude Traunecker, *L'Expedition d'Egypte (1798–1801)*. Editions Armand Colin, Paris, 1989. Chapter 11, "Aspects Scientifiques." Arabic translation, 1996.

20. "Science," in *Academic Press Dictionary of Science and Technology*. Harcourt, Brace, Jovanovich, San Diego, 1992, p. 1926.

21. Commentaire, *Marat homme de science?*, in *Collection: Les empecheurs de penser en rond*, Jean Bernard, Jean-Francois Lemaire, Jean-Pierre Poirier, ed. Synthelabo, Paris, 1993, pp. 151–154.

22. "Un enseignement hégémonique: les mathématiques," in *La formation polytechnicienne, 1797–1994*, B. Belhoste, A. Dahan Dalmedico, A. Picon. ed. Dunod, Paris, 1994, pp. 31–44.

23. Préface, Marcel Reinhard, *Le Grand Carnot*. 2nd ed., 2 Vols. Hachette, Paris, 1994.

24. "Palisot de Beauvois et les Américains," in *Nature,* histoire, société; Essals en hommage a Jacques Roger. C. Blanckaert, J.-L. Fischer, R. Rey, eds. Klinckseick, 1995, pp. 371–389. French original of D45.

25. Preface, Jean-Pierre Poirier, *Lavoisier: Chemist, Biologist, Economist.* University of Pennsylvania Press, 1996.

26. "De l'histoire naturelle à la biologie: Relations entre les programmes de recherche de Cuvier, Lamarck, et Geoffroy Saint-Hilaire," *Le Muséum au premier siècle de son histoire*, pp. 229–239. Éditions du Muséum national d'Histoire Naturelle. Paris, 1997.

27. Préface. *Le Fonds d'Archives* Seguin: Aux origines de la revolution industrielle en France, 1790–1820, Michel Cotte, ed. Archives departementale de l'Ardèche, Privas, 1997.

28. Table ronde, in colloquium "Sciences, Mythes, et Réligions en Europe," dir. Dominique Lecourt, Euroscientia conferences, Royaumont, 14–15 October, 1997, pp. 179–182 and following.

29. "Révolution française et science," in *Dictionnaire critique de la science classique*. Michel Blay et Robert Halleux, ed., Flammarion, 1998, pp. 155–164.

30. "Les polytechniciens face a l'Egypte," in *L'Expedition d'Egypte, une entreprise des Lumieres*, Patrice Bret, ed., Institut de France, Paris, 1999, pp. 43–51.

31. "Science and War in Revolutionary and Napoleonic France," *Form, Zahl, Ordnung, Studien zur Wissenschaft- und Technikgeschichte*. Festschrift für Ivo Schneider zum 65. Geburtstag, Herausgegeben von Rudolf Seising, Menso Folkerts, und Ulf Hashagen. Franz Steiner Verlag, Munich, 2005.

32. "Pierre-Simon Laplace," in *Princeton Companion to Mathematics,* Timothy Gowers, ed., Princeton University Press, 2008, pp. 752–754.

33. "Jean-Paul Marat," Encyclopedia of European Social History, 1789–1914. Scribners, 2001.

34. "Science in the Eye of the Beholder, 1789–1815," in *Advancements of Learning, Essays in Honour of Paolo Rossi*, John L. Heilbron, ed. Bikblioteca di Nuncius, Studi e Testi, LXII. Olschki, Firenze, 2007. Preprinted in A14.

35. "Pierre-Simon Laplace," Routledge On-Line Encyclopedia of Philosophy. Due on line 2007.

36. "A la découverte de Benjamin Franklin," *Benjamin Franklin, Homme de Science, Homme du Monde*. Catalogue, Exposition du 4 decèmbre, 2007 au 30 mars, 2008. Conservatoire des Arts et Metiers, Paris, 2007.

1.3 Contributions to the Dictionary of Scientific Biography

1. Belidor, I (1970) 581–582.
2. Lazare Carnot, III (1971) 70–79.
3. Condillac, III (1971) 380–383.
4. Diderot, IV (1971) 84–90.
5. Alexandre Koyré, VII (1973) 482–490.
6. Voltaire, XIV (1976) 82–85.
7. Laplace, XV (1978) 273–403, with contributions by R. Fox and I. Grattan-Guinness.

1.4 Articles and Review Essays

1. "Physick and Philosophy: A Study of the Influence of the College of Physicians of London upon the Foundation of the Royal Society," *The Journal of Modern History*, XIX, 3 (September 1947), pp. 210–225. Reprinted in A14.
2. "The Work of Élie Halévy: A Critical Appreciation," *Journal of Modern History*, XXII, 3 (September 1950), pp. 232–250. Reprinted in Gerald Wayne Olsen, ed., *Religion and Revolution in Early Industrial England, the Halevy Thesis and its Critics* (Lanham, MD: University Press of America, 1990). Reprinted in A14.
3. "Notice Biographique de Lavoisier par Madame Lavoisier," *Revue d'Histoire des Sciences et de leurs Applications*, IX, 1 (January–March 1956), pp. 52–61.
4. "The Formation of Lamarck's Evolutionary Theory," *Archives Internationales d'histoire des sciences*, IX, 37 (October–December 1956), pp. 323–338. Reprinted in A14.
5. "Perspectives," *The American Scientist*, 45, 2 (March 1957), pp. 169–176. (Review article on Joseph Needham, *Science and Civilisation in China*, II (1956).)
6. "The Discovery of the Leblanc Process," *Isis*, 48, 152 (June 1957), pp. 103–138. Reprinted in A14.
7. "L'oeuvre d'Éie Halévy, appréciation critique," *Revue de métaphysique et de morale*, 2 (avril–juin 1957), pp. 157–186. (Trans, of (4).)
8. "The origin of Lamarck's Evolutionary Views," *Actes du VIIIe Congrès International d'Histoire des Sciences* (Florence, 3–9 September 1956), Florence, 1957, pp. 544–548.
9. "The Natural History of Industry," *Isis*, 48 (December 1957), pp. 398–407. Reprinted in A.E. Musson, ed., *Science, Technology and Economic Growth in the18th Century*, (London: Methuen, 1972), and in A14.
10. "A Physicist Looks at Greek Science," *American Scientist*, 46 (March 1958), pp. 62–74. Reprinted in A14.

11. "Science in the French Revolution," *Behavioral Science*, IV, 1 (January 1959), pp. 67–73; and *Proceedings of the National Academy of Sciences*, 45, 5 (May 1959), pp. 677–687.

12. "Solomon's House: The Tercentenary of the Royal Society," *The Carleton Miscellany* (Spring1961), pp. 3–18.

13. "Galileo and the Law of Falling Bodies," *Science and Mathematics Weekly*, I, 15 (April 19, 1961), pp. 170–173.

14. "Perspectives," *American Scientist*, 50 (December 4, 1962) pp. 626–639. On Dijksterhuis. Reprinted A14.

15. "The Nature of Science," *Science*, 138 (December 14, 1962), pp. 1251–1253. Review article on T.S. Kuhn, *Structure of Scientific Revolutions*. Reprinted in A14.

16. "In Memoriam Alexandre Koyré," with Pierre Costabel, *Archives internationales d'histoire des sciences*, XVII, 67 (avril–juin 1964), pp. 149–156.

17. "Remarks on Social Selection as a Factor in the Progressivism of Science," *American Scientist*, 56 (December 1968), pp. 439–450. Reprinted in A14.

18. "Probability and Politics: Laplace, Condorcet and Turgot," *Proceedings of the American Philosophical Society*, 116, 1 (Feb. 1972), pp. 1–20.

19. "Mertonian Theses," *Science*, 184 (10 May 1974), pp. 656–660. Reprinted in I. Bernard Cohen, ed., *Puritanism and the Rise of Modern Science: the Merton Thesis* (New Brunswick and London: Rutgers University Press, 1990). Reprinted in A14.

20. "A Note on Darwin's Language," *Current Anthropology*, 15 (1974), p. 224.

21. "On Creativity and Science," *University: A Princeton Quarterly*, 63 (Winter 1975), p. 206.

22. "A Note on Prizes," *Isis*, 66 (1975), pp. 473–474.

23. "Eloge: Carl B. Boyer, 1906–1976," *Isis*, 67 (1976), pp. 610–614.

24. "L'Oeuvre scientifique de Sadi Carnot," *Bulletin du Club Français de la Médaille*, 61 (2ème semestre, 1978), p. 1214. A printer's error is corrected in No. 62, p. 43.

25. "Memoires inedits ou anonymes de Laplace sur la theorie des erreurs, l'analyse, et les probabilities", *Revue d'histoire des sciences et de leurs applications*, 33 (1979), pp. 225–265.

26. In Japanese: "Science and Society" the case of Laplace and Probability," *Nature* (Tokyo) (1982–83), pp. 86–92, 94–101.

27. "Due fratelli nel pallone," *Il Mattino* (Naples) (9 June 1983), Cultura, p. 5.

28. "Aloft with the Montgolfiers," *The Sciences*, 23, 4 (July/August 1983), pp. 46–56.

29. "Can the History of Scientific Institutions Replace the History of Scientific Knowledge?" *Minerva*, XX, 1–2 (Spring–Summer 1982), pp. 232–238.

30. "Éloge of Maurice Daumas," *Isis*, 76 (1985), pp. 72–74.

31. "U.S. Flight No. One: January 9, 1793," *American Heritage of Invention and Technology*, I, 2 (Fall 1985), pp. 63–64.

32. "The Idea of Revolution," Review Essay on I. Bernard Cohen, *Revolution in Science*, *Science*, 229 (13 September 1985), pp. 1077–1078.

33. "Le Fonds Seguin & Privas," *Revue d'Histoire des Sciences et de leurs Applications* 39 (1986), pp. 273–275. Also in English in *History and Technology* II (1986), pp. 331–334.
34. "Science and Politics, with special reference to Revolutionary and Napoleonic France," *History and Technology*, IV (1987), pp. 213–223.
35. "History of the Social Sciences," *Revue de Synthèse*, 109 (1988), pp. 379–385.
36. "La Science à l'aube des temps modernes," *Science et Vie* (Hors serie), 168 (Mars 1989), pp. 6–11.
37. "Salomon Bochner as Historian of Mathematics and Science," *Historia Mathematica*, 16 (1989), pp. 316–323. Reprinted in A14.
38. "Scientific Aspects of the French Egyptian Expedition, 1798–1801," *Proceedings of the American Philosophical Society*, 133, 4 (1989), pp. 447–474.
39. "Scienza e istruzione nella Rivoluzione francese," *Intersezioni*, IX, 3 (dicembre 1989), pp. 401–413.
40. "Aux Origines du CNRS," Colloque scientifique sur l'Histoire du CNRS, le 23–24 octobre 1989. *Cahiers del'Histoire du CNRS*, 6 (1990), pp. 9–30.
41. "Scholarship epitomized," Essay review on R. C. Olby, G. N. Cantor, J.R.R. Christie, M.J.S. Hodge, eds., *Companion to the History of Modern Science* (London and New York, 1990), *Isis*, 82 (1991), pp. 94–98.
42. "Chaleur, son, courants éléctriques: De l'astronomie à la physique mathématique," *Cahiers de Science & Vie.*, Hors serie No. 5(October 1991), pp. 6–22.
43. "Palisot de Beauvois on the Americans," *Proceedings of the American Philosophical Society*, 136 (March 1992), pp. 33–50. Reprinted in A14.
44. "Science and Secret Weapons Development in Revolutionary France, 1792–1804: A documentary history." *Historical Studies in the Physical and Biological Sciences*, 23, Pt. 1 (1992), pp. 35–152.
45. "Recent Trends in the Historiography of Science," *Bulletin for the History of Chemistry*, 15–16 (1994), 19–26.
46. "The Scientific Importance of Napoleon's Egyptian Expedition," *Scientific American*, 271 (September 1994), pp. 78–85.
47. "Jerome Blum, 1913–1993," *Proceedings of the American Philosophical Society*, 138, 3 (1994), pp. 408–412.
48. "*L'exposition du système du monde*, Deux cents ans après sa publication, retour sur le célèbre ouvrage de Laplace," *La Recherche*, 292 (Novembre 1996), pp. 76–79.
49. "Charles Scribner, Jr., 1921–1995," *Isis*, 88 (June 1997), pp. 302–303.
50. "The Spirit of Accountancy Raised to Genius," *Chemical Heritage*, 14, 2 (1997), pp. 10–11.
51. "L'Encyclopédie:Vues d'Ailleurs," *Les Cahiers de Science & Vie*, 47 (Octobre 1998), pp. 90–96.
52. Review article. Ken Alder, *Engineering the Revolution: Arms and Enlightenment in France, 1763–1815* (Princeton University Press, 1997). *Technology and Culture*, 39 (October 1998), pp. 733–754.

53. "Des polytechniciens en Égypte," *Sabix*, 20 (Janvier 1999), pp. 39–41. Bulletin des amis de la Bibliothèque de l'École Polytechnique.
54. Review Essay. "Some Recent 'Big Pictures' in the History of Science," *Annals of Science*, 59 (2002), pp. 409–412. Lewis Pyenson and Susan Sheets-Pyenson, *Servants of Nature*. New York and London (Norton), 1999; John V. Pickstone, *Ways of Knowing*. Manchester University Press and University of Chicago Press (2001).
55. "A Professional Life in the History of Science," *Historically Speaking*, 5, 3 (January 2004), pp. 2–6. Reprinted in A14. Chinese Translation, *Science and Culture Review*, 5 (July 2008), pp. 88–97.
56. "Janis Langins on the Corps Royal du Genie Militaire," Essay Review of *Conserving the Enlightenment* (2003). *Annals of Science*, 62, 2 (2005).
57. "The Rare Book Room, A Rare Privilege," *Princeton University Library Chronicle*, LXVII, 1 (Autumn 2005), pp. 95–106.
58. "The Scientific Revolution," *Historically Speaking*, VIII, 1 (2006).
59. "The Distorted Meridian," Focus Section, *Isis*, 98, 4 (December 2007), pp. 788–795.
60. "Atop Mont Blanc," Essay Review of Martin Rudwick, *Bursting the Limits of Time* (Chicago: University of Chicago Press, 2005). *Historical Studies in the Natural Sciences*, 18, 1 (2008), pp. 163–171.
61. "L'École Polytechnique," *SABIX, Bulletin de la Société des Amis de la Bibliothèque de l'École Polytchnique*, 42 (April, 2002), pp. 5–19.
62. "Témoignages," Mémorial René Taton. *Archives Internationales d'Histoire des Sciences*, 57, 159 (Décembre, 2007), pp. 287–288, 567–568, 619–620.

1.5 Reviews

1. Hutchinson, Francis E., *Henry Vaughan, A Life and Interpretation*. Oxford: Clarendon Press, 1947. *Journal of Modern History* 21 (1949), 169–170.
2. George B. Jeffery, *The Unity of Knowledge: Reflections on the Universities of Cambridge and London*. New York: Cambridge University Press, 1950; R. W. Livingstone, *Leadership in Education*. New York: Oxford University Press, 1950; A. H. Smith, *Idleness as a Part of Education*. New York: Oxford University Press, 1950. *The Journal of Higher Education* 22 (November 1951), 450.
3. F.A. Hayek, ed., *John Stuart Mill and Harriet Taylor: Their Correspondence and Subsequent Marriage*. Chicago: University of Chicago Press, 1951. *Journal of Modern History* 24 (1952), 430–431.
4. Aram Vartanian, *Diderot and Descartes*. Princeton: Princeton University Press, 1952. *Isis* 44 (1953), 389–391.
5. Bernard de Fontenelle, *Entretiens sur la pluralité des mondes*. ed. Robert Schackleton. Oxford: Clarendon Press, 1955. *Isis* 47 (1956), 452–453.

6. René Réaumur, *Memoirs on Steel and Iron*. Tr. Anneliese Sisco; Introduction and notes, Cyril Stanley Smith. Chicago: University of Chicago Press, 1956. *Isis* **48** (1957), 499–500.

7. Shmuel Sambursky, *Physical World of the Greeks*. London: Routledge and Kegan Paul, 1956. *Isis* **4** (1958), 356–358.

8. Brooke Hindle, *The Pursuit of Science in Revolutionary America*. Chapel Hill: University of North Carolina Press, 1956. *Pennsylvania History* **24** (April 1957), 167–169.

9. Paul Aubry, *Monge, le savant ami de Napoléon Bonaparte*. Paris: Gauthier-Villars, 1954. *Scripta Mathematica* **22** (September–December 1956, publ. 1957), 245–246.

10. Joseph Needham, *Science and Civilization in China*, Vols. 1 & 2. Cambridge: Cambridge University Press, 1956. "Perspectives," *American Scientist* **45** (March 1957), 169–176. Correspondence in next two issues.

11. Nora Barlow, ed., *The Autobiography of Charles Darwin*. London: Collins, 1956; David Lack, *Evolutionary Theory and Christian Belief*. London: Methuen, 1957. *Victorian Studies* **2** (December 1958), 166–169.

12. Cabanis, Georges, *Oeuvres philosophiques*, ed. Claude Lehec and Jean Cazeneuve. Corpus général des Philosophes Français. 2 vols, Presses Universitaires de France, 1956. *Isis* **50** (1959), 76–78.

13. Charles Singer, E. J. Holmyard, A. R. Hall, Trevor I. Williams, eds., *A History of Technology*, Vol. Ill. Oxford: The Clarendon Press, 1957. *Isis* **50** (1959), 163–165.

14. Gertrude Himmelfarb, *Darwin and the Darwinian Revolution*. New York: Doubleday, 1959. *Isis* **51** (1960), 216.

15. Francis C. Haber, *The Age of the World: Moses to Darwin*. Baltimore: The Johns Hopkins Press, 1959.

16. Gerhard Hennemann, *Naturphilosophie im 19. Jahrhundert*. Munich: Alber, 1959. *Isis* **53** (1962), 273–275.

17. James R. Newman, *Science and Sensibility*, 2 vols. New York: Simon & Schuster, 1962. *The New York Times Book Review* (21 January 1962), 10.

18. Lewis Feuer, *The Scientific Intellectual: The Psychological and Sociological Origins of Modern Science*. New York: Basic Books, 1963. *Science* **141** (19 July 1963), 257–258.

19. Joseph Agassi, *Toward an Historiography of Science*. S-Gravenhage: Mouton, 1963. *Isis* **55** (1964), 97–99.

20. Maurice P. Crosland, ed., *Science in France in the Revolutionary Era, Described by Thomas Bugge*. Cambridge, MA: MIT, 1970. *Nature* **228** (31 October 1970), 479–480.

21. Arthur M. Wilson, *Diderot*. New York: Oxford University Press, 1972. *Historia Mathematica* **2** (1975), 342–344.

22. John Graham Smith, *The Origins and Early Development of the Chemical Industry in France*. Oxford: Clarendon Press, 1979. *Isis* **72** (1981), 133.

23. Jack Worrell and Arnold Thackray, *Gentlemen of Science: Early Years of the British Association for the Advancement of Science*. New York: Oxford University Press, 1981. *American Historical Review* **87** (1982) 1092–1093.

24. Paul T. Durbin, ed., *A Guide to the Culture of Science, Technology, and Medicine*. New York: The Free Press, 1980. Some medical journal.

25. Rene Taton, ed., *Enseignement et diffusion des sciences en France au XVIIIe siècle*, 2nd ed. Paris: Hermann, 1984. *Revue de synthese* **IV**, 2 (avril-juin 1987), 314.

26. Josef Konvitz, *Cartography in France, 1600–1848: Science, Engineering, and Statecraft*. Chicago: University of Chicago Press, 1987. *Isis* **78**:4:294 (1987), 609–611.

27. Harry Paul, *From Knowledge to Power; the Rise of the Science Empire in France, 1860–1939*. New York: Cambridge University Press, 1985. *American Historical Review* **92** (October 1987), 972.

28. Otto Mayr, *Authority, Liberty, and Automatic Machinery in Early Modern Europe*. Baltimore: Johns Hopkins University Press, 1986. *American Historical Review* **93** (February 1988), 136–137.

29. Antoine Laurent Lavoisier, *De la richesse territoriale du royaume de France*, ed. Jean-Claude Perrot. Paris: C.T.H.S., 1988. *Isis* **80** (1989), 184–185.

30. R. Rashed et al., *Sciences a l'epoque de la Revolutionfrancaise*. Paris: Blanchard, 1988. *Archives Internationales d'Histoire des Sciences* **39** (Juin 1989), 167–170.

31. Joachim Fischer, Napoleon und die Naturwissenschaften. Stuttgart: Steiner, 1988. *Archives Internationales d'Histoire des Sciences* **40** (Juin 1990), 100–102.

32. Maurice Crosland, *Science Under Control: The French Academy of Sciences, 1795–1914*. Cambridge, 1992. *Nature* **357**, 6380 (25 June 1992), 652–653.

33. Loren R. Graham, *Science in Russia and the Soviet Union*. New York: Cambridge University Press, 1993. *New York Times Book Review* (21 March 1993), 25.

34. Bernard Gamier and Jean-Claude Hocquet, eds., Genèse et diffusion du système métrique. Paris: Éditions du Lys, 1990. *Archives Internationales d'Histoire des Sciences* **43**, 131 (1993), 419–420.

35. *The Papers of Benjamin Franklin*. Vols. 28, 29 (Nov. 1, 1778–June 30, 1779) ed. Barbara Olberg et al. New Haven: Yale, 1990, 1992. *The Journal of American History* **80** (Sept. 1993), 644–645.

36. Eric Brian, *La Mesure de l'État: Administrateurs et géomètres au XVIIIe Siècle*. Albin Michel, 1994. *Le Monde* (22 July 1994), 7.

37. Carnot, Lazare, *Saggio sulle macchine in generale*. Ed. and tr. Antonino Drago and Salvatore D. Manno. (Classici delle scienza, 3.Naples: Cuen, 1994. *Isis* **86** (1995), 498.

38. Pascal Duris, Linne et la France (1780–1850). Geneva: Droz, 1993. *Archives Internationales d'Histoire des Sciences* **46**, 136 (June 1996), 175.

39. Dora B. Weiner, *The Citizen-Patient in Revolutionary and Napoleonic Paris*. Baltimore: Johns Hopkins Press, 1993. *Archives Internationales d'Histoire des Sciences* **46**, 137 (Dec.1996), 382–385.

40. Jean Dhombres, ed., *La Bretagne des savants et des ingenieurs*. Editions Ouest France: Rennes, 1994. *Archives Internationales d'Histoire des Sciences* **47**, 138 (1997), 228–229.

41. Edward O. Wilson, *Consilience: The Unity of Knowledge*. New York: Knopf, 1998. *American Scientist* 86 (May–June 1998), 280–283. Greek trans. *Eleptherotypia*, 10 September 1999, pp. 6–7.

42. Michael Shortland, ed., *Hugh Miller and the Controversies of Victorian Geology*. Oxford: The Clarendon Press, 1996. *American Journal of Science* **298**, 3 (March 1998), 263–264.

43. Marie-Noëlle Bourguet, Bernard LePetit, Daniel Nordman, Maroula Sinarellis, eds., *L'Invention scientifique de la Méditerranée: Égypte, Morée, Algérie*. Paris: Éditions de l'EHESS, 1998, **53**, 328 pp. *Revue d'histoire des sciences* 2–3 (2000), 308–310.

44. Anne Marie Claire Godlewska, *Geography Unbound: French Geographic Science from Cassini to Humboldt*. 1999. Chicago/London: University of Chicago Press. xii + 444 pp. *Isis* **92** (June 2001), 400–402.

45. Davide Arecco, *Montegolfiere, scienze et lumi nel tardo settecento, cultura accademica e cognoscenze technice della vigilia della Rivoluzione francese all'eta napoleonica*. Bari: Ed. Cacucci 2003. p. 247. Illustrated. *Revue d'Histoire des Sciences* **56**, 1 (In Press).

46. Michel Cotte, *De l'espionnage industriel à la veille technologique*. pp. 289. Belfort-Montbéliard: Presse Univerisaires de Franche-Comté. *Technology and Culture* **47** (Oct. 2006), 27–28.

47. Jonathan Simon, *Chemistry, Pharmacy, and Revolution in France, 1777–1809*, p.189. Aldershot: Ashgate, 2005. In Press, *Medical History* (Wellcome Institute) **51**, 4 (Oct. 2007), 435–582.

48. Maurice Crosland, *The Language of Science: From the Vernacular to the Technical*. Cambridge: Lutterworth Press, 2006. *Scientific Institutions in France and Britain, c. 1700–1870*. Aldershot: Ashgate, 2007. *British Journal for the History of Science* **41**, 4 (2009), 611–613.

49. Fabien Locher, Le savant et la tempête: Étudier l'atmosphère et prévoir le temps au XIXe siècle. Presse Universitaire de Rennes, 2008. Technology and Culture. 2009.

50. Patricia Fara, *Science: A Four Thousand Year History*. New York: Oxford University Press, 2009. *British Journal for the History of Science* **42**, 4 (2010), 613–615.

51. Marie Thebaud-Sorger, *L'Aerostation au temps des Lumieres*. Presse Universitaire de Rennes, 2009. *Technology and Culture* **52**, 2 (April 2011), 394–396.

52. Michael R. Lynn, *The Sublime Invention: Ballooning in Europe, 1783–1820*. London: Pickering and Chatto, 2010. *British Journal for the History of Science* **44**, 1 (2010), 130–131.

Chapter 2
A Professional Life in the History of Science*

Charles Coulston Gillispie

It was with some compunction that I acceded to the flattering invitation from Donald Yerxa, editor of *Historically Speaking,* to write of a professional life in the field of my specialty. Reluctance was the greater in that I had already given an account of that career in *Isis* on the occasion of the 75th anniversary of the History of Science Society in 1999.[1] In all probability, however, there is little if any overlap between subscribers to *Isis* and those to *Historically Speaking.* That such should be the case is one of the situations discussed. Anyone who consults the earlier essay will find that it turns on personal and institutional factors. I tried not to repeat myself more than was necessary to make what follows intelligible, and ventured instead to offer some reflections on the context of my work in relation to the development of the historiography of science.

First of all, a word about the subject. The generation to which I have the good fortune to belong is commonly said to have founded the history of science as a professional field of scholarship in the years after World War II. Marshall Clagett, I. Bernard Cohen, Henry Guerlac, Erwin Hiebert, Alistair Crombie, Giorgio di Santillana, Rupert and Marie Hall, Georges Canguilhem, René Taton, Thomas S. Kuhn—those are among the notable names. Having majored in some branch of science as undergraduates or the equivalent, and gone on to graduate school before or just after the war, all of us had somehow developed a strong ancillary taste for history. We came out of service of one sort or another in 1945, dazzled like everyone else by Hiroshima, the Manhattan Project, sonar, radar, penicillin, and so on. Independently of each other, or largely so, we each harbored a sense that science, even like art, literature, or philosophy, must have had a history, the study of which might lead to a better appreciation of its own inwardness as well as its place in the development of civilization.

---

* Reprinted from *Historically Speaking*: *The Bulletin of the Historical Society*, V:3 (January 2004), pp. 2–6.

C.C. Gillispie (✉)
Department of History, Princeton University, Princeton, NJ, USA
e-mail: gillispie@Princeton.edu

J.Z. Buchwald (ed.), *A Master of Science History*, Archimedes 30,
DOI 10.1007/978-94-007-2627-7_2, © Springer Science+Business Media B.V. 2012

With a few stellar exceptions, the history of science until that time was the province either of philosophers—Condorcet, Comte, Whewell, Duhem, Mach—each adducing exemplary material in service to their respective epistemologies, or of elderly scientists writing the histories of their science, or sometimes all science, in order to occupy their retirement. Though not written in accordance with historical standards, neither of these bodies of literature is to be ignored. The one is always suggestive and sometimes informative, the other often informative, almost always technically reliable, and rarely of much interpretative significance. Of the two notable scholars who flourished in the 1920s and 1930s, George Sarton was a prophet and scholarly bibliographer rather than a historian, while E. L. Thorndike was a devoted, learned antiquarian riding his hobby horse of magic and experimental science through the library of the Vatican. Though much and rightly respected, neither found a following. Nor did E. J. Dijksterhuis, whose *The Mechanization of the World Picture* (1950) is a classic that will always repay study.

Anticipations of a fully historical history of science appeared in the work of Hélène Metzger on 18th-century chemistry and Anneliese Maier on medieval science. Herbert Butterfield's *The Origins of Modern Science, 1300–1800* (1950) was a godsend both in itself and in that it was one of the few things one could expect undergraduates to read. The same was true of Carl Becker's *Heavenly City of the 18th-Century Philosophers* (1932), a supremely literate essay which (unfortunately in my view) has fallen into disfavor among students of the Enlightenment, and also of Arthur O. Lovejoy's *The Great Chain of Being* (1936), a founding work in the modern historiography of ideas. Two ancillary masterpieces, one from the side of sociology, the other from philosophy, were still more inspirational in exhibiting respectively the social and the intellectual interest that the history of science may hold, namely Robert K. Merton's path breaking *Science, Technology, and Society in Seventeenth-Century England* (1938) and Alexandre Koyré's superb *Études Galiléennes* (1939).

I had read none of these works when, safely out of the army in graduate school at Harvard in 1946–47, I thought to find a thesis subject in what to me was the *terra incognita* of the history of science. My scientific and military backgrounds were respectively in chemistry and a 4.2-inch chemical mortar battalion, but I had taken almost all my electives in history as an undergraduate at Wesleyan, graduating in 1940. The emphasis in the excellent department there was on English history, and my instinct was to look to Britain for a subject, rather than to chemistry. I'm not sure I even knew that there had been a chemical revolution centering on the work of Lavoisier. Darwin was the obvious link between science and intellectual history, but, such was my naiveté, it hardly seemed possible that anything new could be said about the theory of evolution, about science and religion, or about social Darwinism, and I elected to look into the background. That turned out to be in geology, whence my first book, *Genesis and Geology:A Study in the Relations of Scientific Thought, Natural Theology, and Social Opinion in Great Britain, 1790–1850* (1951). It has been in print ever since. Harvard University Press saw fit to put it in a new suit of clothes and reissue it in 1996. A foreword by a scholar of the next generation, Nicolaas Rupke, analyzes the way in which it came to mark a new departure in the historiography of science. He credits me with a novel methodology, first, in consulting, not only the original scientific texts, but the general periodical literature of the time; and second in telling not merely of technical discovery, but of the way in which varying religious views of geologists entered into the formation of their theories, and also the way in which the climate of social opinion entered into the discourse of theology as well as science.

I had no notion of anything of the sort. So far as I was aware, my thesis was a new departure for me, but not for a subject of which I was quite ignorant. Nothing was farther from my thoughts than methodology, something fit for Marxists and sociologists. All that we students of history were taught to do was to go look at the sources, all of them. Perhaps it was lucky that I had never taken a course in geology. Though formally trained in science, I wrote my thesis as someone being trained in history. Had I written it as a scientist, it would have been a chronicle of discovery, a sequence of correct theories displacing incorrect theories, the context being the state of knowledge about the earth in the author's time.

This is not to say that persons trained in a science cannot convert their approach so as to treat its development by historical standards. There are distinguished instances in later years. But I am not among them. Nor is it to deny that it is an advantage, if not quite a necessity, for historians of science to have had scientific training. The reasons are not so much technical as psychological. Except for contemporary or highly mathematical topics, one can always inform oneself about the technicalities, as I was able to do with respect to early 19th-century geology. But it is difficult though not impossible—again there are distinguished instances—to appreciate what it is to know something scientifically without having experienced it.

The department of history at Princeton offered me a job in 1947. Harvard granted me the Ph.D. in 1949, and *Genesis and Geology* appeared to almost inaudible acclaim in 1951. There was no question of my teaching history of science at the outset, and I was quite unprepared to propose any such thing. The curriculum there had the advantage for neophyte faculty that they did not have the labor of preparing courses, and instead led freshman classes and preceptorial discussion groups in the courses taught by senior faculty, whatever the subject. Thus one learned a lot of history while having time to develop one's knowledge and scholarship. When as an assistant professor I had a course of my own, it was modern English history. Only in 1956 did I feel ready to offer history of science. In the interval, I had been able to read all the titles mentioned above and many others. I was informed about courses being offered by Henry Guerlac at Cornell, by Marshall Clagett and Robert Stauffer at Wisconsin, and by Bernard Cohen and others under James B. Conant's leadership in the General Education Program at Harvard. Equally important, and in a personal way more so, I had come to know Alexandre Koyré, who spent half the year annually at the Institute for Advanced Study from 1956 until 1962.

The opportunity to offer an undergraduate course in the history of science opened with the inauguration in the curriculum of an interdisciplinary humanities program. The senior faculty responsible accepted my proposal for a course on the history of scientific ideas from Galileo to Einstein. The notion was to present something that might contribute to the liberal education of students of science and engineering while opening to students in the liberal arts an awareness of the place of science in modern history. Enrollment was nothing of a mass movement, but the undergraduates who did participate in discussion of the material throughout the next three years helped me form a sense of the themes that made for viability. I was thus able to develop the lectures into a book, *The Edge of Objectivity, an Essay in the History of Scientific Ideas* (1960).

The time must have been ripe. That book has been translated into half a dozen languages, beginning with Japanese and ending with Greek. In 1990 Princeton University Press issued a second edition, which is still in print. The preface consists of a review of the thematics of the literature in the intervening thirty years. On its first appearance I had ventured to express the hope that my book might contribute to the development of a professional approach to the history of science.

It would have been more seemly to recognize that *The Edge of Objectivity* was an early instance of such a movement already under way at the hands, largely, of the colleagues mentioned above in the second paragraph. Professional graduate study in history of science was then available only at Wisconsin, Cornell, and Harvard. My book was well enough received that Princeton thereupon agreed to my complementing undergraduate instruction with a graduate program that required additional staff.

In point of content, our attention, like that of colleagues elsewhere, was on the ways in which study of nature reciprocally formed and was formed by the world pictures of classical antiquity, the Middle Ages, the Renaissance, the Enlightenment, and modern times. In point of context, the tendency was to look to philosophy in antiquity, to theology in the Middle Ages, to art and humanism in the Renaissance, to secularism and literature in the Enlightenment, and to industrialization and military technology in modern times. With respect to science itself, the seminal transitions were what attracted scholarship: the Scientific Revolution, mechanization, the Chemical Revolution, the Industrial Revolution, Darwinian evolution. Chronologically, the center of gravity tended to be the 17th century. Other than Darwinism, much else in the 19th century and almost everything in the 20th—relativity, quantum mechanics, and genetics—awaited scrutiny. The narrative line throughout followed the route taken by the creation and transformation of scientific ideas and theories. We wrote, in a word, intellectual history of technicalities with important philosophical overtones. If social, economic, or political awareness crept in, it was around the edges.

The publication of the *Dictionary of Scientific Biography* (1970–1980) affords more objective evidence that a fledgling profession had come into existence by the 1960s, when its preparation began under my direction. The initiative came, not from a historian of science, but from the publisher, Charles Scribner, Jr., who had made a hobby of the history of science since his wartime service in cryptography. Soon after *The Edge of Objectivity* appeared, he asked whether I thought a series of books on the history of science would be viable. I had to say that most of the series known to me started off with one good book by the initiator, and then tailed off into mediocrity since few leading scholars were ever willing to write books on commission. Scribner agreed. His firm was publisher of the *Dictionary of American Biography,* however, and he then had the idea that something of the sort might be feasible in history of science. That, I thought, might work. One could probably persuade first-rate scholars to write, not whole books, but authoritative articles about figures known to them from their own studies.

What had not occurred either to Charles Scribner or myself was that preparation of the *Dictionary of National Biography* and later the *Dictionary of American Biography* had come about at a comparable stage in the formation of a professional discipline of historiography in Britain and the United States respectively. Such, quite serendipitously, proved to be the case with the *Dictionary of Scientific Biography (DSB)*. The quality of the board of editors, of the advisory committee, and of the thousand and more contributors whom it proved possible to enlist from every country with a scientific tradition other than mainland China, then incommunicado, not to mention a large grant from the National Science Foundation and sponsorship by the American Council of Learned Societies—all that succeeded, not only in the main purpose of eliciting over 5,000 articles in sixteen quarto volumes, but also in the unforeseen effect of drawing into a sense of common purpose practitioners dispersed among a miscellany of universities, institutes, national societies, and diverse academies throughout the world.

The *DSB* reflects the time in which it was conceived and composed in another way. The emphasis by design is on the content of the science created—one did not then say constructed—by the men and the few women who are subjects of the articles. The instructions requested authors to keep personal biography and extra-scientific context to the minimum required in order to explicate how the work was possible and wherein it contributed to the development of positive scientific knowledge. It is fair to say that the *DSB* was brought into being by a generation of scholars and scientists who, whatever their other differences, believed in the overall beneficence of science, as by and large did public opinion generally.

The climate of opinion changed amid the seismic shifts in cultural attitudes in the late 1960s and early 1970s. Amid the manifold, largely academic, rebellions of those years, authority became suspect everywhere, including the authority of science. In consequence what had been marginal became central, and social history became the approach of choice in historiography generally, and notably so in history of science. That development bore out a prediction by Robert

Merton, to the effect that sociology of science would flourish only if and when the role of science in society should be perceived as problematic.

So it has proved. In consequence, historians of science who came to the forefront in the generation currently in its prime have tended to see sociology, and to a degree anthropology, rather than philosophy as the disciplines with which to link arms. The merit of the approach is not to establish the truism that science is a social and cultural product. No one ever doubted it. But with a few exceptions, the earlier generation never undertook much in the way of analysis of context. We produced little comparable to the fine-grained accounts that distinguish current work by recapturing the actuality of experiment; the life of a laboratory; the labor of field work in natural history and geology: the recalcitrance of instruments; the differences between what scientists say and what they do; the role of research schools; the place of patronage; the occasional cheating; the interplay of professional rivalries, of personal loyalties and hostilities, of institutional standing, of public reputations, of social position, of gender, race, material interest, ambition, shame, guilt, deceit, honor, pride. The practice of scientific research is currently shown to exhibit, in short, the springs of action that make people tick in all walks of life.

All that is to the good. At the same time, the emphasis on the practice, rather than the content, of science may entail certain drawbacks. Current authors often seem to lose interest in science once it is made. Phenomena for which it is difficult to seek any sociological dimension, say the return of Halley's comet, the law of falling bodies, or the fissionability of Uranium 235, are little scrutinized for themselves. What matters is the way they became known. In consequence, or perhaps because of that approach, the fit, if any, with nature is often taken to be ancillary at best, while analysis of the quality of the science under consideration is left aside.

Looking back at my career in the course of writing this essay, I realize that its development might be seen as a set of responses to what was happening in the historiography of science at large. If so, I was a fish in the stream under the impression that the choices were my own. Apart from the *DSB,* an organizational and editorial job, my most considerable effort has been directed toward the material covered in two books, *Science and Polity in France at the End of the Old Regime* (1980) and its sequel, *Science and Polity in France, the Revolutionary and Napoleonic Years* (2004). They are really volumes I and II of a single work. The former is being reissued with the latter, but I did not want to call it Volume I since it could have stood on its own feet if its author had fallen off his in the interval.

That research started, not in response to changing fashion in the historiography of science, but much earlier in consequence of teaching preceptorial discussion groups in Robert Palmer's course on the French Revolution during the academic year of 1951–52. That was the best undergraduate course, including any of my own, in which I have ever participated. *Genesis and Geology* had just appeared. I had begun to feel (no doubt wrongly) that English history, important though it is, held few surprises. It occurred to me that something must have happened to science during the French Revolution, as many things clearly did in this country amid the major events of the last century. The Guggenheim Foundation agreed, and its generosity allowed my wife and me to spend the academic year 1954–55 in Paris, where we have been for part of almost every year until the above work was completed.

That halcyon year was my introduction to archival research. It was clear ahead of time—and this was the attraction of the problem—that the period of French scientific preeminence in the world coincided with that in which political and military events centering in France were a turning point in modern history. The question was: what did these sets of developments have to do with each other? In the process of working that out amid the minutiae of the documents and the magnitude of all that happened in both domains, I came to feel that what I shall call the public history of science may better be elucidated through the medium of events, institutions, and practices than through abstract configurations of ideas and culture. What the relations of science and politics were I shall leave to readers of the books and not attempt to summarize here. Suffice it to say that they turned on the process of modernization in both areas and on the orientation toward the future that is always characteristic of science and was then radically characteristic of politics.

My career, such as it is, has unfolded not in accordance with some agenda, but as a set of responses to a series of lucky accidents—being a historian by nature who happened to study chemistry and mathematics, taking up Charles Scribner's idea for the *DSB,* precepting in Palmer's course on the French Revolution. Personal rather than professional encounters made possible two of the four books that are spin-offs from the research on French science. During our many sojourns in France, my wife and I chanced to meet descendants of two distinguished families, the Carnots and the Montgolfiers. Lazare Carnot has been known to historians only as the "Organizer of Victory" during the revolutionary wars. So he was, but he spent only six years in government during a long life, most of which was occupied with highly original work, not fully appreciated at the time, in mathematics and physics.

Learning of my interest in that aspect of his life, current members of the family arranged for me to spend a summer going through Carnot's papers, which no one had ever seen, in the house in Burgundy where he was born. The result was *Lazare Carnot, Savant* (1971), to which book my esteemed colleague A. P. Youschkevitch of the Soviet Academy of Sciences contributed a chapter. That was another lucky break. He was the only other historian of science who had ever taken an interest in Carnot. In the midst of a discussion about Russian collaboration in the *DSB,* I mentioned a hint in papers I had seen that Carnot had submitted an early draft of his book on the foundations of the calculus to a prize competition set by the Prussian Academy of Sciences. On his way back to Moscow he searched its archives in East Berlin, found it, and contributed a chapter analyzing Carnot's approach.

I knew, of course, that hot-air balloons are called *montgolfières* after the brothers Joseph and Etienne, who invented them in 1783. On meeting Charles de Montgolfier at a wedding reception, I asked whether he was descended from the big balloon. Sure enough, collaterally at least, and since I expressed interest, he invited us to visit in the country house in Annonay, where his ancestors were in the paper business. There he showed me designs, sketches, correspondence, all scattered among drawers and attics in his and his cousins' houses. Thence *The Montgolfier Brothers and the Invention of Aviation, with a Word on the Importance of Ballooning for the Science of Heat and the Art of Building Railroads* (1983). I give the full title (though aeronautics would have been more accurate than aviation) since it suggests, that even like Carnot's work in mechanics, Joseph de Montgolfier's further inventions (which to him were more important than the balloon), along with those of his nephew Marc Seguin, belong to the pre-history of the physics of work and energy.

Two other publications were happenstance in different ways. Firestone Library in Princeton University is fortunate to possess a rare deluxe printing of the *Description de l'Égypte,* this one having been presented by Napoleon to the king of Prussia and bought at auction in 1865 from an impoverished descendant of a Prussian courtier by Ralph Prime of the class of 1843, later one of the founding trustees of the Metropolitan Museum in New York. It had been clear from the outset that a chapter on the scientific component of Bonaparte's Egyptian expedition would be important in my book. While studying the gorgeous plates, I bethought me that a former student who had just started an architectural publishing business might be interested to see them. He turned over a few pages, and said, "Wow, can we do that?" It had never occurred to me to reproduce them, and that was the origin of *Monuments of Egypt, the Napoleonic Edition,* 2 vols. (Princeton Architectural Press, 1987), which I edited in collaboration with Michel Dewachter, an Egyptologist then with the Collège de France.

In like manner, *Pierre-Simon Laplace, a Life in Exact Science* (1997) emerged from an earlier publication, in this case the *DSB*. I had never intended to write a book about Laplace, who lies on the frontier of my ability to follow mathematical reasoning other than qualitatively. Unfortunately, or perhaps fortunately, two colleagues who had successively undertaken to contribute the article on Laplace failed one after the other to keep their commitments. *Faute de mieux* Laplace devolved upon the editor as default author. I worked on him for a year, harder than I have on anything else, and with the collaboration of Robert Fox and Ivor Grattan-Guinness for particular topics, produced a lengthy article, of which the subsequent book is a revision and enlargement.

Thus, exposure to archives and the close-in research required for these books, as well as editing the articles, many of them very technical, in the *DSB*—these were the experiences that led me to think that limiting one's attention largely to the history of scientific ideas and theories was like following the tips of icebergs, except that the history of science is anything but a frigid subject matter. One might perhaps consider that my individual development exemplifies Auguste Comte's dictum to the effect that, just as every discipline passes through theological and metaphysical stages before becoming positive, so every person is a theologian in infancy, a metaphysician in youth, and a physicist on reaching maturity.

However that may be, the discipline of the history of science has reached maturity. The first meeting of the History of Science Society I attended in 1952 comprised thirty or forty persons, for few of whom was the subject a livelihood. The most recent numbered upwards of 600, the great majority of whom are professional scholars in the discipline. The Society has an endowment and an office with an executive officer. A hundred or more books and collections are reviewed in every issue of the quarterly *Isis*. All that spells success. In only two ways do I feel some slight twinge of regret or disappointment, the first with respect to science and the second with history.

The perception of science as socially problematic in the 1970s and 1980s stemmed in some degree, though by no means entirely, from widespread feelings of anti-scientism in academic and literary circles. In consequence, science studies, whether sociological, political, historical, or a mixture, are often perceived by scientists as hostile enterprises. The most obvious complaint is that critics with no technical qualifications to understand the subjects they discuss are violating the precincts of science. The accusation is nonetheless damaging for being usually, though not always, incorrect or irrelevant or both. The second-order concern among scientists is that the image of science is thus tarnished at a time of weakened political support and stringent restrictions on funding. But the sense of offense goes deeper. While willing to agree that questions of power and advantage are factors both in the macro- and micro-politics of science, scientists resent any implication that their work serves no purpose larger than their own, that they are not in the last analysis investigators of the nature of things, that objectivity is an illusion and rationality a sham. There is the counter-cultural *casus belli* of what journalists have called the science wars.

There was, as well as I can recall, no sense of resentment or hostility to the history of science during the time when our discipline was getting into its stride. On the contrary. We met with every encouragement, institutional and moral, on the part of scientific colleagues. We needed it. I doubt that the discipline could have matured in the face of their enmity and contempt. I do not think that any discipline can flourish in a healthy manner in a mood of hostility to its subject matter. Not that one would argue that prudential reasons should lead historians, or social scientists generally, to refrain from critical and even skeptical scrutiny of the objects of their studies. Still, if we are to recreate the past, the essential matter is to see the subject whole. To set out to see through it is to turn the creatures one studies into specimens. By and large, however, I feel optimistic and think the tide of anti-scientism, if that is what it was, has turned. Much of the work of recent years engages science and scientists on their own terms as well as on the author's.

The slight disappointment has to do with history. It was our hope at the outset, even our expectation, that the historical profession would come to accord the role of science in history a place comparable to that of politics, economics, religion, diplomacy, or warfare. Science after all has been a factor shaping history no less powerfully than have those other sectors. That has not happened. A few departments of history—Princeton's among them—do offer undergraduate and graduate work in the field. But at many, and perhaps most institutions, the subject is taught, if at all, in a separate department or under the aegis of a science and technology studies program. Nor are writings in the history of science as widely read as are those in the conventional fields. The best known, unfortunately in my view, are those written in a more or less iconoclastic vein. Perhaps the barrier is psychological. There may be a fundamental divide between temperaments drawn to history and those drawn to science. At Princeton more of our undergraduate students are majoring in science, engineering, and pre-medical programs than in history or literature. The famous, or infamous, two cultures problem may well be real. Still, we work in hopes that it may be abated.

Note

1. "Apologia pro Vita Sua," *Isis*, 90 Supplement (1999): S84–S94.

Chapter 3
A Career in the History of Science as a Student of Charles Gillispie

Seymour H. Mauskopf

Almost fifty years ago to the day, I began my graduate study in the history of science at Princeton University under the guidance of Professor Charles Coulston Gillispie. I would like to share some recollections about my graduate training at Princeton in the early 1960s and some contextualizing of this period in the more general development of the history of science in the 1950s and 1960s. Charles played a very important role in this development, both through his scholarship and his institution-building at Princeton. And, of course, he played a decisive role in my career. I shall include some reflections on his seminal book, *The Edge of Objectivity*, which also celebrates its fiftieth anniversary and conclude with an overview of my career as an historian of science.

I had, in fact applied to graduate school not to study history of science but medieval history. As an undergraduate at Cornell, I had been a pre-med chemistry major for three years. But in my first semester I had become enthralled by a history course in "western civilization," particularly the section on medieval history. I switched from chemistry to history and wrote an honors thesis on medieval canon law. However, I also took the survey history of science course taught by Professor Henry Guerlac as well as his seminar on the eighteenth century. I believe that a factor in my choice of Princeton for graduate study was the knowledge that history of science was also available, but that was not my initial primary interest.

However, when I came to Princeton I was informed by the Director of Graduate Studies that I had been placed in a new program in history and philosophy of science because of my heavy concentration of undergraduate pre-med science courses. I was, of course, given the choice of following my original intention and not joining the program. In a real sense, I was able to do both. As required, I worked up four fields for prelims: two in the history of science (one being history of chemistry) as well as medieval and renaissance history among the historical fields. It was an odd mixture, given the fact that

S.H. Mauskopf (✉)
Professor Emeritus, Department of History, Duke University, Box 90719, Durham, North Carolina 27708-0719, USA
e-mail: shmaus@duke.edu

J.Z. Buchwald (ed.), *A Master of Science History*, Archimedes 30,
DOI 10.1007/978-94-007-2627-7_3, © Springer Science+Business Media B.V. 2012

I opted in my second year of graduate study to pursue a research topic in eighteenth- and nineteenth-century history of science, one that would turn into my dissertation.

There was one other graduate student in history of science my first year, Michael McVaugh. Michael and I shared an interest in medieval history and consequently did similar fields for our prelims (not history of chemistry, however). Michael remained both a medievalist and an historian of science.[1] Michael and I received academic positions in 1964 at the University of North Carolina, Chapel Hill and Duke University respectively. Only eight miles apart, we have had a very productive friendship and joint teaching and research careers.

At this time – and for many subsequent years – I had little knowledge of the context of my being placed in the history and philosophy of science program. But Charles himself provided an account of the genesis of the Princeton program in 1999,[2] and I then learned why I was recruited for the program.

1960 represented something of a watershed for the history of science. The field had begun to professionalize in the 1940s and 1950s in Europe and in the United States, with programs in this country at Harvard, Wisconsin and Cornell. Moreover, a research and pedagogical literature was coming into existence; I still have my copy of A. Rupert Hall's *The Scientific Revolution, 1500–1800: the Formation of the Modern Scientific Attitude,*[3] used in Guerlac's history of science course at Cornell. Thomas Kuhn's *The Copernican Revolution: Planetary Astronomy in the Development of Western Thought*[4] was published in 1957, and, in that year, the program at Wisconsin staged a conference both to take stock of and to prioritize the emerging research field. Published in 1959 as *Critical Problems in the History of Science,*[5] it was retrospectively assessed as "a milestone in the development of the history of science as an organized discipline in the English-speaking world."[6] The following year – 1960, when the Princeton program was inaugurated – saw the publication of Charles' own *The Edge of Objectivity: An Essay in the History of Scientific Ideas.*[7]

1960 not only marked the inception of the Princeton program but also the one at the University of Indiana. These programs, moreover, represented an innovation with respect to previous U.S. programs; at both Princeton and Indiana, history of science was explicitly linked with *philosophy of science.* This innovation

---

[1] There was a third student in philosophy of science: Bill Schuyler.

[2] Charles C. Gillispie, "Apologia pro Vita Sua," *Isis,* 90 (1999) Supplement: Catching up with the Vision: Essays on the Occasion of the 75th Anniversary of the Founding of the History of Science Society," S84–S94.

[3] London: Longmans, Green, 1954.

[4] Cambridge: Harvard University Press.

[5] Marshall Clagett, ed. (Madison: University of Wisconsin Press, 1959).

[6] "A.T." (Arnold Thackray), "Retrospective Review, *Critical Problems in the History of Science.* Review," author[s]: Martin Rudwick; William Coleman; Edith Sylla; Lorraine Daston, *Isis,* 72(2) (Jun 1981), 267 (267–283).

[7] Princeton: Princeton University Press, 1960.

arose naturally at Indiana due to the presence of the philosopher of science, Norwood Russell Hanson, who had received his training in Great Britain, where the fields had been linked, and whose own research was concerned with the origins and genesis of systems of scientific knowledge and, hence, with the history of science. In this respect Hanson was unusual among philosophers of science in the 1950s.[8]

At Princeton, the linkage of the two fields seems to have been more serendipitous. At the end of *The Edge of Objectivity* Charles, a member of the Princeton history department, proclaimed that his goal for the book was to "help win for history of science a place in historiography comparable in interest and professionalism to that which the philosophy of science has for long held in philosophy."[9] At the same time, Charles saw disjunctions between history of science and history (as well as philosophy) and impediments in remaining just a member of a history department, the main one being the difficulty in attracting naturally first-rate graduate students in history of science (felt by many of us subsequently who *have* remained members of history departments).[10]

On the other hand, Charles' initiative had little to do with a felt need on his part to associate with philosophers of science. Nor, in his account, did the distinguished Vienna Circle philosopher of science, Karl Hempel, also at Princeton, play any role in the formation of the HPS program.[11] The resultant program was designed *not* to produce "hybridization" of history and philosophy of science but "reciprocity between them."[12] My own experience and that of many others was that even reciprocity was hard to come by. I was not encouraged to take philosophy of science courses (although I did do one on my own initiative). Reflecting on his days as a history of science graduate student in the Princeton program in the late 1960s, Kenneth Caneva has expressed this absence of reciprocity even more strongly: "I neither knew nor cared where the philosophers were."[13] As the volume from which this quotation comes testifies, there has been much interest recently about how to bring history and philosophy of science into a truly reciprocal relationship, and I have taken part in this enterprise.

---

[8] *Patterns of Discovery: An Inquiry into the Conception Foundations of Science* (Cambridge: Cambridge University Press, 1958).

[9] *Edge of Objectivity*, p. 521.

[10] After characterizing his first graduate seminar in history of science at Princeton (1959), as a "failure," Charles wrote: "As between history proper, philosophy proper, and history of science, we had no stock-in-trade. The disappointment taught me that there was little prospect of recruiting historians of science among graduate students who already knew what they intended to do and that any contribution I might make to graduate education would have to be among people who came to Princeton for that purpose," in "Apologia pro Vita Sua," S89.

[11] Rather, Charles credited the classicist and philosopher (and new Chair of the philosophy department), Gregory Vlastos with helping in the formation of the program.

[12] "Apologia pro Vita Sua," S90.

[13] Kenneth L. Caneva, What in Truth Divides Historians and Philosophers of Science? In *Integrating History and Philosophy of Science*, eds. Seymour Mauskopf and Tad Schmaltz. Dordrecht: Springer (Boston Studies in the Philosophy of Science, Volume 263), 2012, p. 49.

Our first history of science seminar was not with Charles but with John Murdoch, who passed away recently. For me at least, the seminar was something of a baptism by fire. As his seminar topic, John chose Euclid's *Elements* and the classical development of axiomatic methods. The seminar, lasting three hours of an afternoon, was run as an advanced course, with members of the Princeton faculty sitting in. The graduate students (basically Michael McVaugh and I) had do to weekly presentations on a variety of arcane subjects (to us) based on a variety of texts in several languages. I was so much a novice that I never quite knew when I was falling flat uttering some nonsense in the eyes of the scholar – and that probably saved my sanity. But this method of teaching works! Fifty years later, I still can recall all sorts of information about Euclid's *Elements* and its predecessors. I thank John for holding us to such challenging standards.

The second semester, Charles taught his first seminar, on the core of the Scientific Revolution, from Galileo to Newton with something of a focus on Christiaan Huygens. The length and format was similar to John's, and the assignments were similarly "challenging." Other scholars sat in regularly, the most memorable of whom, perhaps, was Father Stanley Jaki, who went on to a very productive and distinguished career in history and philosophy of science. If memory serves, early in the semester, the graduate students were each assigned an entire volume of the Antonio Favaro edition of the works of Galileo, each volume running to hundreds of pages in Latin and Italian. Somehow, we did the assignments. Later in the semester, in what is to me my most memorable seminar report challenge ever, I presented the entire *Horologium Oscillatorium* of Huygens. It was a daunting experience.

In the course of the seminar, we took one short "field trip" out to the Institute for Advanced Study to meet Alexandre Koyré. This was more of a pilgrimage than a field trip, and there was an almost reverential mood to the occasion. The image that sticks in my mind from that meeting is of Koyré showing us a copy of Johannes Kepler's *Harmonices Mundi*, open to a page with musical notation concerning the harmony of the spheres following the statement of the Third Law of planetary motion. I would much like to remember what exactly Koyré said about that.

The summer after my first year, I took a French reading course at Columbia University. I had entered graduate school at Princeton with some Latin and German but no French. However, in keeping with the general rigor of Princeton graduate seminars, one was expected to read French, and I taught myself enough to get through the French translation of Huygen's *Horologium Oscillatorium* for my seminar presentation. I then took the departmental French reading examination at the end of the spring semester – and was passed (with a comment to the effect that "I knew what I knew and would learn what I didn't know"). But I felt that I needed something more systematic; hence the summer course.

In order to test my growing mastery of French, I read (and took notes on) Koyré's already classic *Études galiléennes*. That was followed by my reading Charles' own *Edge of Objectivity*. Reading both these works at leisure enabled me to acquaint myself with the then dominant perspective of the field of the history of science.

The *Études galiléennes* was published in 1939 and came to define the core transformation in the central episode in the history of science, the Scientific Revolution. Koyré focused on the mechanics developed in the seventeenth century, which, as is well known, he characterized as involving a sharp conceptual break with the anterior physics traditions of Aristotle and his medieval commentators – what he indeed termed an "intellectual mutation."[14] This break involved the substitution of a geometric scientific world view (sometimes termed by Koyré "Platonic" and, more appropriately, "Archimedean") for the hitherto dominant Aristotelian one.

The notion that the core of modern science was the formulation of a geometrical vision of the world was hardly new; it had already been articulated by Ernst Cassirer and Edwin A. Burtt.[15] But perhaps never before had it been developed in such a concentrated and elegant form; in particular, never before had the core concepts of the new physics – space, motion, the structure of the universe – been so carefully articulated. Moreover, what gave his work particular cachet was his close reading of texts not only of the admittedly great scientists like Galileo, Kepler, and Descartes but also of the less well known ones, such as Giovanni Battista Benedetti.

It was Koyré's careful historicism combined with a sophisticated "progressivist" analysis of intellectual mutation during the Scientific Revolution that, in my opinion, made Koyré's *Études galiléennes* so powerful a force among the first generation of professional historians of science. Reading it in tandem with Charles' *Edge of Objectivity* during the breaks in my French course in the summer of 1961 gave me a splendid general overview of the development of modern science.

But it was only many years later that I realized it gave me more than that. While meditating on (and writing about) the historiography of science over the past century, I came to the realization that these two works encompassed an intellectual era in the history of science. More particularly, I realized that *Edge of Objectivity* can be seen as a true sequel to Koyré's *Études galiléennes*. Charles practically says as much in the Bibliographical Essay that follows the text of his book:

> I...owe more to him professionally than to anyone else. His writings have revealed to me wherein the intellectual content of the history of science consists. His is by far the greatest influence on this book....He is the master of us all.[16]

---

[14] Koyré himself dedicated the work "À la Mémoire d'Émile Meyerson" and mentioned on the first page of text Duhem, Cassirer and Bachelard, to whom he credited the phrase "mutation intellectualle." *Études galiléennes*, I. "A L'aube de la science classique" p. I-5.

[15] Cassirer, *Das Erkenntnisproblem in der Philosophie und Wissenschaft der neueren* Zeit (3 vols) (Berlin: Bruno Cassirer, 1906–1920). Burtt, *The Metaphysical Foundations of Modern Physical Science: A Historical and Critical Essay* (New York: Harcourt Brace, 1925). It had also been developed in the historical writing of the Dutch scholar Eduard Jan Dijksterhuis (1892–1965). See H. Floris Cohen, *The Scientific Revolution*, p. 59ff.

[16] *Edge of Objectivity*, p. 523. I do not recall reading the Bibliographical Essay when I first engaged the book.

Charles began his narrative where Koyré's theme of the geometrization of nature stopped, for his "objectivity" overlaps Koyré's "geometrization". Charles essayed a definition of this protean theme early in the book:

> Modern science...is impersonal and objective, It takes as its starting points outside the mind in nature and winnows observations of events which it gathers under concepts to be expressed mathematically if possible and tested experimentally by their success in predicting new events and suggesting new concepts. Modern science...is first of all metrical and experiential.[17]

What is "outside the mind in nature" is therefore, for Charles, something very close to Koyré's geometrical nature.

But, though he clearly believes that this is indeed the vision of scientific reality, Charles has not written simply a triumphalist paean to its advance through the different sciences. Ultimately, his is a tragic if also heroic view of the development of science; in his first use of the phrase that served as the book's title, he wrote of the "cruel edge of objectivity." By this he meant what he termed the "'Fatal Estrangement' between science and ethics" – and, indeed, between the human condition and true nature (mechanical, impersonal) – that followed from Galileo's reduction of everything in the material world to the "primary qualities" of the geometry and motions of its constituent particles.[18] This tragic theme shoots through Charles' narrative to be explicitly expressed as such in the book's final sentence, on Einstein's sense of simultaneous liberation and loneliness as he ventured from religious belief to scientific reality.[19]

As the subtitle of his book, "An Essay in the History of Scientific Ideas," indicates, Charles shared Koyré's idealist approach to the history of science. But he was far from Koyré's rigorous exclusion of extra-intellectualist influences. The role of experiment – which Koyré had downplayed in his own work – was prominent in *The Edge of Objectivity* as were, on occasion, even social and cultural contexts, particularly in connection with eighteenth-century French science, an historical domain that was to become Charles' chosen area of research for the rest of his career.

This book was very ambitious; it was the only one by the new breed of professional historians of science to try to provide a comprehensive and coherent account of "modern" (i.e. post-Scientific Revolution) science. Given the almost centrifugal complexity of modern science, some organizing principle like "objectivity" was essential. Significantly, Charles' attempt has not been superseded.

---

[17] *Edge of Objectivity*, p. 10. The close relationship to Koyré's work is shown by Gillispie's use of Galileo as the starting point for his narrative of objectivity. See especially p. 42ff. By the same token, Gillispie is disparaging about Galileo's contemporary, Francis Bacon, pp. 74–82.

[18] *Edge of Objectivity*, p. 44.

[19] "Surely the very generality of his liberation, rendering the perfectly benign perfectly irrelevant to the vast impersonality of nature, invested his inner freedom and security with the loneliness of a Greek tragedy." *Edge of Objectivity*, p. 520.

The *Edge* adhered to what I have termed a progressivist historical view of science, whereby the process of scientific change over time entailed, in the contemporary words of George Sarton, a "purification of its methods" and a "gradual abandonment of ancient errors, poor approximations, and premature conclusions."[20] Whatever tragic implications were contained in modern scientific methods and perspectives, they were, to Charles, undoubtedly truer than what they supplanted. For instance, although he characterized Aristotelian physics as a "serious physics," Charles also identified what to him was its fundamental problem:

> It was wrong .... Nature is not like that, not an enlargement of common sense arrangements, not an extension of consciousness and human purposes ... For the order is mathematical and notes harmonious, Platonic rather than Aristotelian.[21]

The hegemony of the progressivist perspective did not last much beyond the publication of the *Edge*. But the book continues to be esteemed down to the present. As I was working on this essay, I mentioned the work to two fellow historians, one a historian of science in England and the other a European intellectual historian in the United States. I was delighted to find that both knew the book well and testified to its importance in their own professional formation.

Throughout the second year at Princeton, joined by Bob Silliman, I took Charles' seminar, which centered on the history of nineteenth-century science. It was in the course of this seminar – stimulated by one of Charles' observations concerning the significance of Louis Pasteur's molecular images with respect to his first great discovery of crystalline asymmetry and optical activity – that I undertook what became my dissertation project and first monograph, *Crystals and Compounds*.[22]

Just before the start of my third year, Michael McVaugh and I took our preliminary examination. As I mentioned, I had a rather unusual group of fields. The historical ones were in medieval and early modern European history; the history of science fields were medieval history and history of chemistry. The concentration of the Middle Ages reflected both my own original predilection and the presence of John Murdoch at Princeton. It wasn't the best preparation for working on a topic in the eighteenth and nineteenth century and would perhaps not have been permitted once the program was more firmly established.

The field of the history of chemistry was one that I essentially worked up myself, and I have recently come upon my reading notes for this area. Inspecting the notebook confirms the thesis of a paper that I recently gave, namely that the

---

[20] George Sarton, *A History of Science: Hellenistic Science and Culture in the Last Three Centuries B.C.* (New York: Wiley, 1965 [first published 1959]), p. viii.

[21] *Edge of Objectivity*, p. 13.

[22] Crystals and compounds: Molecular structure and composition in 19th-century France. *Transactions of the American Philosophical Society*, 66/3 (1976), 5–82.

history of chemistry had been largely pursued by scholars trained as chemists (and who often continued as practicing chemists) down to the 1960s.[23] Nevertheless, some of them had certainly been pioneers in developing the history of science. An exemplary case here is the British chemist and historian of science, Douglas McKie, whose biography of Lavoisier was on my reading list. Another was James Bryant Conant, whose classic "Overthrow of the Phlogiston Theory"[24] was also on the list. I continued to hone my French by reading the works of that extraordinary scholar whose life ended so tragically, Hélène Metzger. Finally, I read a work that presaged a principal – indeed *the* principal direction that the history of chemistry was to take over the next three decades: Henry Guerlac's *Lavoisier – The Crucial Year*, which had just been published in 1961.[25]

Although we had a reasonably good idea of what the examinations in our fields of European history would be like, neither of us had the slightest notion about those in history of science. We were the experimental animals, after all. We worked together intensely on our joint field of medieval science. We both survived.

I spent my third year at Oxford University, the recipient of an exchange arrangement between the Cornell Branch of Telluride Association, my undergraduate scholarship house, and Lincoln College, Oxford. Thanks to helpful networking by Charles, I was welcomed into the history and philosophy seminar taught by Alistair Crombie and Rom Harré. I also settled into my dissertation research. Again, through Charles' initiative, I was put in touch with Frank Greenaway, of the Science Museum in London. Frank was a very knowledgeable and helpful mentor to me as I began to formulate my dissertation research project. I also contracted double pneumonia and met my future wife in what was certainly an eventful year.

Early in my stay at Oxford, I received by letter an offer from the chair of the history department at Duke University of a tenure-track appointment there in the history of science. Since I had barely settled down in Oxford and had hardly begun research on my dissertation, I wrote back immediately to decline the offer with regret. Virtually by return mail, the chair suggested that the position could be held for an extra year if that would help me reach a positive decision. I could hardly refuse this, and now, forty six years later, I am still at Duke about to retire.

In the fourth year, I was back at Princeton working on my dissertation, supported by a Woodrow Wilson dissertation grant. Most of my research material was in published form but much of it, particularly eighteenth-century

---

[23] Seymour Mauskopf, "Do Historians or Chemists Write Better History of Chemistry?" Delivered in session, "ACS Chemical Landmarks Program: Celebrating the History of Chemistry," March 26, 2007: Published as "Do historians or Chemists write better history of chemistry?" *Bulletin for the History of Chemistry*, 36(2) (2011), 61–67.

[24] Complete title: "Case 2: The Overthrow of the Phlogiston Theory: The Chemical Revolution of 1775–1789." In *Harvard Case Histories in Experimental Science*, eds. James Bryant Conant et al. 2 vols. (Cambridge: Harvard University Press, 1957). Vol. 1, pp. 65–115.

[25] Complete title: *Lavoisier – The Crucial Year: The Background and Origin of His First Experiments on Combustion in 1772* (Cornell: Cornell University Press).

works, was not available at Princeton, and I had to travel to the New York Public Library to read this material.[26] In Princeton's Firestone Library, I did have the kind of fortunate serendipitous experience that researchers like to recount. While searching for something (so totally unmemorable that I have long forgotten what it was), I found next to it J. D. Bernal's *Science and Industry in the Nineteenth Century*, which contained Bernal's essay commemorating the centenary of Pasteur's discovery of crystalline asymmetry and optical activity.[27] That essay gave me the key for which I had been searching. It led me to investigate what became the heart of my dissertation: the fifty-year French tradition that bridged crystallography and chemistry beginning with Haüy's crystal structure theory and climaxing with Pasteur's discovery. Some years later, when developing my dissertation for publication, I was able to mine Bernal's article again – this time for the name of the possessor of the laboratory notebook in which Pasteur recorded his discovery as well as the steps leading up to it, and secure a microfiche copy.

The following year (1964), my bride (and now wife of forty six years) and I moved to Durham, North Carolina where I began my long teaching career. Michael McVaugh also started his long career eight miles away at the University of North Carolina – Chapel Hill. A few years after we arrived, we began to teach a joint UNC – Duke upper-class – graduate seminar in history of science. Joint residency permitting, we have taught this seminar for forty years.

As I implied, I was hired by Duke sight unseen. I have no doubt that Charles gave me a powerful recommendation for which I am extremely grateful. Duke was a fine university when I arrived, and it has only grown better. But this was the period of the great expansion of higher education generally (the 1950s and 1960s), and, in this expansion, history of science had for a time something of a privileged position because, as has been told to me at Duke, the field looked to be the natural "bridge" field between the "two cultures" (or however many there be) of the natural sciences and the other areas of academic pursuit. I have taken this functional view of the field to heart throughout my career and have founded and/or directed at Duke programs of a broad inter- or cross-disciplinary nature.[28] This "bridge" function of the history of science was reinforced for a time by the appearance of Thomas S. Kuhn's *The Structure of Scientific Revolutions* in 1962. It was also exemplified by ambitious narratives like

---

[26] One such work was René-Just Haüy, *Essai d'une théorie sur la structure des cristaux*. This was crucial for my research.

[27] *Science and Industry in the Nineteenth Century* (London: Routledge and Paul, 1953).The essay's title is "Molecular Asymmetry."

[28] Three of these were: interdisciplinary faculty seminars, which the Provost William Bevan asked me to organize in the early 1980s; and the Society of Duke Fellows, for which I served as faculty mentor in the early 1990s. This was the organization of all graduate students at Duke with named fellowships. I was to help facilitate cross-disciplinary discussions and events; FOCUS Interdisciplinary Programs, which I directed from 1995 to 2003.

Charles' *Edge of Objectivity*, and by the founding of graduate programs that featured the history of science. To this list one must certainly add *The Dictionary of Scientific Biography*, of which Charles was Editor in Chief.

These were heady years for the history of science. But, as the 1960s progressed and turned into the 1970s, the social-cultural attitude towards science took a turn towards criticism and even hostility. Within the history of science, the earlier idealist and progressivist perspective gave way to a variety of other approaches and perspectives, some highly critical of the earlier tradition, such as the social constructivist (SSK) and feminist programs. As the exemplary text of the earlier perspective, *The Edge* itself was sometimes used as the point of departure to launch these critical perspectives.[29]

If these approaches have not become entirely hegemonic in the history of science, they have certainly deeply influenced research in the field. Since these perspectives have much in common with post-modernism generally, they have facilitated unprecedented interaction between some historians of science and scholars in the humanities. But, as the "sciences wars" of the 1990s showed, the anti-progressivist perspectives have alienated many scientists from the history of science enterprise, as I have personally witnessed at Duke. History of science has thereby lost much of its original promise as a bridge field between the sciences and the other domains of knowledge.

I shall conclude by relating my own research and publications to the guidance provided by Charles' scholarship and to the some of the changes in historical perspective that have influenced me. My dissertation and subsequent monograph, *Crystals and Compounds*, were cast in what I referred to earlier as an idealist and progressivist mold. They were primarily about the history of scientific ideas, with little social or cultural context. Moreover, I had no doubt – and have none now – that my narrative concerned real advances in natural knowledge. Yet, at the same time, I was concerned with demonstrating that "incorrect" theories (retrospectively evaluated), particularly those of the brilliant chemist, Auguste Laurent, could exert decisively positive influence on scientific discovery.

Although I certainly did not become an anti-progressivist ideologue, my subsequent research topics had very different perspectives than did *Crystals and Compounds*. The next one, jointly carried out with Michael McVaugh, was a history of parapsychology, a subject that had been uniquely pursued at Duke University since the late 1920s by the psychology professor, J. B. Rhine. The intellectual framework behind our taking up this research topic, which occupied us throughout the 1970s, was the work of our first and probably most distinguished graduate student, Betty Jo Teeter Dobbs. Betty Jo had participated in our first joint seminar – on Newton – and had written a research paper on

---

[29] Barry Barnes and Steven Shapin, "Where is the Edge of Objectivity? [Essay review of Mary Douglas, *Implicit Meanings: Essays in Anthropology*], *British Journal for the History of Science*, 10 (1977) 61–66; and Carolyn Merchant, "Critiques & Contentions: *Isis'* Consciousness Raised," *Isis*, 73 (1982) 398–409.

Newton's chemical ideas in the Queries to the *Opticks* and one or two other published papers. This had developed into a dissertation on Newton's *alchemical* studies, where she showed that Newton had spent most of his career involved in them, had brought the same careful methodology and experimental technique to them that he had employed in his much better known work in physics, and finally, that some alchemical ideas appeared to influence those in physics, such as "action at a distance."

Betty Jo's work appeared to have important implications for the traditional narrative of the Scientific Revolution (largely still idealist and progressivist in the late 1960s), and Michael and I were sufficiently intrigued by her work to take up the history of our own local "pseudo-science," parapsychology. The result was *The Elusive Science: Origins of Experimental Psychical Research.*[30] I know that Charles was bemused by our research turn, and, in his own magisterial *Science and Polity in France at the End of the Old Regime,*[31] published the same year as our history of parapsychology, he showed very little sympathy for Mesmerism, an ancestor, as it were, to parapsychology. At the same time, a review of *The Elusive Science* by Harry Collins, then one of the most provocative young social constructivists (and still highly provocative), took us to task for not using "the most recent discussions in the sociology and social history of science, though there material is ideal for such an application."[32]

Since then, my research has returned to the history of chemistry but with a very different turn. I have moved from an idealist focus on chemical theory to the study of chemical technology, specifically chemists and the fabrication and improvement of munitions and explosives. This is, I should say, in line with a more general turn in the orientation of much current history of chemistry towards considering chemistry as much as artisanal practice as natural philosophy. I will end by noting how useful Charles' own discussion of Lavoisier's activities as chief chemist at the *Régie des poudres* (The French gunpowder administration) in *Science and Polity in France at the End of the Old Regime* was to me in getting oriented in this area of research.

---

[30] Baltimore: Johns Hopkins University Press, 1980.

[31] Princeton: Princeton University Press, 1980.

[32] H. M. Collins, review *The Elusive Science: Origins of Experimental Psychical Research* by Seymour H. Mauskopf; Michael R. McVaugh. *Isis*, 72(4) (December 1981) 670–671.

Chapter 4
Charles Gillispie in the Digital Age

Jane Maienschein and Manfred D. Laubichler

> *There are three principal means of acquiring knowledge available to us: observation of nature, reflection, and experimentation. Observation collects facts; reflection combines them; experimentation verifies the result of that combination. Our observation of nature must be diligent, our reflection profound, and our experiments exact. We rarely see these three means combined; and for this reason, creative geniuses are not common.*
> *Denis Diderot*

Denis Diderot was in more than one way a predecessor of Charles Gillispie, and this widely quoted passage does not, of course, apply only to knowledge about nature. It also describes how we acquire our understanding of history. Given the enormity of the world and the almost endless wealth of observations, it is beyond the reach of any single scholar to achieve anything close to a complete overview. Therefore, "profound reflection" and "exact experimentation" are all the more important. And so, we might add, are the organized and guided efforts of large communities of scholars and scientists working towards a common goal. Only then can we even approach collective "creative genius."

Charles Coulston Gillispie has been a worthy successor to Diderot on all these counts. A half century ago, Princeton University Press published *The Edge of Objectivity. An Essay in the History of Scientific Ideas*. This volume cost $7.50 in hardback at the Yale Co-op, where one of us (JM) bought it in 1970 for undergraduate courses taught by historians of science Larry Holmes and Martin Klein. Considered a standard for the sort of survey courses in the humanities for which it was written, the book was also on the reading list for JM's graduate study at Indiana University in 1972. Two decades later, it was still one of the first books that ML (then a biology graduate student at Yale) purchased in the same bookstore and read in what became a five year intensive

J. Maienschein (✉)
Center for Biology and Society, Arizona State University, Tempe, AZ 85287-4501, USA

Marine Biological Laboratory, Woods Hole, MA 02543, USA
e-mail: Maienschein@asu.edu

J.Z. Buchwald (ed.), *A Master of Science History*, Archimedes 30,
DOI 10.1007/978-94-007-2627-7_4, © Springer Science+Business Media B.V. 2012

tutorial in history of science with Larry Holmes and he encountered it, as well as its author, again when he began his graduate studies in history of science at Princeton. The book remains in print, though in 2011 costing $46.95 for the paperback version.

How can one author's book that spans the history of ideas in science from Aristotle into the 20th century remain relevant for so long? And what might we do today with such a project? Our contention is that the lasting value of the approach and the book lies precisely in its focus on ideas and the fact that it is the product of "profound reflection" and "diligent observation" and that the way it probed the transformations of and connections between ideas represents a form of "exact experimentation." But, like his predecessor Diderot, Gillispie realized that such an endeavor can only be sustainable in the long run if it also involves a whole community of scholars (or a network in 21st century-speak). Today such a community would include the kind of digital and computational approaches that some of us, including his long-time Princeton colleague Robert Darnton, have begun to champion (Darnton, Robert, 2008; 2010). In fact, Gillispie's own editing of the marvelous *Dictionary of Scientific Biography* offers an early hint at what is now becoming possible with digital publishing, the "Web 2.0" and beyond.

While Gillispie wrote other books, most notably the magisterial two volumes on *Science and Polity in France*, it is *The Edge of Objectivity* that captures the breadth of ideas as they develop through centuries of individuals, institutions, and intellectual traditions (Diderot, Denis, 1753; Gillispie, Charles Coulston, 1960; 1980). And while some reviewers at the time, and others since, have found things with which to disagree, nearly all have agreed with A. Rupert Hall's review in *Isis*. After fussing about various examples, yet admiring the product as a whole, Hall concluded that "There are in the book perhaps some of the defects as well as the virtues of intellectual brilliance; either way it is an intellectual exercise to read it. And it has much that any historian of science would be justly proud to have written" (Hall, 1960). Indeed, any historian would have been proud to have written such a book a half century ago. We believe that today it is no longer possible to accomplish such a task and that it would be foolish to try. There is just too much scholarship to master, too many additional original sources to consider, and too many enticing new scholarly tools that offer other ways of working. Diligent collection of data has increased dramatically and, in order to accomplish profound refection today we have to explore new forms of exact experimentation enabled by digital and computational approaches, even in the traditional fields of humanistic scholarship.

Diderot and the Enlightenment encyclopedists already recognized that some projects call for collaborations. So did Gillispie with the massive project that emerged over the decade 1970–1980 as the sixteen volumes of the *Dictionary of Scientific Biography*. A considerable army of historians wrote articles that were typically in the five-page range, with a few notable lengthy entries as well. These volumes, plus a two volume Supplement added in 1990, have served as a standard reference source since. The *New Dictionary of Scientific Biography*, edited by Noretta Koertge and published in 2007 included new entries and a set of updated biographies for major figures. That the project has had such staying power and called for an update demonstrates the value of the original project, as well as its capacity for revision and extension. Charles Scribner's and Sons, with funding from the American Council for Learned Societies, helped make the project possible. Again, a large network of contributors contributed their scholarly interpretations, and the work has been recognized as important with awards and enthusiastic reviews.

Scribner's decided to publish the 2007 *New Dictionary* in print, and also with an on-line option that has a comprehensive index. However, what we can see from the vantage point of 2011 is that we need additional new digital and computational ways of working in order to capture the full advantage of the broad vision that Gillispie offered in both *The Edge of Objectivity* and the *DSB*. Digital and computational approaches begin to offer a vast array of new tools that can help with traditional forms of scholarly pursuits—the modern day descendants of *The Edge of Objectivity*, as well as enable new kinds of scholarship that transcend individual projects. We imagine Charles Gillispie as amused and fascinated by such possibilities. And we hope he agrees that it is a new kind of work about which reviewers will say "And it has much that any historian of science would be justly proud to have written."

4.1 Digital History and Philosophy of Science: The Heritage of Charles Coulston Gillispie

In our subsequent discussions we link history and philosophy of science because they are inseparable dimensions of understanding the transformations of our knowledge of nature and the world. It is also a tribute to the original conception that inspired the formation of the History and Philosophy Program at Princeton University, one of the many ways Gillispie has shaped the intellectual landscape of the last decades. Digital History and Philosophy of Science (HPS) has three dimensions based on the kind of digital objects involved and the computational methods used: (1) found objects, that is the published and archived objects of traditional scholarship available in digital form; (2) new publications written specifically for a digital publication environment that include work that is traditionally scholarly but also provides digital links to other work and sources in ways that make each new object part of a growing complex network of digitally available interconnected objects (texts, images, maps, etc.); (3) newly discovered findings that result from use of new tools, including found relationships among items not known to be linked. This requires the use of computational tools and repositories and represents the truly transformative aspects of Digital HPS. We look at each of these and at some sample efforts to develop each.

1. Found Objects would typically start with existing publications and expand from there. Take Gillispie's *Edge of Objectivity*. Right away he introduces

particular *people*, including Galileo, Paolo Sarpi, Machiavelli, Leonardo, Newton, and Einstein—all in the first ten pages. Then there are particular *works* by the individuals. People work in *places*, so we read also of *institutions* such as Oxford, and of *contextual influences* such as the Roman Catholic Church. He introduces algebra, logic, and eventually the other *fields* of science. These are all relatively defined units that can be linked throughout the work and connected with other uses of the same. We can flag such key words and phrases, to be linked with others directly and through more complex relationships (as discussed further below).

Interpretations are harder. There are *ideas* and *concepts* such as that of "objectivity" that run throughout. And Gillispie adds his own interpretation, saying that modern science "is impersonal and objective. It takes its starting points outside the mind in nature and winnows observations of events which is gathers under concepts, to be expressed mathematically if possible and tested experientially by their success in predicting new events and suggesting new concepts. Modern science has not abandoned rationality, but it is first of all metrical and experiential." (p. 10)

Further, Gillispie recognized that science is not something automatic running along on its own, but rather that it involves interpretation by human actors to develop the science. And these actors work in social contexts for particular purposes, and therefore exercise judgment about the use of science in society. That is, "the influence of science is not simply comfortable. For neither in public nor in private life can science establish an ethic. It tells us what we can do, never what we should." (p. 154) That comes from society, and the result changes in different societies.

Capturing such complex interpretations and understanding of the *contexts* digitally is harder, but not impossible. New scholarly tools include annotation approaches to include abstract *concepts* and *interpretations*, linked to the authors, and such tools are providing considerable promise for complex interpretive linking that yield new scholarly findings and make possible new scholarly interpretations because of the new ability to visualize connections.

Gillispie's own *Edge* does not include illustrations, but a 21st century project easily could, if a collaboratively collected database of such images existed in a stable repository with shared protocols and standards for documentation and archiving of metadata to facilitate shared use. The objects in question would each need to have adequate and appropriate metadata identifiers, but standards already exist for such work.

And what if a shared repository included biographies and descriptions of the people, places, institutions, contexts, ideas, concepts, and interpretations. Plus published work included in the references and mentioned in the discussion. If all these objects were available in the same repository, what glorious fun it would be to read the wide-sweeping scholarly interpretation. Instead of being an ill-informed undergraduate student struggling to connect the many pieces and not knowing much about each, the reader could look up additional information, find pictures, see the original texts mentioned, and so on. And this imagined research methodology does not require that all the scholars and all the owners of scholarly materials put them in the same physical space. Certainly not. The repository can be distributed in as many places as desired. What is necessary is shared vocabularies so that different word usages are linked, and shared metadata and descriptors so that difference objects are found.

The nucleus of such a distributed digital repository is already emerging as individual projects within HPS have begun to collaborate, standardize their practices and develop the necessary tools together. These projects—including the Newton projects at Indiana and Sussex, the Embryo Project at Arizona State and the Marine Biological Laboratory (MBL), the Einstein Papers project at Caltech, the History of Quantum Mechanics Project and the many digital projects connected to the Max Planck Institute for the History of Science in Berlin as well as others—have formed a consortium that coordinates these activities and also lobbies for this approach. So far the Marine Biological Laboratory in Woods Hole and the Max Planck Institute for the History of Science in Berlin have stepped up to become the kind of partner that Charles Scribner was to Gillispie's *DSB* project. These two institutions act as hosts for repositories, clearing houses for tools, incubators for projects and, as of Spring 2011 also as an educational and training center. The MBL, with it long tradition of advanced summer courses, is beginning to offer a course in computational methods for digital HPS.

This is being done, and we can and should do much, much more to share resources. If we took all the money that all the scholars have used to visit collections dispersed around the world, and used them to make materials available to everybody, that would be a fine start.

2. New Publications can take traditional forms, of course, and appear in proprietary journals. But we can hope that scholars will increasingly come to feel able to contribute to open access publications, where a work can include links to all the found objects used to generate the results. This is starting to happen. And a new publishing project at the Max Planck Institute for the History of Science in Berlin (MPI)—The Max Planck Research Library for the History and Development of Knowledge—is taking these possibilities in new directions. The goal here is to publish annotated editions of sources and translations as well as edited collections and workshop volumes in an open access environment while maintaining all the standards of peer review. However, historians of science who appreciate the tactile experience of a book need not worry. While all the works are available in an open access digital format, and thus also accessible to scholars from less prosperous countries and universities, they can also be acquired as well produced bound volumes. Print-on-demand technology has developed to such a degree that this is now possible— these volumes can even be ordered on amazon.com.

Digital publications can also take the form of short entries that help to link across a wide range of other contributions. This is an approach we are developing with the Embryo Project Encyclopedia (http://www.embryo.asu.edu), not unlike that of Wikipedia and other similar projects that take the same basic approach (see Laubichler et al., 2009; Maienschein and Laubichler, 2010). The idea here is to provide the complement to Gillispie's breadth and interpretative scope, his "profound reflection." The focus of these projects is a further development of the encyclopedic vision; to take big ideas and complex developments and "atomize" them. Break the big into component pieces, and provide short descriptive articles. What is different from the traditional multi-volume encyclopedia is that this approach is open-ended, that the links and cross-references are easy to follow and can grow dynamically and that interpretative texts can directly link to sources. Such a structure becomes increasingly necessary as the amount of available information begins to threaten its use. Such (short) interpretative essays then serve as the glue that provides links among objects that might not be linked otherwise.

3. Newly discovered findings also can result from the application of new computational tools and methods. For example, such tools enable us to establish relationships among items not known by any one scholar to be linked. It is the small world paradigm of network theory applied to the understanding of the history of science or for Princeton insiders, it is a very complex application of multiple Erdös numbers. To begin with, annotation tools can be used to search texts and other documents and harvest information that is then coded in the form of relationships. Of particular interest are so-called RDF triples, which allow connections such as "Newton was a critic of Hooke" or "Harvey studied at Padua." While some naïve researchers assume that they can find such relationships with tools like Google, just try it yourself and you will immediately see why that cannot work. Only when the exact words appear together will they be found. But sophisticated scholarly annotation tools can let the researcher discover connections. Then these can be captured and stored as new scholarly findings for other scholars to use. To be even more useful, the triples can also be given contextualizing information as well, though tools for such "quadruples" are still in development.

To illustrate how computational methods and approaches can change the practice and methodological foundations of scholarship, we briefly describe the change from traditional interpretive history to a new form of scientific history. What are the methods that allow for detailed historical analysis with data driven computational approaches, what we are calling scientific history? It requires starting with *defining basic informational units* for complex historical processes—what we call a functional relational unit (see below) and *storing* these "atoms of information" in a *centralized repository*, as well as *developing tools that enable queries that can retrieve more information and links than an individual scholar has access to* in his university or research institute. Resulting search outcomes can then reveal aspects of historical developments that were not visible initially: not just more information, but new kinds of information.

The structure of functional-relational units as statements linking two objects creates a highly connected network of knowledge, which, in turn, leads to multiple kinds of synthetic knowledge units that would not have been obvious without the linkages enabled by computational annotation tools and storing information units within a centralized repository.

This means that we have to define historical statements as data so that they can be analyzed with computational methods, and that also requires social cooperation as scholars see the value of this new approach and contribute data. Traditionally, historians have shared their interpretations of data but not the data themselves, which can be found in their notebooks and private annotations of sources. This has made it very difficult and time-consuming to reanalyze most historical works. But, as changed practices requiring that data be shared together with published analyses have transformed the sciences and made them more effective, we argue that scientific history will have a similar transformative effect on understanding of the history and nature of science.

Focusing on textual information we need to extract basic informational statements as well as the relevant contextual information. Our "atom of information" is a functional relational unit that connects two objects through a relevant functional relation; for example "Charles Darwin travelled on *HMS Beagle*." Without additional contextual information such a statement has limited use. We therefore need to be able to connect this simple statement with more layers of contextual information, such as metadata of the text from it was extracted, information about who created the statement (a trusted or less trusted source, for instance, or a machine), and additional layers of information that represent the larger historical context (such as 19th century Britain, or history of biology, the history of exploration). The structure of the statement as well as the layers of contextual information allow us to reconstruct complex networks of events, such as where Darwin's journey took him, who else was on board the HMS Beagle, what publication refer to the *Beagle*, additional information about places Darwin visited, animals he collected and what we know about them today, who he wrote to on his trip, and what the British government felt they got out of their investment in the *Beagle*, and so on.

Now, these are all different sorts of information that require looking in many different places and relying on a certain amount of chance to find and link the "atoms" of knowledge. Historians of science have perfected this approach as part of their craft. However, no one can know it all. Furthermore, the situation is much harder if we are looking at a complex large-scale 20th century science project rather than somebody famous and well-documented like Darwin. For these cases, a data driven social-computational and collaborative approach that is the core of the third dimension of digital HPS is essential.

4.2 Conclusion

Scholarship is changing, and this is good. Gillispie wrote at the end of *Edge* that "History is made by men, not by causes or forces, and I have tried to write with due attention to the intellectual personalities who have borne the battle, and not without sympathy for its casualties. And though I have written as closely to the texts as my competence permits, I want the tale to move unencumbered by the barnacles of scholarly apparatus." (p. 521). In his *Isis* review, Hall reflected that he would have wished for a few more such barnacles to guide his reading. Yet we have appreciated reading without all the footnotes and side trails considered necessary by historians such as Hall himself.

Nonetheless, the new digital HPS can make everybody happy. By including all the barnacles, which are anchored in their own carefully documented archives or libraries and storage facilities but also freed through digital linkages, we have both. Narratives with their own personalities and diversity of directions, yes, and also all the scholarly apparatus. This apparatus can reassure other scholars, but how much more effective that reassurance is when we can all read the same documents and study them, complete with annotations and links to other materials and sources.

Diderot argued that the three means of acquiring knowledge are rarely combined, but if they are we are confronted with a creative genius. Few individual have reached this level. But as cultural evolution progresses, the locus for a creative genius increasingly shifts towards a well connected network of individuals each making their own contributions, but doing so as parts of a whole that is most certainly bigger than the sum of its parts.

Acknowledgment Thanks to the National Science Foundation for support through multiple grants.

References

Darnton, Robert. 2008. "The Library in the New Age." *New York Review of Books*. June 12, 2008.

Darnton, Robert. 2010. "Can we create a national digital library?" *New York Review of Books*. October 28, 2010.

Dictionary of Scientific Biography. 1970–1980. *Supplement*. 1981, edited by Charles Coulston Gillispie; Gillispie and F. L. Holmes et al. New series 2007, edited by Noretta Koertge. New York: Scribner.

Diderot, Denis. 1753. Quoted from *On the Interpretation of Nature, no. 15* in Lester G. Crocker, editor and Derek Coltman translator. 1966. *Diderot's Selected Writings*. New York: Macmillan.

Gillispie, Charles Coulston. 1960. *The Edge of Objectivity. An Essay in the History of Scientific Ideas*. Princeton: Princeton University Press.

Gillispie, Charles Coulston. 1980. *Science and Polity in France Science and Polity in France. The End of the Old Regime.* Princeton: Princeton University Press; 2004. *Science and Polity in France: The Revolutionary and Napoleonic Years.* Princeton: Princeton University Press.

Hall, A. Rupert. 1960. Review of *Edge. Isis* 51: 344–347.

Laubichler, Manfred, Jane Maienschein, and Grant Yamashita. 2009. "The Embryo Project and the Emergence of a Digital Infrastructure for History and Philosophy of Science." *Annals of the History and Philosophy of Biology* 12: 79–96.

Maienschein, Jane and Manfred Laubichler. 2010. "The Embryo Project: An Integrated Approach to History, Practices, and Social Contexts of Embryo Research" *Journal of the History of Biology* 43: 1–16.

Part II
Archaeology

Chapter 5
Peking Man: New Light on an Old Discovery

Tore Frängsmyr

5.1 Background

During the late 19th century and in the beginning of the 20th century Swedish voyages of discovery were directed towards two goals, the North Pole and East Asia. This was partly by chance but also reflected the achievements of certain notable individuals. Swedish arctic research started with Otto Torell's journeys to Iceland, Greenland and the Arctic Ocean during the 1850s. One of his assistants was Adolf Erik Nordenskiöld who was later to lead the first expedition through the North-East Passage with the ship Vega (1878–1880). This success stimulated a series of enthusiastic expeditions supported by The Royal Swedish Academy of Sciences, King Oscar II and private donations, not least from Alfred Nobel (Liljequist, 1993).

This wave of almost intoxicating delirium was terminated abruptly in 1897 with the tragic balloon expedition that Salomon August Andrée undertook together with two other men. The aim was to reach the North Pole with the help of a balloon, but the enterprise failed monumentally. The balloon was forced down onto the ice after just a few days, following which the group drifted on an ice flow fighting ice, cold and polar bears. Three months later, in October 1897, they made landfall on the glacier covered island Vitön in Svalbard where they soon after perished. The expedition was first re-discovered in 1930 when the tragic chain of events was reconstructed with the help of surviving diaries and roles of film. A byproduct of the numerous relief expeditions that were sent out in the years following 1897 to investigate the fate of André's group, was that large areas of the island world of the Arctic Ocean were mapped (Sundman, 1968).

The voyages of discovery to East Asia were also inspired by Nordenskiöld and his expedition. When the vessel Ymer anchored in Stockholm on April 24th 1880 in the presence of the King and thousands of sightseers, a fifteen year old boy, Sven Hedin, stood amongst the crowds on the quay. He dreamed of becoming just as successful an explorer as Nordenskiöld. At the age of twenty

T. Frängsmyr (✉)
Uppsala University, Uppsala, Sweden
e-mail: Tore.Frangsmyr@idehist.uu.se

J.Z. Buchwald (ed.), *A Master of Science History*, Archimedes 30,
DOI 10.1007/978-94-007-2627-7_5, © Springer Science+Business Media B.V. 2012

he took employment with the Nobel brothers in Baku, at the same time learning Russian, Persian and other key languages. At twentyone, in 1886, he made his first journey of discovery in Persia, after which followed a series of expeditions interleaved with studies of geography in Germany. He found ruined cities and "the wandering lake" Lop Nor, reconstructed ancient trade routes and demonstrated the existence of the Transhimalayan mountain chain. As late as 1935, Hedin organized large expeditions with Swedish, Chinese and German participants and published a series of books translated into the leading languages of culture. Through these he made a major contribution to Western knowledge of the geography of China and central Asia (Odelberg, 2008).

Interest for the North Pole and for China was to develop side by side in a man who later played a major role in the discovery of Peking Man, Johan Gunnar Andersson (1874–1960). He was educated as a geologist and already at an early stage in his career took part in expeditions to the North Pole, the first together with Alfred Gabriel Nathorst to Spitsbergen in 1898. During the following year he lead his own expedition to Bear Island in the Barents Sea, and in 1892 he joined the famous expedition to the South Pole under the leadership of Otto Nordenskjöld (nephew to Adolf Erik Nordenskiöld). Andersson and some of his colleagues wintered over under extremely harsh conditions in huts built of stone with skins or canvas as a roof. After two years in the Antarctic they succeeded in making their way back to civilization (Andersson, 1904, 1944, 1954). Andersson made a successful career as a geologist and in 1906 was appointed Director of the Geological Survey of Sweden (SGU). He was Secretary General of the XI International Geological Congress in Stockholm (1910), when Gerard de Geer presented his classic geochronological researches and the development of the last glaciation in Scandinavia (Frängsmyr, 2006, ch. 5).

In 1914 Andersson threw himself into a new adventure. He was granted leave of absence to carry out an assignment for the Chinese government that would eventually lead him into a new area of study. His commission in China was to search for ores, especially iron ore, and to establish the Geological Survey of China, which took place in 1916. In addition to prospecting for ores he became interested in the famous, so-called "dragon bones." In Chinese native medicine it was the custom to prepare medicines from the ground down bones and teeth of "dragons". In contrast to Western notions, dragons were viewed as a positive force in eastern mythology, having a beneficial effect both medicinally and in questions concerning climate. A German doctor, K. A. Haberer, brought back from China a quantity of these dragon bones, and the paleontologist Maximilian Schlosser from Munich determined that they were derived from fossil mammals (bear, hyena, rhinoceros, hippopotamus and deer) (Andersson, 1926, 1932 (English translation 1934), 1933).

Andersson understood the paleontological value of these finds but faced difficulties in undertaking a systematic study. He was not a paleontologist by training and, for commercial reasons, the apothecaries were reluctant to reveal the source of their dragon bones. He established contact with some female

Swedish missionaries who were familiar both with the local geography and the fossil localities and subsequently made an agreement with Dr. V. K. Ting who had been appointed to lead the new Geological Survey of China. Since specialist knowledge was not available in China, Andersson was given permission to send the most valuable material to Sweden for study. Duplicate material and other finds were to remain for study in China. A special publication series, *Paleontologia Sinica*, was established to present the scientific results.

The collected material was sent to Uppsala where Carl Wiman, an old student colleague of Andersson, had been appointed Sweden's first professor in paleontology in 1910. Wiman supervised the time consuming preparation of the material and undertook its scientific description. As the collections grew in size Andersson realized that he needed the services of a specialist paleontologist in China and he turned for help to Professor Wiman (Frängsmyr, 2006, ch. 6; Wiman, Carl, ms, unpublished autobiographical notes). It was here that Otto Zdansky made his appearance, but first some words are due about Andersson's own activities. He raised money in Sweden through a special China Committee to undertake archeological excavations. He was successful in locating an entire Chinese stone age settlement with stone axes and ceramic vessels, and it has been said that he "discovered" the Chinese prehistory by this work during the 1920s. He collected and purchased numerous objects, magnificently painted vessels, urns, vases and water pots. He undertook excavations in the settlement Yangshao cun in the Henan republic in eastern China where his discoveries resulted in the recognition of a unique culture between the stone age and the bronze age: the Yangshao culture. The unique and extensive collections were transported home to Stockholm. Andersson returned to Sweden after eleven years in China and was appointed Professor of Geology in 1925. But already in 1929 he became Professor in East Asian Archaeology and superintendent for the east Asian collections he had himself brought home (Niklas Jacobsson, manuscript, "Johan Gunnar Andersson," unpublished master's dissertation, p. 186).

5.2 Zdansky's Secret

In the spring of 1920 a group of Austrian students arrived in Uppsala. The years following the First World War were exceptionally difficult in Vienna and the students starved and froze in unheated rooms. Uppsala Students' Union responded by inviting some twenty students to stay in Uppsala for the remainder of the year. Amongst these was Otto Zdansky (1894–1988) and his brother Erich. After unsuccessful studies in engineering, Otto had switched to paleontology and managed to start a thesis study under Professor Othenio Abel of a collection of fossil turtles located in the museum in Vienna. In Uppsala he made contact with Professor Carl Wiman for whom his work as assistant provided a small sum for his keep. Amongst other duties, he participated in the sorting and analysis of the rich material that Wiman had received from China. At the close of the year, however, Zdansky returned to Vienna to complete his thesis (Zdansky, manuscript "Hågkomster", unpublished autobiography).

When Carl Wiman received the request from Johan Gunnar Andersson to send a suitable paleontologist to China his thoughts turned to Zdansky. Already at the beginning of 1921 he wrote to him about the offer to work with Andersson in China. Zdansky would receive free travel and subsistence under three years, but he should not expect a salary. These were not exactly princely terms but he had no alternative offers. Zdansky had, however, the presence of mind to request that he should be granted the right to describe and publish scientific studies of his own discoveries. In the long term, this was naturally a stroke of fortune (Zdansky ms, Wiman ms).

After six weeks travel by ship Zdansky arrived in Peking where he met Johan Gunnar Andersson and V. K. Ting, the director of the Geological Survey of China, together with the vice-director, Dr. Weng Wenhao (W. H. Wong). Following Andersson's directions he started work at first near the town of Zhoukoudian where there were several caves with fossil bones. Conditions in the town were rather primitive. Zdansky set up camp, with a camp bed, in the town temple. Local men helped with the excavation. Andersson paid him a visit after several days, accompanied by Walter Granger from the American Museum of Natural History in New York who was recently employed as chief paleontologist on the American expedition to Mongolia lead by Roy Chapman Andrews. The most productive cave was called *Dragon Bone Hill*, in English translation, but "Chou K'ou Tien" became the most used designation following Zdansky's use of the name in his first report in 1923. In modern orthography this is Zhoukoudian (Zdansky ms; Andersson, 1932, pp. 122, 125).

Zdansky carried out excavations all through the summer of 1921. The material was uncovered and sorted on site, with individual bones being cleaned and labeled. Part of the material was left unprepared in blocks and packed into crates since there was insufficient time to examine all samples in detail and prepare out the individual fragments. It was the material in these crates that was later to be examined in Carl Wiman's laboratory in Uppsala. During the excavations Zdansky found a molar from a hominid. He discovered the tooth while sieving material from the cave. He knew at once that it came from a hominid, but he just slipped it into his pocket without saying anything to anyone. It is not mentioned with a single word in his report of 1923. Naturally, he was aware that this was a sensational discovery, and that it could be a trace after the oldest fossil man yet found. He was to remain silent for five whole years, and declared later that he understood what a fuss the discovery would have caused. He had priority to work with his own discoveries, and wished to complete his studies in peace and quiet.

After Zhoukoudian Zdansky devoted himself to other excavations in consultation with Andersson. He made expeditions to many different provinces during the late autumn of 1921, all of 1922 and a large part of 1923. He found remains of camels, saber-toothed tiger, rhinoceros, giraffe and several species of dinosaur. In the bone caves of Chi Chia Kou (Jijiakou), which were mined in small tunnels into the mountain, he was able to buy material from the workers as well as find material himself. On occasions his activities were interrupted on account of groups of brigands and soldiers passing through the district. The times were troubled, to say the least (Frängsmyr, 2006, ch. 6).

Andersson financed the project with money from the China Committee in Stockholm where he received strong support from the Crown Prince, afterwards King Gustaf VI Adolf. The grants, however, were exhausted by the end of 1923, and Zdansky started to prepare for his journey home. He was offered paid work with the American expedition, but in the absence of any guarantee of more long term employment, he felt obliged to decline. During the journey home his diaries and field notes were stolen, with the result that it is sometimes difficult to precisely date his discoveries and various expeditions. In the middle of January 1924 Zdansky arrived back in Uppsala where he once more started to work with Carl Wiman in order to describe parts of the Chinese material. Wiman was clearly supportive of Zdansky and helped him find cheaper and better accommodation (Zdansky ms; Wiman ms).

Zdansky found another hominid tooth while he sat and worked with the material from Zhoukoudian; it was from the same locality and with the same state of preservation as the first, but this time a premolar. Not even now did he inform those around him. The two teeth were kept in a small glass jar in a bookshelf in his work room. It was no more dramatic than that. He knew the secret he alone was privy to, but he was in no hurry. First priority lay with the scientific work ahead.

5.3 The Discovery Becomes Public

In 1926 it was planned that Crown Prince Gustaf Adolf would make an extended foreign journey together with his new wife Louise (they had married in 1923), almost a round-the-world trip across North America and Asia. Johan Gunnar Andersson offered to meet them in Peking and, together with his colleague in Stockholm, Axel Lagrelius, to arrange a symposium to present the excavations and collecting activities in Kina. As chairman, the Crown Prince was a key person in the China Committee and Andersson wished to arrange a formal scientific gathering in his honor (Lewenhaupt, 1928).

With this in mind, Andersson wrote to his colleague Carl Wiman in Uppsala to enquire if his researchers had made new discoveries during their study of the Chinese material. Wiman sent him reports concerning the dinosaur *Helopus* (now called *Euhelopus*), the giraffes in the Hipparion beds and a long snouted, three-toed horse. Wiman also asked Zdansky is he had anything interesting to present – and he did! Zdansky himself writes that this was the first time that an outsider was given information about the two teeth. Since he had recently completed his manuscript description of the Zhoukoudian fauna he removed the section concerning the teeth and edited it slightly "so that it could be printed and presenting with the meeting in Peking as a preliminary communication" (Zdansky ms, fol. 57).

The formal scientific symposium was held on October 22nd 1926 in the auditorium of the College of Medicine in Peking. The royal guests were formally welcomed and the Crown Prince lectured on the traditions of archeological research in China. Amongst other presentations was a lecture by the French priest Pierre Teilhard de Chardin. The highlight of the evening was Andersson's report on the extensive research in progress at Professor Wiman's laboratory in Uppsala. As a conclusion, he presented an illustration of the teeth and read out Zdansky's short written description. Andersson emphasized that "this in itself highly incomplete find may well come to be recognized as the most important discovery resulting from the Swedish collecting in China" (Andersson, 1932, pp. 124–128, quotation p. 125 f.; Zdansky ms, p. 57).

The presentation of the discovery was greeted with complete silence. The sensation was total. Andersson had skillfully anticipated the effect of the public announcement. It was clear to all that this discovery could spring from the oldest human ever found. As Andersson commented, "the far-reaching significance of the communication was only fully appreciated by the scientific leaders in Peking, namely Doctors Ting, Wong, Black and Grabau." It may be recalled that Ting and Wong comprised the leadership of the Geological Survey of China. The Canadian anatomist Davidson Black was employed by Peking Union Medical College in Peking, and the American Walter W. Grabau worked as a paleontologist at Peking University. The activities of the latter two institutions were based on grants from the Rockefeller Foundation.

The news spread around the world by cable. Two days later it was presented in the Swedish newspapers. Grabau immediately gave the discovery the name Peking Man. Black submitted an article to *Nature* where it was published in November 1926, and somewhat later in *Science* (December 1926) (Black, 1926a, 1926b). Andersson noted that Black in his article spoke of "Wiman's pictures and text", despite the fact that it was Zdansky that delivered both. The same tendency to marginalize Zdansky is clearly visible in Black's article in *Nature* which fills little more than a column. He claims there that Andersson discovered the locality which was subsequently "partly excavated" by Zdansky. The material from the excavations has been prepared, it is said, in Wiman's laboratory "and afterwards studied there by Dr. Zdansky" (Black, 1926a, p. 734). Thus, it was not even mentioned that Zdansky found the first tooth at the locality and the second in the laboratory. Black dated the find to the late Tertiary or early Quaternary, indicating an age of about 1 million years. He made comparisons with a tooth which Haberer had obtained from a Chinese apothecary which had been described by Schlosser in 1903. The latter characterized the tooth as "*Homo? Anthropoid?*" and pointed out that some kind of pre-human might be found in China in the Tertiary or Quaternary.

Black also stressed that the find showed great similarities with Java Man and Piltdown Man (which at this time was still taken seriously), and the somewhat younger Heidelberg Man. All these discoveries were made in different directions and at relatively great distances from the Central Asiatic Plate which had probably been the centre of the radiation. The new discoveries, writes Black, give us additional evidence for the hypothesis that Man originally came from central Asia: "The Chou Kou Tien discovery therefore furnishes one more link in the already strong chain of evidence supporting the hypothesis of the central Asiatic origin of the Hominidae." (Ibid.)

The only one to express concern about the derivation of the discovery from a human was Father Teilhard de Chardin but Andersson defended Zdansky's conclusion. Clearly, it was possible to confuse worn down carnivore teeth and early hominid teeth but he had "complete confidence in Zdansky's critical acuity" and his experience with studies of China's fossil carnivores (Andersson, 1932, pp. 126–128). Zdansky kept a low profile, both in the preliminary report and the final printed version. The information he compiled in 1926 was printed in a revised version in the *Bulletin of the Geological Survey of China* (submitted in April 1927 and printed in autumn of the same year). He establishes without doubt that the teeth had belonged to Man and not an ape, but he is skeptical about some of the conclusions in the paleontological literature and meant that finds of this kind often give rise to unrestrained speculation. He had no wish to contribute to the formulation of far reaching conclusions on the basis of the small amount of material he had presented and which he, until further notice, will designate as "? Homo sp." (Zdansky, 1927, 284).

Zdansky disagreed with the opinion that the find of Peking Man should be of Tertiary age (at least a million years old following the accepted chronology of the time). Until a complete description of the fauna from Zhoukoudian was available he would only say that "these teeth (which are of Quaternary age) should be regarded as decidedly interesting but not of epoch-making importance". He takes the same wait and see attitude when he writes about the teeth in his own larger study of the cave which was essentially completed during the autumn and published in 1928, *Die Säugetiere der Quatärfauna von Chou K'ou Tien*. Clear criticism of Andersson, Black and others who speculate too eagerly on the basis of the available material can be read between the lines. Consequently, the question mark accompanying the determination *Homo* remained (Zdansky, 1928, 140 f).

5.4 A New Discovery

During his 1926 presentation Andersson announced that the Swedish expedition did not plan to continue its excavations at Zhoukoudian, but would welcome that others completed the project. In particular he had in mind collaboration between the Geological Survey of China, Peking Union Medical College and the Rockefeller Foundation. And so it proved to be. Supported by Rockefeller finances, systematic excavations of the cave were to be undertaken, and it was natural that Zdansky was asked to lead these. In the mean time, Zdansky had accepted a position in Cairo, Egypt, and did not wish to change this decision. Thus, the offer was passed on to Birger Bohlin, another of Carl Wiman's younger colleagues, who was on site in Zhoukoudian in April 1927 when excavation recommenced. Bohlin had received his doctoral degree for a thesis concerning the Chinese material, in particular giraffes from the Hipparion epoch. The Geological Survey of China was formally responsible for the project, and as a result, it was decided that all material should be considered its property and not allowed to be transported outside of China. The official leaders of the project were Dr. V. K. Ling and geologist Li Ji, but Bohlin was responsible for the actual excavations. All discoveries of hominid material were to be studied by Black at the medical college's anatomical department (Bohlin, 1944, pp. 9–20; Lanpo and Weiwen, 1990, ch. 5).

Accordingly, work in the cave made a new start with great resources. About 3,000 cubic meters of deposits were quarried and yielded large quantities of bones and other remains from various mammals. Politically, at this time, a regular civil war raged between the nationalists and the communists. The political troubles had gradually increased and various combatant groups passed back and forth through the landscape. Boxes with quarried material were often searched by guerillas or soldiers in search of weapons.

Three days before work was planned to finish for the year, on October 16th 1927, Birger Bohlin found a tooth from a hominid. He followed Zdansky's example and put it into his pocket. This was probably the safest place taking into account all the current strife and the wandering bands of thieves. Bohlin managed to return to Peking three days later and immediately sought out Black with his discovery. He was so eager that he went directly to Black without first having contacted his own wife. Both Bohlin and Black knew at once that they had further evidence for the existence of Peking Man.

The new discovery was fully described by Black in a report later during the autumn of 1927 within the series *Palaeontologia Sinica* (Black, 1927). He compared the tooth in detail with the two earlier finds of Zdansky, and even stressed that it had been found in the same area of the excavation as the two others. Following comparisons with living material, he concluded that the recently discovered tooth was more primitive than a human tooth but more specialized than that of a chimpanzee. He drew comparisons with photographs of Zdansky's discoveries and with copies of these made under the direction of Wiman. He considered it established that the teeth came from two different individuals of the same kind, the pre-human known as Peking Man. He deduced also that this creature did not live during the Tertiary as previously supposed, but during the Quaternary (about 500,000 years old), as Zdansky had always claimed. In conclusion, Black considered that the material in question represented a new species of hominid, and he gave it the name *Sinanthropus pekinensis* (Black and Zdansky) (Ibid., p. 21).

5.5 Scientific Priority

We have now arrived at a key point in the narrative. Peking Man has been discovered and accepted by the scientific community; he has received a scientific name and been placed within the hominid family tree. In terms of age, he is described as contemporaneous with Java Man and both are considered to be definitely older than Piltdown Man and Heidelberg Man, not to mention the Neanderthals which were clearly later. Some considered Peking Man to be the oldest. We will now briefly examine how this tale of discovery was recorded in the contemporary world and by posterity. Zdansky's discovery of the two teeth can be regarded as beyond discussion. He found the first tooth at the excavation site in 1921 and the second tooth in Uppsala while working through the material some time between 1924 and 1926. The third tooth was discovered by Bohlin at the excavation site in October 1927. The account of the three teeth and presentation of the new species published by Davidson Black in 1927 contains some amazing errors. In the preface, signed by W. H. Wong (Weng Wenhao), it is stated that Zdansky made his excavation in 1920(!) and that he subsequently found both teeth during his examination of the collections in Uppsala. In his introduction Black writes that Zdansky found the two teeth in Wiman's laboratory during the summer of 1926(!). Thereafter he relates that Bohlin found the third tooth and admits that he had not participated in the excavations personally but had just visited the locality on several isolated occasions. The nonchalant recording of the factual data concerning Zdansky's discoveries is astounding, taking the importance of the scientific text into account (Ibid., espec. foreword and p. 1).

Johan Gunnar Andersson held an ambivalent opinion of Zdansky. He pointed out willingly that he had personally taken the initiative for the excavation at Zhoukoudian, which was correct, but also maintained that he had predicted what Zdansky would find. His letters indicate a noticeable irritation concerning Zdansky and imply that he worked too slowly and avoided making obvious conclusions. At the same time as he expresses his great respect for Zdansky's knowledge, he shows a tendency to want to marginalize him. In his book *The children of the yellow earth* he writes that Zdansky "with extraordinary skill excavated our most common vertebrates and thereafter described large parts of this unprecedented material" (Andersson, 1932, p. 112).

In the preface to the same book, however, he thanks some twenty colleagues and sources of inspiration without any mention of Zdansky. Instead he thanks Davidson Black whose "greatest fame is the brilliant feat of research through which he re-awakened Peking Man from the dead". In so far as Black never made any discoveries of his own from the locality, this statement stands out as greatly exaggerated, if not directly erroneous. In another of his books, *Chinese and Penguins*, Andersson speaks of "our discoveries" in the cave of Zhoukoudian: "Together with a prehistoric animal world of rhinoceros, deer, tiger and the like, we chanced upon a pair of teeth that were very human-like. Chinese scientists have continued the excavations after we finished and as a result found skulls, jaws and loose teeth of a human-like creature that came to be called Peking Man, *Sinanthropus pekinensis.*" Truly a remarkable description (Ibid., forward; Andersson, 1933, quotation p. 393).

Andersson and Black have influenced all other historians. Even Jia Lanpo, himself a paleontologist who worked at Zhoukoudian in the early part of the 1930s, followed this version of history in his various accounts, the latest in *The Story of Peking Man* (with Huang Weiwen, 1990) (Lanpo and Weiwen, 1990, ch. 2). He cites long passages from Andersson, not least where he claims to have predicted what Zdansky should find in the cave. Ralph von Königswald, also a paleontologist who visited Zhoukoudian, follows Andersson's path in his book on prehistoric man (*Begegnungen mit dem Vormensch*, 1955; in English as *Meeting prehistoric man*, 1956). To be sure, he describes Zdansky as the actual discoverer, but it is Andersson's contribution that is given prominence, often in Andersson's own words. Additionally, he describes how Bohlin found his tooth and how Black on the basis of this single tooth (!) identified a new prehistoric man (Königswald, 1955, pp. 43–57). But Black made use of all three teeth, considering them to be derived from two individuals of the same kind. His attribution of the species description to "Black & Zdansky" shows that he will emphasize the connection between Zdansky's find and the new tooth, although not to the extent that Bohlin's name was included. Königswald recounts that Black had a large gold chain with a small hollow charm to contain the tooth specially made. And so he travelled around the world with the gold chain and the fossil around his neck "in order to show it to his colleagues everywhere and hear their opinions" (Ibid., pp. 47f).

The same or similar errors are repeated in all the books written about Peking Man. Ruth Moore thinks it was Zdansky who presented his discovery in Peking in 1926. Herbert Wendt confuses Zdansky's discovery with finds that should have been made by Haberer and Andersson (Frängsmyr, 2006, ch. 8). Christopher G. Janus and William Brashler describe Bohlin as a German paleontologist and claim that Davidson Black "would contribute more than any other person to the excavations at Choukoudian" (Janus and Brashler, 1975, quotation p. 19).

The examples could be repeated. In summary, it can be stated that the most common error is that Davidson Black is given as the discoverer of Peking Man, as for example in *Encyclopedia Britannica* (Encyclopaedia Britannica, 1987, vol. 12). The next most common error is that Andersson is given as the discoverer. Both these individuals have personally contributed to this creation of the myth. In an article in a Swedish journal from 1928 Andersson writes about the locality "where I and my assistant paleontologist, Dr. O. Zdansky, started excavations in summer 1921." Subsequently he writes "When he (Zdansky) occupied himself with processing of this cave fauna in Uppsala in 1926, he found amongst a wealth of other material two teeth that were very hominid-like. It was this find that I communicated in autumn 1926 in Peking" (Andersson, 1928, 19/2 1928).

The description is directly misleading, perhaps because Andersson wishes to cover up the five years where Zdansky had kept him ignorant of the first discovery. In addition he praises Davidson Black who dared to identify a new hominid species and "in this way was less cautious than Dr. Zdansky", since the latter "only diffidently" placed a query with the species. Another article in the same journal was attributed to Black, although probably translated and expanded by Andersson. The title of the article was "Definitive clarity achieved concerning Professor J.G. Andersson's prehistoric man from Peking." (Andersson, 1929, 15/5 1929; Black, Davidson, 1929) When the first complete cranium of Peking Man was found in the late autumn of 1929, Andersson wrote yet another article where he discusses the excavations that under the summer of 1921 were made "by Dr. Zdansky under my leadership". Additionally, he says that he encouraged Zdansky to continue digging during summer 1923 "and he found then two teeth of a human-like creature" (Andersson, 1929, 29/12 1929). Not even the date is correct. Andersson's portrayals of the earliest discoveries have also formed the basis for accounts on the international stage. In reality, no new facts have been added to the literature since the 1930s.

5.6 Conclusion

Honor to whom honor is due. Andersson and Black made great contributions but neither of them was a paleontologist. They did not take part in the excavations and neither of them made any discovery. It was Zdansky and Bohlin who made the discoveries, thanks to their scientific competence. They recognized immediately that the teeth belonged to a hominid species. If Zdansky had not discovered the first two teeth it is likely that excavation at Zhoukoudian would not have continued. Some might think that Zdansky's behaviour was not scientifically correct. He kept quiet about his discovery for several years without even communicating the news to his employer Andersson. He had not finished his research when he was given the offer to lead the new excavations. He wished to complete his survey of the cave at Zhoukoudian in peace and quiet, and he did not want to make too far reaching conclusions (Zdansky, 1928; ms, passim). Behind this austere attitude lay a scientist's skeptical conviction of not accepting that which is not yet proven.

Zdansky was employed at the university in Cairo where he built up a paleontological department. His recently formed family, a wife and son, accompanied him during the first years. Afterwards they remained in Uppsala and Zdansky returned home every year on vacation. The rising tide of nationalism in Egypt became ever stronger and was to reach its culmination in 1952 with Colonel (later President) Nasser's coup d'état. It became increasingly difficult for Europeans to live and work in the country, and in August 1950 Zdansky left Cairo where he had lived for twenty three years. He had no position in Uppsala University and had to make do with short periods as a teacher of elementary courses. He resumed his research into the collections brought home from China and in 1952 found yet another tooth. This tooth was described by Zdansky the same year in an article in *Acta Zoologica* (Zdansky, 1952). All told, this was thus the fourth tooth. Together with the first two teeth, it remains in safe keeping in Uppsala. That found by Bohlin stayed in China.

The excavations at Zhoukoudian continued for a decade after Bohlin's 1927 discovery, at times on an almost industrial scale (Lanpo and Weiwen, 1990; Lanpo, 1980; Black, 1929; Wenzhong, 1929). A variety of finds of greater and lesser significance were made but the political disturbances made the work increasingly difficult. A state of civil war existed with numerous political fractions involved. Furthermore, China was infiltrated by the Japanese which resulted in a full scale Japanese occupation in 1937. Excavations ceased on July 9th 1937; it was no longer possible to work in a meaningful manner. By that time, the remains of about forty individuals of Peking Man had been found. They were well taken care of at Peking Union Medical College.

The Japanese entered the Medical College on the day after the attack on Pearl Harbor on December 7th 1941 and demanded the keys to the strong room where the remains of Peking Man were stored. The room was empty. No one could say where the remains could be found. It is most likely that they were placed in packing cases and placed on board of an American ship with the intention of transporting them to the USA, but that the ship was sunk by the Japanese. Before this happened, however, the institute's last director, Franz Weidenreich, had replicas made of all the finds (Weidenreich, 1935, 1936, 1939. For later literature, see also Shapiro, 1974; Rukand and Shenlong, 1983, 1985; Mateer and Lucas, 1985; Boaz and Ciochon, 2004). The fate of the original material is still unknown.

Until now, the three teeth in Uppsala have been all that remains of the original material of Peking Man. Since this article was written, however, a sensational new find has been made. Some of the original sample boxes, untouched since the 1920s, were finally opened in March 2011 and another tooth was found. This canine tooth shares the same derivation and time of collection as the earlier teeth. Since Bohlin's tooth has disappeared we can talk therefore about four original teeth of Peking Man, all of which are still in Uppsala.

Acknowledgements The translation was made by John S. Peel, Professor of Historical Geology and Paleontology at Uppsala University and, for many years, responsible for the care of the three teeth of Peking Man as Director of the Paleontological Museum (now a part of the Museum of Evolution). I am grateful for his support and interest, and advice in certain scientific matters.

Under my visit to China in 1995 I learned that the Chinese were well aware of the role of Otto Zdansky (1894–1988) in the discovery of Peking Man. This awareness is much less in Sweden and Western countries, where Zdansky's contribution is often overlooked. I met Zdansky, probably in 1963, while I was a young student, but without being aware of his discoveries.

After returning to Sweden in 1995 I made contact with his son Göran Zdansky, a University Lecturer in Chemistry. He loaned me photographs, notes and letters, together with un unpublished autobiography referred to herein as "Zdansky, ms".

Carl Wiman (1867–1944) was appointed Sweden's first professor in paleontology at Uppsala University in 1911. His granddaughter Mrs Görel Oscarsson kindly loaned me Carl Wiman's unpublished autobiographical notes, referred to in the following text as "Wiman, ms".

Birger Bohlin (1898–1990) was the paleontologist who succeeded Zdansky in China and who found a third tooth of Peking Man. His son Staffan Bohlin has kept his notes and some drawings, which I could use.

This article is mainly based on my book written in Swedish, *Pekingmänniskan: En historia utan slut* (Stockholm: Natur och Kultur, 2006).

Bibliography

Andersson, Johan Gunnar, 1904. *Två år bland Sydpolens isar.*
— 1944. *Antarctic.*
— 1954. *Sydpolens hjältar.*
Andersson, Johan Gunnar, 1926. *Draken och de främmande djävlarna.*
— 1928. *Sinanthropus pekinensis,* nydöpt urgammal medlem av människornas familj, *Svenska Dagbladet* 19/2 1928.
— 1929. En fullständig skalle av *Sinanthropus pekinensis, Svenska Dagbladet* 29/12 1929.
— 1932. *Den gula jordens barn.*
— 1933. *Kineser och pingviner: En naturforskares minnen från jordens fyra horn.*
— 1934. *Children of the Yellow Earth.*
Black, Davidson, 1926a. Tertiary Man in Asia: The Chou K'ou Tien Discovery, *Nature* 20/11 1926.
— 1926b. Tertiary Man in Asia: The Chou K'ou Tien Discovery, *Science* LXIV:1668 1926.
— 1927. *On a Lower Molar Hominid Tooth from the Chou K'ou Tien Deposit, Paleontologia Sinica, ser D* VII:1.
— 1929a. Definitiv klarhet vunnen om prof. J.G. Anderssons urmanniska vid Peking, *Svenska Dagbladet* 15/5 1929.
— 1929b. Preliminary Note on Additional Material Discovered in Chou K'ou Tien in 1928, *Bulletin of the Geological Survey of China* 1929:8.
Boaz, Noel T. and Ciochon, Russell L., 2004. *Dragon Bone Hill: An Ice-Age Saga of Homo erectus.*
Bohlin, Birger, 1944. *På jakt efter urmänniskans förfäder.*
Encyclopaedia Britannica, article Peking Man, 12 (1987) unpublished.
Frängsmyr, Tore, 2006. *Pekingmänniskan: En historia utan slut.*
Janus, Christopher G. and Brashler, William, 1975. *The Search for Peking Man.*
Königswald, G.H.R. von, 1955. *Begegnungen mit dem Vormenschen.*
— 1957. *Möten med urtidsmänniskan.*
Lanpo, Jia, 1980. *Early Man in China.*
Lanpo, Jia and Weiwen, Huang, 1990. *The Story of the Peking Man.*
Lewenhaupt, Sten, 1928. *Axel Lagrelius' Kinaresa.*
Liljequist, Gösta, 1993. *High Latitudes: A History of Swedish Polar Travels and Research.*
Mateer, Niall J. and Lucas, Spencer, 1985. Swedish Vertebrate Palaeontology in China: A History of the Lagrelius Collection, *Bulletin of the Geological Institutions of the University of Uppsala,* new series, 11.
Moore, Ruth, 1953. *Man, Time and Fossils.*
— 1954. *Felande länken.*
Odelberg, Axel, 2008. *Äventyr på riktigt: Berättelsen om Sven Hedin.*
Rukand, Wu and Shenlong, Lin, 1983. Peking Man, *Scientific American* 248:6.

— 1985. Chinese Palaeanthropology: Retrospect & Prospect, *Palaeoanthropology and Palaeolithic Archaeology in the People's Republic of China,* eds. W. Rukand and John W. Olsen.

Shapiro, Harry L., 1974. *Peking Man.*

Sundman, Per Olof, 1968. *Ingen fruktan, intet hopp: Ett collage kring S.A. Andrée, hans följeslagare och hans polarexpedition.*

Weidenreich, Franz, 1935. The *Sinanthropus* population of Choukoutien (Locality 1), with a preliminary report on New Discoveries, *Bulletin of the Geological Survey of China* 1935:14.

— 1936. The Mandibles of Sinanthropus pekinensis: A Comparative Study, *Palaeontologica Sinica* 1936:7.

— 1939. *Sinanthropus* and His Significance for the Problem of Human Evolution, *Bulletin of the Geological Survey of China* 1939:19.

Wendt, Herbert, 1953. *Ich suchte Adam.*

— 1955. *I urmänniskans spår.*

Wenzhong, Pei, 1929. An Account of the Discovery of an Adult Skull in the Choukoutien Cave Deposit, *Bulletin of the Geological Survey of China* 1929:8.

Zdansky, Otto 1923. Ueber die Säugerknochenlager von Chou K'ou Tien, *Bulletin of the Geological Survey of China* 1923:5.

— 1927. Preliminary Notice on the Two Teeth of a Hominid from a Cave in Chihli (China), *Bulletin of the Geological Survey of China* 1927:5.

— 1928. *Die Säugetiere der Quartärfauna von Chou K'ou Tien, Paleontologia Sinica, Series C* V:4.

— 1952. A New Tooth of *Sinanthropus pekinensis* Black, *Acta Zoologica* 1952:XXXIII.

Chapter 6
The Puzzle Picture of Lucretius: A Thriller from Herculaneum

Knut Kleve

There is an engraving that portrays the grave of Napoleon. Two large trees overshadow the grave. There is nothing else to be seen in the picture, and the immediate spectator will see no more. Between these two trees, however, is an empty space, and as the eye traces out its contour Napoleon himself suddenly appears out of the nothingness, and now it is impossible to make him disappear. The eye that has once seen him now always sees him with anxious necessity.
S. Kierkegaard, The Concept of Irony (1841, p. 56)

6.1 A Non-believing Believer

Titus Lucretius Carus wrote *On the Nature of Things (De Rerum Natura)* in the first half of the first century BC in six books and (at least) 7409 hexameters to free his friend Memmius[1] and everybody else who happened to read his poem, from the fear of gods and death, and so be able to live a happy life. *On the Nature of Things* is both a didactic poem[2] and an epic,[3] hailing Lucretius' philosopher hero Epicurus[4] as the first man (or rather god!) to give a complete, atomistic and irrefutable explanation of everything going on in the world.

---

[1] Friendship was highly valued in Epicureanism: "Friendship goes dancing round the world proclaiming us all to awake to the praises of a happy life" (Epicurus, Bailey 1926, p. 115). On the shithead Memmius see Smith, pp. xlvi ff.

[2] Kleve (1979). Cf. Kubbinga pp. 19ff.

[3] Kleve (1999).

[4] On Epicurus in general cf. Rist. Short, but modern and instructive is Kubbinga, pp. 8–13.

K. Kleve (✉)
University of Oslo, Oslo, Norway
e-mail: knut.kleve@ifikk.uio.no

J.Z. Buchwald (ed.), *A Master of Science History*, Archimedes 30,
DOI 10.1007/978-94-007-2627-7_6, © Springer Science+Business Media B.V. 2012

Lucretius' literary model was Quintus Ennius, "father of Roman poetry," and his verses have a weight and majesty and a depth of passion, which have caused critics to rank him as the equal of Virgil, if not his superior.[5]

St. Jerome tells that Lucretius committed suicide in his forties after being driven mad by a love-potion (a statement cherished by the Church).[6]

6.2 Mad Ideas

Among Lucretius' crazy Epicurean thoughts may be mentioned: the cruelty of religion (1.80-101), nothing is created out of nothing (1.205-07), nothing destroyed into nothing (1.262-04), invisible atoms (1.265-270) moving in void (1.329-69) and having parts (1.599-634), infinity of the universe (1.958-1001), absolute velocity (2.142-64), swerving atoms (2.216-93), infinity of worlds (2.1048-76), birth and decay of worlds (2.1105-74), the soul as a part of the body (3.94-416), refutation of teleology (4.823-57), invisible matter in space (5.717-19), beginning of life (5.783-820), survival of the fittest (5.855-77), primitive man (5.925-87), early civilisation (5.1011-27), origin of language (5.1028-90), discovery of fire (5.1161-1240), origin of religion (5.1161-1240).

6.3 Defeat and Triumph

In late Antiquity Lucretius and materialistic philosophy went out of fashion. Idealism and religion prevailed. In the Middle Ages Aristotle became *the* philosopher of the Church. *On the Nature of Things* was only just saved in a few manuscripts into the Age of Enlightenment.[7]

Numerous modern scientists and thinkers are influenced and inspired by Lucretius, among them Copernicus, Bruno, Bacon, Galileo, Kepler, Hobbes, Gassendi (atomic theory), Bayle, Beeckman (first molecular theory),[8] Newton, Voltaire, Goethe, Darwin,[9] Marx, Nietzsche.[10] In 1924 Albert Einstein wrote a preface to the famous German classicist Hermann Diels' translation of *On the Nature of Things*[11] where he expressed his astonishment that Lucretius, who had

---

[5] Formulations borrowed from Bailey in *The Oxford Classical Dictionary*, 1950, s. v. "Lucretius." For Cicero's high valuation of the poem cf. Smith p. xi.

[6] On St. Jerome's statement cf. Smith pp. xviii ff.

[7] On the manuscripts of Lucretius cf. Smith pp. liv ff.

[8] On Isaac Beeckman see Kubbinga pp. 33ff.

[9] Kleve (1976).

[10] On the aftermath in general see Boyancé pp. 316ff. More literature in *Der kleine Pauly* s. v. "Lucretius," section VI. On atomism in the tradition see Kubbinga pp. 29ff.

[11] Diels pp. VIa-b.

a knowledge of science far below the level of an ordinary school pupil today could present such a mature world picture. Einstein also found the Roman Lucretius' admiration for Epicurus and Greek culture in general "moving" and very different from his own homeland's attitude to other nations. (Einstein left Germany for America in 1933 because of the persecution of the Jews.)

6.4 The Papyrus Villa

The Papyrus Villa (Fig. 6.1), north of Herculaneum, was buried with the city and the city of Pompeii during the eruption of the volcano Vesuvius in 79 AD. The villa contained a library of about two thousand book scrolls, rediscovered in a heavy state of carbonization in the 1750s (Fig. 6.2). Two hundred scrolls were cut open by knife, six hundred unrolled by means of the fish glue method of

Fig. 6.1 Papyrus Villa, inner garden (peristylium). Reconstruction, Getty Museum, Malibu, California

Fig. 6.2 Charred book scrolls

Antonio Piaggio.[12] The legible papyri are for the most formerly unknown works on rhetorical, ethical and aesthetical themes written by the Epicurean philosopher Philodemus,[13] a contemporary of Lucretius. Fragments from several books of Epicurus' main work *On Nature* were also found. Remains from Lucretius' *On the Nature of Things* were just identified as late as 1989.[14]

In the lifetime of Philodemus and Lucretius the villa was probably one of the land-houses of the Roman nobleman Lucius Calpurnius Piso Caesoninus, the father-in-law of Julius Caesar and political adversary of Cicero. It served as a meeting-place for a philosophical circle around Philodemus.

6.5 Big Ego

In 1984 Brynjulf Fosse, Fredrik Störmer and I were invited by the National Library in Naples to resume the unrolling of the Herculaneum papyri. The fish glue method of Piaggio had been abandoned for more than a century. We made a glue of gelatine and acetic acid adaptable to different degrees of carbonization.[15] To our assistance we had a team of technicians from the library and scholars from Marcello Gigante's International Centre for Herculaneum Studies. Among the scholars were Gigante's favourite students Mario Capasso and Anna Angeli. About two hundred scrolls were unrolled with the new method.

Two papyrus scrolls donated to Napoleon in 1802 by the King of Naples were returned from Paris to be opened by the team. One of the scrolls unrolled by Tommaso Starace turned out to be a work *On Slander* by Philodemus, dedicated to Virgil, the final proof that the famous poet was a member of the

---

[12] On the method of Piaggio cf. Sider pp. 46ff.

[13] Gigante (1990).

[14] Kleve (1989).

[15] Kleve 1991. Before his death Fosse started experiments to read scrolls without actually opening them, by means of roentgen and ultra sound. D. Delattre (Paris) is conducting similar experiments today.

Fig. 6.3 PHerc.Paris. 2, fragment 279A, with dedication to Virgil (in vocative) visible on the next last line: ΚΑΙ ΟΥΕΡΓΙ[ΛΙΕ] ("and you, Vergil!")

circle in the Papyrus Villa (Fig. 6.3). In 1989 Gigante and Capasso wrote an article about the sensational discovery,[16] and Gigante had by and by a plaque from the Mommsen Society hung up on the premises, announcing him as the very first to read the name of Virgil in a Herculaneum papyrus.

The relationship between Gigante and Capasso got sour. Capasso was showing up before the plaque growling: "Wrong! *I* read it, not he." And Gigante began warning about Capasso, urging me to remove him from the team. But I had no intention to deprive myself of a valuable partner, and told Gigante so.

6.6 Big Surprise

One day Capasso came to me with a handful of papyrus pieces he had found in a forgotten drawer. They turned out to be fragments of Lucretius' *On the Nature of Things*! Or so I thought. I suggested a common edition, but Capasso would not believe that the fragments were of Lucretius. Instead I dedicated my edition to him, Gigante and Angeli. Gigante's old dream that Lucretius should one day be found in Herculaneum had at last been fulfilled, and he enthusiastically published

---

[16] Gigante-Capasso.

my results in his Naples journal *Cronache Ercolanesi* (note 14). The discovery was
a sensation and the readings were included in the forthcoming Lucretius editions.[17] The piece now numbered *PHerc.* 1829 was the very first to be identified
(Fig. 6.4a,b):

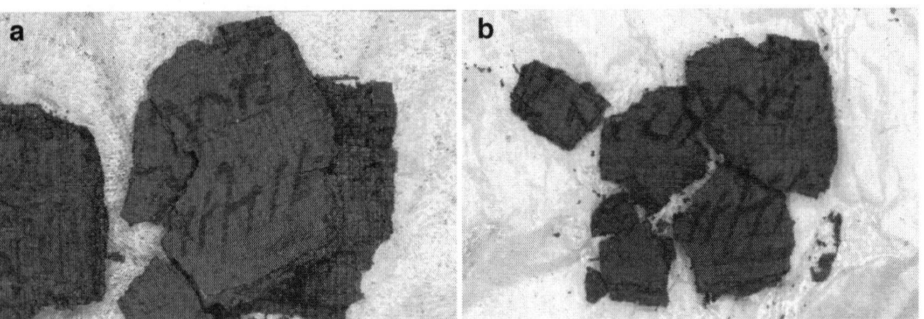

Fig. 6.4 (**a**) PHerc. 1829 before breakage. (**b**) PHerc. 1829 after breakage

There are two layers; the upper one is lower (a so called *sottoposto*), with a
fragment of Lucretius, book 5, 1301 (on the development of warfare: ordinary
chariots with two or four horses came before chariots with scythes on the
wheels):

et quam falciferos armaTUM · ESCendere currus

"and before *climbing up armed into* chariots set with scythes" (For the use of *interpunctio* (·) cf. below.)

The second layer is higher (a *sovrapposto*) and contains a fragment from the
same book, verse 1409 (on the development of music, a development that
according to Lucretius has made nobody any happier):

et numerum servare genus didicere neqUE HILO

"and they have learnt to keep to a special rhythm, *but* (are) *not a whit* (more pleased)"

A Herculaneum session was being announced in Lille (France), and Capasso
asked if I was coming. I told him no, and then he read a paper with a total
rejection of my readings by reducing the fragments to insignificance. Conclusion: can *not* be Lucretius, author unknown. In elevated style he likened himself
to the protagonist in Ibsen's *An Enemy of the People* (who reveals that the public
baths are polluted). Capasso's trump card, however, was a note he had found
later on in the drawer with the information that the pieces originally belonged to
the larger *PHerc.* 395. And that papyrus was absolutely *not* Lucretius![18]
    I answered his criticism during the Vienna Papyrologist Congress in 2001,
with Capasso in the audience. One picture of *PHerc.* 395 sufficed to show that

---

[17] Smith; Flores.

[18] Capasso (2003).

the scroll contained the second book of Lucretius (below Fig. 6.13a, b). In an appendix I added some extra twenty readings from the same book and in conclusion remarked that I did not see much likeness between Capasso and Ibsen's protagonist in *An Enemy of the People*. He rather reminded me of the same writer's *Master Builder* who, in an attempt at climbing his highest tower, falls down to smash his head (censored in the printed version).[19]

6.7 Lucretius Herculanensis

As already suggested, the fragments of *On the Nature of Things* are contained in seven papyrus pieces, now numbered *PHerc.* 1829-31, and in the large *PHerc.* 395, with seventeen spacious papyrus flakes. The fragments are heavily carbonized and torn up, catalogued as illegible, looking like patchwork. However, by means of microscopy, photography, image processing and search programs it has been possible to decipher the ruined text. The pieces are from Lucretius book 1, 3, 4 and 5, while the large papyrus contains book 2.

The middle "patch" (Fig. 6.5, surrounded by "patches" from other parts of the book) contains fragments of Lucretius, book 2, 206-9 (on the universal motion downwards):

> *nocturnasque faces caeli sublime vOLAntis*
> *nonne vides longos flammarum ducerE TRActus*
> *in quascumque dedit partis natura MEAtum*
> *non cadere in terram stellas et sidERA Cernis*

Do you not see that the nightly torches of the sky *flying on* high, *trail* long *tracts* of flames behind towards whatever side nature has set them *to travel*? Do you not *see* stars and *constellations* drop to the earth?

I had difficulties reading *R* until I realized that our small *r* (often written like a *Λ*) was already used in the time of Lucretius!

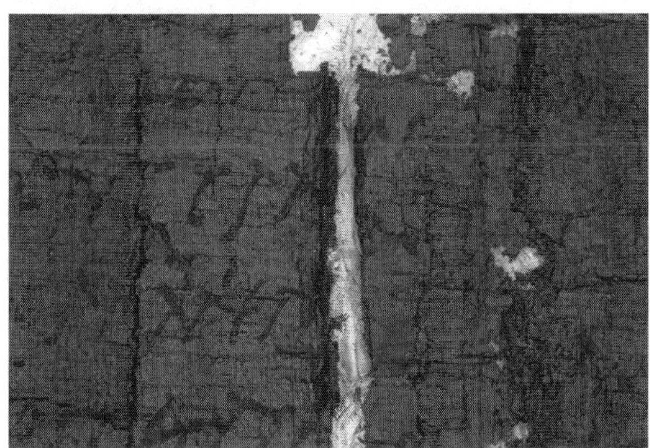

Fig. 6.5 PHerc. 395.13. "Patchwork"

---

[19] Kleve (2007), cf. Delattre.

There is also a drawing from the year 1805 (before the age of photography) of fragments from Lucretius' book 1, the original papyrus text being lost (Fig. 6.11). Thus all the books except book 6 of *On the Nature of Things* are represented among the Herculaneum papyri. There can be no doubt that the whole poem was once present in the library of the Papyrus Villa.

The script is an irregular Roman cursive, very much like the one used in the poet Ennius (Fig. 6.6), the comedian Caecilius Statius (Fig. 6.7) and graffiti from Pompeii and Herculaneum (Fig. 6.8). The cursive was easy to write, but difficult to read and for that reason abandoned as a book script in the second half of the last century BC and replaced by the Roman capitals still familiar to us today (Figs. 6.9–6.10). However, the older script continued to be used for daily purposes, letters and documents, eventually to pop up in the Carolingian renaissance as our lower-case letters.[20]

Examples of Latin script types in the Papyrus Villa and Pompeii:

Fig. 6.6 PHerc. 21. Ennius, Annales (Yearbooks). On the Pyrrhus war Kleve (1990)

Lucretius is the first author in the Papyrus Villa whose text is known beforehand. This gives a unique opportunity for observing in detail how flood,[21] compression, heat, carbonization, decomposition and unrolling can affect a papyrus text. But knowing Lucretius in advance also offers an opportunity for determining the original dimensions of a scroll. A stipulation of letter sizes, number of verses within a column, distances between columns etc. gives a preliminary estimate of 0.30 m in height and 25 m in length for a book of Lucretius, an increase on modern estimates of Herculaneum papyri.[22]

The Herculaneum fragments mostly support the readings of the oldest and best Middle Age manuscripts, now in Leiden, O(blongus) and Q(uadratus), called so after their shape (note 7). The Herculaneum papyri are written eight hundred years before O and Q and can confirm lacunas in the younger

---

[20] Tjäder pp. 124–127 follows the cursive until the fourth century AD. For Roman script types in Herculaneum see Kleve (1994 p. 316); Sider p. 65.

[21] Lead from the water pipes in the Papyrus Villa has been found on the scrolls, cf. Störmer.

[22] Capasso (1991 pp. 223ff).

Fig. 6.7 PHerc. 78.8. Caecilius Statius, Faenerator (The Money-Lender), end of comedy.
Kleve (1996)
Note: (Two last lines:)
saVIARE IUS / (stroke marking end of verse)
coronis (sign marking end of book)
(Priest:) "You *may kiss* the bride." (Happy ending!)

Fig. 6.8 Graffito in verse (distichon) outside an inn in Pompeii (the letter e carved // on the
concrete wall, backhand writing):
miximus in lecto fateor peccavimus
hospes si dices quare nulla matella fuit
"I pissed in bed, sorry to say, sir. If you ask why: there was no chamber pot!"

Fig. 6.9 PHerc. 817. Hexameter poem on the war against Antonius and Cleopatra. Author
unknown. Imperial time

Fig. 6.10 PHerc. 1475.
Senate speech. Author
unknown. Imperial time

manuscripts assumed by modern scholars and even give glimpses of their contents.[23]

However, despite their age the extant papyri are not Lucretius' original manuscript. This is shown by metrical (but not grammatical) errors in the texts.[24] These same errors make clear that the Herculaneum texts were not dictated, but transcribed, which was the common way of copying in the Papyrus Villa.[25]

Interpunctiones or word dividers (·) are sparingly used in the oldest Latin papyri (and totally absent in the Greek ones). In Lucretius they seem to have been used for metrical reasons, in order to stress special pauses within the verse, which indicates that the poem was supposed to be read aloud, say, during the yearly memorial ceremony on Epicurus' birthday.

Use of *interpunctio* (·) (Fig. 6.11):

Fig. 6.11 Oxford drawing no. 1615 from the year 1805. Original papyrus lost. Kleve (2010)
Note: There are both Oxford and Naples drawings of the papyri, but of *PHerc. 395* only a tiny Oxford drawing exists.

The drawing contains a fragment of Lucretius, book 1.983 (on the infinity of the universe):

effugiumQUE·FUGAE pRolatet copia semper

(no boundary can be set for the universe) *and* the opportunity *for escape* always *postpones* the escape

---

[23] Kleve (1989 p. 11); Kleve (2007, pp. 350, 352).

[24] Kleve (2007, pp. 349, 351, 352).

[25] Capasso (1991, pp. 219f.).

The letter *A* has disappeared in a compression of the original papyrus leaving only the left part visible, looking like an *I*. The *interpunctio* (·) is placed in a metrically "forbidden" pause[26] to illustrate the impossibility of ever crossing any boundary of the universe. This trick is combined with one of Lucretius' numerous word jingles *(effugium. . . fugae)*.[27]

Lucretius is often regarded as an erratic Epicurean, a lone wolf, outside the school, because of his alleged pessimism.[28] He seems to be obsessively occupied with decomposition and decay and ends his poem with the most miserable description of the great pestilence in Athens. Lucretius also chose to give his message a poetical presentation. Epicurus on the other hand had an optimistic view of life: "Pleasure is easy to obtain, pain easy to endure!"[29] He despised culture, disliked poetry and rhetoric, and recommended plain prose for philosophical writing.[30]

Philodemus, however, Lucretius' older contemporary, represented a new type of Epicureanism. Besides being a philosopher he was a valued epigrammatist[31] and recognised the charm and force of special types of rhetoric and poetry. He further realized that to meet the requirements of his Roman pupils he had to offer a broad introduction to Greek culture.[32]

Lucretius' pessimism may after all be just apparent, as he evidently left his poem unfinished. He had probably no intention of restricting himself to physics or ending up with a description of pestilence.[33]

The Papyrus Villa must have been a perfect place for Lucretius to dwell. There he had easy access to the works of his two heroes: Epicurus for philosophy and Ennius for poetry. There he found friends and partners of discussion. And there was a circle of young Romans eager to read and listen to his great poem. That transcripts of his poem circulated on the premises seems as natural as there being several copies of Epicurus' *On Nature* in the Villa (Fig. 6.12a–c).[34]

---

[26] Crusius pp. 49f.

[27] Bailey (1950) ad loc.

[28] Cf. *Der Kleine Pauly* s. v. "Lucretius," section V.

[29] The last part of the "fourfold remedy" (τετραφάρμακος), the first part being: "God is not to be dreaded/death not to be feared," Arrighetti [196].

[30] Bailey (1926, pp. 21, 139, 165); Kleve (1997 p. 55, n. 53).

[31] Gigante (2002).

[32] Kleve (1997).

[33] Sedley.

[34] Gigante (1979, p. 53).

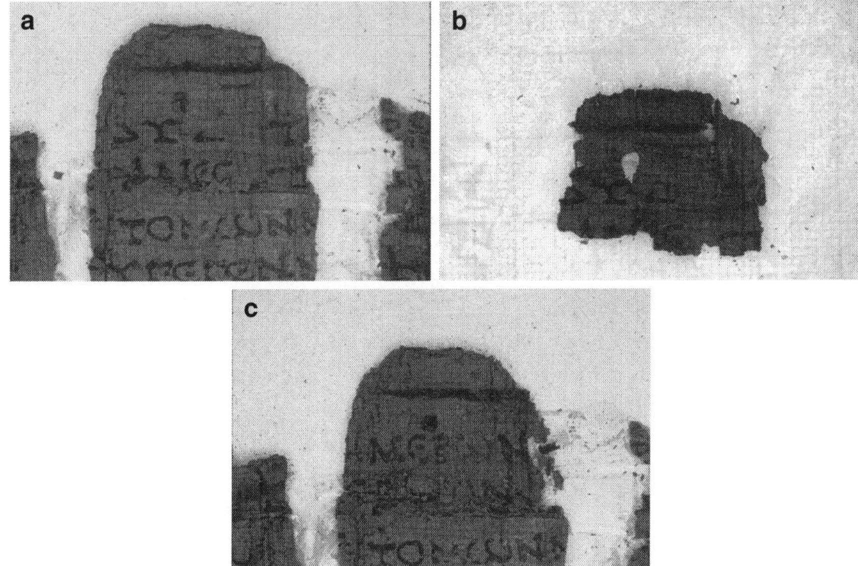

Fig. 6.12 (**a**) PHerc. 697. Epicurus, On Nature. (**b**) PHerc. 697. Epicurus, On Nature. Sovrapposto removed with ethanol solution. (**c**) PHerc. 697. Epicurus, On Nature. Another sovrapposto visible. More to be removed

6.8 Bad Eyesight

"He learnt a lesson!" somebody remarked after my Vienna encounter with Capasso. He did not. Capasso came to me in the shape of a young woman. By rearranging fragments and restricting rules for how letters can look in *PHerc.* 395 Beate Beer reaches the conclusion: can *not* be Lucretius, author unknown. Her trump card, however, is to make the crucial part of my Vienna picture disappear. It is neither visible in the official images[35] nor in the original papyrus, and has so to be dismissed.[36] But a Neapolitan colleague of mine just read it on the spot, and I am still keeping my old slide. There is every reason to show it again (Fig. 6.13a–b):

---

[35] Internet: http://www.herculaneum.ox.ac.uk/Lucretius-Herculaneum/PHerc0395 (The pictures in the present article are my own. They are mostly clearer than the official ones because they are in colors and taken at a closer range.)

[36] Beer pp. 76f.

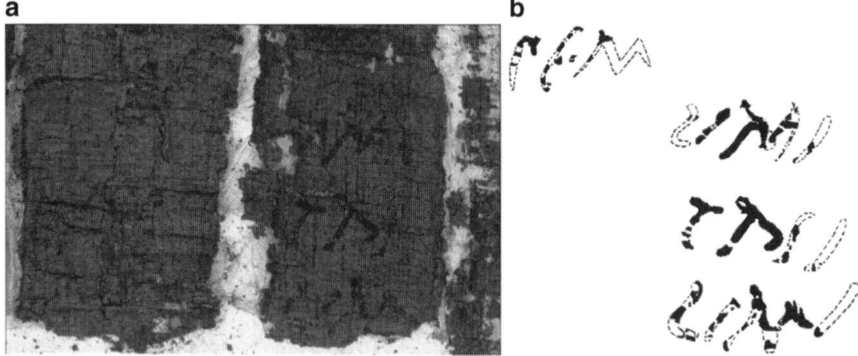

Fig. 6.13 (a) PHerc. 395.17. (b) Reconstruction, PHerc. 395.17

The picture presents three verse endings followed by the usual strokes: UM/ TAS/ UM/, plus a fragment of a verse above with the letters CE·M (non existent according to Beer). The *interpunctio* (·) indicates a word division. This exact combination is found only once in Latin literature: Lucretius, book 2, 1080-83 (ironically enough on nothing in the world being unique):

> *sint genere in primis animalibus indiCE·Mente*
> *invenies sic montivagum genus esse ferarUM/*
> *sic hominum geminam prolem sic denique muTAS/*
> *squamigerum pecudes et corpora cuncta volantUM/*

… (there) are (innumerable specimens of the same) kind. Begin with the animals, *let reason be your guide,* and you will find that this is so with the race of *wild beasts* that haunts the mountains, so with the twofold breed of men, so also with *the dumb* scaly fish and all the bodies of *flying* (fowls).

6.9 Stale Beer

Beer draws her main argument from *PHerc.* 395.5 (Fig. 6.14a–b):

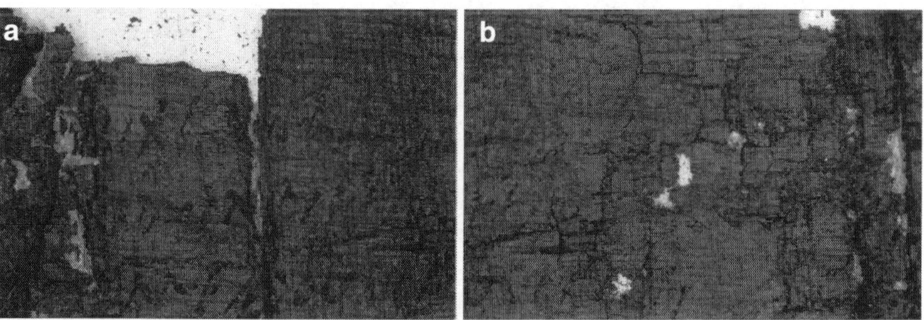

Fig. 6.14 (a) PHerc. 395.5. (b) PHerc. 395.5. Continuation

Beer reads[37]:
]s·graeco[
]rise[
pr]o duci[bus
...Greek...before the leaders...

If this is right, it is certainly not Lucretius. "Greek" is called "Graius," not "Graecus" in Lucretius, and the form "ducibus" does not occur in his poem.

Beer thinks that the three lines are written on the same papyrus layer and so belong together, but this is hardly right. If the printer is kind to my pictures it will be seen that the lines are written on different layers. Further, remnants of ink between the lines show that text we cannot read any more has preceded and followed the visible lines.

Beer takes only the text on the right side of the vertical cleft in the papyrus[38] into account, on the assumption that the two sides belong to different layers (*sottoposto/sovrapposto*). But that is not right either. Taken together the two sides will bring us straight back to Lucretius.

The first fragment is of Lucretius book 2.660 (on atoms of different shape and bulk moving in one conglomeration):

saepe itaque ex uno tonDENtES·GRAEMina campo[39]

for that reason (several animals) *are* often *cropping grass* from one field

The left part of the fragment is covered by a *sovrapposto* (with the letters *UN?*). Then a *D* is partly visible (written as our small *d*), followed by a fragmentary *EN*. The letter *t* has disappeared in the cleft, and the following *E* is just partly preserved. The fragmentary *M* was written as our cursive, or more like a *Π*.[40]

The second fragment is of Lucretius book 2.705 (on non-existent fantasy creatures):

tum flammam taetro spirantis orE CHIMAERas

further *chimeras* breathing flame from monstrous *mouths*

The papyrus cleft divides the *M*.

The third fragment is of Lucretius book 2.748 (verse of transition):

quod quoniam vinco fieri tuM·ESSE DOCEBo[41]

since I have succeeded in showing that this is so (that things in darkness have no color),
I *will next explain that* (also atoms) *are* (without color)

---

[37] Beer pp. 76, 82.

[38] The vertical clefts in the papyri are a result of compression from rubble, ashes and lava during the eruption of Vesuvius. On how they can be helpful in text reconstruction see Nardelli.

[39] On the spelling *graemina* for *gramina* see Kleve (2007, pp. 349, 352).

[40] For another *M* of this type see Kleve (2010). Cf. also *M* in the second line fig. 6.13a–b.

[41] On the metrically impossible *tum* see Kleve (2007, pp. 349, 352); Flores ad loc.

And since *I* have succeeded in showing that this is so, I will next explain that my readings are also verifiable by simple arithmetic:

The columns of *PHerc.* 395 contained 18-22 verses, which can be calculated from fragments of intervals where ends and beginnings of verses are readable.[42]

The fragments of verse 660, 705 and 748 stand in a vertical row and make up two combinations of *sottoposto/sovrapposto:* 660/705 and 705/748. These verses, then, were originally written on a level plus one line in the length of the scroll.

The distance between verse 660 and verse 705 is 705 – 660 = 45 verses or two columns, both 22 verses, plus one. The distance between verse 705 and verse 748 is 748 – 705 = 43 verses or two columns, both 21 verses, plus one, or one column with 20 and another with 22 verses, plus one.

6.10 Moral of the Story

From my student days I remember an Austrian professor who proved that Linear B could *not* be Greek. Beware of obscurantism! *Eppur si muove!*

Bibliography and Abbreviations

Arrighetti = Epicuro, Opere, ed. G. Arrighetti, Torino: Einaūdi 1967.

Bailey 1926 = Epicurus, The Extant Remains, ed. C. Bailey, Oxford: Clarendon 1926.

Bailey 1950 = Titi Lucreti Cari De Rerum Natura Libri Sex, ed. C. Bailey, Vol. I–III, Oxford: Clarendon 1950.

Beer, B., Lukrez in Herculaneum? – Beitrag zu einer Edition von PHerc. 395, *Zeitschrift für Papyrologie und Epigraphik* 168, 2009, 71–82.

Boyancé, P., Lucrèce et l'épicurism, Paris: Loūvain 1963.

Capasso 1991 = M. Capasso, Manuale di papirologia ercolanese, Congedo Editore 1991.

Capasso 2003 = M. Capasso, Filodemo e Lucrezio: due intellettuali nel patriai tempus iniquum, in: A. Monet (ed.), *Le jardin romain. Epicurism et poésie à Rome,* Lille 2003, 77–107.

Crusius, F. – Rubenbauer, H., Römische Metrik, Munich: Hùeber 1955.

Delattre = D. Delattre, Présence ou absence d'une copie du De Rerum Natura á Herculanum? (Réponse à M. Capasso) in: A. Monet (ed.), *Le jardin romain. Epicurism et poésie à Rome,* Lille 2003, 109–116.

Der Kleine Pauly. Lexikon der Antike, Drückenmüller 1964–75.

Diels = Lucretius, De rerum natura : lateinisch und deutsch von Hermann Diels, Bd. 2, Berlin: Weidmann 1924.

Flores = Titus Lucretius Carus, De rerum natura, ed. E. Flores, Vol. I-III, Naples: Bibliopolis 2002–2009.

Gigante 1979 = M. Gigante (ed.), Catalogo dei Papiri Ercolanesi, Naples: Bibliopolis 1979.

Gigante 1990 = M. Gigante, Filodemo in Italia, Florence 1990. Felice le Monnier. Trans. D. Obbink: Philodemus in Italy. Michigan U.P.

---

[42] Kleve (2007, pp. 351, 352, 353); Kleve (2010).

Gigante 2002 = M. Gigante, Il libro degli epigrammi di Filodemo, Naples: Bibliopolis 2002.

Gigante-Capasso = M. Gigante – M. Capasso, Il ritorno di Virgilio a Ercolano, *Studi italiani di filologia classica,* Terza Serie, Volume VII, Fasc. 1, 1989, 3–6 (plus picture).

Kierkegaard, S., The Concept of Irony with Constant Reference to Socrates. Trans. L. M. Capel, London: Collins 1966.

Kleve 1976 = K, Kleve, Antikk og moderne utviklingslære [Ancient and Modern Theory of Evolution], *Samtiden* 85, Oslo 1976, 625–635.

Kleve 1979 = K. Kleve, What Kind of Work did Lucretius write? *Symbolae Osloenses* 54, 1979, 81–85.

Kleve 1989 = K. Kleve, Lucretius in Herculaneum, *Cronache Ercolanesi* 19, 1989, 5–27.

Kleve 1990 = K. Kleve, Ennius in Herculaneum, *Cronache Ercolanesi* 20, 1990, 5–16.

Kleve 1991 = K. Kleve, A. Angeli, M. Capasso, B. Fosse, R. Jensen, F. C. Störmer, Three Technical Guides to the Papyri of Herculaneum, *Cronache Ercolanesi* 21, 1991, 111–124.

Kleve 1994 = K. Kleve, An Approach to the Latin Papyri from Herculaneum, in: *Storia poesia e pensiero nel mondo antico. Studi in onore di M. Gigante,* Naples 1994, 313–320.

Kleve 1996 = K. Kleve, How to read an illegible papyrus. Towards an edition of PHerc. 78, Caecilius Statius, Obolostates sive Faenerator, *Cronache Ercolanesi* 26, 1996, 5–14.

Kleve 1997 = K, Kleve, Lucretius and Philodemus, in: K. A. Algra, M. H. Koenen and P. H. Schrijvers (edd.), *Lucretius and his Intellectual Background,* Amsterdam 1997, 49–66.

Kleve 1999 = K. Kleve, Lukrets' "Om verdens natur" som heltedikt [Lucretius' "On the Nature of Things" as Epic], *Romansk forum* 10, Oslo 1999, 53–64.

Kleve 2007 = K. Kleve, Lucretius Book II in PHerc. 395, *Akten des 23. internationalen Papyrologenkongresses 2001,* Vienna 2007, 347–354.

Kleve 2010 = K. Kleve, Lucretius Herculanensis, PHerc. 395, and disegno oxoniense 1615, *Cronache Ercolanesi* 40, 2010 (in press).

Kubbinga, H., The Molecularization of the World Picture, or the Rise of the Universum Arausiacum I–II, Groningen U.P. 2009.

Nardelli, M. L., Ripristino topografico di sovrapposti e sottoposti in alcuni papiri ercolanesi, *Cronache Ercolanesi* 3, 1973, 104–115.

PHerc. = Papyrus Herculanensis (Naples).

PHerc.Paris. = Papyrus Herculanensis Parisinus (Paris).

Rist, J. M., Epicurus: An Introduction, Cambridge 1972. CUP Archive.

Sedley, D., How Lucretius Composed the De rerum natura, in: K. A. Algra, M. H. Koenen and P. H. Schrijvers (eds.), *Lucretius and His Intellectual Background,* Amsterdam 1997, 1–19.

Sider, D., The Library of the Villa dei Papiri at Herculaneum, Los Angeles: Getty 2005.

Smith = Lucretius, De Rerum Natura (Loeb Classical Library), trans. W. H. D. Rouse, rev. M. F. Smith, London: Harvard U.P. 1992.

Sottoposto = lower papyrus layer (belonging nearer the beginning of the scroll).

Sovrapposto = higher papyrus layer (belonging further inside the scroll).

Störmer, F. C., Kleve, K., Fosse, B., What happened to the papyri during the eruption of Vesuvius? *Cronache Ercolanesi* 16, 1986, 7–9.

Tjäder, J. O., Skrift, skrivande och skrivkunnighet i det romerska världsriket, in: *Kungl. Humanistiska Vetenskaps-Samfundet i Uppsala,* Årsbok 1981–82, 83–127 (Italian version: Considerazioni e proposte sulla scrittura latina nell'età romana, *Palaeographica diplomatica et archivista, Studi Battelli I,* Roma 1977, 31–60).

Part III
Astronomy

Chapter 7
Urania Propitia, Tabulae Rudophinae
faciles redditae a Maria Cunitia
Beneficent Urania, the Adaptation
of the *Rudolphine Tables* by Maria Cunitz

N.M. Swerdlow

Maria Cunitz's *Beneficent Urania*, published in 1650, has the distinction of being the earliest surviving scientific work by a woman on the highest technical level of its age, for its purpose was to provide solutions to difficulties in the most advanced science of the age, the mathematical astronomy of Kepler's *Rudolphine Tables*. Her work is at once original and the product of a long history. In 1577 Tycho Brahe began his program of observations while constructing the Castle of Uraniborg on the island of Hven in the Danish Sound, granted him by King Frederick II, with the object of a complete reform of astronomy, to produce new and accurate tables of the motions of the sun, moon, and planets, which he had envisioned years earlier. The observations, by Tycho and his many assistants, with the finest and largest new instruments, exceeding all previous observations in quality and quantity, many, many thousands, continued through twenty years on Hven, two years of travel through Germany, and, after Tycho entered the service of Rudolph II in 1599, at Prague and the estate of Benatky granted him by the Emperor. It was now for Rudolph that Tycho intended to complete his great reform of astronomy. In 1600 Johannes Kepler, about to lose his position as district mathematician in Graz because of the persecution and expulsion of Protestants by Archduke Ferdinand of Styria, and having no alternative, became an assistant to Tycho and began his investigations of the motion of Mars by which, as it turned out, planetary theory, mathematical and physical, was transformed. In September of 1601, Tycho introduced Kepler to the Emperor, who promised him the title of Imperial Mathematician to collaborate with Tycho on the new tables, which Tycho requested permission to call the *Tabulae Rudolphinae*, as was only reasonable as the Emperor was paying for them. The very next month, on 24 October, Tycho died, and responsibility for the tables fell to Kepler, although with interference from Tycho's heirs, their control of Tycho's observations and publication of the tables. But eventually Kepler had full access to the

N.M. Swerdlow (✉)
Division of Humanities and Social Sciences, California Institute of Technology,
Pasadena, CA, USA
e-mail: swerdlow@caltech.edu

J.Z. Buchwald (ed.), *A Master of Science History*, Archimedes 30,
DOI 10.1007/978-94-007-2627-7_7, © Springer Science+Business Media B.V. 2012

observations and full responsibility for the tables, and even after devising his new, physical planetary theory in the *Astronomia nova*, completed about 1605 and published in 1609, his work on this immense obligation was only beginning.

He began deriving the elements of the orbits of the planets from the observations while working on the *Astronomia nova*, and continued in the years following, after which computation of the tables, and checking the computations, was a labor of the sheerest tedium. During this time, Kepler wrote the *Epitome of Copernican Astronomy*, in principle a less technical exposition of the entirety of his new mathematical and physical astronomy, upon which the tables were based, finally published, with later revisions, in 1618, 1620, and 1621. By 1616 the tables were sufficiently advanced to compute the first of the *Ephemerides novae*, 'New ephemerides of the celestial motions from the year 1617 of the common era, above all from the observations of Tycho Brahe, physical hypotheses, and the *Rudolphine Tables*.' But then something happened he had not anticipated, a 'fortunate calamity for my *Rudolphine Tables*', as he described it in 1618, logarithms. John Napier had published his *Description of a Wonderful Table of Logarithms* in 1614, with a table at 1' intervals of sines and natural logarithms of trigonometric functions, with instructions for their use but no explanation of their computation, which appeared only in his posthumously published *Construction of a Wonderful Table of Logarithms* in 1619. Kepler, who first saw the *Description* in 1617, knew a good thing, a great thing, when he saw it. For logarithms were the most important innovation in computation since Hindu-Arabic numerals, even since place-value notation of the Babylonians. The effect was immediate. Kepler replaced many of the tables he had earlier computed with tables using logarithms, and the replacement tables were *computed* with logarithms. He was entranced with logarithms, introduced them in explaining calculations in later parts of the *Epitome*, began to use them in ephemerides for 1620, dedicated to Napier, and wrote two fundamental works on logarithms, *Chilias logarithmorum* (One Thousand Logarithms, 1624), an explanation of their computation with his own table, and *Supplement to Chilias logarithmorum* (1625), an explanation of their use. The time required for revising, really for recomputing, the *Rudolphine Tables* to incorporate logarithms, and for rewriting their very lengthy instructions, was extensive, and the result was virtually a new work, not completed until 1624, eight years after the earlier version, which also gave Kepler the opportunity to introduce yet more revisions and corrections. And there still remained the task of setting type and reading proof of a highly complex manuscript containing thousands of numbers, each of which would have to be checked and rechecked. The printing was finally completed in September 1627 and the tables dedicated to Emperor Ferdinand II, the very one whose earlier forced conversion of Protestants had Kepler expelled from Graz in 1600 and from Linz in 1626, to whom Kepler presented a copy in 1628, which the Emperor graciously received although Kepler was never paid what he was owed for all his years of work. Kepler took copies to the Frankfurt fair where the price was set at 3 gulden and sales were not particularly brisk.

The *Rudolphine Tables* were recognized as a contribution to mathematical astronomy of singular importance and *assumed* to be more accurate than any other since they were based upon Tycho's observations. But they were not widely used. Kepler and his son-in-law Jacob Bartsch computed from them ephemerides for 1621–28 and 1629–36, which Kepler himself had printed and published in 1630. But few had the patience for such tedious calculation, and the principal reason was the logarithms. Using tables of the traditional sort, as those of Reinhold (1551), Stadius (1560), Longomontanus (1622), and Landsberg (1632), with practice a planetary position in longitude and latitude can be computed in, say, twenty minutes, in a series of calculations less, with nothing more than a sexagesimal multiplication table for multiplication and division of at most three-place numbers. I have found that, using the logarithms in the *Rudolphine Tables*, computation of a planet's geocentric longitude and latitude takes *at least* three times as long, and the chance of error in numerous direct and inverse interpolations, especially of logarithms and in tables of non-integer entries, is not small, as I can also attest. And I cannot believe that many in the seventeenth century found it easier or faster. Consequently, there was interest in converting the tables to a form easier and faster to use, and some number of these were published, not as revisions of the *Rudolphine Tables*, but as new tables that happened to borrow from them. The tables we are considering, however, *Beneficent Urania* (*Urania propitia*) of Maria Cunitz, were published explicitly as a handier, easier version of the *Rudolphine Tables*, and for ingenuity in adaptation and ease of use may be, indeed, are the most successful, the most admirable of all.

Maria Cunitz (Maria Cunitia, 1610?–1664, see additional biographical information in the Appendix), the 'Silesian Pallas', was a wonder of learning who already in her fifth year could read, through constant reading of both sacred and secular histories, was proficient in seven classical and modern languages— according to a frequently repeated tradition going back to her own lifetime, Latin, Greek, Hebrew, and the vernaculars German, Italian, French, Polish— and learned in history, mathematics, and medicine, as well as cultivating the arts 'suitable to persons of her sex' of painting, poetry, and music, of playing musical instruments and embroidering. The contemporary account, by Johannes Herbinius, who knew her, is translated in the Appendix. She came of distinguished Silesian Protestant families on both sides, of her father, Heinrich Cunitz (1580–1629), who had studied at universities in Rostock and Frankfurt an der Oder, a physician learned in astrology and astrological medicine, on which he published, the natural sciences, and mathematics, and her mother, Maria Schultz (ca. 1584–1634), herself educated, daughter of Anton Schultz, who published on practical mathematics. Her parents, originally from Liegnitz, were married in 1603 in Wohlau, where she was born, the eldest of four daughters and one son; in 1615 the family moved to Schweidnitz (all within 50 km. of Breslau). Her education, and self-education, were supervised and encouraged by her parents. In 1623 she married David Gerstmann, a lawyer, who died in 1626. When Wallenstein began to enforce the Catholic faith in the region in 1629, the family fled to Liegnitz, where her father died on 5 August leaving his widow. The following year she married Elias von Löwen (Elias a Leonibus, formerly Elias Kretschmar, ca. 1602–1661), also a physician, proficient in astronomy and astrology, who may have come to Schweidnitz about 1626 and accompanied the family to Liegnitz. Not long after their marriage, they were forced to flee again to Pitschen, to the east near the Polish border. Elias had studied medicine at Frankfurt an der Oder and also astronomy with David Origanus, who published ephemerides for 1595 to 1630, later extended to 1654, as well as some number of annual calendars with prognostications. Elias also contributed to astrology, and before he received a title of nobility, under the name Elias Crätschmair published in Breslau in 1626 *Horologium zodiacale*, perpetual tables for finding the planetary hours, and in following years calendars with prognostications for 1627–1629.

Much of what is known about Cunitz, her interest in astronomy and astrology, and her work on the tables is found in Elias's 'Husband (*Maritus*) to the reader' in Latin and her own 'Answer to probable objections of the reader' in German in *Beneficent Urania*. Her principal concern is to answer the doubt, not without reason, of such ability as hers in the highest of the human sciences because of the infrequency of such accomplishments by 'Weiblichen Geschlechts Personen', which is also Elias's concern in greater detail. When he first came to Schweidnitz, he writes, there was constant report that the elder daughter of Heinrich Cunitz is not only devoted to the study of languages and histories, but also diligently pursues astrology, or natural predictions sought from the configuration of the stars, and, by erecting a nativity horoscope from ephemerides, somehow prepares a common judgment not at all badly. Of the former, he had no doubt, since he knew other women distinguished to some degree in studies of this kind, but of the latter he was undecided, being somewhat doubtful that a thorough knowledge of this science could belong to her sex. He found the reports were not exaggerated, indeed, her learning and virtue exceeded all that was said. Initially, their bond was their interest in astrology, for which he instructed her in all the complexities and refinements of chronocrators, commonly called directions, profections, and such, which determine at what age the events already known from a common judgment, a horoscope, will occur, for the most part sought in vain, for which the ephemerides she had been using, published before Kepler's, were insufficiently accurate. Directions are the most complex part of the judgment from a horoscope, so he was teaching her the most advanced astrology requiring the lengthiest calculations. He also instructed her in the correct calculation of the motions of the planets, which would free her from the use of ephemerides, requiring more advanced arithmetic and trigonometric computation, all of which she mastered very quickly and asked to learn more rapid methods of computation. Within four months she could not only compute perfectly the motions of the planets in longitude and latitude from Longomontanus's *Astronomia Danica* (1622), which Elias had praised highly, and solve plane and spherical triangles, but also after exactly one year, wearied by the length of those calculations, advised by a few things in letters from Elias, she reduced the calculation to a shorter method, putting the equations of the eccentric and of the sphere together in a single correction, which she then used for some time for her pleasure. What this means is that she computed double-entry tables combining the equations of the first and second inequalities, presumably for all the planets, a colossal amount of computation.

This took place before the death of Maria's father and before her marriage to Elias in 1630, after which, he says, he could advise her in person. It is said that after her marriage, she applied herself so completely to astronomy, devoting the greater part of the night to observations or calculations, and the day to sleeping, that she neglected her domestic affairs, meaning, if true, that she was remiss in giving orders to the servants. Still, there are records of her having three sons. When Elias learned that the calculation of the *Astronomia Danica* was often shown to be incorrect by observations of his own and others, and the *Rudolphine Tables* always approached closer to the heavens, he suggested she become acquainted with them, to which she consented. If this took place not long after their marriage, she was among the first to adopt this most innovative and advanced astronomy; it was in 1630 that Kepler published, in Sagan in Silesia, his *Ephemerides* for 1621–1636 based upon the *Rudolphine Tables*. Even if she began some years later, she was still early to use the tables. The tables presented formidable difficulties, including the use of logarithms, iterative calculations, and interpolations, so that computation appeared too lengthy and tedious, and she asked whether a shorter computation was possible. Elias explained at length the difficulties of adapting the tables and the effects of approximations, in particular on motions over a long period, and when she understood all of this, she resolved to prepare a handy abridgement of easier tables although the labor to do so was immense. But the beginnings of the work came in dangerous times, for the devastation of warfare intervened and they were again forced to leave their home and flee, from Pitschen to the north into neighboring Poland, where they were received by Sophia Lubinska, the Abbess of a convent in Olobock, who showed them every kindness, benevolence, and favor, and a peaceful and safe stay, in the village of Lubnitz just across the border from Pitschen, and after she departed from the living, by her successor Ursula Kobiersicka, to both of whom Elias expresses the greatest gratitude. Under their care and protection, from year to year, my Cunitz steadily made further progress with these tables, so that the first and second parts were completed in 1643 and the third part in 1645, and by an exertion, so arduous, tenacious, and of long duration, triumphing over all difficulties by the gracious aid of God, she at last applied the finishing touch. This history has been set out at length for the reader, he writes, lest anyone falsely think the work perhaps not of a woman, pretending to be of a woman, and only thrust upon the world under the name of a woman. The instructions for the tables, in Latin and German, written after the computation, contain later dates of observations of Jupiter in 1649, just before the beginning of type-setting and printing. By this time they had returned to Pitschen in Silesia, where they lived when the tables appeared and for the rest of their lives. During these years Cunitz's accomplishment and learning were recognized, indeed famous, in Northern Europe, and she corresponded with Ismaël Bullialdus (Boulliau) and Johannes Hevelius, the most distinguished astronomers, theoretical and observational, of her day. In 1655 a fire in Pitschen consumed her and her husband's library, instruments, and pharmaceutical supplies, and probably for this reason only a few scraps of paper by Cunitz survive.

The result of Cunitz's great labor, *Urania Propitia*, printed by Johann Seyffert in Oels and published in 1650, was a radical transformation of the *Rudolphine Tables*:

> BENEFICENT URANIA, or wonderfully easy (*faciles*, handy) astronomical tables, comprehending the essence (*vim*) of the physical hypotheses brought forth by Kepler, satisfying the phenomena by a very easy, brief way of calculating without any mention of logarithms; the concisely taught use of which for the period of time, present, past, and future—adding as well a very easy correction of the superior planets, Saturn and Jupiter, to a more accurate computation and improved agreement with heaven— MARIA CUNITZ imparts to devotees of this science in a two-fold idiom, Latin and vernacular. (German:) That is, new and long-desired, easy astronomical tables, through which the movement of all planets, with regard to longitude, latitude, and other phenomena in all moments of time, past, present, and future, is supplied in an especially handy method, presented here for the benefit of the devotees of science of the German Nation.

Urania is the muse of astronomy, after whom Tycho named the Castle of Uraniborg. The elimination of logarithms (*sine ulla logarithmorum mentione*), commended as a virtue in the title no less, might not have pleased Kepler after all his labor to include them, but it certainly shows that not everyone welcomed them with delight. The last phrase in the German is literally 'the science-loving German Nation' (*den kunstliebenden Deutscher Nation*).

Urania propitia is a substantial publication, longer than the *Rudolphine Tables*, containing 286 pages of tables, 144 pages of instructions in Latin, for 'necessitas', 'Nothwendigkeit', 118 pages in German, for 'amor patriae', 'Liebe deß Vaterlandes'. The complete explanation is quite interesting:

> It remains here to forestall suspicion, for some will ask why it is that against the custom of all expert practitioners I have written the introduction in a two-fold idiom; and it will be suspected (for men are prone to suspicion) that either through eagerness for novelty or display of Latinity I seek after a little empty glory. To these I answer that I have done this neither heedlessly nor without careful thought, but persuaded by the encouragement of two weighty authorities, of which the first is LOVE OF FATHERLAND, the other NECESSITY itself. I believe it is unknown to no one that the German Nation abounds in those of abilities suited to astronomical practice although often lacking knowledge of Latin, to whom I do not see by what pretense of duty I, being of the German Nation, could refuse or take away the benefit of my labors since it is agreed by the opinion of foreigners also that the Fatherland claims a great part of our labors as its own. And we see that most other Nations publish their writings, in particular their discoveries, in the vernacular idiom familiar to them (German: *in ihrer Muttersprache*). But because this work, in as much as it is intended for the universal improvement of the republic of letters, should also be imparted to other nations not proficient in our idiom, necessity itself ordained to entrust this introduction to the Latin language as being more universal and familiar to most Nations. In addition, it is also to be feared, lest

perhaps kidnappers (*plagiarii*)—for by the wickedness of our age such shamelessness has come about that very many are found who have no scruples to translate works of some note into another language, with the authors not knowing, not consulted, not willing, rather with their names entirely suppressed, and maliciously take credit for themselves—seizing this opportunity for conveyance in the open market of their own desire, ambition, and arrogance, expose for sale, enveloped in a veil of Latinity, this my authentic progeny, honorably and freely born and raised.

The work is dedicated by 'Maria Cunitia together with her husband Elias a Leonibus' to Emperor Ferdinand III, son of Ferdinand II to whom Kepler dedicated the *Rudolphine Tables*, and a version of the dedication, written by Cunitz in the customary excessively florid style in praise of the Emperor, worthy to be revered as divine and numinous, survives in her own hand. The instructions, in both languages, are a model of clarity and show a complete understanding of all aspects of mathematical astronomy, including its history from antiquity to the time of composition. The writing is both learned and elegant, with classical orthography and idioms, a large vocabulary, including of technical terms, which are explicitly defined in both Latin and German, and a fondness for metaphors drawn from the law. Although one hardly thinks of Silesia as a center of humanism and erudition, especially during the recurrent depredations of the Thirty Years War, Cunitz was part of a cultured circle of physicians, academics, lawyers, and ecclesiastics, as shown by thirteen pages of complimentary verses in *Urania Propitia, ad aevi nostri eruditam Palladem*, that make present-day professors of Latin appear schoolboys, not to mention the ingenious acrostics on her name and a final Italian sonnet. And as noted, Kepler published his *Ephemerides* in Silesia. Lest anyone think Elias wrote the instructions, one need only compare his 'husband to the reader' with the instructions; and it is not a matter of the ornate style of prefaces, the writers cannot be the same. Although the use of the tables, like all tables of their time, was primarily astrological, the instructions are entirely astronomical, without applications such as houses and directions, which are, however, referred to and Kepler does include in a supplement to the *Rudolphine Tables*. The Latin instructions are longer and contain technical and historical discussions of considerable interest not present in the German, which are more strictly practical, although they have their own interesting differences from the Latin.

The tables are in three parts. Part I: sexagesimal sines, solutions of small right triangles in minutes and seconds, tables for spherical astronomy for degrees of the ecliptic of: declination, right ascension, oblique ascension for latitudes 0° to 72° at 2° intervals, angle of the ecliptic with the horizon, angle of the ecliptic with vertical circles of altitude. These are followed by the equation of time, cities with latitude and difference of time from the meridian of the tables, and historical epochs, from Kepler's date of Creation to Anno Hejirae. There are also tables for sexagesimal multiplication and interpolation in double-entry tables. Part II: Tables of mean motions and corrections for the sun, planets, and moon in longitude and latitude. Part III: Tables for computation of eclipses: day of the year of the conjunction of the sun with the ascending and descending node; 'astronomical golden number' for finding the date of new and full moon, near conjunction and opposition; latitude of the moon in eclipses; true hourly motion, semidiameter, parallax of the sun and moon, semidiameter of the shadow; parallax of the moon in altitude, longitude, and latitude; parallactic correction between the true and apparent hourly motion of the moon for geographical latitudes 36° to 60° at 3° intervals; illuminated portion of the moon, also applicable to observing the libration; refraction. These are followed by a catalogue of longitude, latitude, and magnitude of brighter and more notable fixed stars.

Our study is concerned entirely with the tables of Part II, the most innovative, their form, use, and computation. Part II also contains tables for the additional inequalities of the moon and for lunar latitude according to both Tycho and Kepler, which differ slightly. We shall not analyze these, as doing so would require explanations of Tycho's and Kepler's lunar theories, which would add many pages, or the tables for the equation of time in Part I, also part of lunar theory, or the tables for eclipses in Part III, a study in themselves. The instructions also contain much of interest, including corrections of ancient observations of Mercury reported in the *Almagest*, with an examination of errors in using the apparent diameter of the moon to estimate distances, and computations of historical eclipses, as the lunar eclipse of 20 September 331 BC, eleven days before Alexander's defeat of Darius III at the Battle of Gaugamela. But these too would add many pages to this paper, and surely, none would ever wish it longer than it is.

Although Cunitz describes the arrangement of the planetary system according to the Tychonic theory, she also describes Kepler's model, with an elliptical orbit physically produced by a libration. 'The orbit of a planet (*orbita planetae*) is not a mathematical circle but a kind of natural revolution (*gyrus*) that the planet, the sun and moon about the earth, but Saturn, Jupiter, Mars, Venus, and Mercury about the sun, describe in the universe (*in universo*) by a nonuniform motion and libration in certain and fixed periods.' (German:) 'The orbit of the planets (*der umbkrais des Planetens*) is a slightly elongated (*etwas ablänglichter*) circle, which, the sun and moon about the earth, the other five, Saturn, Jupiter, Mars, Venus, Mercury, about the sun, describe in endless space (*unendlichen raum*), moving around nonuniformly in a certain period and at the same time by a nonuniform approaching and withdrawing.' The 'libration', a motion toward and away from the sun on either side of the apsidal line as the planet moves about the sun, in Kepler's physics caused by alternately attracting and repelling magnetic forces between the planet and the sun, makes the planet depart inward from a circle and describe an ellipse. Cunitz also refers to the orbit as elliptical, calls the direction of the planet from the sun heliocentric, and measures the mean and true anomalies of the first inequality from the greatest distance from the sun, which, following Kepler, she calls the aphelion, still an uncommon term. There is more, and it is particularly interesting. She tabulates Kepler's physical equation of time, described as a variation in the length of the day due to the nonuniform addition to 360 days of $5\frac{1}{4}$ days in each year depending upon the variation in the distance of the sun from the earth, Kepler's own description, which only makes sense physically as a variation in the speed of the diurnal rotation of the earth, as Kepler explained it, since one could hardly imagine the heavens doing such a thing, especially with *in universo* translated as *in dem unendlichen raum*, perhaps a clue for the discerning. And she reports that many times it was observed by my husband as also by me that this motion is subject to exceptional physical intensions and remissions, that is, observational confirmation of a nonuniform diurnal rotation with a physical cause, of Kepler's physical equation of time. But she refers only to the apparent diurnal rotation of the heavens, never to a true rotation of the earth. This is, to say the least, puzzling. Why she should have adopted Tychonic theory is hardly clear, and she offers no explanation. (It has been suggested that her father knew Tycho in Rostock in the summer of 1597. David Origanus, with whom Elias studied astronomy, held the Tychonic theory with the diurnal rotation of the earth, as did Longomontanus and any number of others, but not Cunitz.) Yet, the single sentence in Latin and German quoted here is the only place, to my knowledge, Tychonic theory is so much as mentioned, and the instructions contain not one word of opposition to Copernican theory. We would hardly object to the suggestion that her own opinion was in fact Copernican.

One of the clearest statements of Cunitz's adherence to Kepler's planetary theory occurs in the Latin version of Chapter 15, on Mars, in which earlier theories are criticized and Kepler's commended:

This is that fiery star, the motion of which, not only did the ancients call unobservable (see Pliny 2.17; now 2.15 (77)), but also among the moderns, Maestlin, an eminent astronomer most deserving of praise in this science, attests in the preface to his *Ephemerides* (1580) in 1577 that errors of this (motion) cannot be confined within the limits of two degrees (assuming the customary models and the common ancient hypothesis). And there was in every case agreement of observation (*experientiae*) to the extent that in the year 1625 in the autumn, the difference of calculation from heaven increased to 3°, 4°, and nearly 5°. The reason was:

1st. Ignorance of the true center of motion, which (our) predecessors considered to be a certain point, empty and devoid of all moving force, they called the center of the Great Sphere (*Orbis Magnus*).
2nd. Ignorance of the causes of motion, which they assumed to be in the circular motion (*circulatione*) of the centers of epicycles.
3rd. Ignorance of the true planetary path or figure of the orbit, which, by the revolutions (*circuitibus*) of so many centers in circles subordinated to one another, they made puff-cheeked (*buccosam*), and from that cause the distances (were) faulty.
4th. Ignorance of the true (anomaly of) parallax (*commutationis*), from which there results a deformation (*depravatio*) of all angles of the last triangle.

Kepler, to whom access to the Brahean observations was freely available, was the first who examined these very matters by carefully investigating and, supported by the invincible strength of demonstrations, found:

1st. The true and genuine center of the moveable world, namely, the sun itself, not some empty point in its vicinity.
2nd. The true cause of motion, namely, the rotating motion of the sun about its own axis, which, by motive fibers sent out through an image (*fibris motoriis per speciem*

emissis), causes the bodies of the planets arranged in order about it to turn in a revolution (*gyrum*).

3rd. The genuine figure of the orbit, namely, elliptical, which the planet describes, by librating itself (*sese librando*), approaching and fleeing the sun, in accordance with the measure of half the eccentricity, in the first and third quadrant from the aphelion in the proportion of the versed sines, but in the second and fourth (quadrant in the proportion) of the right sines, and from this (results) the true and genuine distances, of the planet as well as of the earth, from the sun.

4th. The true (anomaly of) parallax, arising, not from some fictitious place, but from the apparent and true place of the sun joined to the place of the planet, and from this the true and demonstrative given quantities (*didomena*) of the last triangle, granting which, by virtue of necessary demonstration, the true measure of the motions is determined.

Now at last, under the authority of this Master (Kepler), Mars not only finds for itself a calculation agreeing with heaven, but also shows how, by the bisected eccentricity of the sun being known—without which the motion of no planet can be computed agreeing everywhere with heaven—all astronomy is transformed from a useless, monstrous farrago (*farragine*) of fictitious circles to natural causes and demonstrative and undoubted calculation.

The criticism here is of the hypothesis, the model, for the planets, of both the first and second inequalities, in the theory of Copernicus and, following him, Longomontanus and Landsberg. It is notable that the highly compressed descriptions, both of the flaws of this model and the virtues of Kepler's, are in heliocentric form, which is an indication of the way Cunitz actually thinks. The defective model has for its center the mean sun, the center of the Great Sphere, of the sphere carrying the earth about the sun, which is only near the sun. Kepler's has the sun itself, the true sun, as the center of the moveable world, of the system of planets. The defective model uses two epicycles, or the equivalent, for the first inequality, and the result is a path that is puff-cheeked (*buccosam*, Kepler's term in the *Astronomia nova*, although for a different model), extending outward from a circle at the sides, with false distances. Kepler's model has the correct, physical cause of motion, the rotation of the sun sending out an image of itself causing the planets to move around the sun and, on either side of the apsidal line, to 'librate', to approach or flee the sun, producing the true elliptical form of the orbit, in which the planet moves inward from a circle at the sides and the distances are correct. And the correct location of the center in the true sun rather than the center of the earth's sphere gives the correct anomaly of parallax, of the second inequality, and the correct solution to the 'last' triangle, determined by the distance from the sun to the planet, from the sun to the earth, and the angle between the two distances, which gives the true (geocentric) position of the planet. Finally, it is noted that only with Kepler's bisection of the eccentricity of the sun, compared to the undivided eccentricities of Ptolemy, Copernicus, Tycho, and tables based upon a simple eccentric circle for the sun or earth, can the (geocentric) motions of the planets be determined correctly. Elsewhere in the instructions, there are other statements of this kind, criticizing traditional planetary theory and commending Kepler's, with comparisons of computations from other tables showing their errors—especially of Philip Landsberg, a rude, arrogant

Dutchman, who claimed his 'perpetual' tables, 'agreeing with observations of all times,' were superior to all others, which he disparaged at every turn, Kepler's in particular—from which there is no doubt Cunitz is in every way a Keplerian. And since it is hard to imagine how Kepler's astronomy, physical and mathematical, which she undoubtedly follows, indeed, insists upon, can be reconciled with anything but Copernican theory, as Kepler himself believed, we presume her also to be a Copernican, even if she never says so.

Before explaining the form, use, and computation of the tables in Part II, we shall set out a description of Kepler's planetary theory, shown in Fig. 7.1, upon which the tables are based. The planet moves in an ellipse with the sun at a focus such that the line joining the planet to the sun describes an area proportional to time, although this area is not measured in the ellipse and is only a method of computation for the underlying physics, that the time to traverse each small arc is proportional to the distance of the planet from the sun. On the apsidal line AB, we describe an ellipse and its major auxiliary circle of radius R and center M, with the sun S at a focus with eccentricity $e = SM$, and let the planet be at P in the ellipse and projected to P' in the major auxiliary circle on the ordinate $P'PN$. Drawing SP and SP', measured from the aphelion A, areas ASP in the ellipse and ASP' in the circle are proportional to the time since passage of the aphelion. Area ASP' is the mean anomaly M, and drawing MP' from the center of the circle, area ASP' is divided into sector $AMP' = E$, the eccentric anomaly, and triangle $SMP' = e \sin E$, the physical equation. The mean anomaly is the sum of the eccentric anomaly and the physical equation, in the circle, area $ASP' = $ sector $AMP' + $ triangle SMP', or $M = E + e \sin E$, which is measured in degrees where $e° = (180°/\pi)e/R$. And in the ellipse, angle $ASP = \upsilon$ is the true anomaly and SP the distance of the planet from the sun. Given the eccentric anomaly E, one may easily find the mean anomaly M, the true anomaly υ, and the distance SP; thus,

$$M = E + e \sin E, \qquad \cos \upsilon = \frac{e + R \cos E}{R + e \cos E}, \qquad SP = R + e \cos E.$$

But to find E from M, the eccentric anomaly from the mean anomaly, which is what always must be done as M, not E, is a linear function of time known from tables of mean motion, is quite another matter. For the relation $M = E + e \sin E$,

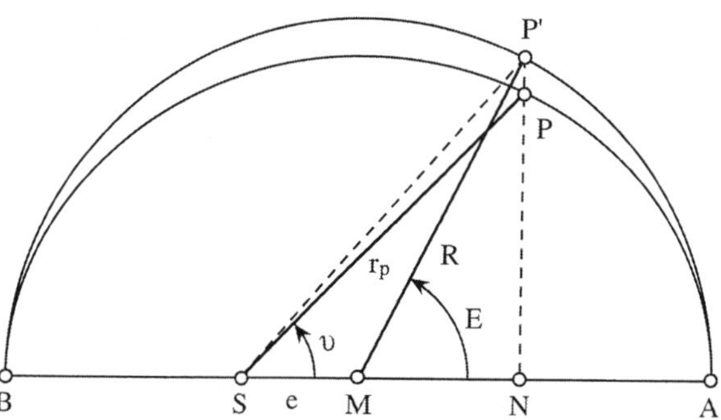

Fig. 7.1

known as 'Kepler's equation', has no geometrical or algebraic solution for E. If we are given arc $AP' = E$, proportional to sector $AMP' = E$, and point S on diameter AB, we may construct triangle $SMP' = e \sin E$, and the sum $E + e \sin E = M$. But if we are given an area M of part of a circle, as proportional to the time since P' passed A, and point S on diameter AB, we must construct a line from S to P' such that SA, SP' and the arc of the circle contain the area M. For then we could divide sector $ASP' = M$ into E and $e \sin E$ by drawing MP' from the center such that the difference $M - e \sin E = E$. However, there is no geometrical construction for line SP', as Kepler suspected and as Newton, among others, was later to show; but there are ways of finding E from M to any degree of precision by iteration or interpolation, as Kepler himself was the first to do. The *Rudolphine Tables* contain interpolations for finding E from M and v from M, but they are difficult to use and replacing them with an easy, direct calculation of v from M, of the true anomaly from the mean anomaly, without having to use E or a coefficient of interpolation, is one of the major accomplishments of Cunitz's tables.

Through a truly colossal labor, greatly simplifying computation for the user, Cunitz computed: (1) tables for mean anomaly M in days since aphelion passage, longitude of the aphelion λ_A, longitude of the ascending node Ω, for single years of 1600–1700 with centennial additions and subtractions for ±5600 years; (2) tables for the first inequality, entered with M in days, giving the true anomaly v and the true daily and hourly motion Δv; (3) double-entry tables for the equation of the second inequality c, reduction of heliocentric to geocentric longitude, entered with the true anomaly v and the anomaly of parallax α at intervals of $5°$, sixteen pages for each planet; (4) double entry tables for geo-centric latitude β, entered with the distance of the planet from the ascending node ω and the anomaly of parallax α at intervals of $10°$, three pages for each planet. In this way, Maria Cunitz's tables are as efficient an adaptation of the *Rudolphine Tables* as can be imagined, and there is not a logarithm in sight.

The computational procedures of Cunitz's tables, (1–2) using time in days directly as an entry rather than motion in degrees and (3–4) large double-entry tables for each planet, are not in themselves new although the way she applies them definitely is. In the mid-fifteenth century Giovanni Bianchini used both in tables based upon the *Alfonsine Tables* but rearranged to facilitate calculation. The procedure is that one first computes the time since the planet's passage of the apogee of the epicycle, equivalent to the mean anomaly of the epicycle, with the mean longitude and mean eccentric anomaly, and then in double-entry tables, entered with the time and mean eccentric anomaly, reads corrections to the mean longitude at apogee with the true motion for days and hours, which together give the true longitude, and in the same table the latitude. Bianchini's tables were printed in 1495 and again in 1553, so they must have had some use although it is not clear how much. And Giovanni Magini's *Tabulae secundorum mobilium coelestium* (1585) contain enormous double-entry tables for the cor-rection of both inequalities of longitude, entered with the mean eccentric and epicyclic anomalies, as Cunitz had computed for Longomontanus's tables, and likewise of latitude, entered with the argument of latitude and the epicyclic anomaly, although again, how much they were used is not certain, especially since Magini already did all the computing one could wish in his *Ephemerides of the Celestial Motions*, eventually for all of 50 years, 1581–1630. But in basing her tables on the *Rudolphine Tables*, Cunitz left these far behind, left all earlier tables far behind, for the principles underlying her tables, Kepler's mathematical and physical astronomy, are completely different and far superior to anything that came before. And this includes the contemporary, and thoroughly disagreeable, astrologer Jean Baptiste Morin's *Rudolphine Tables computed for the meridian of Uraniborg...reduced to an accurate and easy shorter way* (1650), which convert the mean anomaly to integer entries in the tables and substitute common logarithms for Kepler's natural logarithms, but leave the computation of geocentric longitude and latitude entirely to the user. Indeed, Maria Cunitz's *Beneficent Urania*, distinguished by the ingenuity of the tables and great proficiency of the instructions, is also, as noted above, the earliest surviving scientific work by a woman on the highest technical level of its age, for in the generation after Kepler, no contemporary was more accomplished in mathematical astronomy, and as such, she and her work are of enduring interest, of enduring importance, to the history of science.

What she intended to accomplish is explained clearly at the beginning of the instructions to Part II.

> Having thus far explained in the first part the properties of the first motion, by which, according to even the rudest perceptions of any man, the stars apparently move from east to west in a diurnal revolution (*circumgyratione*), and the considerations arising from that, I now undertake the second motion, which may GOD cause to succeed, or the motion belonging to the planets, by which (they move), not from east to west, but clearly in the opposite direction, slowly from west to east, and according to the sense perception only of assiduous observers, each one of them born in its own degree of slowness and swiftness. Which (second motion), since it is subject to so many and such great irregularities in longitude and latitude, and the calculation of it subject to the perplexities of so many cautions, to the accumulations, prone to error, of so many corrections, mean motions, anomalies, is so exceedingly difficult that it has deterred very many of ability, otherwise well versed in those things which take place by the first motion, from further progress in this science. I was entirely under the guidance of my husband in this: in order that the greatest part of those difficulties be removed as far as was possible, I would devise a particular method, the most concise and clear, for discovering those insidious digressions in the motion of the planets. Therefore, preserving the given quantities (*didomenis*) of the RUDOLPHINES in all the planets, by concentration to the point of exhaustion, I computed in an entirely new form this second part of the tables that I now impart to the astrophile.

And succeed she did. We shall explain Cunitz's tables for the planets, their use, and the methods, unsurprisingly not disclosed, by which they were calculated, taking as our example the tables for Mars. Our concern is strictly with the tables and their relation to the *Rudolphine Tables, not with their accuracy compared to modern planetary theory*, about which Cunitz could know nothing. We can say simply that, as we shall see, they are close enough to the *Rudolphine Tables* to have nearly the same accuracy, and are thus comparable to the best tables of their day.

1. Mean Motions

'Table I of Mars, the lowest of the superior planets, showing the time from aphelion, the position of the aphelion, and the ascending node.' These are epochs of the mean anomaly M in time since aphelion passage in years, days, hours, minutes, and of tropical longitudes of the aphelion λ_A and ascending node ☊, which include both their independent sidereal motions and the constant of precession, for single years 1600–1700. The epochs are for noon of 1 January of current years in the Julian calendar, 1 year later than the corresponding epochs of the *Rudolphine Tables*, which are for completed years, e.g. the epoch for 1601 corresponds to the epoch for 1600 in the *Rudolphine Tables*, which is noon of 1 January 1601, and the meridian of the tables is Uraniborg, just as the *Rudolphine Tables*. In the table, B indicates a bissextile year, y a year of 365 days, s signs of 30°. In notating sexagesimal numbers, a comma separates places and a semicolon separates integers and fractions, e.g. $12;22,30 = 12 \cdot 1 + 22/60 + 30/60^2$. The first number in the table under λ_A, $4^s\ 28;57,47 = 4 \cdot 30 + 28 \cdot 1 + 57/60 + 47/60^2$ or $148;57,47$.

Year	M			λ_A		☊	
B 1600	1^y	257^d	$23;40^h$	4^s	$28;57,47°$	1^s	$16;43,52°$
1601	0	301	23;44	4	28;59,54	1	16;44,32
1602	1	301	23;44	4	29; 1, 1	1	16;45,12
…		…			…		…
1653	0	59	1;27	4	29;57,55	1	17;18,59
1654	1	59	1;27	4	29;59, 2	1	17;19,39
1655	0	102	1;31	5	0; 0, 9	1	17;20,18
…		…			…		…
1698	0	7	2;56	5	0;48, 8	1	17;48,48
1699	1	7	2;56	5	0;49;15	1	17;49,27
B 1700	0	50	3; 0	5	0;50,22	1	17;50, 7

There are also small tables of the cumulative number of days in calendar months and of λ_A and ☊ in calendar months and days. The use of the tables is straightforward. To give an example, for 1601 February 12 3;30 AM, which is February 11 $15;30^h$, we have:

Date and Time		M		λ_A		☊	
1601 Jan 1	0^y	301^d	$23;44^h$	4^s	$28;59,54°$	1^s	$16;44,32°$
Feb 1		31			0; 0, 6		0; 0, 3
Feb 11		10			0; 0, 2		0; 0, 1
$15;30^h$			15;30		0; 0, 0		0; 0, 0
Feb 11 $15;30^h$	0y	343^d	$15;14^h$	4^s	$29; 0, 2°$	1^s	$16;44,36°$

The values are adapted from the *Rudolphine Tables* as follows: For the year 1600 completed of the *Rudolphine Tables*, corresponding to 1601 January 1, the mean heliocentric longitude $\bar{l}_p$, longitude of the aphelion λ_A, and mean anomaly in degrees M° are

$$\bar{l}_p = 307;14,47^\circ, \quad \lambda_A = 148;39,54^\circ, \quad M^\circ = \bar{l}_p - \lambda_A = 158;14,53^\circ.$$

The period of the mean anomaly P_M and motion of the mean anomaly per day $M^{\circ/d}$ are

$$P_M = 1^y \, 321^d \, 23;56^h, \quad M^{\circ/d} = 360^\circ/P_M = 0;31,26,28,14^{\circ/d}.$$

Hence, the mean anomaly M expressed as the time elapsed since passage of the aphelion,

$$M = M^\circ/M^{\circ/d} = 0^y \, 301^d \, 23;43,53^h \approx 0^y \, 301^d \, 23;44^h,$$

in agreement with the tables. The following values of M are computed by adding 365 days, or 366 days in a bissextile year, modulo the period of the mean anomaly P_M. For example, for 1602

$$0^y \, 301^d \, 23;44^h + 365^d = 666^d \, 23;44^h = 1^y \, 301^d \, 23;44^h,$$

which does not exceed P_M. For 1603, since M for $1602 + 365^d > P_M$, we have

$$1^y \, 301^d \, 23;44^h + 365^d = 1^y \, 666^d \, 23;44^h - 1^y \, 321^d \, 23;56^h = 0^y \, 344^d \, 23;48^h.$$

The tables give $0^y \, 344^d \, 23;47^h$ because the additions of single years have been computed to one more place and then rounded. To compute back to 1600, a bissextile year, since $366^d > M$ for 1601, we add P_M and then

$$0^y \, 301^d \, 23;44^h + 1^y \, 321^d \, 23;56^h - 366^d = 1^y \, 623^d \, 23;40^h - 366^d = 1^y \, 257^d \, 23;40^h.$$

The values for λ_A and Ω are copied directly from the *Rudolphine Tables* for the following year, e.g. the value for 1600 for 1601, etc.

We have recomputed M for 1601 for the sun, planets, and moon, and summarize the results in the table below. The columns are P_M, the period of the mean anomaly, which is given in the text; $M^{\circ/d} = 360^\circ/P_M$, the mean anomaly per day; from the *Rudolphine Tables* $M^\circ = \bar{l}_p - \lambda_A$, the mean anomaly in degrees for 1600 completed; then $M = M^\circ/M^{\circ/d}$, the mean anomaly expressed as time since passage of the aphelion (apogee for the sun and moon). These are followed by M in the tables in hours since the days are the same, the difference in time Δt, and the difference in mean anomaly $\Delta M^\circ = \Delta t \cdot M^{\circ/h}$.

Body	P_M		$M^{\circ/d}$	M°	M		M Tables	Δt	ΔM°
Sun	0^y 365^d	$6;13,58^h$	$0;59, 8, 9,37^\circ$	$195;11,15^\circ$	0^y 198^d	$0;56,31^h$	$0;55,59^h$	$-0;0,32^h$	$-0;0,1^\circ$
Saturn	29 180	14;57, 0	0; 2, 0,22,59	302;29, 3	24 285	15;36, 9	15;36	$-0;0,9$	0;0
Jupiter	11 317	11; 2, 0	0; 4,59, 8,14	333;52,34	11 3	1;39,21	2; 0	$+0;20,39$	$+0;0,4$
Mars	1 321	23;56, 0	0;31,26,28,14	158;14,53	0 301	23;43;53	23;44	$+0;0,7$	0;0
Venus	0 224	16;52,25	1;36, 7,36,41	51; 8,49	0 31	22;11,30	22;11,10	$-0;0,20$	$-0;0,1$
Mercury	0 87	23;16,44	4; 5,32,18,10	173;57,31	0 42	12;12,31	12;12,32	$+0;0,1$	0;0
Moon	0 27	13;18,35	13; 3,53,56,14	150;30, 0	0 11	12;27,51	12;27,52	$+0;0,1$	0;0

The differences Δt are very small except for Jupiter, which appears to be a (rather severe) rounding, and the resulting differences $\Delta M°$, which affect the longitude, are minute.

The table of M, λ_A, and $☊$ for 1600 to 1700 is followed by a table of 'secular reduction', to be added to the epochs for later centuries, or subtracted for earlier centuries, which also contains a column of the complement of the period of the anomaly P_M, that is $P_M - M$, for which the addition or subtraction is reversed, to be used if an addition gives M greater than P_M or a subtraction a negative M, and also for computing M for negative dating, $-y$ AD $= y$ BC $- 1^y$.

Cent.	M Past −	Fut. +	$P_M - M$ Past +	Fut. −	λ_A Past −	Fut. +	$☊$ Past −	Fut. +
100	0^y 114^d	$3;17^h$	1^y 207^d	$20;39^h$	0^s	$1;51,35°$	0^s	$1; 6,15°$
200	0 228	6;33	1 93	17;23	0	3;43, 9	0	2;12,29
...	...		...		...		...	
1600	1 87	4;32	0 234	19;24	0	29;45,14	0	17;39,52
1700	1 201	7;49	0 120	16; 7	1	1;36,48	0	18;46, 7
...	...		...		...		...	
2700	0 333	16;43	0 353	7;13	1	20;12,34	0	29;48,32
2800	1 82	19;59	0 239	3;57	1	22; 4, 8	1	0;54,46
...	...		...		...		...	
5500	0 94	12;45	1 227	11;11	3	12;16,43	2	0;43,18
5600	0 208	16; 1	1 113	7;55	3	14; 8,18	2	1;49,32

The limit of 5600 years was chosen to extend the tables back to AD $1600 - 5600^y = $ AD -4000, just prior to Kepler's date of Creation 24 July 3993 BC, which Cunitz gives in her table of historical epochs. An example of computing forward to 1854 January 1 is as follows:

Date	M			λ_A		$☊$	
1654 Jan 1	1^y	59^d	$1;27^h$	4^s 29;59, 2°		1^s 17;19;39°	
$+200^y$	0	228	6;33	3;43, 9		2;12,29	
1854 Jan 1	1	287	8; 0	5 3;42,11		1 19;32, 8	

Months, days, hours and their motions are then added as in the previous example for 1601. To compute back to the beginning of the Christian Era, AD 1 January 1 at 12 PM, we *add* $(P_M - M)$ for 1600 years, converting 365^d to 1^y, and subtract λ_A and $☊$.

Date	M			λ_A	$☊$
1601 Jan 1	0^y	301^d	$23;44^h$	4^s 28;59,54°	1^s 16;44,32°
$-1600^y + (P_M - M)$	$+0$	234	19;24	− 29;45,14	− 17;39,52
AD 1 Jan 1	1	171	19; 8	3 29;14,40	0 29; 4,40

From $M° = 281;17,30°$ in the *Rudolphine Tables*, $M = 1^y\ 171^d\ 19;6,21^h$, corresponding to a difference of $-4''$ in $M°$. The computation of M using $P_M - M$ is equivalent to the longer calculation,

$$M + P_M = 0^y\ 301^d\ 23;44^h + 1^y\ 321^d\ 23;56^h = 2^y\ 258^d\ 23;40^h,$$

$$(M + P_M) - M(1600^y) = 2^y\ 258^d\ 23;40^h - 1^y\ 87^d\ 4;32^h = 1^y\ 171^d\ 19;8^h.$$

For negative AD or BC years, which are counted backward, there is a special rule: Since in the table of epochs, years are counted forward from 1600 to 1700, subtract the centuries from the complement of the years of the century to 100 years. Thus, for 357 BC, which is AD $-356 =$ AD $-400 + 44^y$, one subtracts $1644 - 2000^y = -400 + 44^y =$ AD $- 356$ or 357 BC. The example given in the instructions is for that very year, of the birth of Alexander the Great. Again, we *add* $(P_M - M)$ for the interval of 2000 years and subtract λ_A and Ω.

Date	M			λ_A		Ω	
1644 Jan 1	0^y	206^d	$1;\ 9^h$	4^s	$29;47,53°$	1^s	$17;13,\ 1°$
$-2000^y + (P_M - M)$	$+1$	100	$6;13$	-1	$7;11,32$	$-$	$22;\ 4,50$
AD -356 Jan 1	1	306	$7;22$	3	$22;36,21$	0	$25;\ 8,11$

From $M° = 351;46,41°$ in the *Rudolphine Tables*, $M = 1^y\ 306^d\ 7;22,9^h$, an excellent agreement.

2. Correction of the First Inequality, True Heliocentric Longitude

'Table II of Mars showing the true distance from the aphelion and also the true diurnal and hourly motions' The table for the first inequality, entered with M in years and days since aphelion passage, gives the true anomaly v and the true motion Δv per day and per hour for interpolation between the entries for days. The table for Mars is at intervals of 2 days.

M		v		Δv^d	Δv^h
0^y	0^d	0^s	$0;\ 0,\ 0°$	$0;26,14°^{/d}$	$0;\ 1,\ 6°^{/h}$
0	2	0	$0;52,27$	$0;26,14$	$0;\ 1,\ 6$
0	4	0	$1;44,54$	$0;26,14$	$0;\ 1,\ 6$
	$\cdots$		$\cdots$	$\cdots$	$\cdots$
0	340	5	$27;46,53$	$0;38,\ 1$	$0;\ 1,34$
0	342	5	$29;\ 2,56$	$0;38,\ 2$	$0;\ 1,35$
0	344	6	$0;18,59$	$0;38,\ 2$	$0;\ 1,35$
	$\cdots$		$\cdots$	$\cdots$	$\cdots$
1	316	11	$27;22,42$	$0;26,14$	$0;\ 1,\ 6$
1	318	11	$28;15,\ 9$	$0;26,14$	$0;\ 1,\ 6$
1	320	11	$29;\ 7,36$	$0;26,14$	$0;\ 1,\ 6$

Where d is the interval of M in days, the differences $\Delta v^{\mathrm{d}} = (v_{\mathrm{n}+1} - v_{\mathrm{n}})/d$ and $\Delta v^{\mathrm{h}} = \Delta v^{\mathrm{d}}/24$. In using the table, one first reads v for years and days of M and then multiplies Δv^{d} and Δv^{h} by any remaining days and hours and adds. From our previous example for 1601 February 12 3;30 AM, which is February 11 15;30^{h}, let $M = 0^{\mathrm{y}}\ 343^{\mathrm{d}}\ 15;14^{\mathrm{h}} = 0^{\mathrm{y}} + 342^{\mathrm{d}} + 1^{\mathrm{d}} + 15;14^{\mathrm{h}}$, from which

$$v = 5^{\mathrm{s}}\ 29;2,56° + 1 \cdot 0;38,2° + 15;14 \cdot 0;1,35° = 6^{\mathrm{s}}\ 0;5,5°.$$

Then the heliocentric longitude,

$$l_{\mathrm{p}} = \lambda_{\mathrm{A}} + v = 4^{\mathrm{s}}\ 29;0,2° + 6^{\mathrm{s}}\ 0;5,5° = 10^{\mathrm{s}}\ 29;5,7°,$$

and the argument of latitude, the distance from the ascending node,

$$\omega = l_{\mathrm{p}} - \Omega = 10^{\mathrm{s}}\ 29;5,7° - 1^{\mathrm{s}}\ 16;44,36° = 8^{\mathrm{s}}\ 12;20,31°.$$

The table is computed from Kepler's table of equations, and the method of computation is our next concern. As noted, the most obvious feature distinguishing the *Rudolphine Tables* is the use of logarithms. The word 'logarithm', invented, although not explained, by Napier, means 'ratio-number' or 'calculational number'. It is the exponent m relating a number N to a base a, $a^{\mathrm{m}} = N$ or $m = \log_a N$. Most logarithmic tables are common logarithms to base 10, $\log_{10} N$, first published in 1617 by Henry Briggs in consultation with Napier in *Logarithmorum Chilias prima* (First Thousand Logarithms), of the numbers 1–1000; common logarithms of trigonometric functions computed by Edmund Gunter followed in 1620. Kepler's logarithms, like Napier's, are natural logarithms to the base $1/e$, $\ln N = \log_{1/e} N$, where $1/e = 0.36787\ldots$ is the reciprocal of the natural or Napierian base $e = 2.71828\ldots$. Consequently, they are inverse to N, decreasing as N increases, and are positive for $N < 1$ and negative for $N > 1$, the opposite of later tables, and are multiplied by 10^5 so are of up to six integer places and two decimal fractional places, for which a decimal point is used. Below is a specimen of Kepler's equation table for Mars using logarithms and corresponding to the model shown in Fig. 7.1 for an eccentric anomaly of 42°–45°.

1. Ecc. An. Phys. Eq.	2. Log. Intercol.	3. True Anom.	4. Dist. Log.	5. Int.
42°	13800		162841	
3;33,10°	0;52,16	38;33,39°	48761	16
43	13590		162674	
3;37,15	0;52,23	39;29,35	48658	16
44	13390		162504	
3;41,16	0;52,29	40;25,34	48554	17
45	13180		162331	
3;45,13	0;52,35	41;21,37	48448	17

We show the equation table in five columns—Kepler counts them as three—of which three have two lines per entry. Column 1, 'anomaly of the eccentric with the physical part of the equation,' contains in its first line the eccentric anomaly E in integer degrees and in its second the physical equation $e° \sin E$, where $e° = (180°/\pi)e/R$. For Mars, where $R = 100{,}000, e = 9265$ and $e° = 5;18{,}30°$. The mean anomaly $M = E + e° \sin E$ is thus not an integer, which makes interpolation in the table difficult. Column 3, 'coequated anomaly', is the true anomaly v computed from $\cos v = (e + R \cos E)/(R + e \cos E)$. Column 2, 'intercolumniation with logarithm,' is for interpolation in computing v directly from M. The second line is the coefficient $k = (v_n - v_{n-1})/(M_n - M_{n-1})$, applied where M lies between M_n and M_{n+1}, neither integers, and the first is the logarithm $\ln k$. The interpolation of v from M is $v = v_n + k (M - M_n)$. (Note that k is computed from lines n and $n-1$, but applied between lines n and $n+1$. This is in fact a systematic error in the application of k, for a more accurate computation of v, closer to a strict trigonometric computation, results from using k in line $n+1$. I am at a loss to explain why Kepler did this, and I do not know whether any later user of the tables noticed and corrected it; Cunitz did not.) If $M > 180°$, one uses $360° - M$ to find $360° - v$ and then v. The first three columns suffice to find the heliocentric longitude of the planet in its own orbit, $l_p = \lambda_A + v$. Column 4, 'distances with logarithms,' is the distance from the sun to the planet, $r_p = \bar{r}_p (1 + e \cos E)$, with e a decimal fraction of 1, and the mean distance $\bar{r}_p$, which appears in the column at $E = 90°$, for Mars $\bar{r}_p = 152{,}350$, is in units in which the mean distance of the sun from the earth $\bar{r}_s = 100{,}000$. The logarithm is $\ln (r_p/\bar{r}_s)$, which is positive for $r_p < \bar{r}_s$ and negative for $r_p > \bar{r}_s$; thus it is positive for inferior planets, negative for superior planets, and both for the sun. Column 5 is an interpolation factor for finding the logarithm for a given M. The distance r_p or the logarithm $\ln (r_p/\bar{r}_s)$ is used in reducing the heliocentric longitude of the planet to geocentric longitude.

Cunitz's table for finding the true anomaly v from the mean anomaly M is computed from Kepler's using k. For a mean anomaly M in years and days, compute the mean anomaly in degrees by $M° = M \cdot M^{°/d}$, and from column 1 of Kepler's table find the closest mean anomaly $M_n = E + e° \sin E$ less than $M°$ and take the difference $dM = M° - M_n$. In the same line with M_n, find in column 2 the coefficient of interpolation k, in column 3 the true anomaly v_n, and form the interpolated true anomaly $v = v_n + k \cdot dM$. If $M° > 180°$, use $360° - M°$ to find $360° - v$ and then v. (The computation can be done using $\ln k$, but is actually more trouble and Cunitz probably did not use it. Many of the logarithms, for k and for $r_p/\bar{r}_s$, are not accurate, and give slightly different results.) The table gives specimen computations of v for M from 88^d to 94^d, corresponding to Kepler's table given above.

Table for Mars

M	$M°$	M_n	dM	k	$k \cdot dM$	v_n	v	v Table
88^d	46; 6,49°	45;33,10°	0;33,39°	0;52,16	0;29,18,46°	38;33,39°	39; 2,58°	39; 3, 0°
90	47; 9,42	46;37,15	0;32,27	0;52,23	0;28,19,50	39;29,35	39;57,55	39;57,55
92	48;12,35	47;41,16	0;31,19	0;52,29	0;27,23,36	40;25,34	40;52,58	40;52,58
94	49;15,28	48;45,13	0;30,15	0;52,35	0;26,30,38	41;21,37	41;48, 8	41;48, 8

Table for all bodies

Body	M	$M°$	M_n	dM	k	$k \cdot dM$	v_n	v	v Table
Sun	46^d	46;19,23°	45;43,45°	0;35,38°	0;58,30	0;34,45°	44;16,32°	44;51,17°	44;51,17°
Saturn	3^y 330^d	47;39, 6	47;18,33	0;20,33	0;55,21	0;18,57	42;44,12	43; 3, 9	43; 3, 8
Jupiter	1^y 204^d	47;16,49	46;57,11	0;19,38	0;55,59	0;18,19	43; 4,43	43;23, 2	43;23, 9
Mars	94^d	49;15,28	48;45,13	0;30,15	0;52,35	0;26,31	41;21,37	41;48, 8	41;48, 8
Venus	29^d	46;27,40	46;17, 7	0;10,33	0;59,24	0;10,27	45;42,58	45;53,25	45;53,25
Mercury	13^d	53;12, 0	52;21,30	0;50,30	0;44,12	0;37,12	36; 9,33	36;46,45	36;46,45
Moon	3^d 14^h	46;48,58	46;42, 2	0; 2,56	0;56,23	0; 2,45	43;15,32	43;18,17	43;18,16

We have computed $k \cdot dM$ to three places to check for rounding in v; agreement with the table is perfect if we allow a rounding for 88^d. In the following table we compute one position for each body, with $M°$ and v on either side of $45°$ for all but Venus, which Kepler uses in testing computations of the first inequality; here $k \cdot dM$ is rounded to two places.

The only notable difference, of $0;0,7°$, is for Jupiter; Mercury is definitely misprinted, ...54 for ...45. We have checked other values of v and found differences of a few seconds, but no larger errors that are not typographic. The number of calculations is impressive. For the anomalistic period P_M of each planet, the intervals of M are: Saturn 30^d and 5^d at the end of each year; Jupiter 12^d and 5^d; Mars 2^d, Venus, Mercury, and the sun 1^d; for a total of 1361 calculations. The moon is at intervals of 1^h and has 638. Hence, calculations of the kind shown here were carried out nearly 2000 times, not counting checking and correction of errors. In using the *Rudolphine Tables*, each step is required for *every* computation of the true anomaly from the mean anomaly, and logarithms make the computation longer. Cunitz has saved all of that work by computing the true anomaly directly along with the additions for single days and hours, for which anyone using her tables would be grateful.

3. Correction of the Second Inequality, True Geocentric Longitude

'Table (III) of the difference of the geocentric longitude (*loci*) of Mars from the heliocentric longitude.' The table for the equation of the second inequality c, the reduction of heliocentric to geocentric longitude, is double-entry, the horizontal entry the true anomaly v, the vertical entry the anomaly of parallax α, both at intervals of $5°$. Our specimen, drawn from the part of the table for $\alpha < 180°$, shows v at intervals of $1^s = 30°$ up to $6^s \, 0°$; for α we begin and end at $5°$ and $175°$, since c for $0°$ and $180°$ is zero, and include the range $125°–135°$ in which the maximum equation for Mars occurs; the field contains the equation c. The parts of the table for v and $\alpha > 180°$ differ slightly because of the displacement of the apsidal lines of the planet and sun, changing the relation of v for each, as will be explained. Where the heliocentric longitude of the planet is l_p and the geocentric longitude of the sun l_s, for a superior planet $\alpha = l_s - l_p$ and for an inferior planet $\alpha = l_p - l_s$. Column headings, not shown, indicate where c is positive or negative; in the part of the table shown, for $\alpha < 180°$, c is positive; for $\alpha > 180°$, c is negative.

α		v	$0^s \, 0°$	$1^s \, 0°$	$2^s \, 0°$	$3^s \, 0°$	$4^s \, 0°$	$5^s \, 0°$	$6^s \, 0°$
0^s	5		$1;53°$	$1;54°$	$1;56°$	$1;59°$	$2; 3°$	$2; 4°$	$2; 5°$
0	10		3;46	3;47	3;50	3;56	4; 2	4; 7	4;10
	...		...	...	...	...	...	...	...
4	5		36; 9	36;48	38;37	41; 8	43;40	45;33	46;15
4	10		36; 3	36;47	38;45	41;28	44;14	46;15	46;59
4	15		35;36	36;24	38;32	41;28	44;27	46;37	47;23
	...		...	...	...	...	...	...	...
5	20		13;53	14;36	16;19	18;54	21;48	24; 3	24;35
5	25		7;13	7;37	8;33	10; 0	11;39	12;58	13;16

A problem with this table is that within the intervals of 5°, interpolation is not necessarily secure, especially where c changes rapidly, and not linearly, on either side of $\alpha = 180°$, where a precision of $1'$ would be uncertain. This, and the inconvenience of double interpolation, are the limitations of double-entry tables unless the entries are very close together. Special tables are provided to aid with the double interpolation. There is another limitation to this table. Kepler makes two corrections depending upon the heliocentric latitude, the reduction to the ecliptic and the shortening of the distance from the sun to the planet in the plane of the ecliptic, *before* computing the geocentric longitude. These corrections, which are considered in the treatment of latitude below, are not made in Cunitz's tables.

The computation of the table is straightforward in principle, but laborious in practice. The configuration of the sun, earth, and planet upon which it is based is shown in heliocentric form in Fig. 7.2a for a superior planet and 7.2b for an inferior planet, in which the sun is S, the earth O, the planet P, the distances are r_p and r_s, the angle of parallax α, and the correction in the table c. Eccentricities, from which r_p and r_s are variable, are not shown. For a superior planet, c corrects

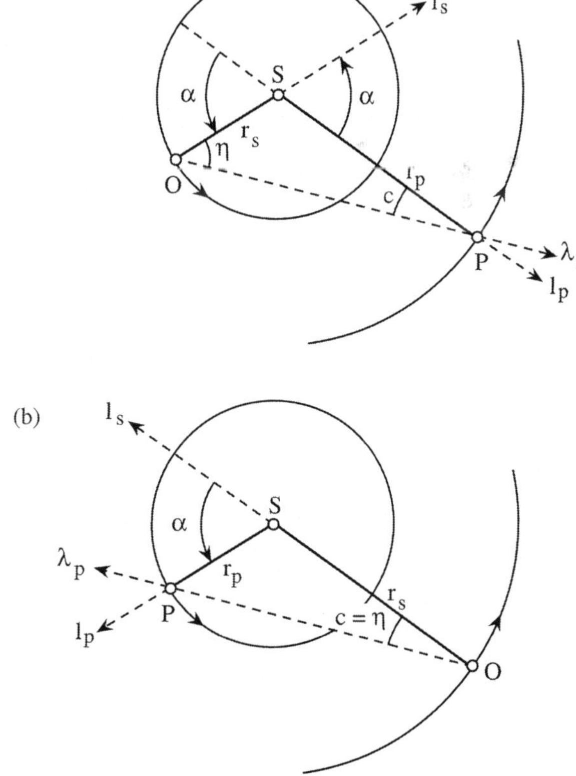

Fig. 7.2

from the heliocentric direction of the planet $l_p = \lambda_{Ap} + \upsilon_p$ to the geocentric direction $\lambda_p = l_p + c$; for an inferior planet, c corrects from the geocentric direction of the sun $l_s = \lambda_{As} + \upsilon_s$ to the geocentric direction of the planet $\lambda_p = l_s + c$; in both cases $c > 0°$ for $\alpha < 180°$ and $c < 0°$ for $\alpha > 180°$. The correction can also be applied to the elongation η of the planet from the sun as $\lambda_p = l_s \pm \eta$. For an inferior planet, directly $\eta = c$ and $\lambda_p = l_s + \eta$, where $\eta > 0°$ for $\alpha < 180°$ and $\eta < 0°$ for $\alpha > 180°$; for a superior planet, measured continuously from l_s westward, in the negative direction, $\eta = \alpha + c$, where $c < 0°$ for $\alpha < 180°$ and $c > 0°$ for $\alpha > 180°$, and then $\lambda_p = l_s - \eta$, as shown in the figure and for all values of α.

The correction c is found as the solution to a triangle, with two sides and the included angle given, for the angle opposite the shorter of the given sides. Thus, in the simplest solution,

$$\text{Superior planet:} \quad \tan c = \frac{r_s \sin \alpha}{r_p + r_s \cos \alpha}, \quad \text{Inferior planet:} \quad \tan c = \frac{r_p \sin \alpha}{r_s + r_p \cos \alpha}.$$

What makes the computation laborious is that r_p and r_s vary as functions of true anomalies υ_p and υ_s, and since the apsidal lines of the planet and the sun do not coincide, these are completely independent, which is also why every c in the table must be computed separately with no symmetries for α and υ on either side of $180°$. There are thus three variables, r_p, r_s, and α, and a double-entry table can use only two. Kepler provides a table, of just two pages, in which the entries are α and the logarithms of the ratios of (r_p/r_s) for a superior planet or (r_s/r_p) for an inferior planet, which are found first, and this is one way of reducing the entries to two. However, the table is very difficult to use and requires an iterative calculation, with logarithms no less, that seems more trouble than solving the problem directly. But there is an easier way of reducing the computation to two variables because α itself includes υ_s, which determines r_s, and this is implicit in using the table and may be used in computing the table. Thus, for a superior planet

$$\alpha = l_s - l_p = (\lambda_{As} + \upsilon_s) - (\lambda_{Ap} + \upsilon_p) = \lambda_{As} - \lambda_{Ap} + \upsilon_s - \upsilon_p, \quad \upsilon_s = \upsilon_p + \alpha + (\lambda_{Ap} - \lambda_{As}),$$

for an inferior planet,

$$\alpha = l_p - l_s = (\lambda_{Ap} + \upsilon_p) - (\lambda_{As} + \upsilon_s) = \lambda_{Ap} - \lambda_{As} + \upsilon_p - \upsilon_s, \quad \upsilon_s = \upsilon_p - \alpha + (\lambda_{Ap} - \lambda_{As}),$$

so that for any value of α, r_s can be found from υ_s just as r_p is found from υ_p. And Kepler's correction tables give r_s and r_p corresponding the eccentric anomaly E, from which r_s and r_p corresponding to the true anomaly υ may be found by interpolation. Although Table 3 is only to minutes, the following calculations are to seconds in order to avoid cumulative rounding errors in the minutes—a result within $\pm 0;1°$ of the table should be considered agreement—and we measure r_p and r_s where the mean distance of the sun $\bar{r}_s = 1$. For a superior planet, let us take Mars. For the epoch 1600 in the *Rudolphine Tables* and 1601 in Cunitz's tables, $\lambda_{Ap} = 148;59,54°$ and $\lambda_{As} = 95;44,8°$, so that

$$\lambda_{Ap} - \lambda_{As} = 148;59,54° - 95;44,8° = 53;14,46°, \quad \text{and thus} \quad v_s = v_p + \alpha + 53;15,46°.$$

First, let $v_p = 0^s\,0°$, for which $r_p = 1.66465$. Then, taking α from 5° and using v_s to find r_s, from the formula given for c:

α	v_s	r_s	c	c Table
$0^s\ 5°$	$58;15,46°$	1.00924	$1;53,12°$	$1;53°$
$0\ 10$	$63;15,46$	1.00784	$3;46,\ 6$	$3;46$
$0\ 15$	$68;15,46$	1.00639	$5;38,30$	$5;39$

Now let $v_p = 1^s\,0°$, for which $r_p = 1.64217$, and beginning α at $2^s\,0°$, we have:

α	v_s	r_s	c	c Table
$2^s\ 0°$	$143;15,46°$	0.98546	$21;47,21°$	$21;48°$
$2\ 5$	$148;15,46$	0.98460	$23;26,19$	$23;26$
$2\ 10$	$153;15,46$	0.98386	$25;\ 2,38$	$25;\ 2$

For an inferior planet, we take Venus. For the epoch 1600 in the *Rudolphine Tables* and 1601 in Cunitz's tables, $\lambda_{Ap} = 301;14,22°$ and $\lambda_{As} = 95;44,8°$, so that

$$\lambda_{Ap} - \lambda_{As} = 301;14,22° - 95;44,8° = 205;30,14°, \quad \text{and thus} \quad v_s = v_p - \alpha + 205;30,14°.$$

Let $v_p = 0^s\,0°$, for which $r_p = 0.72914$, and using v_s to find r_s, taking α from 5°:

α	v_s	$360° - v_s$	r_s	c	c Table
$0^s\ 5°$	$200;30,14°$	$159;29,46°$	0.98311	$2;\ 7,44°$	$2;\ 8°$
$0\ 10$	$195;30,14$	$164;29,46$	0.98265	$4;15,27$	$4;16$
$0\ 15$	$190;30,14$	$169;29,46$	0.98229	$6;23,\ 3$	$6;23$

Next, let $v_p = 2^s\,15°$, for which $r_p = 0.72540$, and taking α from $3^s\,0°$:

α	v_s	$360° - v_s$	r_s	c	c Table
$3^s\ 0°$	$190;30,14°$	$169;29,46°$	0.98229	$36;26,42°$	$36;27°$
$3\ 5$	$185;30,14$	$174;29,46$	0.98208	$38;10,37$	$38;10$
$3\ 10$	$180;30,14$	$179;29,46$	0.98200	$39;50,44$	$39;50$

Hence, within the rounding error of $\pm 0;1°$, the computations agree with the tables for Mars and Venus. We have found that other values are as close. Of course, we have tested only a very small sample of the $72^2 = 5184$ calculations for each planet, a total of 25,920 for all five planets, but from what we have checked, it appears that this was (probably) the method of calculation—an alternative is the law of tangents, using the law of cosines to find the distance from the earth to the planet requires a far more laborious computation—and that the accuracy of the calculations is quite good.

Note that these calculations of Table 3 depend upon $\lambda_{Ap} - \lambda_{As}$ for the epoch 1600 of the *Rudolphine Tables*. Since λ_{Ap} and λ_{As} move independently, because of the different sidereal motions of the apsidal lines, although both contain the same constant of precession, $\lambda_{Ap} - \lambda_{As}$ slowly changes. For example, for Mars in 1600 $\lambda_{Ap} - \lambda_{As} = 53;15,46°$ and in 2000 $\lambda_{Ap} - \lambda_{As} = 53;51,13°$, a difference of $0;35,27°$, and over longer periods the differences will be greater. Since for a superior planet $v_s = v_p + \alpha + (\lambda_{Ap} - \lambda_{As})$ and for an inferior planet $v_s = v_p - \alpha + (\lambda_{Ap} - \lambda_{As})$, for any value of α the relation of v_p and v_s will change with $\lambda_{Ap} - \lambda_{As}$ and so too will the relation of r_p and r_s. How much effect does this have on the computation of c? Does Table 3 remain accurate, or at least useful, over extended periods? These questions are not addressed in the instructions to the tables, but are in Elias von Löwen's account of the origin of the tables, where he states that over a period of one thousand years the differences are minimal and so the tables, although not perpetual, are useful for a long period. His account is rather obscure, and I have not been able to recompute the numbers he gives aside from the changes of $\lambda_{Ap} - \lambda_{As}$ from the *Rudolphine Tables*. But I have recomputed the examples for Mars and Venus given above for one thousand years after 1600, and found the differences from Table 3 very small, less than the rounding error of $\pm 0;1°$. However, this excellent agreement breaks down for Mercury, for which in 1000 years the change of $\lambda_{Ap} - \lambda_{As}$ is nearly $12°$. Near superior and inferior conjunction, the differences are still less than $\pm 0;1°$, but near greatest elongation the differences becomes notable, as much as $+0;9°$ in the examples I have computed. This, however, affects only Mercury, while for the other planets, and even for Mercury far from greatest elongation, Table 3 is safely useable for ± 1000 and even ± 2000 years from 1600.

4. Geocentric Latitude

'Table IV of Mars, latitudes (*latitudinaria*).' The table for latitude is entered horizontally with the argument of latitude ω, the true heliocentric distance of the planet from the ascending node, $\omega = l_p - \Omega$, and vertically with the anomaly of parallax α. The geocentric latitude β in the field is positive, north, for $\omega < 180°$ and negative, south, for $\omega > 180°$. The heliocentric latitude that underlies the computation of the table does not appear. Here too there are imprecisions of interpolation within the intervals of $10°$, particularly near $\alpha = 180°$ where β reaches its maximum and its rapid change is far from linear. We give entries at intervals of 1^s and 2^s and for $\omega = (10°, 190°)$ since at $\omega = (0°, 180°)$, the nodes, $\beta = 0°$; the limits are at $\omega = (90°, 270°)$ where the greatest latitudes occur at opposition, $\alpha = 180°$, and are $+4;32°$ and $-6;51°$.

α	ω	$0^{\mathrm{s}}\,10°$	$1^{\mathrm{s}}\,0°$	$3^{\mathrm{s}}\,0°$	$5^{\mathrm{s}}\,0°$	$6^{\mathrm{s}}\,10°$	$7^{\mathrm{s}}\,0°$	$9^{\mathrm{s}}\,0°$	$11^{\mathrm{s}}\,0°$
0^{s}	$0°$	$+0;11°$	$+0;33°$	$+1;\,8°$	$+0;34°$	$-0;11°$	$-0;33°$	$-1;\,5°$	$-0;32°$
1	0	0;12	0;35	1;11	0;36	0;12	0;35	1;\,7	0;33
3	0	0;16	0;46	1;35	0;47	0;16	0;46	1;29	0;45
5	0	0;35	1;40	3;14	1;38	0;35	1;43	3;34	1;44
6	0	0;55	2;30	4;32	2;26	0;58	2;58	6;51	3;\,4
7	0	0;35	1;40	3;13	1;40	0;36	1;44	3;33	1;45
9	0	0;16	0;46	1;35	0;47	0;16	0;46	1;30	0;45
11	0	0;12	0;35	1;11	0;35	0;12	0;35	1;\,7	0;34

The method of calculation of the table is in Fig. 7.3, which shows in 7.3a the plane of the orbit of the planet intersecting the plane of the ecliptic with an inclination i at the nodal line $\Omega\mho$ passing through the sun S. The eccentricity of the planet's orbit is not shown. The heliocentric latitude b is found from the argument of latitude ω, the true distance of the planet P from the ascending node Ω, $\omega = l_{\mathrm{p}} - \Omega$, and the inclination i of the plane of the orbit to the ecliptic, $b = \sin^{-1}(\sin\,\omega\,\sin\,i)$, where $b > 0°$, to the north, for $\omega < 180°$ and $b < 0°$, to the south, for $\omega > 180°$. Kepler tabulates b at intervals of $1°$ for ω from $0°$ to $90°$, which can serve for all four quadrants. The heliocentric latitude affects the longitude of the planet in two ways through the projection of P into the ecliptic at P'. The first is the reduction δ, the difference between the distance from the node of P in the orbit and of P' in the ecliptic, shown by perpendiculars to the orbit and ecliptic, which directly affects the heliocentric longitude and indirectly the geocentric longitude. The second is the curtation (*curtatio*), the 'shortening' of the distance of the planet from the sun, $r_{\mathrm{p}} = SP$ in the orbit to $r'_{\mathrm{p}} = SP'$ in the plane of the ecliptic, which affects the geocentric longitude since correctly r'_{p} is used in its computation. The effects of both are very small for every planet except Mercury, and Kepler notes that the reduction and curtation can for the most part be omitted.

Kepler tabulates the reduction and curtation with the latitude for the planets and the reduction alone for the moon. Cunitz tabulates the reduction only for the moon, reaching $0;6,33°$. Whether the reduction is applied can be investigated for Mercury, for which $i = 6;54°$ and the reduction reaches $0;12,30°$, which would change α by that amount and affect c in Table 3. The maximum reduction $\delta_{\mathrm{m}} = \sin^{-1}(\tan^2\tfrac{1}{2}i)$ occurs at $\pm(45° + \tfrac{1}{2}\delta_{\mathrm{m}}) \approx \pm45°$ from each node, so if we compute c without the reduction for nodal distances of $\pm45°$ to compare with c in Table 3, any difference should be evident if the reduction were applied. Now, in computing the geocentric latitude below, we use the relation $\upsilon_{\mathrm{p}} = \omega + (\Omega - \lambda_{\mathrm{Ap}})$ in order to find υ_{p}, and in computing the geocentric longitude of an inferior planet, we used $\upsilon_{\mathrm{s}} = \upsilon_{\mathrm{p}} - \alpha + (\lambda_{\mathrm{Ap}} - \lambda_{\mathrm{As}})$. Combining

Fig. 7.3

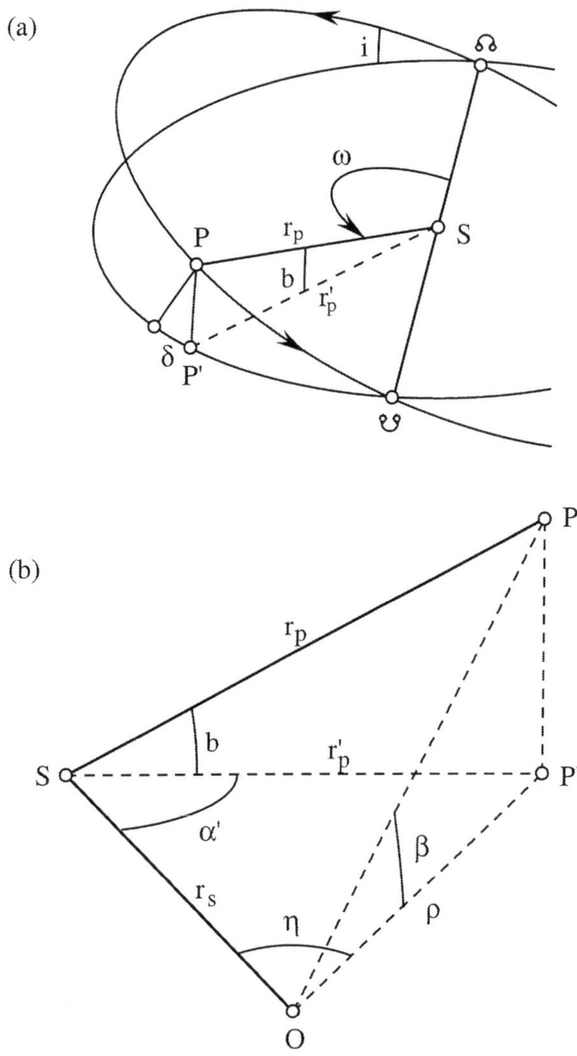

the two, $v_s = \omega + \Omega - \lambda_{Ap} - \alpha + \lambda_{Ap} - \lambda_{As} = \omega - \alpha + (\Omega - \lambda_{As})$. Thus, for any value of ω we may find v_p and r_p, v_s and r_s, and solve for c as we did before in checking Table 3. Computing c in this way, we find no difference from Table 3 greater than $\pm 0;1°$, which shows that the reduction was not applied. We also investigate the curtation by its effect on c. The curtation of r_p to $r'_p = r_p \cos b$ has its greatest effect at the limits of latitude, $\omega = (90°, 270°)$, and for Mercury reduces r_p by $0.0073 r_p$, which can reduce the maximum value of c, near greatest elongation, by about $0;10°$. We therefore test the curtation for $\omega = (90°, 270°)$ and near greatest elongation, with α in the range of $105°$ to $115°$. In this test, we find no difference in c greater than $0;2°$, which shows that the curtation was not applied.

The geocentric latitude differs from the heliocentric because of the variable distance between the earth and the planet. In 3(b) the sun is S, the earth O, the planet P is projected into the plane of the ecliptic at P', the heliocentric latitude is b, and the geocentric latitude β. Let $\alpha' = 180° - \alpha$, the elongation η, and the distances $SP = r_p$, $SP' = r'_p$, $OS = r_s$, $OP' = \rho$. In plane triangles SPP' and OPP', $\tan \beta / \tan b = r'_p/\rho$, and since $\sin \alpha' = \sin \alpha$, in triangle SOP', $r'_p/\rho = \sin \eta / \sin \alpha$. Hence,

$$\frac{\tan \beta}{\tan b} = \frac{\sin \eta}{\sin \alpha} \qquad \text{or} \qquad \tan \beta = \tan b \left(\frac{\sin \eta}{\sin \alpha} \right),$$

which is simple enough as all distances are eliminated, although the effects of r_p and r_s are present in η, and the effect of r_p in place of r'_p is minute. Kepler uses logarithms with the cotangent of b and β. This method, he says, is usable where α and η are 'moderate', for if α is (very) close to $0°$ or $180°$, $\sin \eta / \sin \alpha$ approaches a zero divisor, and Kepler has other methods, including using distances, $\tan \beta = \tan b \left(r'_p/\rho \right)$, which is straightforward at $\alpha = (0°, 180°)$ where for a superior planet $\rho = r'_p \pm r_s$ and for an inferior planet $\rho = r_s \pm r'_p$, and r_p will do for r'_p.

Just as in Table 3 for finding geocentric longitude, a function of three variables, ω, α, η, must be reduced to two. We have earlier seen that the elongation η is computed from the angle of parallax α: for an inferior planet $\eta = c$ and $\eta > 0°$ for $\alpha < 180°$ and $\eta < 0°$ for $\alpha > 180°$; for a superior planet, measured continuously from l_s westward, in the negative direction, $\eta = \alpha + c$, where $c < 0°$ for $\alpha < 180°$ and $c > 0°$ for $\alpha > 180°$. Hence, we must find c, which we do in Table 3 from υ_p and α, so we must first find υ_p for a given value of ω. Since $l_p = \lambda_{Ap} + \upsilon_p = \Omega + \omega$, so $\upsilon_p = \omega + (\Omega - \lambda_{Ap})$. Then, using υ_p and α, we find c from Table 3, from which we find η and solve for β from b using the preceding formula. Here is an example for Mars. From the *Rudolphine Tables* at the epoch 1600 and Cunitz's tables at 1601, $\Omega = 46;44,32°$ and $\lambda_{Ap} = 148;59,54°$, so that

$$\upsilon_p = \omega + (\Omega - \lambda_{Ap}) = \omega + (46;44,32° - 148;59,54°) = \omega + 257;44,38°.$$

Let $\omega = 30°$, for which, from Kepler's table of heliocentric latitude $b = 0;55,15°$, and let $\alpha = 30°$. Thus,

$$\upsilon_p = \omega + 257;44,38° = 287;44,38°, \quad c = 11;47°, \quad \eta = \alpha - 11;47° = 18;13°.$$

Then from the formula,

$$\tan \beta = \tan 0;55,15° (\sin 18;13° / \sin 30°) = 0.010049, \quad \beta = 0;34,32°. \text{ Table: } \beta = 0;35°.$$

The following table shows solutions for different values of ω and α:

ω	b	υ_{p}	α	c	η	β	β Table
30°	0;55,15°	287;44,38°	60°	23; 3°	36;57°	0;38,21°	0;38°
60	1;35,42	317;44,38	30	11;27	18;33	1; 0,53	1; 1
60	1;35,42	317;44,38	60	22;17	37;43	1; 7,36	1; 7
90	1;50,30	347;44,38	150	31;13	118;47	3;13,33	3;14
90	1;50,30	347;44,38	170	13;52	156; 8	4;17, 4	4;17

For an inferior planet, we take Venus, for which at the epoch 1600 in the *Rudolphine Tables* and 1601 in Cunitz's tables, $\Omega = 73;0,45°$ and $\lambda_{\mathrm{Ap}} = 301; 14,22°$, so that

$$\upsilon_{\mathrm{p}} = \omega + (\Omega - \lambda_{\mathrm{Ap}}) = \omega + (73;0,45° - 301;14,22°) = \omega + 131;46,23°.$$

Let $\omega = 30°$, for which, from Kepler's table of heliocentric latitude $b = 1;41,0°$, and let $\alpha = 30°$. Thus,

$$\upsilon_{\mathrm{p}} = \omega + 131;46,23° = 161;46,23°, \quad \eta = c = 12;22°,$$

and from the formula,

$$\tan\beta = \tan 1;41°(\sin 12;22°/\sin 30°) = 0.012587, \quad \beta = 0;43,16°. \quad \text{Table: } \beta = 0;44°.$$

The following table again shows solutions for different values of ω and α:

ω	b	υ_{p}	α	$\eta = c$	β	β Table
60°	2;54,56°	191;46,23°	60°	24;21°	1;23,20°	1;23°
60	2;54,56	191;46,23	90	35;27	1;41,31	1;42
90	3;22, 0	221;46,23	120	43;49	2;41,33	2;42
90	3;22, 0	221;46,23	150	43;38	4;38,29	4;38
90	3;22, 0	221;46,23	170	23;38	7;44, 1	7;44

It can be seen that the comparisons for both Mars and Venus are all within the rounding error of $\pm 0;1°$, and the same is true using $\tan\beta = \tan b \left(r_{\mathrm{p}}/\rho\right)$ at $\alpha = (0°, 180°)$, so the computation of the tables appears to be excellent. The number of calculations for Table 4 is nowhere near as large as Table 3, but is large enough, $12 \cdot 36 = 432$ for each planet and thus 2160 for all five planets.

Table 4 for geocentric latitude has the same condition as Table 3 for geocentric longitude, that the underlying parameters for the epoch 1600 change slowly with time, for longitude $\lambda_{\mathrm{Ap}} - \lambda_{\mathrm{As}}$ increases—except for Jupiter—increasing υ_{s} for any value of υ_{p}, for latitude $\Omega - \lambda_{\mathrm{Ap}}$ decreases for all the planets, reaching nearly $-11°$ for Jupiter in 1000 years, decreasing v_p for any value of ω. In order to check whether Table 4 is usable for other epochs, it is first necessary to find $v_p = \omega + (\Omega - \lambda_{Ap})$ and, for a superior planet $v_s = \omega + \alpha + (\Omega - \lambda_{As})$ or for an inferior planet $v_s = \omega - \alpha + (\Omega - \lambda_{As})$, to find r_p and r_s to compute c and η. Or one can trust Table 3 for c since we have seen that the changes over 1000 years are insignificant except for Mercury near greatest elongation. One can then take the heliocentric latitude b from the *Rudolphine Tables* and use the same procedure as above to find the latitude β and compare with Table 4 to see if the table remains applicable. We have carried out specimen calculations for 1000 years after 1600 using the same values of ω and α for all five planets, including greatest latitude at $\omega = 90°$, with additional calculations for Mercury, and have used both Table 3 and direct computation to find c and then η. The results are that for every planet except Mercury, the differences from Table 4 are mostly less than $\pm0;1°$, in some cases less than $\pm0;2°$; for Mercury at $\omega = 90°$ the differences do not exceed $-0;6°$, at least in the tests we have made. Hence, with the exception of Mercury near greatest latitude, Table 4 appears to be safely useable for at least ±1000 years from 1600 and probably for ±2000 years.

We conclude this examination of the tables with a summary of the procedure for computing the position of a planet in longitude and latitude.

(1) For the sun, in Table 1 find the mean anomaly M and longitude of the apogee λ_A.

(2) In Table 2, with M find the true anomaly v_s and form the true longitude $l_s = \lambda_A + v_s$.

(3) For the planet, in Table 1 find the mean anomaly M, the longitude of the aphelion λ_A, and the longitude of the ascending node Ω.

(4) In Table 2, with M find the true anomaly v_p and form the heliocentric longitude $l_p = \lambda_A + v_p$ and the anomaly of parallax α, for a superior planet $\alpha = l_s - l_p$, for an inferior planet $\alpha = l_p - l_s$.

(5) In Table 3, with v_p and α, find the correction c. The geocentric longitude λ_p is then, for a superior planet $\lambda_p = l_p + c$ and for an inferior planet $\lambda_p = l_s + c$, where $c > 0°$ for $\alpha < 180°$ and $c < 0°$ for $\alpha > 180°$.

(6) Form the argument of latitude $\omega = l_p - \Omega$, and in Table 4, with ω and α, find the geocentric latitude β, where $\beta > 0°$, north, for $\omega < 180°$ and $\beta < 0°$, south, for $\omega > 180°$.

Corrections of Saturn and Jupiter

Cunitz explains that although she is aware of no tables more perfect than the Rudolphine, nevertheless, no one should think there is nothing in them worthy of correction. The corrections of Saturn and Jupiter, promised in the title of the book, are intended as the first corrections of imperfections concealed in the remaining planets, the sun and moon, and in many fixed stars, on the basis of numerous, trustworthy observations through 28 years by her husband (and

also, for the most recent observations of Jupiter, by her). But these corrections
will only be published if the ones given here as specimens are received with
gratitude and not, by virtue of the wickedness of the age, with hatred and
calumnies, in which case all the rest will be suppressed. It is later reported
that the greatest discrepancy of Venus, which occurs rarely, nowhere exceeds
$\frac{1}{4}^{\circ}$. The corrections of Saturn and Jupiter are supported by selected observations
made with a telescope, using the aperture of the instrument as a field of known
angular diameter and estimating the distance of the planet from a star as a
fraction of that diameter. The coordinates of the star are from Tycho's catalogue in the *Progymnasmata* or the expanded version in the *Rudolphine Tables*, for
1600 completed, 1 January 1601, precessed in longitude to the date of the
observation using Tycho's rate of precession, 0;0,51° per Julian year, 1;25°
per Julian century, which Cunitz tabulates with the mean anomaly of the sun,
as Kepler does with the mean longitude. The latitude is unchanged. There are
not that many stars of known coordinates close enough to the ecliptic to be seen
along with a planet in the field of view of a telescope—in the examples given,
about $\frac{1}{4}^{\circ}$, about the semidiameter of the moon—so the opportunities for such
observations are not frequent.

Now, Cunitz writes, from 29 observations of Saturn in very diverse points of
the eccentric and of the anomaly of parallax, which is necessary if one wishes
this matter to be accomplished favorably, very carefully compared among
themselves, my husband found the greatest excess of the *Rudolphine Tables*
from heaven of $4\frac{1}{2}'$, and this in the first semicircle of the anomaly of parallax. In
order to correct the position of Saturn, the calculation is to be modified as
follows:

(1) In Table 1, from the mean anomaly *M subtract* 1 day $1\frac{1}{2}$ hours.
(2) In Table 3, from the correction *c*, for each 1° of *c subtract* 20″ and for each 3′
    subtract 1″.

An auxiliary table is provided for the subtractions from *c*. The subtraction of
1 day $1\frac{1}{2}$ hours from *M* is a change in epoch, reducing the mean anomaly and
also the mean longitude by −0;2,8°. In the *Rudolphine Tables*, the mean distance
of Saturn from the sun $\bar{r}_p = 9.51$ where the mean distance of the sun from the
earth $\bar{r}_s = 1$, from which the maximum correction at mean distance $c_m = \sin^{-1}(1/9.51) = 6;2,9°$. The maximum subtraction from c_m is $6;2,9 \cdot 20″ = 0;$
$2°$, reducing c_m to $6;0,9°$, equivalent to increasing the mean distance to
$\bar{r}_p = 9.5628$, which holds very nearly for all values of *c*.

In order to support this correction, of many examples of our observations,
Cunitz selects one from 1627 on 4/14 December (Julian/Gregorian) in the
morning, taken as 6 AM. No location is specified, but it must have been in
Schweidnitz, in the Catalogue of Places $+0;13^h$ east of Uraniborg, taken here
as on the same meridian since there would be no difference for Saturn. Saturn
was seen to have covered (*texisse*) the penultimate star of the left wing of Virgo
(44 Virgo, of the sixth magnitude; Ptolemy's star with the same description is
46 Virgo), but applying the telescope (*perspicillum*), the star was perceived

standing to the left and upwards by about a sixth part of the instrument, $2'$ or $3'$; hence the total aperture was from $12'$ to $18'$. From Tycho's catalogue precessed to the date of the observation, the longitude of the star is $\underline{\Omega}9;28,30° + 0;22,53° = \underline{\Omega}9;51,23°$, the latitude $+2;23\frac{1}{2}°$, and from the observation, the longitude of Saturn is taken as the longitude of the star, or a little more. But according to Kepler's *Ephemerides*, the longitude of Saturn at that time was $\underline{\Omega}9;56°$ and latitude $+2;21°$. And computation with Cunitz's tables gives the longitude $6^s\ 9;56,20°$ and latitude $+2;20\frac{1}{2}°$. Hence the computed longitude exceeds the observed by about $4\frac{1}{2}'$ and the latitude, Cunitz says, agrees excellently with the observed. When, as is also shown, the corrections are applied to the computation, the longitude of Saturn is $6^s\ 9;52,28°$, 'agreeing closely enough with heaven, which showed Saturn conjoined to the star in longitude or in advance of it by 1 minute.' Although we have deliberately not compared Cunitz tables with modern theory, we shall compare the observation and computations here, which turns out to be of interest. (In all the following computations of positions, because of interpolations in Tables 3 and 4, which are to minutes, the seconds are not secure, and any comparison beyond minutes is false precision. For the modern computation we have used the Alcyone Ephemeris.) For 1627 4/14 December 6 AM at the meridian of Uraniborg, we have:

Body	Observed	Kepler *Eph.*	Tables	Tables Cor.	Modern
Saturn λ	189;51,23+°	189;56°	189;56,20°	189;52,28°	189;59,15°
β	+2;23,30+	+2;21	+2;20,30	+2;20,30	+2;20,37
44 Vir λ	189;51,23				190; 0,46
β	+2;23,30				+2;22,32

What we have here is an example of bad luck, that the longitude of 44 Virgo in Tycho's catalogue is in error by $-9'$, and so too is the precessed longitude, when most of the stars in the catalogue have much smaller errors, $\pm1'$ or $\pm2'$. Consequently, the observed longitude of Saturn is also taken too low and shows an error of $-8'$ from modern computation. And the uncorrected computations from the *Rudolphine Tables*, in Kepler's *Ephemerides* and Cunitz's tables, agree better with the modern computation than the correction to fit the observation. Hence, were it not for the large negative error in the longitude of the star in Tycho's catalogue, observation would have shown that a smaller positive correction was needed, so the correction is both too large and in the wrong direction. How 28 other observations of Saturn could have supported this correction, I do not know. Before investigating this observation, I was curious to see whether perhaps it would show detection of an effect of the great inequality of Jupiter and Saturn, the largest perturbation in the planetary system, but it shows nothing of the kind, merely the effect of the error in longitude of 44 Virgo in Tycho's catalogue.

The error for Jupiter is much larger than for Saturn, as Cunitz reports her husband found from 87 observations very carefully and accurately examined, reaching $13\frac{1}{2}'$ from heaven in the second semicircle of the anomaly of parallax, and the same error will appear in these tables as long as they are not corrected. In the first semicircle of the anomaly of parallax, the errors of the *Rudolphine Tables* for the most part cancel each other, so it is no wonder that Kepler, although a very diligent and skilled master, as is clear from his written account, was deceived in this matter. For three elements of the theory of Jupiter suffer from error, the longitude of the aphelion, the mean motion, and the proportion of the radius of the orbit of Jupiter to the distance between the sun and earth. Of all these, although no one taken by itself and separately causes much (error), nevertheless, combined they accumulate a deviation—in the first and third quadrant of the eccentric and in the second semicircle of the anomaly of parallax—as was just described, which, in order that we may remove it entirely, we shall make use of this very easy computation:

(1) In Table 1, from the mean anomaly M *subtract* 8 days 5 hours and to the longitude of the aphelion λ_A *add* $0;45°$. This corrects two of the elements.
(2) In Table 3, from the correction c, for each $1°$ of c *subtract* $36''$ and for each $5'$ *subtract* $3''$. This corrects the third element.
(3) In Table 4, to the geocentric latitude β *add* 1/60th part.

An auxiliary table is provided for the subtractions from c. The subtraction of 8 days 5 hours from M is a change in epoch, reducing the mean anomaly by $-0;40,55°$, and the addition of $0;45°$ to λ_A increases the mean longitude by $\bar{l}_p = \lambda_A + M° = 0;45° - 0;40,55° = +0;4,5°$, less than one day's motion. The mean distance of Jupiter from the sun $\bar{r}_p = 5.2$ where the mean distance of the sun from the earth $\bar{r}_s = 1$, from which the maximum correction at mean distance $c_m = \sin^{-1}(1/5.2) = 11;5,15°$. The maximum subtraction from c_m is $11;5,15 \cdot 36'' = 0;6,39°$, reducing c_m to $10;58,36°$, equivalent to increasing the mean distance to $\bar{r}_p = 5.2518$, which holds very nearly for all values of c. The addition to β of 1/60th part implies an increase of the inclination i and the maximum heliocentric latitude b in the same proportion, from $1;19,20°$ to $1;20,39,20°$.

Three observations are given to confirm the correction for Jupiter. The first was in 1627, after two and a half days of rain during which Jupiter could not be seen, following 25 Apr/5 May scarcely 1 hour after midnight, taken as $12;54^h$ after noon. The calendar date, also given, and time are thus 26 Apr/6 May 12;54 AM. Jupiter was located to the right (west) of the higher star in the forehead of Scorpio (β Sco) in a telescope (*perspicillo*), the aperture of which was $14'$, by a fourth part, that is, by $3\frac{3}{4}'$, and a little lower (*paulo humiliore*). The observation and computations are shown in the following table. Cunitz gives the coordinates of β Sco precessed from Tycho's catalogue, and places Jupiter at a longitude lower by $-0;2,30°$ and an unspecified lower latitude.

We have interpolated the position of Jupiter in Kepler's *Ephemerides* and computed from Cunitz's tables without correction; the corrected computation is given by Cunitz, and this is followed by the modern computed positions.

Body		Observed	Kepler *Eph.*	Tables	Tables Cor.	Modern
Jupiter	λ	237;56°	237;49,14°	237;49, 1°	237;55,43°	237;55 30°
	β	+1; 5 –?	+1; 3	+1; 3	+1; 2,49	+1; 4, 0
β Sco	λ	237;58,30				237;59,37
	β	+1; 5				+1; 3,17

Note that according to the report of the observation, Jupiter was a little lower than the star, as it is in Kepler's *Ephemerides* and Cunitz's tables, both uncorrected and corrected. But by modern computation, Jupiter was higher than the star, which, although the difference is barely 1′, would be evident in the telescope since even its altitude from the horizon was higher. An explanation we can suggest is that when the observation was made, Jupiter was recorded as higher than the star, but then 20 or more years later when this section of the instructions was written, the position was computed, the latitude found to be lower, and the earlier report altered in the belief it must be in error. However, the longitude of Jupiter is improved in the corrected computation, by +6′, as Cunitz mentions, for in the uncorrected computation it is much farther from the star than the separation seen in the telescope. So this does appear to be a correction of the *Rudolphine Tables*, in longitude if not in latitude.

The next observation is altogether different, and hardly what one would expect. In October 1629 from 7/17 to 19/29, thanks to mostly clear weather, a series of observations made in the evening of Jupiter with fixed stars exceeded the Rudolphine calculation by +9′ more or less. A single example is given for 18/28 October exactly 1 hour after apparent sunset, that is, at an apparent time of 5;56 hours after noon, 'in the presence and with the help of friends not inexperienced in this matter.' Using a *radius astronomicus*, a cross-staff for measuring the tangent of arcs, ten Roman feet long, which is quite large for such a thing, perhaps the reason for the help of friends, and provided with a sight (*dioptra*) of narrow aperture at the eye, Jupiter was observed five times ahead (*in antecedentia*, to the west) of the preceding (western) star in the tail of Capricorn (γ Cap) by 17;17° refracted, which was truly 17;18°, and exactly in a straight line with the two stars in the preceding (western) horn (α & β Cap). Without explaining the computation of the longitude from the observation, for which there is insufficient information, it is concluded that Jupiter was in Ⅵ29;25$\frac{1}{2}$°, and no latitude is given. In the following table, we give the observed longitude of Jupiter and the precessed longitude of γ Cap from Tycho's catalogue, then Jupiter interpolated in Kepler's *Ephemerides*, the uncorrected computation from the tables, which is not given, Cunitz's corrected computation, and last the modern positions of Jupiter and γ Cap.

Body	Observed	Kepler *Eph.*	Tables	Tables Cor.	Modern
Jupiter λ	299;25,30°	299;15,44°	299;16, 4°	299;24,51°	299;10, 2°
β	?	–0;45	–0;45,46	–0;46,41	–0;43,59
γ Cap λ	316;38,28				316;36,23
β	–2;26				–2;31,20

Here it is evident that the observation itself is in error, by $+15'$, because measurement with a radius is not reliable, particularly at a separation as great as 17°. Consequently, what would have been an error of $+6'$ in the uncorrected computation, compared to the modern longitude, is increased to $+15'$ in the corrected computation. Of course, Cunitz could not know this as she assumed the observation to be correct. But it shows the danger of relying upon a single observation, or even many observations, of unknown accuracy to correct, or verify the correction, of any element, let alone of three elements, of a planetary theory.

The last observation is actually a set of two of Jupiter passing the same star twice in retrograde and direct motion. 'I add for the sake of astrophiles two noteworthy very recent observations, which I obtained with my husband.' The first is in 1649 on 1/11 May in the evening 10;55 hours after noon, using a telescope (*tubum opticum*) with an aperture of $14'$ to estimate the distance of Jupiter from the first, higher star in the left wing of Virgo (η Vir), finding the planet a little more than half the aperture, $7'$ or rather $8'$, to the north, and its center a little west of a perpendicular to the star by the width of its own small disc, that is, about $1'$. (Jupiter had passed the star in retrograde motion.) Then, just before the beginning of type-setting and printing (*sub initium operarum typographicarum*), on 30 May/9 June in the evening 9 hours after noon, Jupiter had not yet reached the same star in longitude, the centers being separated less than $2'$ but more than $1\frac{1}{2}'$, and it was a little more northern than the star. The following day 31 May/10 Jun, at about the same time, Jupiter had now passed over the star (in direct motion), the centers being separated by $1'$, from which it is gathered that at about noon of that day they were in conjunction in longitude and that Jupiter with its southern edge touched the star, and thus it had a northern latitude greater than the latitude of the star by $\frac{1}{2}'$. From these two observations, it is concluded that, not only is a correction required in the Rudolphine calculation for Jupiter, but also in the position of the fixed star. For in 1629 my husband, as yet unmarried and more attentive to this research, began to examine carefully the fixed stars in relation to each other whenever the occasion allowed, and he found the longitude and sometimes also the latitude of some in the catalogue were not accurate to the least degree. Among them, in the longitude of this fixed star $2\frac{1}{2}'$ is lacking, but the latitude is in excess by $1'$, and it could be shown in this configuration; but not wishing to make trouble for himself, he added only $1'$ to the star so that it is in ♍29;17°, for which the catalogue has ♍29;16°, and left the latitude, $+1;25°$ in the catalogue, untouched. For the observation of 31 May/10 Jun, the star is precessed to

♍29;57° and the latitude reduced to +1;24° In the comparisons below, we have computed the uncorrected positions for both dates and the corrected position for 1/11 May; the corrected position of 31 May/10 Jun is computed by Cunitz.

Date & Time	Body		Observed	Tables	Tables Cor.	Modern
1/11 May	Jupiter	λ	179;56°	179;46,46°	179;59,50	179;56,32°
10;55^h		β	+1;33	+1;30,24	+1;31,58	+1;30,45
31 May/10 Jun	Jupiter	λ	179;58	179;45,52	180; 0, 4°	179;56,58
noon		β	+1;25,30	+1;22,42	+1;24, 6	+1;23,21
	η Vir	λ	179;57			179;56,38
		β	+1;24			+1;22,36

It is notable that the observations agree very well with the modern computation; that the corrected computations are better than the uncorrected is the purpose of the corrections. (Cunitz computed the uncorrected position for 1/11 May as 179;46° and we find the corrected position for 31 May/10 Jun 180;0,16°.) That the observations are so good shows that the method, estimating the distance from a star as a fraction of the aperture of a telescope, is good in principle to about 1′ provided the coordinates of the star are accurate, as they are in the case of η Virgo. What then is the reason for the problems of the uncorrected positions? We have checked the underlying mean and true longitudes of Jupiter and the sun and the longitudes of the aphelia in Cunitz's tables and the *Rudolphine Tables*, and found the discrepancies arise from a combination of factors: errors of +0;6° in Jupiter's mean longitude and −1;10° in its aphelion, errors reaching −0;10° in Jupiter's true heliocentric longitude and +0;6° in the sun's true longitude. What I have not been able to isolate, which would have been of interest, is the great inequality of Jupiter, which probably does make some contribution, but is hidden within the more obvious errors. In any case, the sort of corrections to the tables made here to fit these observations, or some number of observations, cannot hold everywhere, so, like the corrections for Saturn, are not an improvement of the *Rudolphine Tables*, even if the tables can use improvement.

Appendix

Account of Maria Cunitz in Johannes Herbinius's

On the Learning of Distinguished Women

But two Muses of this age and order of the learned will surpass the admiration of future generations, CUNICIA and SCHURMANNIA. The former, MARIA CUNICIA, named from the paternal family from Cunitz, but from bond of matrimony called a Leonibus (von Löwen), Mistress of and heir to estates (*Domina et haereditaria*) in Kunzendorff and Hohengiersdorff of our

Silesia; although thus far she has dwelt in the small town of Pitschen, by a fortunate destiny she has achieved a distinction in honor and glory to which now the world does not suffice. This noble heroine, as she was born to illustrious and learned parents, so not except wisely did she consent to companionship and marriage, as wife of the distinguished and accomplished man, Elias a Leonibus (formerly named Kretschmer), doctor of medicine, successful and worthy practitioner of medicine throughout Silesia and neighboring Poland. The skill of this Muse in languages, both of those called learned and of other foreign languages, German, Italian, French, Polish, and such, in works of elegance fashioned with the needle and painted with the brush, in sweetness of playing music, and above all her modesty and gentleness in so great a display of learning, I have not experienced in one of the learned acknowledged far and wide. But lest we speak of her insufficiently, and by our fault detract from the virtues of her genius, it is preferable to be silent and listen to the opinions and approbations of the world. This so very illustrious woman has become famous from having some time ago published a large mathematical book with the title MARIAE CUNICIAE URANIA PROPITIA, which book, just as it is of assured benefit to its readers, so is it destined to undying fame and glory in histories. For it is necessarily deserving of the recognition and commendation of future generations unless the praise of women appear the censure of men. For example, Strada, *De bello belgico*, lib. 5.

Translated from *Dissertatio historica I, De foeminarum illustrium eruditione* (Historical disquisition I, On the learning of distinguished women). Johannes Herbinius (ca. 1632–3–ca. 1676–9), who knew Cunitz in Pitschen in Silesia, studied at Wittenberg, where this curious work was presented on 20 April 1657. It is set out as a disquisition on the remark in Tacitus's *De moribus Germanorum* (*Germania*) 19.1, *Literarum secreta viri pariter ac foeminae ignorant* (The men and women alike have no knowledge of intimate correspondence), meaning they are virtuous. Somehow, this leads to a consideration of the erudition of women, first in antiquity, and then in the modern period, in which the two outstanding examples are Maria Cunitz and Anna Maria van Schurman (1607–78), artist, poet, philosopher, theologian, then in Utrecht, who knew even more languages than Cunitz and is today more famous. Herbinius himself is best known for *Dissertationes de admirandis mundi cataractis supra et subterraneis* (Disquisitions on wonderful waterfalls on and below the earth, 1678), which also considers the cause of the tides and the location of paradise. But more to our purpose, years earlier he wrote *Famosae, de solis vel telluris motu, controversiæ, examen theologico-philosophicum* (Theological-philosophical consideration of the famous controversy concerning the motion of the sun or the earth, Utrecht, 1655), which concludes (pp. 313–27) with what amounts to a second letter of dedication to Maria Cunitz after a lapse of four years in their communication, perhaps after he left Pitschen. The letter is written from Utrecht, where he knows Schurman, and some phrases from the letter appear in the later *Dissertatio*. The *Examen* itself argues that Scripture cannot be used to refute or defend the motions of the earth, which are held for good astronomical reasons by mathematicians. Herbinius is not a mathematician and remains undecided, although so much of the treatise is devoted to showing that Scripture cannot be used to refute the mobility of the earth that he seems inclined to accept it. Perhaps this was also Cunitz's opinion. What is essential is that he knew Cunitz, which is obvious where he writes in the first person, so his account, even if appearing conventional praise, in translating which, we have reduced superlatives to positives, must be assumed correct. Two small points: Kunzendorff is a village near Liegnitz and Hohengiersdorff a village near Schweidnitz in estates Cunitz inherited from her father. The reference to Faminio Strada's *De bello belgico* concerns Margaret, Duchess of Parma, regent to Philip II in the Netherlands, formidable in learning but unsuccessful in governing.

Additional biographical information: The year of Cunitz's birth is a matter of uncertainty and curiosity. The most thorough discussion is by Liwowsky (2010). Since her parents were married in 1603 and she was the eldest of five children, a probable year would be 1604, which is frequently given. This would make her nineteen at the time of her first marriage in 1623, discovered by Guentherodt (1991b), which seems reasonable. However, the year 1610 is also given, and it is noted in a letter of 1651 from Elias von Löwen to Johannes Hevelius that she was married at the age of $13\frac{1}{2}$, which, as strange as it appears, confirms 1610. What are we to think of this? Liwowsky notes such an uncommonly early marriage would have been arranged by her father and cleverly suggests that perhaps to him his highly gifted daughter had become so unheimlich that he decided her new roll as Mutter und Hausfrau would put her back on the 'right path'. Fortunately, it didn't. (May we hope she was born in 1604 after all?) The German note to the reader in *Urania propitia* is signed: 'In the year 1650 after the birth of the Son of God on my birthday on Sonntag *exaudi*.' Assuming her birthday on Sonntag *exaudi*, the sixth Sunday after Easter, is for 1650, her birth date is Julian 26 May or Gregorian 29 May, the Gregorian date one week *before* the Julian. The evidence for her children is in Pitschen church records cited by Scheibel (1798). A son, Franz Ludwig, was baptized on 19 November 1647 and died 26 January 1648. The eldest son, Elias Theodatus, eloped with the daughter of an apothecary in Pitschen and was secretly married in Poland before a Catholic priest on 6 June 1657. Another married son, Henricus Antonius, is also mentioned in 1657. According to the same records, Elias von Löwen died on 27 April 1661 and Maria Cunitz on 22 August 1664. Liwowsky has assembled a great deal of information concerning Cunitz's extended family and others in her life.

References

Urania Propitia is available on line from the libraries of Wolfenbüttel, the University of Wrocław, and the University of Florida. In a copy at the Adler Planetarium in Chicago, the Latin instructions are underlined and annotated, showing it was at one time used; the copy from Wolfenbütel has only marks of notice (") in the German instructions. My original study of the work some years ago was from the Adler copy since supplemented by Wolfenbüttel. For the *Rudolphine Tables*, I have used the edition with notes by Franz Hammer in *Johannes Kepler Gesammelte Werke* 10, Munich, 1969. The principal study of the *Rudolphine Tables* is Volker Bialas, *Die Rudolphinischen Tafeln von Johannes Kepler, Mathematicsche und astronomische Grundlagen, Nova Kepleriana*, Neue Folge, Heft 2, *Abhandlungen der Bayerische Akademie der Wissenschaften*, Mat.-Naturwiss. Kl., Neue Folge, Heft 139, Munich, 1969. The literature on Cunitz goes back to the enthusiastic appreciation of her virtues, written in her own lifetime and translated here in the Appendix, *Dissertatio historica I, De foeminarum illustrium eruditione*, presented at the University of Wittenberg in 1657 by Johannes Herbinius, who knew Cunitz in Pitschen. A rather trivial source that appears to stand behind later accounts is Johann Eberti, *Eröffnetes Cabinet dess gelehrten Frauen-Zimmers, darinnen die Berühmtesten dieses Geschlechtes umbständlich vorgestellet werden*, Franckfurth und Leipzig, 1706, 116–18; and *Schlesiens hoch- und wohlgelehrtes Frauenzimmer, nebst unterschiedenen Poetinnen, so sich durch schöne und artige Poesien bey der curiesen Welt bekandt gemacht*, Breslau, 1727, 25–28 (which I have not seen). There is a more interesting memoir by Alfonse des Vignoles, president of the Berlin Academy, in an Eloge des Madame Kirch (Maria Winckelmann Kirch), à l'occasion de laquelle on parle de quelques autres Femmes & d'un Paisan Astronomes, *Bibliotheque Germanique*, Anne'e MDCCXXI, v. 3, Amsterdam, 1722, 155–83 at 163–68. Vignoles's memoir is translated with extensive notes and corrections by Johann Ephraim Scheibel, *Astronomische Bibliographie*, Abt. 3,2, Breslau, 1798, 361–78, following a notice at 355–58. Scheibel's is the most thorough of the early accounts. Abraham Kästner, *Geschichte der Mathematik*, v. 4, Göttingen, 1800, 430–38 contains a summary of the notes to the reader in *Urania Propitia* along with information from the earlier secondary sources. There are notices by Lalande and, depending in part on Lalande, by Delambre, *Histoire de l'astronomie moderne*, 1821, v. 2, 323–26. Delambre examined *Urania Propitia* with technical understanding, although his description is brief and dismissive.

Of recent scholarship, the most substantial is by Ingrid Guentherodt, who has greatly enlarged the biographical information on Cunitz and written extensively on her contribution to scientific literature in German. Among her many publications, which we have found very helpful, are: Maria Cunitz und Maria Sibylla Merian: Pionierinnen der deutschen Wissenschaftssprache im 17. Jahrhundert, *Zeitschrift für germanistische Linguistik* 14, 1986, 23–49. Urania Propitia (1650)—*in zweyerly Sprachen*: lateinisch- und deutschsprachiges Compendium der Mathematikerin und Astronomin Maria Cunitz, *Res Publica Litteraria, Die Institutionen der Gelehrsamkeit in der frühen Neuzeit*, eds. S. Neumeister, C. Wiedemann, *Wolfenbüttler Arbeiten zur Barockforschung* 14, Wiesbaden, 1987, Teil II, 620–40. Maria Cunitia: Urania Propitia, Intendiertes, erwartetes und tatsächliches Lesepublikum einer Astronomin des 17.

Jahrhunderts, *Daphnis* 20 (1991a), 311–53. Frühe Spuren von Maria Cunitia und Daniel Czepko in Schweidnitz 1623, "Homines quoque si tacent, vocem invenient libri", *Daphnis* 20 (1991b), 545–84. Kirchlich umstrittene Gelehrte im Wissenschaftsdiscurs der Astronomin Maria Cunitia (1604–1664): Copernicus, Galilei, Kepler, *Religion und Religiosität im Zeitalter Barock*, ed. D. Breuer, *Wolfenbüttler Arbeiten zur Barockforschung* 25, Wiesbaden, 1995, 857–72. Cunitz, Merian, Leporin: das Wagnis der Erkentnissuche, Kosmos, Tierwelt, Menschenwelt, *Frauen in der Aufklärung*, eds. I. Bubenik-Bauer, U. Schalz-Laurenze, Frankfurt, 1995, 173–93. Zum Briefwechsel des schlesischen Gelehrtenehepaars Cunitia/de Leonibus um 1650 mit den Astronomen Hevelius, Danzig und Bullialdus, Paris, *Kommunikation in der Frühen Neuzeit*, eds. K.-D. Herbst, S. Kratochwil, Frankfurt, 2009, 171–88. The most current and detailed biographical information is by Klaus Liwowsky, *Einige Neuigkeiten zur Familie der Schlesierin Maria Cunitz*, 3. Auflage, 2010, who has kindly sent me his fine study and in an exchange of letters provided advice and corrections for this paper, for which I am very grateful.

Chapter 8
Simplicity in the Copernican Revolution: Galileo, Descartes, Newton

David B. Wilson

"Copernican Revolution" is too simple. The shift from an earth-centered to a sun-centered system required a *century* to occur. It was not "revolutionary" in the swift sense, though it was a huge conceptual change. Moreover, Copernicus was hardly alone in causing the change. Those who supported Copernicus (Kepler, Galileo, Descartes, Newton) did so dramatically differently. Indeed, for those four, Copernicus arguably provided only superficial similarity. This history is complicated, serving as one of many examples as to why historians seek and prefer complicated historical explanations. Nevertheless, they habitually employ simple phrases to state the gist of those complexities, examples being the Scientific Revolution and the Industrial Revolution.

Physical scientists, by contrast, seek and prefer simple explanations. Many different phenomena yield to the same principle or theory. That simple theory is not perceived as an imperfect-but-helpful summary of complexity, but as a valid or true explanation of those different phenomena.

The resultant query, therefore, is: what is historians' simple representation of their complex history of scientists' simple theories? This chapter is not so complicated as to cover the whole history of science, rather focusing on the history of "Simplicity in the Copernican Revolution: Galileo, Descartes, and Newton."

In 1600, there existed no unequivocal, empirical evidence for Copernicanism, by then more than a half century old. The principal observational astronomer of the previous century, Tycho Brahe, rejected Copernicus' conclusions, proclaiming instead that the planets orbited the sun while the sun orbited the stationary earth. Exceedingly few astronomers endorsed Copernicus in 1600.

In 1700, there existed no unequivocal, empirical evidence for Copernicanism. Different Copernican theories certainly offered explanations of phenomena. However, it was the discovery of stellar aberration in the late 1720s that established empirically that the earth did move.

D.B. Wilson (✉)
Department of History, Iowa State University, Ames, IA 50011, USA
e-mail: davidw@iastate.edu

J.Z. Buchwald (ed.), *A Master of Science History*, Archimedes 30,
DOI 10.1007/978-94-007-2627-7_8, © Springer Science+Business Media B.V. 2012

Against this background of the uncertainty of empirical evidence during the seventeenth century, this chapter considers the roles of non-empirical considerations, especially that of simplicity. Galileo, Descartes, and Newton all employed simplicity, but, of course, in different ways.

8.1 Galileo

As early as the 1590s, Galileo accepted Copernicanism, well before his telescopic observations from around 1610. Evidence from the tides convinced him. In the Copernican system, the annual and daily motions of the earth combined, so that any point on the earth's surface was always either accelerating or decelerating. That change in speed produced tides, just as water on the floor of a moving barge would surge in one direction or the other if the barge accelerated or decelerated. By 1616, Galileo had written a long, but unpublished, treatise on the tides, and they filled the fourth (and final) chapter of his 1632 *Dialogue on the Two Chief World Systems*.

His telescope encouraged Galileo to support Copernicus publicly. His 1610 *Starry Messenger* contained a few Copernican statements, but his 1613 *Letters on Sunspots* vigorously defended Copernicus. However, although Galileo's telescopic discoveries answered significant objections to Copernicanism, those discoveries were also consistent with both the Tychonic and Ptolemaic systems.

Those observations involved the sun, the moon, Jupiter, and Venus. Sun spots appeared, disappeared, moved, and changed in size and shape. The lunar surface exhibited earth-like mountains. Such observations undermined the long-held dichotomy between the celestial and terrestrial realms, thus rendering it physically more plausible for the earth to orbit the sun. But the reverse was still physically possible. Jupiter's moons demonstrated the physical possibility of a moon's orbiting a planet which itself was in orbit, thus removing a difficulty facing the Copernican earth-moon arrangement – but without *requiring* that arrangement. Venus' phases showed that Venus orbited the sun, but a sun-orbiting Venus fit with all three of the competing universes. Obviously compatible with Copernicus' and Tycho's universes, Venus' phases meant that Ptolemy's supporters had merely to put Venus in orbit around the sun while the sun orbited the earth. In that *Ptolemaic* sense, Galileo's observations of Jupiter and Venus actually reinforced one another.

Then there was the issue of the Bible and God's relationship with both man's intelligence and nature's structure. Because the biblical God created the universe, Galileo explained that natural philosophy and astronomy allowed man's God-given mind to improve man's comprehension of God. Moreover, because the Bible was accommodated to mankind's understanding of nature at the time it was written, any disagreements between the Bible and modern astronomy were no problem. Famously, the Bible taught man how to go to heaven, not how the heavens go. Even so, Galileo argued, a literal interpretation of the controversial passage in Joshua fit better with a sun-centered than with an earth-centered universe. God's bringing the sun to a standstill "in the midst of the heavens" agreed with Galileo's conclusion that a *rotating* sun resided at the center – in the *midst* – of the Copernican universe. No doubt reflecting Galileo's high regard for mathematical knowledge as well as the biblical truth that God created man in his image, Galileo concluded that whatever mathematics man knew he knew just as well as God did. This human-Divine convergence undoubtedly provided the theological context for Galileo's search for nature's mathematical structure and its "simplifications and conveniences" (Galilei, 1632, 123). Though that search's conclusions could be overturned by future empirical evidence, they nevertheless substantially increased the probability of the Copernican system (Galilei, 1632, 122).

Here are four of five examples. First, in an earth-centered universe, an astonishingly large number of objects had to move at astonishingly high speeds to circle the earth every twenty-four hours. In Copernicus' universe, these were only *apparent* motions that resulted from the much slower rotating speed of the vastly smaller earth. The latter agreed with the concept of nature "which by general agreement does not act by means of many things when it can do so by means of a few" (Galilei, 1632, 117). Second, in the stationary-earth universe, celestial objects – in addition to their twenty-four-hour periods – rotated in opposite directions. The stars moved in one direction and the planets in the opposite. For Copernicus, however, "the contrariety of motions is removed, and the single motion from west to east accommodates all the observations and satisfies them all completely" (Galilei, 1632, 117). One motion was "much simpler and more natural" than two (Galilei, 1632, 118). Third, in the Copernican system there was better "order" for objects circling the center. Faster planets were nearer the sun and slower ones farther away, proceeding logically to the absolute slowness of the stationary stars (Galilei, 1632, 118). A fourth example followed from Venus' phases, disclosing it to be a non-luminous planet. In the Copernican system, the non-luminous earth was in motion like Venus, and the luminous sun was at rest like the luminous stars, providing simple consistency (Galilei, 1632, 267).

Fifth and most important were the tides. Providing Galileo's best argument for Copernicus to begin with, they now filled the fourth and concluding chapter of his *Dialogue*. As previously mentioned, Galileo explained that the earth's two motions caused tides. His mathematically elegant theory joined two circular motions. A simple, mathematical structure underlay complicated tidal phenomena. At great length, Galileo's *Dialogue* explained how these simple causes led to their complicated results. He hardly convinced all. Relying upon tidal *observations*, the Catholic Church pointed out that the basic interval between high and low tides was only half that required by Galileo's theory. Despite such empirical challenges, Galileo's *Dialogue* persisted with his mathematically simple, Copernican theory of the tides.

8.2 Descartes

Descartes' Jesuit teachers taught him about recent telescopic discoveries. Galileo's findings clearly contributed to Descartes' cosmology. Sunspots may have played the most crucial role, but Venus' phases, the moon's mountains, and Jupiter's moons all contributed. Descartes' mature Copernican cosmology followed from two, basic considerations: his *rationalist* argument for the existence of mind, God, and matter and his *empirical* awareness of the properties of light. Descartes' *Meditations on First Philosophy* (1641) presented his well-known, rationalist, "*Cogito ergo sum*" argument for a non-deceiving God, guaranteeing the validity of his own "clear and distinct" idea of the material world's existence. His empirical knowledge of light was epitomized in the subtitle of the book that he withdrew from publication in 1633 after Galileo's trial: *Le monde, ou traité de la lumière*. Descartes eventually published his Copernican arguments in 1644 in his *Principles of Philosophy*.

It was not that Descartes was forming a physical explanation of an already accepted Copernican system, but that he was formulating a rationalist-empiricist account that required Copernicanism to be true, that decisively selected Copernicanism above any competitors. Descartes did declare that Venus' phases eliminated Ptolemy's universe but also that Tycho Brahe's and Copernicus' universes were equally consistent with astronomical observations. That is, neither Galileo's nor anyone else's observations decided between them.

Conceptual relations between God, mind, and matter established necessary truths about that material world. Mind clearly and distinctly perceived the impossibility of void space, and, therefore, matter filled the universe. Hence, motions within the universe had to be closed motions, void space being the impossible alternative. Also, only matter was spatially extended, not mind or God. Finite, created entities obviously had to be preserved – or continually recreated – by their infinite creator. Hence, a constant God assured the constancy of the essential properties of the universe – that is, the total amount of motion of matter would remain the same. Such necessarily true principles would underlie any specific explanations of specific empirical findings.

Universe-filling light provided the key empirical insight into the structure of Descartes' matter-filled universe. That insight did depend upon the denial of an Aristotelian celestial-terrestrial dichotomy – a dichotomy undermined by Galileo's telescopic observations but rejected by Descartes' quite different argument (described below). Nearby terrestrial phenomena now provided valid, or at least analogical, understandings of distant celestial events. Hence, the celestial sun was like terrestrial fire, even though the sun required no fuel. Fire, and the light emanating from it, thus allowed the entire universe to be comprehended accurately.

Light related to matter in three ways. Rarefied material flames produced light. Densely opaque matter halted light. A middle matter transmitted light. That transmission was infinitely swift, just as the impulse from one end of a whacked stick could be immediately felt at the other end. Hence, nearby optical phenomena plus luminous stars meant that the universe consisted of three basic kinds of material particles – small, medium, and large. Small ones emitted light, medium ones transmitted it, and large ones stopped it.

All this justified Copernicanism. Vortex-like rotational motion in the universe was obvious. Planets circled some center, after all, wherever that center was. Whatever the initial distribution of those particles in such a vortex, eventually the smallest ones would be pressed to the center – to form the sun. Even now the different sized particles were not totally separated from each other. Wood burns. There would also be some variation of size within each group. Mid-sized particles filled most of the space around the sun, but somewhat smaller and therefore faster moving mid-sized particles would be closer to the sun. Descartes also speculated that each planet's surface consisted mainly of large, opaque particles while their interiors contained mainly small particles. Various combinations of the two produced planets of various, overall densities, thus placing them at different distances from the sun.

Consequently, although Tycho's and Copernicus' systems agreed equally well with astronomical observations, Descartes' theological-optical argument forced the sun to the center of planetary motions, thus winning the day for Copernicus. That is, in one way or another, Descartes' metaphysical plenum and his empirically fast light combined to entail Copernicanism. But how did simplicity influence this argument?

Like Galileo, Descartes did link "simple" to the Copernican universe. Copernicus' was "somewhat simpler and clearer" than Tycho's (Descartes, 1644, 90). Unlike Galileo, Descartes did not explicitly employ an overall simplicity as evidence for Copernicanism. Also, unlike Galileo, Descartes was contrasting Copernicus with Tycho, not Ptolemy. Moreover, Descartes did not explain *why* Copernicus' system was simpler than Tycho's. It may have been because Copernicus' had only one major center of rotation, whereas Tycho's had two. Or, more likely, Descartes envisioned even simpler simplicities – both metaphysical and physical.

Metaphysical simplicities concerned the possibility and basis of knowledge itself. In his letter to the translator of *Principles*, for example, Descartes wrote of "Metaphysics, which contains the principles of knowledge; among which is the explanation of the principal attributes of God, of the immateriality of our souls, and of all the clear and simple notions which are in us" (Descartes, 1644, xxiv). In the *Principles*, he presented his *Cogito* as something simple and known of itself (Descartes, 1644, 6). Moreover, though we finite beings could not fully comprehend the infinity of God's perfections, "we can however understand them more clearly and more distinctly than any corporeal things; because they fulfil our mind more, and are more simple, and are not obscured by any limitations" (Descartes, 1644, 10).

A critic might argue that all this was merely another example of a Cartesian circular argument. Simplicity implied truth; mind and God being simple ideas were therefore true – meaning that Descartes claimed to know something before he had any knowledge. I would not agree. Rather, Descartes could legitimately reply that he knew mind and God independently of their simplicity. Their being simple ideas, however, provided exemplars for other truths, especially about the corporeal world.

That is, Descartes' evident, implicit, guiding conviction was that the perfect God created a material world wherein simplicity implied essential truth. Here are some examples of *physical* simplicities. Depending upon nothing else than God for its existence, a "substance" had only one "principal attribute" – just as thought was that attribute for the substance of mind, extension was that for matter (Descartes, 1644, 23). Descartes presented his cosmological "hypothesis" as the "simplest and most useful of all" (Descartes, 1644, 91). He continued to declare that "if we can devise some principles which are very simple and easy to know" in order to explain how our present universe developed from a quite different one, then that would be a better explanation than just describing ours (Descartes, 1644, 105). Three entities met that simplicity criterion. First, because space was the same everywhere, extension being the defining property of matter meant that only one kind of matter filled the entire universe, not two as with the celestial-terrestrial dichotomy (Descartes, 1644, 49–50, 106–107). Second were Descartes' laws of motion about which he stated that "I do not think it possible to devise any simpler, more intelligible or more probable principles than these" (Descartes, 1644, 107). Third, "because no proportion and no order is simpler or easier to know than that which consists in equality of all kinds: I am accordingly supposing here that all particles of matter were, in the beginning, equal to one another both in size and motion" (Descartes, 1644, 107). Their interactions over time would have produced the current, different-sized particles.

Vortex-enclosed stars followed directly from God's simply created, evolving universe. Indeed, more stars once existed than do now. Succumbing to their surface, sunspot-like scum, some of those stars became the planets currently orbiting the sun. Their exterior third element and interior first element combined in different average densities, thus defining each planet's proper place in the solar vortex. The sun appeared first. Planets, comets, moons, tides, magnets, and so on arrived later, with all their complexities. Perhaps surprisingly but evidently convincingly, Cartesian simplicity rendered Copernicanism – after matter and light themselves – the easiest and most obvious aspect of God's universe to explain.

8.3 Newton

On the one hand, Newton may not have been part of the Copernican Revolution. My surmise is that astronomers and natural philosophers began generally accepting Copernicus during the 1650s, the decade between Descartes' death and Newton's matriculation at Trinity College. On the other hand, Newton certainly was essential to the Copernican Revolution because it was his version that survived in the long run.

Newton matriculated at Cambridge in 1661 and by 1664 was immersed in Descartes' works. Cambridge's official curriculum may still have embraced Aristotelian natural philosophy, but some important Cambridge minds were reading Descartes closely. As an undergraduate, Newton was a Cartesian Copernican, and he would continue to be something of a Cartesian for several years. He rejected much but by no means all.

Newton's rejections agreed with his English-Cambridge conceptual context. Trinity man Francis Bacon's staunch empiricism from earlier in the century undoubtedly encouraged Newton's denial of Descartes' rationalist approach to knowledge. Empirical results during the century undermined – for an empiricist – Descartes' idea of a plenum, based as it was on a clear and distinct idea. Mental ideas disclosed God's existence to Descartes, whereas revelation and nature's design combined to do so for Newton. But it was absurd for an infinite God to be confined to a geometrical point. He filled all space. Newton thus rejected Descartes' basic metaphysical and physical simplicities.

Newton did accept three important ideas. First, consistent with his connections with Isaac Barrow, Cambridge's first Lucasian Professor of Mathematics, Newton studied Descartes' pure mathematics and would, of course, soon invent the calculus. Second, the world was mechanical, though it contained atoms within void space. Third, in that mechanical context, Descartes' sun-centered, material vortex made sense.

In accepting the Cartesian vortex, however, Newton had implicitly abandoned Cartesian simplicity. That is, he rejected Descartes' basic conclusions resting upon his metaphysical and physical simplicities. In accepting the physically appealing, Cartesian solar vortex, Newton endorsed an *overall* Copernican system which had *not* attracted from Descartes an emphasis on its simplicity. But Newton would still argue for simplicity in his eventual, anti-Cartesian version of Copernicanism.

His two-decade path from his undergraduate study of natural philosophy to his *Mathematical Principles of Natural Philosophy* involved fundamental changes. Unlike Descartes, but again consistent with Cambridge's mathematical tradition, Newton sought a mathematical, Copernican natural philosophy. Though ignored by Galileo and Descartes, Kepler's three, mathematical laws of planetary motion thus captured Newton's attention. But he could not get a mathematical vortex to agree with Kepler. Moreover, especially in the 1670s, Newton pursued alchemy – which included non-mechanical, action-at-a-distance forces. Kepler and alchemy then seemingly combined in Newton's proposal of gravitational attraction to explain motions in the Copernican universe. Evidently, not until the 1680s when he began dealing with comets did gravity become universal for Newton.

It was certainly universal in his 1687 *Principia* which espoused a highly mathematical account of the universe. Gravity at the earth's surface was the same kind of force as lunar gravity, their actions being understood within the context of Newton's three laws of motion. In accord with his third law of motion, for example, the earth and a rock mutually attracted each other, the forces being equal in magnitude and opposite in direction. Same with the moon. The earth's much greater quantity of matter meant that the rock was observed to fall to the earth – not vice versa – and the moon orbited the earth – not vice versa. This theory accurately explained motions of planets and comets and was also applicable to stars, even though there were no observable stellar motions as a consequence of their mutual gravitational attractions. Nevertheless, universal gravitation was obviously an empirical-mathematical concept of enormous insight and success.

But puzzles – or fatal flaws, opponents would argue – did remain. If the stars attracted each other, why did they not move? If planets attracted each other, why did they not alter one another's orbits to undermine the solar system's long-term stability? As a non-mechanical force, was not gravity inconceivable?

Simplicity guided much of Newton's defense of his gravitational theory, especially in what became his *Principia's* four, well-known "Rules for the Study of Natural Philosophy." According to the first rule: "No more causes of natural things should be admitted than are both true and sufficient to explain their phenomena." That was because "Nature does nothing in vain" and "nature is simple" (Newton, 1729, 320). Hence, according to the second rule, causes of the same kind of natural effect were the same, meaning that stones fell in Europe and America for the same reason. These two briefly-stated rules entailed the third, that constant qualities belonging "to all bodies on which experiments can be made should be taken as qualities of all bodies universally." After all, "nature is simple and ever consonant with itself" (Newton, 1729, 320). Properties of sensible objects could legitimately be attributed to objects that were either invisibly small or remotely distant. Those objects were all extended, hard, impenetrable, moveable, and subject to the forces of inertia. Of course, this third rule's climax was universal gravitation: "it will have to be concluded by this third rule that all bodies gravitate toward one another" (Newton, 1729, 321).

Consequently, because gravitational theory worked so well for so much, nature's simplicity established gravity's universality. Simplicity provided Newton a successful defense of universality, a defense required despite the great successes of gravitational theory. That is, resultant problems were insufficient to undermine the theory and remained merely to be solved. But *how* did Newton know that nature was simple? A question not answered by his four rules.

Theology did provide the answer, as disclosed in Newton's Queries in his *Opticks*. Query 28's several rhetorical questions had the same answer. What occupied empty space? Why do the sun and planets "gravitate towards one another?" Why do the planets move in the same direction? Why are animals' bodies so well "contrived?" And to the point of simplicity: "Whence is it that nature does nothing in vain . . . ?" (Newton, 1730, 369). The answer: "And these things being rightly dispatch'd, does it not appear from Phaenomena that there is a being incorporeal, living, intelligent, omnipresent, who in infinite Space, as it were in his Sensory, sees the things themselves intimately, and thoroughly perceives them, and comprehends them wholly by their immediate presence to himself" (Newton, 1730, 370, cf. 403).

But in Newton's argument for simplicity, did God's omnipresence actually reflect Newton's conviction that universality was logically prior to simplicity? And was revelation epistemologically prior to empiricism in linking God and universality? Cited in manuscript by Newton, one biblical passage – Acts 17:28 – confirmed about God that it was "in him we live and move and have our being" (cf. Brooke, 1991, 139). Such a God would surely be consistent, acting the same everywhere throughout the universe. Hence, a biblically established, universal God may well have convinced Newton of the validity of his own third rule which, in turn, justified his first-rule simplicity.

8.4 Conclusion

In conclusion, I would first raise two, "what if" questions – one short, one long. I am not advocating hypothetical history. However, we have been challenged by no less than Francis Crick who asks: if historians cannot answer "iffy" questions then what is the point of historical analysis? (Crick, 1988, 75). My first question is, what if opponents of Copernicus also relied on simplicity? If so, we would be cautioned not overly to credit simplicity's role in the Copernican Revolution. Second, what if eighteenth-century empirical results had been different?

What if stellar aberration, as the prime example, had *not* been observed in the 1720s.[1] James Bradley was apparently surprised by the unexpected observation that led to the aberration explanation. But what if Bradley had predicted the observation, had searched for it, and had found it non existent? That is, the earth was evidently not moving! Combined with an analysis of simplicity's different roles, this iffy question does help illustrate astronomical uncertainties in, say, 1720. This was not an impossible empirical result, and would it not arguably have demolished both Cartesian and Newtonian Copernicanisms – even given all the theological simplicity they had in their favor?

Being present-day Copernicans ourselves, we might initially perceive such a hypothetical, empirical result from the perspective of our own Kuhnian paradigm. Placing ourselves in the context of the 1720s, we could see it as simply a Kuhnian "puzzle" to be solved. Absence of aberration could result

---

[1] In addition to the Simplicity Workshop, I also presented this paper at Caltech and, in addition, posed the stellar-aberration question to students in a class that I was teaching there. This part of the chapter has thus been revised and extended in response to several insightful remarks from many, especially Zagid Abatchev, Diana Buchwald, Jed Buchwald, Paul Gebhart, Kathryn Olesko, Robert Shimizu, Noel Swerdlow, and Caleb Ziegler.

from either a faulty telescope or light speeding much faster than previously mis-measured. If true, those possibilities could successfully solve the puzzle facing our Copernican paradigm.

But I am supposing that we have reached the next stage. That is, the above possibilities have been resolved. We were correct about light's speed and the telescope's accuracy. The puzzle has thus become an "anomaly," posing a serious "crisis" for our Copernican paradigm. I am proposing that within that hypothetical context, with empirical evidence revealing a non-orbiting earth, a Ptolemaic or Tychonic universe seems rather obvious.

This brief hypothetical example thus provides *historical* insight into the *non-*hypothetical situation circa 1720. An earth-centered universe was still *empirically* possible. Newtonians would argue that the sun must be at the center because its mass made it the center of gravity. Cartesians would argue that the sun must be at the center because it is composed of the smallest particles. Their unsettled, contentious disagreement resulted from the fact, a supporter of Ptolemy could argue, that they were both wrong. We could even imagine an astronomer immersed in perpetual meteorological uncertainties observing absolutely predictable planetary motions. Why could he not have argued both for an earth-centered universe and also for a somewhat modified Aristotelian-like dichotomy between the celestial and terrestrial realms – along with an Aristotelian-Ptolemaic account of planetary motions? Surely the Christian God could have provided an Aristotelian-like final cause of planetary motions, and the astronomer could attempt a mathematical analysis of those final causes. His would then have been a *mathematical* universe pervaded by *non-material* causes. Much like that of Isaac Newton! Surely someone actually made this argument, given that British universities were teaching Aristotelian natural philosophy into the second half of the seventeenth century.

If such had transpired, we would be writing not about the Copernican Revolution but the Copernican Diversion – and perhaps the Cartesian Diversion. Descartes had explicitly stated that empirical negation of Copernicanism would undermine his natural philosophy. Could not the Christian God of Galileo, Descartes, and Newton have been once more recognized as the creator of a somewhat modified Aristotelian world? If so, I would be writing about how the ideal of simplicity led astronomers and natural philosophers astray, away from Aristotle. We undoubtedly would pay much more attention to Gian Domenico Cassini (1625–1712), a supporter of Tycho's earth-centered system well into the second half of the seventeenth century and, evidently, never more than a lukewarm Copernican. The non-hypothetical, historical point here is that as influential as simplicity may have been, neither it nor existing empirical evidence had unequivocally established Copernicanism.

But back to non-hypothetical history and the issue of simplicity, which was obviously significant for Galileo, Descartes, and Newton. However, in quite different ways. Galileo's accommodated Bible did support a similar human-Divine appreciation of mathematical simplicity. For Galileo, the overall structure of the Copernican universe was appealingly simple, and his simple explanation of the tides provided crucial support for Copernicus. However, both Descartes and Newton, in effect, agreed with the Catholic Church in rejecting that tidal theory. For Descartes, God not the Bible was necessary to the argument. Simplicity lay in the metaphysical and physical foundations of his natural philosophical case for Copernicanism. But, unlike Galileo, Descartes did not emphasize the overall simplicity of the Copernican system. The young Newton, persuaded by Descartes' *physical* conclusions, denied his metaphysical and physical simplicities. With respect to simplicity, the older Newton was more concerned with gravity's universality. More of a biblical literalist than either Galileo or Descartes, Newton invoked God's omnipresence to justify universality – which dovetailed with simplicity, in the sense of fewer causes.

Finally, what about a simple summary of simplicity's complicated role in the Copernican Revolution? How about: The absence of conclusive empirical evidence greatly strengthened the importance of theologically based simplicity arguments, providing some with compelling reasons to accept Copernicus, but in the end *physical* appeal is what triumphed. That is, whether simplicity appealed to one or not, Descartes' physical insights did. His was an ingenious, insightful physical mechanism that required a central sun. Descartes' serious employment of simplicity produced a simplicity-independent Copernicanism. In that sense, it is perhaps misleading to speak of a Copernican Revolution, because the eventual shift to a sun-centered system was merely one part of the more comprehensive *Cartesian* Revolution. That is, our analysis of simplicity's role in the Copernican Revolution arguably exposes the very concept of a *Copernican* Revolution to be a misleading, lingering vestige of now long-discredited Whiggish history of science.

Acknowledgements I am grateful to Evelyn Fox Keller for encouraging me to write this essay, for a Workshop on Simplicity as an Epistemological Value in Scientific Practice that she and Karine Chemla organized. And thanks to Julie for insightful comments

Bibliography

Brooke, J. H., *Science and religion: some historical perspectives*, Cambridge: Cambridge University Press, 1991.

Crick, F., *What mad pursuit*, Basic Books, 1988.

Descartes, R., *Principles of philosophy*, trans. V. R. Miller and R. P. Miller, Reidel: Dordrecht, 1983 [1644].

Galilei, G., *Dialogue concerning the two chief world systems*, trans. S. Drake, Berkeley: University of California Press, 1962 [1632].

Newton, I., *The principia*, trans. A. Motte, 3rd edn., Amherst: Prometheus Books, 1995 [1729].

——, *Opticks*, 4th edn., New York: Dover, 1952 [1730].

Part IV
Chemistry

Chapter 9
The Weekday Chemist: The Training
of Aleksandr Borodin

Michael D. Gordin

On 3 July 1877 (N.S.), Aleksandr Porfir'evich Borodin (1833–1887), chemistry professor at the Medico-Surgical Academy in St. Petersburg, found himself on a scientific trip near Weimar, Germany, and wanted to pay homage to a great mentor he had never met. The object of his aspirations was not the local chemistry doyen, but Franz Liszt (1811–1886), piano virtuoso and mainstay of avant-garde musical composition. He managed to locate Liszt's house with some difficulty, and while waiting a few hours to be received he wandered around the local monuments to German cultural supremacy: the domiciles of Goethe, Schiller, and Herder. He was finally ushered in to see the Hungarian-born master, and his reception exceeded his wildest fantasies:

> The majestic lively figure of the old man, with an energetic, attractive face, moved before me and spoke unceasingly, tossing questions at me. The conversation was now in French, now in German, skipping from one to the other each minute. When I told Liszt that I am properly a *Sonntagsmusiker* [Sunday musician], he even quipped: ["]*aber Sonntag ist immer ein Feiertag*["] [but Sunday is always a holiday], and that "you have a complete right to '*Feiern*["], i.e., to celebrate.[1]

This episode quickly became legend. Borodin came to the master of modern composition (for the Russians scorned the alternative, Richard Wagner),

---

Abbreviations: *BorP*: A. P. Borodin, *Pis'ma: Polnoe sobranie, kriticheski sverennoe s podlinnymi tekstami*, ed. S. A. Dianin, 4 v. (Moscow: Gos. izd. muzykal'nyi sektor, 1927–1950); TsGIASPb: Central State Historical Archive of St. Petersburg. All dates are given in the old style Julian calendar, which lags 12 days behind the new-style Gregorian calendar in the nineteenth century. Exceptions are indicated by (N.S.). Transliterations follow a modification of the standard Library of Congress format, with the exception of Cui. All unattributed translations are mine. I would like to thank Michael S. Mahoney, Caryl Emerson, and Simon Morrison for valuable comments on an earlier draft of this essay.

[1] Borodin to E. S. Borodina, 3 July [1877] (N.S.), *BorP*, II, 133.

M.D. Gordin (✉)
Department of History, Princeton University, Princeton, NJ, USA
e-mail: mgordin@princeton.edu

J.Z. Buchwald (ed.), *A Master of Science History*, Archimedes 30,
DOI 10.1007/978-94-007-2627-7_9, © Springer Science+Business Media B.V. 2012

confessed his amateur status, and was welcomed.[2] For the remainder of his life, Liszt was an active supporter of the so-called "New Russian School" of music, arranging for concerts of the work of Borodin and his like-minded peers across Western Europe. Propagandist for Russian art Vladimir Vasil'evich Stasov (1824–1906) insisted that Borodin take precious time away from composition and write up his encounter for a Russian journal.[3] This was the stuff of mythmaking, and Stasov was not about to let it slip.

Several features of the Borodin-Liszt encounter have made it a mainstay of the collective hagiography of Borodin, especially the enthusiastic reception of Russian music by elite foreigners and Borodin's casual attitude towards his craft. One might just as well stress other features of the encounter: that Borodin was abroad on a chemist's errand; the profusion of foreign languages and the play of national identity, central for the Hungarian-born and French-educated composer of German ancestry ("He speaks both languages [French and German] excellently, loudly, in lively fashion, with excitement, quickly, and a great deal;—one might think that he is a Frenchman")[4]; and the vital role of Vladimir Stasov in shaping the account. What follows does not pretend to be a comprehensive biography; it is, rather, a focused depiction of the central role of *training* as a category to reformulate the central dilemma that has perpetually obsessed writers on Borodin: how does one reconcile the fact that he was both a scientist and an artist, a chemist and a composer? A profusion of articles, primarily in medical and chemical journals, portray Borodin's life as a problem of "double vocation": his biographers must decide which career—chemistry or music—was the "true" one, and which merely a distraction.[5]

---

[2] Liszt was the undisputed leader of the avant-garde in the 1850s, but the torch passed quite decisively to Wagner a decade later. See Alan Walker, *Franz Liszt: The Weimar Years, 1848–1861* (Ithaca: Cornell University Press, 1989), 336.

[3] Borodin wrote the first draft of the piece as "Moi vospominaniia o Liste" in June–July 1878, reproduced in *BorP*, III, 13 ff. It was printed in *Iskusstvo* in 1882 as "List u sebia v Veimare (iz lichnykh voospominanii A. P. Borodina)," reprinted in *BorP*, IV, 14 ff. On Stasov's role, see Stasov to Borodin, 6 February [18]78, reproduced in S. A. Dianin, *Borodin: Zhizneopisanie, materialy i dokumenty* (Moscow: Gosudarstvennoe muzykal'noe izd, 1960), 212. See also David Lloyd-Jones, "Borodin on Liszt," *Music and Letters* 42 (1961): 117–126; Alfred Habets, *Borodin and Liszt*, tr. Rosa Newmarch, 2d. ed. (London: Digby, Long & Co., [1895]); and Louise Cruppi, "Borodin and Liszt," *The Living Age* 312 (11 March 1922): 600–605. For Liszt's musical influence on the "New Russian School," see Gerald E. H. Abraham, *On Russian Music* (New York: Charles Scribner's Sons, 1939), Chapter 7.

[4] Borodin to E. S. Borodina, 3 July [1877] (N.S.), *BorP*, II, 135.

[5] All of the medical and chemical articles on Borodin, even the best, consist to at least some degree of read-write evidence: an author reads a story in the secondary literature and promptly writes it down without trying to check its veracity against the historical evidence. A selection of the historiography, in no particular order, is: "A. P. Borodin (1834–1877)," *Nature* 134 (1934): 727; "Professor Borodin," *Lancet* (19 March 1887): 601; F. William Sunderman, "Alexander Porfirivich Borodin: Physician, Chemist and Composer," *Annals of Medical History* 10 N.S. (1938): 445–453; Torstein Vik, "Aleksandr Borodin—lege, kjemiker, vitenskapsmann, lærer og komponist," *Tidsskrift for Den norske lægeforening* 30 (1998): 4693–4696; Peter J. Davies, "Alexander Porfir'yevich Borodin (1833–1887): Composer, Chemist, Physician,

By the end of this account I propose to explain *why* this particular question dominates the literature. The root cause of this skewed focus on this single issue can be laid at the door of Vladimir Stasov. For Stasov, Borodin was above all the composer of two symphonies, twelve songs, a symphonic poem, two string quartets, and an unfinished opera, *Prince Igor*—the last of which accounts for the bulk of his reputation both in Russia and in the West (where his music formed the score of the hit musical *Kismet*, which won the Tony Award in 1954). For Stasov, and for historians since, the main issue to be resolved is why Borodin wrote so little, why he sacrificed his "true vocation" of music to his chemistry. But a glance at his chemical productivity also shows a paucity of publications, with a grand total of about 20 works, almost all pre-dating 1872, leaving the last fifteen years of his life unfettered by chemistry. The problem here stems from essentializing the notion of "vocation," as if there were some

---

and Social Reformer," *Journal of Medical Biography* 3 (1995): 207–217; Igor E. Konstantinov, "The Life and Death of Professor Alexander P. Borodin: Surgeon, Chemist, and Great Musician," *Surgery* 123 (1998): 606–616; E. Lee Strohl, Robert W. Jamieson, and W. G. Dieffenbaugh, "Physician-Musicians," *Perspectives in Biology and Medicine* 17 (Winter 1974): 267–285; George Sarton, "Borodin (1833–1987)," *Osiris* 7 (1939): 225–260; K. N. Zelenin and N. I. Liashenko, "Aleksandr Porfir'evich Borodin (K 150-letiiu so dnia rozhdeniia)," *Voenno-meditsinskii zhurnal* 11 (1983): 66–69; Iu. V. Ionov and A. Iu. Ionov, "A. P. Borodin—vrach, khimik, pedagog (K 150-letiiu so dnia rozhdeniia)," *Sovetskoe zdravookhranenie*, 1 (1984): 61–64; Clive B. Hunt, "Aleksandr Borodin: Chemist and Composer," *Chemistry in Britain* 23(6) (1987): 547–550; Walter Kwasnik, "Der Komponist Alexander Borodin (1834–1887) als Chemiker: Zur Wiederkehr seines Todestages am 28.2.1967," *Chemiker-Zeitung/Chemische Apparatur* 91 (1967): 312–313; George B. Kauffman, "Syntheses and Symphonies," *The World & I* 3 (January 1988): 206 211; idem, "Russia's Alcksandr Borodin: Many Gifts, Many Callings," *Industrial Chemist* 8(1) (1987): 40–43; George B. Kauffman, Yurii Ivanovich Solov'ev, and Charlene Steinberg, "Aleksandr Porfir'evich Borodin (1833–1887)," *Education in Chemistry* 24 (September 1987): 138–140; Martin Sherwood, "A Russian of Many Octaves," *New Scientist* 100 (10 November 1983): 424; Jan Smaczny, "Alexander Borodin," *BBC Music Magazine* 5 (May 1998): 46–49; Maurice Schofield, "Borodin—Chemist and Composer," *Chemistry* 49(8) (October 1976): 13–14; Desmond O'Neill, "…aber Sonntag ist immer ein Feiertag: Alexander Borodin, MD, 1833–1887," *Journal of the Royal Society of Medicine* 81 (October 1988): 591–593; Charlene Steinberg, "The Scientific Activities of Aleksandr Borodin," *CHEM TECH* 1 (August 1971): 473–475; Hope Stoddard, "Borodin, Genius in Double Harness," *Musical Opinion* 57 (1934): 502–503; Edmund Yochum, "Symphonies and Syntheses," *The Science Counselor* 7 (1942): 42–43, 59–60; A. D. White, "Alexander Borodin: Full-Time Chemist, Part-Time Musician," *Journal of Chemical Education* 64 (April 1987): 326–327; Erica Kidson, "Alexander Borodin, 1833–1887," *Canon* (June 1955): 423–426; Jerzy Chodokowski, "Aleksander Borodin jako chemik," *Wiadomości Chemiczne* 8 (1954): 369–373; James C. Cole, "Alexander Borodin, the Scientist, the Musician, the Man," *Journal of the American Medical Association* 208(1) (7 April 1969): 129–130; Frederick H. Getman, "Alexander Borodin—Chemist and Musician," *Journal of Chemical Education* 8(9) (1931): 1763–1780; Susan E. Harman, "Alexander Borodin: Medical Educator, Chemist, Composer," *Maryland Medical Journal* 36 (1987): 445–450; Harold B. Friedman, "Alexander Borodin—Musician and Chemist," *Journal of Chemical Education* 18 (1941): 521–525; and William B. Ober, "Alexander Borodin, M.D. (1833–1887): Physician, Chemist, and Composer," *New York State Journal of Medicine* 67 (15 March 1967): 836–845.

kind of mark of Cain on Borodin's forehead that could have told him what his true career was. Borodin was employed as a chemist, taught at a medical school, wrote music, organized women's medical education, and in general partook of the vibrant intellectual life of post-Great Reforms Petersburg. Instead of accepting the categories established in Borodin's obituary by Stasov, we should return to his voluminous and beautifully styled correspondence to observe how he self-consciously understood his unusual life.[6] Borodin's tubercular wife spent her winters in Moscow while her husband worked in Petersburg, and he wrote to her several times a week chronicling his activities. These letters form an amazing panorama of the musical, scientific, and other cultural spheres in which the man moved—and one in which he steadfastly refused to make precisely the "vocational" distinctions Stasov later imposed on him. Instead, one finds the leitmotif of *training*—the process which creates vocations—as a way of understanding Borodin's cultural peregrinations.

9.1 The Random Walk from Chemistry to Music

Borodin's life was colorfully atypical from the moment of his birth. He was born on 31 October 1833, in St. Petersburg, a royal bastard. His father, Luka Stepanovich Gedianov (1772–1843), was an Imeretian prince from Transcaucasia, suitably Russified and living in the center of Petersburg, who sired young Aleksandr with his maid, Avdot'ia Konstantinovna Antonova, a soldier's daughter from Narva who was 24 at the time. In order to establish legitimacy, Borodin was registered as the son of Gedianov's valet, Porfirii Ionovich Borodin, and his wife Tat'iana Girgor'evna Borodina—which technically meant that the boy was a serf. His biological mother—whom he called "auntie (*tetushka*)" for the rest of her life—took charge of his education at home, having him tutored in German (by Fräulein Luischen, a housekeeper), French (by Béguin, who taught at the Lycee), and in English (by John Roper, who served as a governor at a commercial school).[7] He was registered as a free serf on 3 November 1849, and the next year, at age 17, his mother attempted to register him as a student at St. Petersburg University. This proved abortive, but she managed to enroll him as a student at the Medico-Surgical Academy on the Vyborg Side of Petersburg—largely because her current beau, F. A. Fedorov, knew the inspector, Il'inskii, who directed admissions there. Antonova, married

---

[6] The original Borodin template essay is V. Stasov, "Aleksandr Porfir'evich Borodin," *Istoricheskii Vestnik* 28 (1887): 137–168; reprinted in V. Stasov, *Izbrannye sochineniia*, 3 v. (Moscow: Iskusstvo, 1952), 329–365. All quotations come from the original published article. Stasov had already begun to shape the story in the days after Borodin's death with a brief obituary in a newspaper: V. Stasov, "Aleksandr Porfir'evich Borodin: Nekrolog," *Novoe Vremia*, 17 February (1 March) 1887, #3940: 3. Many later biographies betray explicitly or through their footnotes that they are entirely derived from this one ur-source for biographical information.

[7] This information is heavily emphasized by Stasov in "Aleksandr Porfir'evich Borodin," 138–139.

as of Spring 1839 to retired physician Kh. I. Kleineke, moved her family to Aleksandr's new neighborhood.[8]

The Medico-Surgical Academy turned out to be a propitious choice. It was, primarily, a military medical school, and this is the training Borodin received. The Academy was founded by Imperial charter on 18 December 1798, under the reign of Tsar Paul, and in its early years it mostly enrolled the children of foreigners. It continued in a kind of administrative limbo, shuttling between the Ministry of Internal Affairs, the Ministry of Popular Enlightenment, and the Army (its final home), during its first half century.[9] In 1854 construction began on a new chemistry laboratory to be stocked with imported instruments. There were still a great many foreigners at the moment when Borodin enrolled. According to his half-brother, in an account he gave to Stasov, Borodin's "closest comrades [at the Academy] were, for the most part, all German students, which was especially strongly facilitated by the antipathy of our mother to Russians, whom she (although also Russian, but originally from Narva) did not approve of 'for the rudeness of [their] manners.' "[10]

Certainly the most significant move Borodin made at the Academy was to approach his chemistry professor, Nikolai N. Zinin, and ask to perform experiments in his laboratory as training for a career in chemistry. Zinin's imprint on the young man—from taking him in, directing his specialization in precisely the same areas of experimental organic chemistry he himself studied, sending him off to Heidelberg for further study, and even apparently controlling issues of his personal toilet—is hard to overemphasize.[11] The role of Zinin was so prominent that several psychobiographers since have cast Zinin as the first of a series of "father figures" by which Borodin sought to replace his biological father— absent by death, illegitimacy, and his mother's dominance.[12]

---

[8] The biographical particulars here and in what follows are drawn from the most reliable Soviet-era biographies of Borodin: N. A. Figurovskii and Yu. I. Solov'ev, *Aleksandr Porfir'- evich Borodin: A Chemist's Biography*, tr. Charlene Steinberg and George B. Kauffman (New York: Springer-Verlag, 1988); Dianin, *Borodin*; A. P. Zorina, *Aleksandr Porfir'evich Borodin* (Moscow: Muzyka, 1987). On issues of interpretation, however, all these sources follow the basic structure offered by Stasov. A recent Dutch dissertation has attempted to fill in many of these lacunae: Willem Vijvers, "Alexander Borodin: een biografische studie" (Ph.D. dissertation, University of Amsterdam, 2007).

[9] G. Skorichenko, "Mediko-Khirurgicheskaia Akademiia, v vedenii Meditsinskoi kollegii i Ministerstva Vnutrennikh Del, ot osnovaniia do vvedeniia pervago ustava eia, 1798–1808 g.," in Ivanovskii, ed., *Istoriia Imperatorskoi Voenno-Meditsinskoi (byvshei Mediko-Khirurgiches- koi) Akademii za sto let, 1798–1898* (St. Petersburg: Tip. Ministerstva Vnutrennikh Del, 1898): 41–154.

[10] Quoted in Stasov, "Aleksandr Porfir'evich Borodin," 142.

[11] On the micromanagement of Borodin's personal life, see A. P. Dianin, "Aleksandr Porfir'evich Borodin: Biograficheskii ocherk i vospominanii," *Zhurnal Russkogo Fiziko-Khimicheskogo Obshchestva* 20, khim. ch. (1888): 367–379, on 369.

[12] A. Sokhor, *Aleksandr Porfir'evich Borodin: Zhizn', deiatel'nost', muzykal'noe tvorchestvo* (Moscow: Muzyka, 1965), 45; Bärbel Zaddach-Dudek, "A. P. Borodin—russischer Musiker und Naturwissenschaftler im 19. Jahrhundert," in Aloys Henning and Jutta Petersdorf, eds.,

According to an account which is presented only by Stasov, and appears nowhere else in Borodin's writings—including a lengthy obituary for Zinin that he penned in 1880—Zinin once tried to dissuade Borodin from his musical activities, encouraging him to focus on his chemical studies, declaring: "Mr. Borodin, busy yourself less with [musical] romances—I am placing all my hopes in you, that you will be my deputy, and you all the time think about music and two hares [i.e., try to catch two hares simultaneously and you will end up with neither—MG]."[13] Absolutely no contemporary refers to this anecdote in any context except by citing Stasov, and therefore it is highly likely that this is an apocryphal embellishment—a supposition also suggested by Stasov's use of the unusual idiomatic expression in his correspondence.[14] This quotation is the sole source of evidence for the claim that chemists disapproved of Borodin's musical activities. On no other occasion did any chemist—either to Borodin or to a third party—suggest he give up his music or any of his other activities. They were, of course, enthusiastic about Borodin's chemical research, and lamented that he did not complete more of it.

Borodin graduated on 25 March 1856 and served briefly as a physician at the Second Infantry Hospital (where he happened across Modest Musorgskii as an army officer, an inconsequential encounter at the time that was later exaggerated by Stasov), but he preferred to pursue a career in chemistry and not medicine. Borodin returned to work with Zinin, and defended his dissertation—the first in the history of the Academy written and defended in Russian (not Latin)—"On the Analogy of Arensous Acid with Phosphoric Acid in Its Chemical and Toxicological Relations," on 3 May 1858. (He later would obtain a master's in chemistry from St. Petersburg University while working in Zinin's lab at the Medico-Surgical Academy.)[15] Borodin had already been abroad once, escorting the distinguished oculist Ivan Ivanovich Kabat to an international ophthamological congress in Brussels, and which he used to visit

---

Wissenschaftsgeschichte in Osteuropa: Europa literarum artiumque scientiam communicans (Wiesbaden: Harrassowitz, 1998), 87–100; and R. P. LaCombe's theory as reported in George B. Kauffman and Kathryn Bumpass, "An Apparent Conflict between Art and Science: The Case of Aleksandr Porfir'evich Borodin (1833–1887)," *Leonardo* 21 (1988): 429–436, on 434. Stasov stoked these flames: "[Zinin] considered [Borodin] his spiritual son, and Borodin from his side considered him a second father." Stasov, "Aleksandr Porfir'evich Borodin," 149.

[13] Quoted in Stasov, "Aleksandr Porfir'evich Borodin," 144.

[14] In a discussion of César Cui, Stasov snapped: "Go after two hares—and you won't catch one." Stasov to Balakirev, 3 September 1886, in M. A. Balakirev and V. V. Stasov, *Perepiska*, 2 v., ed. A. S. Liapunova (Moscow: Muzyka, 1970–1971), II, 88.

[15] See the correspondence regarding permission to take his master's exam: A. Borodin to rector of St. Petersburg University Aleksandr Pletnev, 23 March 1859, TsGIASPb f. 14, op. 1, d. 5983, l. 1; Pletnev to Dean of Physico-Mathematical Faculty of St. Petersburg University Emilian Khristianovich Lenz, 29 May 1859, TsGIASPb f. 14, op. 3, d. 14709, ll. 51–51ob.

chemistry laboratories (such as Marcellin Berthelot's) in Paris.[16] Zinin believed that a postdoctoral trip to study abroad would be beneficial for the development of young Borodin's chemical career, and he arranged for him to embark on a subsidized three-year stay in Heidelberg (and incidentally also in Paris and Pisa).[17] Upon his return a position at the Medico-Surgical Academy quickly materialized. Borodin was appointed an adjunct at the Academy in October 1862 and was already promoted to adjunct professor on 8 December 1862, and to full professor on 15 April 1864. There he remained until his death twenty-three years later.

The account above portrays a career that is primarily chemical. What about the music? One does not, of course, become a renowned composer overnight, and Borodin had—according to reliable evidence beyond Stasov's obituary—plenty of exposure to music in his youth. He learned how to play piano quite young, at the insistence of his mother, and by 1850 he and his close friend in all things musical, Mikhail Shchiglev, were brought by a violinist friend, Petr Ivanovich Vasil'ev, into amateur chamber musician I. I. Gavrushkevich's circle.[18] Borodin's musical activities, essentially confined to romances—some of them published, probably through a family friend, in 1849—and halting ventures into chamber music, continued into his stay at Heidelberg. He even wrote four songs while a student at the Academy from 1852 to 1855. Much of this juvenilia remains unpublished and Borodin himself almost never referred to these pieces.

His musical career began at a chance meeting at a *kruzhok* (discussion circle) of his colleague at the Medico-Surgical Academy, Sergei Petrovich Botkin, soon to become one of the most distinguished Russian clinical physicians of the second half of the nineteenth century. In late November or early December 1862, at one of the regular Saturday night meetings around 9 pm, Borodin met Milii Alekseevich Balakirev (1836–1910), a local amateur musician who was also a rather hypochondriac patient of Botkin's.[19] Balakirev would form the fulcrum around which Borodin pivoted into the musical world, and it is to this world that we now follow him (although it behooves us to keep in mind that this connection was made by virtue of the Medico-Surgical Academy).

Borodin entered Balakirev's circle at a time of tremendous ferment in musical Petersburg, developments that were intimately tied to the liberalization and

---

[16] Borodin to his mother (Avdot'ia Konstantinovna Kleineke), 15 August 1857, *BorP*, I, 27–28; Dianin, *Borodin*, 40.

[17] For more on the circle of Russian chemists in Heidelberg, see Michael D. Gordin, "The Heidelberg Circle: German Inflections on the Professionalization of Russian Chemistry in the 1860s," *Osiris* 23 (2008): 23–49.

[18] Zorina, *Aleksandr Porfir'evich Borodin*, 23.

[19] Balakirev thought very highly of Botkin, and mentioned him in his correspondence with Stasov on several occasions. (He did not, however, remark on meeting Borodin there; the Botkin historiography maintains a similar silence on the matter.) See Balakirev to Stasov, 31 March 1862 and 29 April 1862, in Balakirev and Stasov, *Perepiska*, I, 184–185.

professionalization that characterized the Great Reforms.[20] In the 1850s there were two opera companies in Petersburg and several concert series, most of them administered by the Imperial Theater Directorate, a department in the Ministry of the Imperial Court, which had held the official monopoly on public entertainment in the winter season since Alexander I instituted the system in 1803 (it was abolished in 1882 by Alexander III). The central feature of the Petersburg music scene in the 1850s was the dominance of foreign musicians and musical instructors, mostly Italians (vocal) and Germans (piano).[21] In 1859, pianist and composer Anton Rubinstein (of Jewish heritage but a baptized Orthodox Christian) formed the Russian Musical Society with several associates. Rubinstein's goals were threefold: an annual concert series; civil status to be granted to musicians under the "free artists" clause of the Table of Ranks; and a Western-style conservatory in St. Petersburg. These goals might seem uncontroversial from our perspective, but they were all positions that called for the establishment of music as a *profession* in Petersburg molded on the prevailing German standards of the musical world, and thus drew fire from self-styled amateur composers: first from the Wagnerian Aleksandr Serov (1820–1871), and then from the circle surrounding Balakirev, dubbed the "New Russian School" and somewhat ironically as the "Mighty Little Heap (*moguchaia kuchka*)" at home and the "Mighty Five" abroad, terms which I shall use interchangeably.[22]

[20] Much of what follows is drawn from the excellent studies: Robert C. Ridenour, *Nationalism, Modernism and Personal Rivalry in Nineteenth-Century Russian Music* (Ann Arbor: UMI Research Press, 1981); Richard Taruskin, *Defining Russia Musically: Historical and Hermeneutical Essays* (Princeton: Princeton University Press, 1997); idem, *Opera and Drama in Russia: As Preached and Practiced in the 1860s* (Rochester, New York: University of Rochester Press, 1993 [1981]); Sigrid Neef, *Die Russischen Fünf: Balakirew—Borodin—Cui—Mussorgski—Rimski-Korsakow: Monographien—Dokumente—Briefe—Programme—Werke* (Berlin: Ernst Kuhn, 1992); and A. Gozenpud, *Russkii opernyi teatr XIX veka (1857–1872)* (Leningrad: Muzyka, 1971).

[21] This foreign dependence persisted into the 1860s, as noted by music critic G. A. Larosh: "Musical Russia even now is nothing but a colony of Germany: all our musical activity, with the exception of the most insignificant in volume—composition—is in the hands of Germans; our pianists, a class of musicians predominant in our times, are almost exclusively Germans; kappelmeisters, teachers, finally instrumental masters and sellers of notes are again Germans, seizing among us all the steps of the musical hierarchy from brilliant virtuosity to humble craftsmanship. However, it is more influential than one usually thinks. Germans don't only materially control our music, but they morally influence it in terms of its national character." Larosh, "Glinka i ego znachenie v istorii muzyki (1867–1868)," in Larosh, *Izbrannye stat'i*, 4 v. (Leningrad: Muzyka, 1974–1977), I, 33.

[22] On Rubinstein's stance toward professionalism and its casting by his opponents as sycophancy to the Germans, see Taruskin, *Defining Russia Musically*, 123. For a sympathetic interpretation of Serov, see Taruskin, *Opera and Drama in Russia*.

The Mighty Five defined the first decade of Borodin's musical life, a decade centered on the person of Milii Balakirev.[23] Balakirev had moved to St. Petersburg from his native Nizhnii Novgorod in 1855, not yet nineteen years old and by training a mathematician. The following year he met the Stasov brothers, Vladimir and Dmitrii, and the three of them formed a *kruzhok* that centered around their shared passion for the music of Mikhail Glinka (1804–1857). They were soon joined by César Cui (1835–1918; of French and Lithuanian parentage), a fortification engineer and composer who in 1857 brought a young military officer—Modest Musorgskii (1839–1881)—into the fold, whom he had in turn met at the house of Glinka's *Doppelgänger* in Russian musical composition, Aleksandr Dargomyzhskii (1813–1869). In November 1861 a piano teacher introduced his student, a naval cadet named Nikolai Rimskii-Korsakov (1844–1908), to Balakirev for further training. They were joined by Aleksandr Borodin in 1862, the oldest by far at the ripe age of 28. This was a group built around the institutions of the *kruzhok*, and was characterized by the fact that all of the members initially earned their livings from careers other than music. By the mid-1870s, the only members of the rapidly dissolving Five who were still engaged in "double careers" were Cui and Borodin—the latter often marked by his career among the group through references to him as "the alchemist," "the chemical gentleman," and "the chemical brigand."[24]

Musical relations among the group were far from equal. Balakirev was clearly the leader, treated by all as the undisputed authority on issues of orchestral music, while Cui was deemed the master of opera.[25] Stasov—the only close member of the group who did not compose original music—assumed the role of chief ideologue, having already shaped many of Balakirev's views, although the public dissemination of those views devolved to Cui, who served as the music critic for the *St. Petersburg News*. (It was only after Cui ceased to perform this role in the mid-1870s that Stasov assumed the mantle).[26] The content of the group's musical ideology has long been a subject of some dispute. Officially, it was couched in the language of "nationalism"—although that

---

[23] On Balakirev, see E. Frid, "Milii Alekseevich Balakirev (1837–1910)," in E. L. Frid, ed., *Milii Alekseevich Balakirev: Issledovaniia i stat'i* (Leningrad: Gos. muzykal'noe izd., 1961), 5–75; Edward Garden, *Balakirev: A Critical Study of His Life and Music* (London: Faber and Faber, 1967); and Evgenii Gippius, "M. Balakirev—sobiratel' russkikh narodnykh pesen [I-II]," *Sovetskaia muzyka*, 4 (1953): 69–76; 5 (1953): 61–67.

[24] See, respectively, V. V. Stasov, *Pis'ma k rodnym*, 3 v. (Moscow: Gos. muzykal'noe izd., 1954–1962), I(ii), 78; and Modest Petrovich Musorgskii, *Literaturnoe nasledie*, eds. A. A. Orlova and M. S. Pekelis (Moscow: Muzyka, 1971), 101 and 118.

[25] The infantilization of the other three was deeply resented by Rimskii-Korsakov in his highly subjective account, *Letopis' moei muzykal'noi zhizni (1844–1906)*, 3d. ed. (Moscow: Gos. izd. muzykal'nyi sektor, 1926), 79.

[26] A. K. Lebedev and A. V. Solodovnikov, *Vladimir Vasil'evich Stasov: Zhizn' i tvorchestvo* (Moscow: Iskusstvo, 1966), 39. Borodin stepped in thrice and wrote anonymous concert reviews in the late 1860s in Cui's stead. These are reprinted in A. P. Borodin, *Kriticheskie stat'i*, 2d. rev. ed. (Moscow: Muzyka, 1982).

would seem to preclude Cui's position on the side of the angels and the placement of Petr Il'ich Chaikovskii (1840–1893) on the Rubinstein side.[27] The Five ostensibly based their music on folk tunes, but so did Chaikovskii; meanwhile they waxed enthusiastic for Franz Liszt or Hector Berlioz.[28] I will postpone the problem of teasing out the content of the ideology for now, to return to it later in one specific aspect which reflects many of its other features: the issue of musical training.

Training was in fact a central feature of the public institution associated with the Five (much as the Russian Musical Society was associated with Rubinstein): the Free Music School, created by Balakirev and his associate Gavriil Lomakin on 18 March 1862. Lomakin had originally managed to constrain Balakirev's propagandistic use of the School, but ended up retiring on 28 January 1868. This left Balakirev in charge of almost all non-state musical events in the capital. For on 16 July 1867 Rubinstein, due to conflicts with the Musical Society's primary patron, Grand Duchess Elena Pavlovna, resigned as director of the Society's concert series, and Balakirev—his arch-nemesis—was appointed in his stead. Balakirev's reign over both the Free Music School and the Russian Musical Society did not last long (he was dismissed in 1869). From 1872 to 1877 Balakirev withdrew into increasingly archaic religiosity, cutting off contact with all of his former musical friends, with the partial exception of Borodin.

Borodin saw this dissolution of the Five to be a natural process. In the beginning, he recalled, when they were all "in the position of eggs under a brood hen (I mean under Balakirev), we were all more or less close."[29] Writing to reassure a friend, he argued that the separation of the group was more an expression of the success of the initial group than its complete disintegration:

> We do not understand the words "the collapse of the circle" entirely identically. After all, you also find among us great differences, and you even say that the works of each of the circle's members are so different and varied in character and spirit, and so on, but isn't this what the fact of "collapse" expresses[?]. . . And if I find such a collapse to be natural, then that is only because this is how it always happens in all areas of human activity In the degree of the development of activity, individuality begins to take

---

[27] Cui would not always be considered fully of the group. After the collapse of the Five, Borodin would write of him: "Tell the truth, for all his advantages, he is still not a *Russian* person and not a *Russian* composer; he doesn't understand properly *Russian* music, he likes it only insofar as it is *good* music in general; he doesn't at all feel, value, or understand the national streak." Borodin to E. S. Borodina, 20 November 1886, *BorP*, IV, 217. Emphasis in original. For more on the rise and fall of Cui—certainly the most obscure member of the group to concert-goers today, although perhaps the most visible and popular of them in terms of opera composition in the 1860s, see Taruskin, *Opera and Drama in Russia*, Chapter 6. On the peculiarity of excluding Chaikovskii, see Vladimir Fédorov, "Čajkovskij, Musicien Type du XIXe siècle?" *Acta Musicologica* 42 (1970): 59–70; and Georg Knepler, "Čajkovskij, Musicien Type du XIXe siècle?" *Acta Musicologica* 43 (1971): 205–235.

[28] These issues are usefully discussed in Ridenour, *Nationalism, Modernism, and Personal Rivalry in Nineteenth-Century Russian Music*.

[29] Borodin to Liubov' Ivanovna Karmalina, 15 April 1875, *BorP*, II, 89.

precedence over the school, over what a person inherited from others. *Eggs which a hen carries all resemble each other; the chickens which come forth from the eggs are already less similar, and they grow up, such that they don't at all resemble each other—from one emerges an impassioned black rooster, from another a peaceful white hen.* Same thing here. The general musical stamp, the general manners, proper to the circle remained, as in the above example the general generic and visible signs of the species of chickens remains, and then each of us, as every grown-up rooster or hen, has his own *personal* character, his individuality. And thank God![30]

To put it another way, their training was complete.

How he used that training was another issue entirely. Issues of both the quantity and the quality of Borodin's musical writing were live ones for his contemporaries, as they have been for scholars since. Borodin's few compositions were across the map in terms of genre, but in each genre—particularly the string quartet and the symphony—Borodin clung doggedly to old models, such as the sonata form of symphonic structure. In terms of innovations in both structure and music theory—strong points for Russian music in general—Borodin was (in all but his musical politics) among the conservatives.[31] While his friends may have looked askance at the lack of novelty in his compositions, they were quick to defend him in terms of quantity. Less, they claimed, was certainly more for the New Russian School. The less Russian composers wrote—and Borodin was their prime example—the better each individual work was.[32]

While Borodin left quite a limited legacy, it was essentially all of high quality.[33] The quantity, however, was meager, and this was attributed to lack of time. Borodin himself lamented the scarcity of time for composition: "In winter I can only write music when I am so ill that I don't give lectures, don't go to the laboratory, yet all the same can work a little. For this reason my musical friends, contrary to universal custom, always wish me not health, but sickness."[34] That said, he undertook no efforts to rearrange his commitments to allow more time for composition. It remained something he did when the occasion presented itself. What resulted were brief ventures in almost every

---

[30] Borodin to Liubov' Ivanovna Karmalina, 1 June 1876, *BorP*, II, 107–108. Ellipses added; emphasis in original.

[31] M. Ivanov, "Muzykal'nye nabroski," *Novoe Vremia*, 23 February (7 March) 1887, #3946: 2. On innovation in theory, see Gordon D. McQuere, *Russian Theoretical Thought in Music* (Ann Arbor: UMI Research Press, 1983).

[32] Cui, "A. P. Borodin," *Nedelia*, 1 March 1887, #9: 287–293, on 289.

[33] Understandably, most biographies of Borodin emphasize his music. The best of these are the two produced for the Grove Dictionary: Gerald Abraham and David Lloyd-Jones, "Alexander Borodin," in *The New Grove Russian Masters 1* (New York: Macmillan, 1986): 43–74; and the extremely thorough and authoritative Robert William Oldani, "Borodin, Aleksandr Porfir'yevich," *Grove Music Online*, ed. L. Macy (Accessed 5 December 2003), <http://www.grovemusic.com>. See also Igor' Belza, *A. P. Borodin* (Moscow: Muzgiz, 1944); and Iv. Remezov, *A. P. Borodin: K 125-letiiu so dnia rozhdeniia* (Moscow: Gos. muzykal'noe izd., 1958).

[34] Borodin to Liubov' Ivanovna Karmalina, 1 June 1876, *BorP*, II, 108.

genre of music explored by the New Russian School. His most numerous works are his twelve mature songs, six of whose words he composed himself, and eight of which were published in his lifetime.[35] Much more widely known are his two symphonies—in E-flat (composed 1862–1867) and in B (1869–1876)—accompanied by a third symphony (in A), only two movements of which were completed at his death and were later orchestrated by Glazunov.[36] Borodin, alone of the Five, took chamber music seriously even in his mature period, and composed two string quartets (in A [1874–1879] and D [1881]), the second of which is still widely played in repertoires both in Russia and abroad.[37] Of his orchestral music, however, the most widely appreciated—both at the time and since—is his symphonic poem, "The Steppes of Central Asia," with its peaceful musical confrontation and synthesis of both Russian folk themes with Oriental ones. This piece was one of twelve commissioned to accompany a production of tableaux vivants celebrating a quarter century of Tsar Alexander II's reign in 1880.[38]

By far Borodin's most famous composition, and the one that has drawn the greatest amount of attention from musicologists, is his posthumous opera, *Prince Igor*, in particular the Polovtsian Dances drawn from it.[39] The idea for turning the unique twelfth-century poetic epic *Tale of Igor's Campaign*—which chronicles the defeat in battle of Russian troops against Polovtsian forces—into an opera, as well as a plan of scenes, clearly lay with Stasov. Borodin's only worry was his competence to undertake the project: "Will I have the strength for

---

[35] For discussion, see César Cui, *Russkii romans: Ocherk ego razvitiia* (St. Petersburg: N. F. Findeizen, 1896), 69–76; Gerald Abraham, "Borodin's Songs," *Musical Times* 75 (November 1934): 983–985; Terence Kelly, "The Songs of Aleksandr Borodin," *Journal of Singing* 524 (March–April 1996): 3–12.

[36] See Gerald Abraham, "Borodin as a Symphonist," *Music & Letters* 11 (1930): 352–359.

[37] See Albrecht Gaub and Melanie Unseld, *Ein Fürst, zwei Prinzessinnen und vier Spieler. Anmerkungen zum Werk Aleksandr Borodins* (Berlin: Ernst Kuhn, 1994), 109–111; and Edward Garden, "The 'Programme' of Borodin's Second Quartet," *Musical Times* 128 (1728) (February 1987): 76–78.

[38] For discussion of these pieces, as well as their "program" of a meeting of caravans in the desert, see Oldani, "Borodin"; and Neef, *Die Russischen Fünf*, 88. The issue of Borodin's Orientalism is a vast one. See Gerald Abraham, "Arab Melodies in Rimsky-Korsakov and Borodin," *Music and Letters* 56 (1975): 313–318; idem, *On Russian Music*, Chapter 6; and Willi Kahl-Köln, "Die russischen Novatoren und Borodin," *Die Musik* 15(1923): 733–738, on 737.

[39] On the composition of the opera, see: Harlow Robinson, "'If You're Afraid of Wolves, Don't Go into the Forest': On the History of Borodin's *Prince Igor*," *Opera Quarterly* 7 (1990/1991): 1–12; Marek Bobéth, *Borodin und seine Oper "Fürst Igor": Geschichte—Analyse—Konsequenzen* (Munich: Musikverlag Emil Katzbichler, 1982); Gerald E. H. Abraham, *On Russian Music* (New York: Charles Scribner's Sons, 1939), Chapter 12; "Borodin, Alexander Porfir'yevich," *The New Grove Dictionary of Opera*, ed. Stanley Sadie (4 vol). (New York: Macmillan Press, 1992): I, 560–561; and Ludolf Müller, "Fürst Igor im altrussischen Heldenlied und in der Oper Borodins," in Ernst Kuhn, ed. *Alexander Borodin: Sein Leben, seine Musik, seine Schriften* (Berlin: Verlag Ernst Kuhn, 1992), 414–422.

it? I don't know. If you are afraid of wolves, don't go into the forest. I'll try."[40] Almost immediately, Borodin began researching the history of twelfth-century Russia so he would be able to draw from what he considered appropriate present-day folk music, historical costumes, and scenery.[41] Borodin would, over the years, rewrite the libretto entirely, removing much of the coherence that had made Stasov's version attractive.[42] The lack of a complete libretto before beginning composition was more responsible than any other factor for the delay of the opera from its genesis in 1869 until Borodin's death in 1887—making it the gold standard for disorganized operatic composition by which other fiascos are measured.[43]

Borodin spent as much time deciding to quit the opera as actually writing it. His first demurral came in March 1870, when he wrote his wife that he had hardly the time and the patience to deal with all the little details, that the public would not like the story due to lack of drama, that "it is no joke to make a libretto which satisfies both the musical and the scenic demands," that he has no experience for it, that he was drawn to symphonic forms, and, in the end, that "opera (not dramatic in the strict sense) seems to me an unnatural thing."[44] In short, whatever excuse came most handy. Intriguingly, an interview with a former medical student, V. A. Shonorov, who lamented the abandonment of the opera, triggered Borodin to get back to work on it. When Borodin told Stasov about his return to *Igor* in 1874, the latter was positively ecstatic.[45] Of all his associates—even more than Stasov—Rimskii-Korsakov was constantly agitated that the opera was still so far from completion, a point he reiterated not just to Borodin ("Write more, using the summer, write as abbreviated, as dirty as possible, but only more quickly"), but to a large number of his correspondents and visitors.[46] Rimskii-Korsakov also volunteered to assist Borodin with orchestration, editing, and copying out parts.

---

[40] Borodin to Stasov, [20 April 1869], *BorP*, I, 142. On the limited links between the *Igor Tale* itself and Borodin's libretto, see Zsuzsa Domokos, "The Epic Dimension in Borodin's Prince Igor," *Studia Musicologica Academiae Scientaurum Hungaricae* 33 (1991): 131–149; and T. Cherednichenko, "Borodin kak poet," *Sovetskaia muzyka* (8) (1978): 94–100.

[41] V. V. Mainov reported to Borodin in the mid-1870s that he had contacted the famous Hungarian traveler Hunfalvi to find out information about the Polovtsy tribes and their possible connection to the Magyars and Pechenegs. See letters reproduced in Dianin, *Borodin*, 200 and 338; and Gerald E. H. Abraham, *Studies in Russian Music* (New York: Charles Scribner's Sons, 1936), Chapter 7.

[42] Kathryn Bumpass and George B. Kauffman, "Nationalism and Realism in Nineteenth-Century Russian Music: 'The Five' and Borodin's *Prince Igor*," *Music Review* 48 (1988): 43–51, on 48; and Taruskin, *Defining Russia Musically*, 154.

[43] David Brown, *Mikhail Glinka: A Biographical and Critical Study* (London: Oxford University Press, 1974), 110; and Abraham and Lloyd-Jones, "Alexander Borodin," 69.

[44] Borodin to E. S. Borodina, 4 March 1870, *BorP*, I, 200.

[45] Stasov to D. V. Stasov, 18 October 1874, in Stasov, *Pis'ma k rodnym*, I(ii), 222.

[46] Quotation from Rimskii-Korsakov to Borodin, 10 August [1879], reproduced in Dianin, *Borodin*, 231. See also Rimskii-Korsakov, *Letopis' moei muzykal'noi zhizni*, 217, 223–224;

After Borodin's death, *Prince Igor* was further delayed both because it was unfinished and due to legal battles over authorship rights following the death of Borodin's wife a few months after his own.[47] Given the repeated foot-dragging Borodin had displayed for over a decade on the completion of the opera, his musical allies already had contingency plans prepared. As Rimskii-Korsakov had noted to a mutual friend in 1884: "Borodin is somehow more and more approaching a collapse in general; there can't even be any talk of composition; when he hears music, then he sleeps; he has completely dropped behind in musical affairs; he uselessly participates in an innumerable quantity of committees. If I survive him, I'll finish 'Igor.' "[48] It is clear that the notion that Borodin abandoned chemistry so he could work on music, or vice versa, rings hollow. Borodin simply was not completing *anything*. Rimskii-Korsakov ended up having to make good on his pledge. He went with Stasov to Borodin's apartment immediately after hearing of the latter's death and seized all his musical manuscripts, and finally completed the opera together with A. K. Glazunov.[49] Borodin's most famous work was thus not, strictly speaking, his.

9.2 Conservatories, Conservatives, and Consternation

The specifics of Borodin's life seem to demand a reckoning. What was he: a composer or a chemist? Why did he shift from one topic and mentor to the next so readily? His music is now considered to be among the greatest produced in Russia, so surely his chemistry must have been equally promising? This last question is one that has occupied a large number of those who write on Borodin. The intrinsic oddities of Borodin's case have prompted many to argue that his status as a chemist was equal to those of his more well-known countrymen— such as D. I. Mendeleev (of the periodic system of chemical elements) and A. M. Butlerov (of the structure theory of organic compounds)—but that foreigners stole all his credit, or that perhaps his musical activities "distracted"

Rimskii-Korsakov to S. N. Kruglikov, letters of 12 August 1879, 23 September 1880, and 28 March 1882, in idem, *Polnoe sobranie sochinenii*, v. 8a (Moscow: Muzyka, 1981), 20, 50, 87; and M. M. Kurbanoff, "A Few Reminiscences of Borodin (1884–1887)," tr. Alfred J. Swan, *Chesterian* 16 (1933): 96–99, on 98. Stasov and Balakirev repeatedly bemoaned Borodin's lack of progress: Balakirev to Stasov, 10 August 1882, and Stasov to Balakirev, 20 June 1884, in Balakirev and Stasov, *Perepiska*, II, 44 and 63.

[47] On the legal dispute, see Stasov to Balakirev, 20 August 1887, in Balakirev and Stasov, *Perepiska*, II, 120.

[48] Rimskii-Korsakov to Kruglikov, 23 February 1884, in Rimskii-Korsakov, *Polnoe sobranie sochinenii*, 130.

[49] Rimskii-Korsakov, *Letopis' moei muzykal'noi zhizni*, 281. For an enumeration of precisely which features of the opera were created from scratch and which were original with Borodin, see Glazunov's account published as V. Stasov, "Redaktsiia 'Kniazia Igoria' Borodina," *Russkaia muzykal'naia gazeta* 3(2) (1896): 153–160.

him from pursuing what *could have been* a career of the same grandeur as his illustrious peers. Both of these approaches are supported by the evidence either weakly or not at all. Although Borodin began the 1860s as one of the more promising young chemists of his generation, for a variety of reasons his research programs petered out largely on their own, not because of subterfuge by foreigners. Borodin seems to have left active chemical research due to a diminution of interest and an increasing desire to devote his time to other issues. As I have described elsewhere, the claims for Borodin's titanic status as a chemist end not with a bang, but a whimper.[50]

Instead of trying to shoehorn Borodin into the role of original chemist, we should recognize that this was not a laurel he coveted for himself. He was happy to devote the last decade-and-a-half of his life not to the production of science, but to the production of *scientists*. The vast majority of Borodin's attention to science concerned not the content of the material but the issue of how to properly *train* practitioners. (He was particularly devoted to the topic of higher education for women, a major effort of the Russian intelligentsia that would occupy too much space to treat satisfactorily here.)[51] It is only by stepping back from Borodin's completed works that one begins to see a larger pattern in which the two separate strands become to cohere; instead of looking at chemistry and music, one should look at *chemists* and *musicians*, and specifically the process by which those two types creatures are consciously formed. It is in the necessity of formalized training of the former and the vital importance of *not* training the latter, for Borodin, that provides a strong link between the Mighty Five, the Russian Chemical Society, and the generalized anxiety about training in the professionalizing and modernizing culture of post-Reforms Petersburg.

Anxieties about the training of musicians abounded in virtually every activity of the Balakirev *kruzhok*. Each member came to the group with another career already underway, and they worked cooperatively to try to develop each other's talents and capabilities so that they could produce mature "Russian" compositions. In fact, the mania for realism among members of the Five was a sign of their dilettante status; one of the main charms of realism for the autodidact is that it proclaims as useless foppery precisely those aspects of counterpoint technique that were the hallmark of formal training.[52] Not having any formal training himself, Balakirev's teaching methods were unorthodox.

---

[50] Borodin's chemical researches are discussed in Michael D. Gordin, "Facing the Music: How Original Was Borodin's Chemistry?," *Journal of Chemical Education* 83 (2006): 561–566. See also Ian D. Rae, "The Research in Organic Chemistry of Aleksandr Borodin (1833–1887)," *Ambix* 36 (1989): 121–137.

[51] On this important topic, see Christine Johanson, *Women's Struggle for Higher Education in Russia* (Kingston, ON: McGill-Queen's University Press, 1987; and Richard Stites, *The Women's Liberation Movement in Russia: Feminism, Nihilism, and Bolshevism, 1860–1930* (Princeton: Princeton University Press, 1978). Unfortunately, no secondary literature yet treats Borodin's role in this movement satisfactorily.

[52] Richard Taruskin, "Realism as Preached and Practiced: The Russian Opera Dialogue," *Musical Quarterly* 56 (1970): 431–454, on 436.

When the *kruzhok* of musicians met, each would bring a piece that he had been working on, and then it was publicly performed for the group and communally criticized—with Cui, Balakirev, and Stasov most prominent in suggesting changes.[53] Stasov encouraged Balakirev's anti-academic stance, urging him "to learn directly in practice, directly in action, and not from textbooks, the entire system of *Russian* music, church and popular, just as you have up till now learned without textbooks one of its members, one of its scales. This learning (*uznavanie*) is truer and more stable than what you would get from [Adolph] Marx's book or any other."[54] Balakirev made a point teaching other members of the Five and his auxiliary students without textbooks.[55]

Of course, not everyone was pleased with Balakirev's teaching methods. Most of the prominent critics of the New Russian School—with the exception of the equally autodidact Aleksandr Serov—felt that the fault of this cadre of musicians was not their lack of talent, but the inferiority of their training. Not surprisingly, many of these contrary voices, such as music critic G. A. Larosh, were conservatory trained, and thus became in turn prime targets for the wrath of Cui and Stasov.[56] The "junior" members of the Five—including the eldest, Borodin—also found some of Balakirev's strictures to be too severe, and one of the most common criticisms of his teachings, and a central reason why the *kuchka* fell apart by the 1870s, was the widespread perception among them that Balakirev was simply too "despotic" to bear.[57] They did not crave the academicism of professionals, but rather a kinder, gentler Balakirev patterned on Franz Liszt. Liszt, like Hector Berlioz, was strongly opposed to what he saw as pedantry and academicism.[58] This glorification of Liszt's anti-academicism was somewhat overstated, since the first thing Liszt did when he arrived in Paris from Budapest for the first time as a piano prodigy was to attempt to enroll in the Paris Conservatory. He was turned away because the institution

---

[53] Stasov, "Modest Petrovich Musorgskii: Biograficheskii ocherk (1883)," in *Izbrannye sochineniia*, II, 184.

[54] Stasov to Balakirev, 20 August 1860, in Balakirev and Stasov, *Perepiska*, I, 115. Emphasis in original.

[55] Garden, *Balakirev*, 61.

[56] As Larosh put it: "With us, in Russia, where the public is small, where traditions are not established, where specialists are very few, it is especially easy to fall into that *kruzhok* manner, to take a half-dozen of one's friends for the Russian people, an anthill for the globe, and examples of such sad confusions in our tiny musical world as more common and usual than in any other." From his "['Demon' A. Rubinshteina] (1877)," in Larosh, *Izbrannye stat'i*, III, 227.

[57] See, for example, Cui to M. S. Kerzina, 23 May [1910], in Cui, *Izbrannye pis'ma*, ed. I. L. Gusin (Leningrad: Gos. muzykal'noe izd., 1955), 404; Rimskii-Korsakov, *Letopis' moei muzykal'noi zhizni*, 283; and. As Borodin wrote to his wife in October 1871: "[Balakirev] is such a despot by nature, that he demands complete subservience up to the tiniest infinitesimals. He can in no way understand and recognize freedom and equal rights. ... He wants to impose his yoke on everyone and everything." Letter of [24–25 October 1871], *BorP*, I, 311. Ellipses added.

[58] Alan Walker, *Franz Liszt: The Virtuoso Years, 1811–1847*, rev. ed. (Ithaca: Cornell University Press, 1988 [1983]), 182.

excluded foreigners, but he quickly found other teachers in their stead.[59] As a private teacher of piano, however, he had essentially no teaching method, eschewing an analytic approach in favor of allowing students to play freely, and only intervening with minor and gentle suggestions. As Borodin observed during his visit to Liszt: "In general between him and his students relations are terribly simple, familiar, and heartfelt, not at all reminiscent of the relations of students to a professor, but more like children to a father, or grandchildren to a grandfather."[60]

This model was self-consciously designed as an alternative to the Conservatory, which was erected under the auspices of the Russian Musical Society by Rubinstein. By the late 1850s, Russia was the only major European country that had no dedicated musical educational establishment, and scattered comments in various newspapers and journals began to call for such an organization. In December 1859 the Russian Musical Society began preparatory courses, and in Spring 1860 courses were offered (some for a fee, others at no cost). On 17 October 1861, the "Musical Academy (*Uchilishche*)" was opened, renamed the St. Petersburg Conservatory in 1866. A Moscow Conservatory followed, quickly supplanting the virtual forest of piano and voice schools in the two metropoles.[61] Rubinstein's goal, as mentioned earlier, was to establish civil status and formal training for musicians on the German model precisely to *eliminate* dependence on Germans for the future of music in Russia. Rubinstein was concerned with establishing music in Russia, not Russian music. Music was international, much like science, and formal training was needed in both.[62]

Stasov would have none of this. At precisely this moment, he was encountering hostility from the Academy of Arts in St. Petersburg against the brand of realist painting he most favored—that of the Wanderers (*Peredvizhniki*)—and he felt that any art-training institute would always be compromised by its connection to the state, *by virtue of being an institution*, from truly supporting original Russian art.[63] The Conservatory would inevitably generate two undesirable features: "Germanness" and "craft"—the former was a nationalist nightmare that would perpetuate perceived German dominance, the latter was the related routinization of musical education and a stifling of creativity. Stasov decried the Conservatory as a bastion of ungifted elites, even twenty-five years after its creation:

---

[59] Paul Metzner, *Crescendo of the Virtuoso: Spectacle, Skill, and Self-Promotion in Paris during the Age of Revolution* (Berkeley: University of California Press, 1998), 138.

[60] Borodin to E. S. Borodina, 12 July 1877 (N.S.), *BorP*, II, 146. See Alan Walker, *Franz Liszt: The Final Years, 1861–1886* (Ithaca: Cornell University Press, 1996), 228.

[61] L. Z. Korabel'nikova, "Muzykal'noe obrazovanie," In Iu. V. Keldysh, et al., eds., *Istoriia russkoi muzyki*, v. 6 (Moscow: Muzyka, 1989): 134–187.

[62] For Rubinstein's clearest expression of his case, see his article, "The State of Music in Russia (1861)," as translated in Stuart Campbell, ed. and tr., *Russians on Russian Music, 1830–1880: An Anthology* (Cambridge: Cambridge University Press, 1994), 65–73.

[63] Lebedev and Solodovnikov, *Vladimir Vasil'evich Stasov*, 66.

> [The Conservatory] wanted only to plant among us German musical routine and the parochial craft stamp; it haughtily ignored Russian music and with scorn looked at Russian composers, whom they all labeled without exception "dilettantes." The new conservatory fully justified its name: it was in the highest degree conservative, more than anything it recognized only the generally accepted "classics" and didn't want to know anything new.[64]

One issue was that the study of formal counterpoint and harmony would turn individuals into pseudo-Germans; another was the very real fact that most of the instructors of music were in fact German and that the language of instruction (certainly for the first decade of the St. Petersburg Conservatory) was German—points criticized even by generally satisfied graduates of the institution like Larosh.[65] Of course, neither Cui nor Stasov thought that one could become a composer with absolutely no training (e.g., the ability to read notes); they objected instead to slavish devotion to a specific body of knowledge— which Borodin himself ridiculed in the case of Nikolai Zaremba's classes at the Conservatory.[66]

Stasov et al.'s hostility to formal training was not merely a characteristic of a group that also happened to be vociferously advocating a self-consciously "national" form of art music; it was constitutive of what they thought "Russian music" meant: the more you were trained, the less Russian your music. As Carl Dahlhaus has noted in his landmark study of "neo-Romanticism" in late-nineteenth-century music, nationalism can best be interpreted not as an issue of music's substance, but of the music's function.[67] That is, composers' claims to "nationalism" and "national styles" of music should not be sought—or not sought exclusively—in the tropes and techniques they employed, their acknowledged influences, and so on; rather, one should look at those claims in the context of the entire spectrum of musical politics, such as local personal conflicts, attitudes to foreign music, and instantiations of pedagogy. In the case of

---

[64] Stasov, "Dvadtsatipliatiletie Besplatnoi muzykal'noi shkoly (1887)," in *Izbrannye sochineniia*, III, 79. Cui also lamented the "craft (*remeslo*)" of Conservatory-trained musicians like Chaikovskii in Cui to Stasov, 3 May [1880], in Cui, *Izbrannye pis'ma*, 102. This emphasis on the Conservatory was the central feature of the Five's critique of Chaikovskii. For his part, it was precisely the *lack* of training that Chaikovskii lamented in Borodin, whom he saw as possessed of "talent, a very great talent which, however, has come to nothing for want of teaching...." Quoted in Victor I. Seroff, *The Mighty Five: The Cradle of Russian National Music* (Freeport, NY: Books for Libraries Press, 1970 [1948]), 206. Ellipses added.

[65] Larosh, "Muzykal'nye pis'ma iz Peterburga: Pis'mo pervoe (1871)," in Larosh, *Izbrannye stat'i*, III, 65.

[66] See, for example, V. Stasov, "Nasha muzyka na posledniia 25 let," *Vestnik Evropy* 5 (October 1883): 561–623, on 565; and Borodin to E. S. Borodina, [21 September 1871], *BorP*, I, 294.

[67] Carl Dahlhaus, *Between Romanticism and Modernism: Four Studies in Music of the Later Nineteenth Century*, tr. Mary Whittall (Berkeley: University of California Press, 1980), 80–81 and 91. For a comparable instance, see Michael Beckerman, "In Search of Czechness in Music," *19th-Century Music* 10 (1986): 61–73.

the New Russian School, these positions were all conflated, as in this (ironically polyglot) letter from Musorgskii to Rimskii-Korsakov in 1868:

> Now again about *symphonic development*. For you it's as if it's frightening that you write like Korsakov and not like Schumann. And I tell you (you scorn fear—*vous êtes brave*), that *okroshka* [Russian bread soup] is horrible for a German, but we eat it with pleasure (*point de comparaison, s'il vous plait, comparaison n'est pas raison*). German *Milchsuppe* or *Kirschensuppe* is horrible for us, and a German is in raptures from it. *Bref, symphonic development, technically understood*, is worked out by a German like his philosophy—at present destroyed by English psychology and our own [M. M.] Troitskii. A German, when he thinks, first *reconnoiters*, and then *proves*, our brother first proves, and then pacifies himself with reconnaissance. . .[68]

Musorgskii took national character as primary, and then constructed differences around it, from taste in soup to symphonic composition. A less essentialist tack would consider the concatenation of tastes as primary and then look at them as *constitutive* of national character. One of the most prominent features of the "Russian" musical style was precisely this failure to obtain adequate training before undertaking composition.

This impulsiveness—or, to use a term its proponents would prefer, spontaneity—was central in shaping reactions to the music of the Five both at home and abroad. For Borodin in particular, his qualities as a *national* composer were rarely spoken of without in the same breath dealing with his qualities as an *untrained* composer. For his critics, such as Larosh, "Borodin, entirely infected with dilettantism, never suffered from *unmusicality*. On the contrary, with him everything is interesting; the interest which attracts [one] to his works comprises at the same time their Achilles's heel. They are *only* interesting,. . . they are not balanced out by any simplicity. . ."[69] The fault here was that Borodin's education was too weak to balance the deleterious influence of the *kruzhok*'s training. Lack of training was both the consequence and the cause of its national qualities. There is an undercurrent in the elitist self-representation of the Balakirev circle and its supporters that while the musicians did not in fact need training to be national, the audiences in Russia *did* need to be trained so they could more appropriately mimic German audiences. As Glinka's sister, L. N. Shestakova, wrote to Borodin after a performance of one of his symphonies was greeted with scorn:

> You see, they didn't understand [Glinka's opera] "Ruslan [and Liudmila]," exactly as they didn't understand your glorious symphony, and I don't at all think that it was an agreed-upon booing, it seems to me simply that they didn't understand and were naughty; their ears are not grown up enough for this symphony, they are such a brilliant piece of jackassery (*osliatina*). . . . Before and even during the performance itself, the public makes noise and disperses. After all it is only possible to do this in Russia. They should try to pull a similar prank in Germany. . . .[70]

---

[68] Letter of 15 August 1868, reprinted in Musorgskii, *Literaturnoe nasledie*, 106–107. Emphasis in original. Ellipses added.

[69] Larosh, "Muzykal'naia khronika," *Russkii Vestnik* (10) (1887): 823–854, on 850. Emphasis in original. Ellipses added.

[70] Letter of 27 February [1877], reproduced in Dianin, *Borodin*, 208. Ellipses added.

Germans were ideal audiences for music; they should just refrain from composition.

Proof of the suitability of Western European audiences for supposedly "purely national" music was evident abroad, particularly in Belgium, where—due largely to the good offices of the Countess Louisa de Mercy-Argentau—Borodin soon became the most prominent of the New Russian School in foreign climes.[71] Even in the heart of the beast, musical Germany, the results were encouraging, as Borodin wrote to Balakirev after a very well-received performance of his second symphony in Baden-Baden: "It is especially pleasing to me personally that this thing had success precisely in Germany...."[72] Truly, for Borodin to be recognized by the Germans on their home soil was the nationalist victory the group as a whole savored in Petersburg—and was substantially more important than the enthusiasm his quartets received in Buffalo.[73] Even domestically, although Stasov would later claim that Borodin had never been appreciated in his own country, he was lauded by large numbers of music aficionados. In 1879, for example, he traveled on business to Odessa and was greeted on arrival as a famous composer.[74] And while it was true that Borodin received substantial criticism in the musical press, it was not the case—as Stasov reported after Borodin's death—that the first symphony had been poorly received, as Balakirev hastened to correct.[75] And all of this without the slavish devotion to training that former Conservatory students like Chaikovskii insisted upon.

Interestingly, it was exactly such a devotion that Borodin himself was trying to establish for chemical education in the predominantly medical Academy. The Medico-Surgical Academy in the 1860s underwent deep transformations in its attitude towards the rigors of medical pedagogy under the presidency of P. A. Dubovitskii (1857–1867). Nikolai Zinin in particular, as Secretary of the Academy during the first eight years of this period, was Dubovitskii's central aide in grounding science more deeply in the medical curriculum, primarily through building a Natural History Institute (in 1863, with a grant of 45,000 rubles to start and 2,000 more annually) and traveling abroad under Dubovitskii's direction to import foreign teaching methods. The natural science chairs were expanded from two (chemistry, physics, and mineralogy on the one

---

[71] On the Countess's very colorful life, see Carlo Bronne, *La Comtesse de Mercy-Argentau*, 2d. ed. (Liège: Soledi, 1945), esp. 65–74 on Borodin. Borodin and the Countess corresponded often, and she translated the lyrics for several of his works into French. See the letters of October 1884 in *BorP*, IV, 92 and 105.

[72] Borodin to Balakirev, 17 May 1880, *BorP*, III, 99. This was also true in Paris. Borodin wrote to A. P. Dianin on 6 November 1877 that he had heard from Turgenev that the second symphony was a tremendous success there: *BorP*, II, 191.

[73] Borodin to E. S. Borodina, 30 November 1885, *BorP*, IV, 99.

[74] See Sokhor, *Aleksandr Porfir'evich Borodin*, 282.

[75] Balakirev to Stasov, 8 December 1888, in Balakirev and Stasov, *Perepiska*, II, 141.

hand; and natural science on the other) to five (chemistry; physics, geography, and climatology; zoology and comparative anatomy; botany; and geology, mineralogy, and paleontology).[76] Borodin later recalled Zinin's three fundamental transformations as the introduction of fresh new teachers, establishing facilities for applied medical and scientific work, and building an "Institute for Young Doctors."[77] These transformations continued after Zinin formally resigned his chair in 1874 to move full-time to the Imperial Academy of Sciences, and Borodin was promoted from Zinin's apprentice to spearheading his own set of academic changes at the Medico-Surgical Academy.[78]

Borodin was already thinking of his brief time abroad at Heidelberg, Paris, and Pisa primarily in terms of pedagogical reform as early as 1863:

> Filled with the conviction that only a scientist who is completely possessed by this subject can be a really good teacher, I tried above all to develop myself from this point of view. This is accomplished first: by mastering what was done by others, and second: by independent research, helping the advancement of science. Without these conditions it is impossible to obtain an accurate, critical outlook in science and to stand at the level of contemporary direction. But this is still not enough for the activity of a teacher: it is necessary to be able to teach others; it is necessary to be able to transmit science to audiences, conforming to their degree of development and to their future purpose. This is achieved, on the one hand, by the study of different methods of teaching others, and on the other hand, by independent training.[79]

Borodin's reforms of chemical pedagogy were perhaps the most time-consuming of his many activities of the 1870s, along with his involvement in committees for higher education for women (themselves emblematic of his concern for proper training). As he wrote to his wife in early September 1869: "I am busy up to here with the construction of the laboratory, the receipt of things, and the organizing of laboratory property." There was no gas or running water in the laboratory buildings yet, and they were filthy.[80] Defending his delay in finishing *Prince Igor* to Liubov' Karmalina, Borodin wrote in 1876— after he had ceased to publish scientific papers—in defense of his expenditure of time on pedagogy: "I love my work, and my science, and the Academy, and my students; my science is practical in the character of the studies, and thus consumes a great deal of time; my male and female students are close to me

---

[76] P. Belogorskii, "Preobrazovaniia shestidesiatykh godov," in Ivanovskii, ed., *Istoriia Imperatorskoi Voenno-Meditsinskoi (byvshei Mediko-Khirurgicheskoi) Akademii za sto let* (1898), 523–579.

[77] Borodin's funeral oration for Zinin, 9 February 1880, reproduced in *BorP*, III, 87–88.

[78] On these later transformations, including the change of name to the Military-Medical Academy, see D. Kodorotov, "Perekhodnoe vremia," in Ivanovskii, ed., *Istoriia Imperatorskoi Voenno-Meditsinskoi (byvshei Mediko-Khirurgicheskoi) Akademii za sto let* (1898), 581–683; and N. Kul'bin, "Imperatorskaia Voenno-meditsinskaia akademiia, 1881–1898 g.," in ibid.: 685–828.

[79] Borodin's final report on his trip abroad, dated 31 January 1863, reproduced in Figurovskii and Solov'ev, *Aleksandr Porfir'evich Borodin*, 143.

[80] Borodin to E. S. Borodina, [8 September 1869], *BorP*, I, 147.

even in other respects than as studying youth, which doesn't limit itself to listening to my lectures, but also needs practical exercises, etc. The interests of the Academy are dear to me."[81]

Assembling the remnants of proper instruction out of the mess he inherited was the task of the next several years. His son-in-law and eventual successor, A. P. Dianin, saw 1874 as the watershed of Borodin's work at the Academy. With Zinin's retirement, Borodin made practical laboratory instruction a requirement in the chemical education of physicians. Since the medical students had different schedules, Borodin had to keep the laboratory open almost all day, every day, so that 300–400 students could conduct experiments.[82] The point of these efforts, as Borodin had articulated in an essay review on pharmacy as early as 1863, was to show students who only dealt with the applied sciences (such as medicine and pharmacy) the kind of strict logic that is possible in the "pure sciences."[83] Here, therefore, unlike in the Balakirev circle, Borodin was adamant about the importance of proper (read: formal) instruction. Guidance under the hands of a master like Balakirev without routine would just be insufficient for the sciences.

The connection and juxtaposition between formal training in the sciences and informal training in the arts (particularly music) was noted by members of the Five as being almost constitutive of the difference between these two domains of human activity. As Balakirev wrote to Stasov, referring to the property dispute at the Medico-Surgical Academy over the disposition of the Borodins' estate, in August 1887:

> All that you have written me concerning the ridiculous orders, concerning the property of the late Borodins, confirms my opinion of the fact that so-called specialists of sciences, especially medical sciences, are very stupid folk, of the sort like teachers of harmony, cobblers, and other workmen. The exceptions are only representatives of the humanities or such luminaries as, for example, Botkin. To the rest all of life appears in the narrow little confines of pedestrian concepts about the elevation or reduction of temperature, diarrhea, constipation, etc., and for the rest, no matter how important, they don't care, but their self-confidence is so high, that in another specialized affair, such as the arranging of spiritual testament, he is not afraid to consider himself competent, believing in his supposedly "bright head."[84]

Arrogance—that was the fundamental problem of a formal education, and it led to mistaking one's competence in one area with talent in another. The Conservatory merely perpetuated this when dealing with the fragile flower of creativity. Stasov had already mooted this point in his initial response to Rubinstein's call for conservatories:

> It is possible that Rubinstein is not aware of the opinion now deep-rooted in the greater part of Europe that holds that academies and conservatories serve only as breeding

---

[81] Letter of 1 June 1876, *BorP*, II, 109.

[82] Dianin, "Aleksandr Porfir'evich Borodin," 373.

[83] A. Borodin, "Referat ob uspekhakh farmatsii v 1861 godu [I-III]," *Voenno-Meditsinskii Zhurnal* 88(10–11) (1863): 220–234, 289–306, 371–403, on 220–221.

[84] Letter of 22 August 1887, in Balakirev and Stasov, *Perepiska*, II, 122.

grounds for tasteless people and aid in the establishment of harmful ideas and tastes. Therefore the best minds search in the sphere of artistic education for means of doing without *higher* educational establishments. *Higher* educational establishments for art are a completely different thing from their counterparts in the sciences, and the two categories should never be confused. There is a vast gulf between the two types. A university and a conservatoire are completely different things. The former communicates only *knowledge*; the latter is not content to do just that and interferes in the most dangerous way with the *creative process* of the artist in training, and extends a despotic power (from which nothing can protect him) over the shape and form of his works. . . .[85]

Along these lines, Balakirev was not entirely pleased with the way his pedagogy was portrayed in Stasov's Borodin biography. As he wrote to Stasov in a comment on the draft:

You wrote that Borodin, thanks to his acquaintance with me, understood that one must relate to authorities critically, that they are not infallible, etc. I could have influence on him only in the *specifically musical* sphere. The question about authorities you touch on is not a specific question but a general intellectual one, and in this sphere Borodin, being not only excellently educated (*obrazovannym*), but even a scientist, had no need to be enlightened by me, who had received only a boy's schooling.[86]

What differentiated Borodin from Balakirev was formal education—but only in the sense that one was a scientist and the other not. In music, all were equal before the Russian spirit.

Training seems so essential to Borodin's self-conception, that one might wonder at how much excavation and pruning of the historiography had to be undertaken to document this connection. Why, indeed, has the importance of training as not only a bridge between the two cultures of Borodin's world, but as a means of denying any direct homology between the two, been left out of the standard account? There are two main strands to this "sidetracking" of the Borodin legacy: the first the shaping of Borodin by Stasov into a posthumous spokesman for the greatness of Russian (read: anti-German) music; and the cooptation by historically-interested chemists of the man in attempt to transcend the debate about the "two cultures." In the process, both traditions only inscribe that divide more deeply.

9.3 Conclusion: Vladimir Stasov and the Two Cultures

Aleksandr Borodin died at an Academy fancy-dress party celebrating carnival at 11:40 PM, 15 February 1887. He was immediately enveloped by his physician colleagues who attempted to revive him, but the heart attack proved fatal. Almost instantly, Vladimir Stasov began to collect his letters and unfinished

---

[85] Stasov, "Conservatories in Russia: Comments on Mr. Rubinstein's Article (1861)," in Campbell, *Russians on Russian Music*, 78. Emphasis in original. Ellipses added.

[86] Balakirev to Stasov, 22 February 1887, in Balakirev and Stasov, *Perepiska*, II, 102. Emphasis in original.

musical manuscripts, solicit reminiscences from his friends, and write obituaries for him that hit the same notes central to the Mighty Five's message: Borodin was a *composer*, at best distracted by chemistry; he was resolutely *nationalist*; and the central tragedy of his brief life is that he was not given enough *time* to devote to finishing his opera. All three of these points are generally accepted by writers on Borodin both at the time and since, and all three of them are equivocal, to say the least. The first point (his status as a composer) assumes some kind of Platonic "vocation" which is imprinted on the soul of an individual and defines the essence of his or her life. The second point (his nationalism) was certainly a factor in his thought, but nowhere near as central as his emphasis on proper forms of training. The third point (time), is the most problematic, since often Borodin did have the time but *chose* to spend it on other matters. Just because Stasov and Rimskii-Korsakov wanted *Prince Igor* to be Borodin's highest priority does not mean that Borodin himself did. Recall that when he died, he had essentially no reputation as an opera composer at all, having only displayed some extracts of *Igor* to the public. The grand reputation Borodin developed was posthumous, and the credit for it lies not only with the composer but with his unbidden publicist—Stasov.

I have mentioned Stasov's imprint so frequently because it is literally inescapable. Stasov, more than any other individual, shaped how people both at the time and since interpreted Borodin, and since Stasov clearly had a very articulate partisan agenda, one needs to be intensely critical of the nuances he imposed on his subject.[87] In his writings on Borodin, Stasov emphasized three narrative points: that he had been a cosmopolitan as a child, speaking a great many languages, who later became a nationalist; that the meeting with Balakirev was the decisive shift in his life; and that while he had worked in chemistry and music simultaneously for most of his life, it was really in music where he fulfilled his destiny.[88] Stasov left a very strong impression with this final point that Borodin was always a genius *manqué*, a promise deferred because of too many alternative commitments. One could always defend Stasov by pointing out that he was there and therefore knew the state of affairs. But, as Sigrid Neef has recently pointed out, the discrepancies between Stasov's version and the surviving historical record best call to mind the common Russian proverb: "He lies like an eyewitness."[89]

---

[87] For Stasov's biography, see Vlad. Karenin, *Vladimir Stasov: Ocherk ego zhizni i deiatel'-nosti*, 2 v. (Leningrad: Mysl', [1927]); and E. G. Salita and E. I. Suvorova, *Stasov v Peterburge* (Leningrad: Lenizdat, 1971).

[88] Cui held the same line in his obituary for Borodin: his science is "fruitful and distinguished; but his musical, compositional activity has a still greater, perfectly outstanding significance." Cui, "A. P. Borodin," 288.

[89] Sigrid Neef, "Wladimir Stassow und das Mächtige Häuflein," in Wladimir Stassow, *Meine Freunde Alexander Borodin und Modest Mussorgski: Die Biographen*, ed. Ernst Kuhn (Berlin: Verlag Ernst Kuhn, 1993), 11–24.

Stasov, of course, did not devote his myth-making attention exclusively to Borodin. In fact, he wrote biographical essays, monographs, or compilations of source material on *all* major Russian composers—including the prior generation of Dargomyzhskii and Glinka—with the exception of Chaikovskii and Balakirev. Borodin simply happened to be the second of the Mighty Five to die, following the more dramatic case of historiographical manipulation he bestowed on Modest Musorgskii—recast by Stasov as a radical populist driven to an alcoholic death spiral by a repressive world. (The historiographical distortion there has been well noted and corrected.[90]) Stasov wanted to build up a legend of the New Russian School, and Borodin and Musorgskii were apposite means to that end.[91] He often had to defend his rush into biographical print, thus consciously shaping the historical memory of the figure in question: "And, in the main, [my critics say] it would have been better if I hadn't written, but someone else. And what? Twenty years have passed—but not a single soul has thought, not a single hand has written since then a single letter. It was exactly the same with Musorgskii, Repin, with Borodin and with everyone, everyone about whom I happened to write. Balakirev rebuked me for Musorgskii, Turgenev for Repin, various others for Borodin."[92] In a sense, Stasov has needed little defense since, since with limited exceptions, like Musorgskii's, most commentators have endorsed his evaluations—both within the Soviet Union, where he fit with the dominant trend to identify Soviet patriotism with Great Russian nationalism, and by anti-Soviet Russian nationalists abroad.[93]

---

[90] Stasov, "Modest Petrovich Musorgskii: Biograficheskii ocherk (1881)," in Stasov, *Izbrannye sochineniia*, II: 161–213. For the revision, see Richard Taruskin, *Musorgsky: Eight Essays and an Epilogue* (Princeton: Princeton University Press, 1993); Caryl Emerson, *The Life of Musorgsky* (Cambridge: Cambridge University Press, 1999); and Francis Maes, "Modern Historiography of Russian Music: When Will Two Schools of Thought Meet?" *International Journal of Musicology* 6 (1997): 377–394. For more general revisions, see idem, *A History of Russian Music: From* Kamarinskaya *to* Babi Yar, tr. Arnold J. Pomerans and Erica Pomerans (Berkeley: University of California Press, 2002 [1996]); and Richard Taruskin, "Some Thoughts on the History and Historiography of Russian Music," *Journal of Musicology* 3 (1984): 321–339.

[91] This perhaps explains the intense interest in trying to find parallels between the two composers: Igor Glebow, "Borodin und Mussorgski (Versuch einer Parallele) (1930)," in Kuhn, *Alexander Borodin* (1992): 342–347; and Kremlev, *A. P. Borodin*, 24 and 83.

[92] Stasov to V. D. Komarova, 25 August 1899, in Stasov, *Pis'ma k rodnym*, III(i), 316.

[93] For a selection of Stasov's writings on music in English, see Vladimir Vasilevich Stasov, *Selected Essays on Music*, tr. Florence Jonas (New York: Da Capo Press, 1980). For an anti-Soviet attempt to rescue Stasov's historiography from Soviet clutches for the greater glory of the Russian nation, see Yuri Olkhovsky, *Vladimir Stasov and Russian National Culture* (Ann Arbor: UMI Research Press, 1983). For other Western, essentially Stasovian interpretations of Russian music, see: Donald N. Ferguson, *A History of Musical Thought*, 2d. ed. (New York: Appleton-Century-Crofts, Inc., 1948); Gerald R. Seaman, *History of Russian Music: From Its Origins to Dargomyzhsky*, v. 1 (Oxford: Basil Blackwell, 1967); Richard Anthony Leonard, *A History of Russian Music* (New York: Macmillan, 1957); James Bakst, *A History of Russian-Soviet Music* (New York: Dodd, Mead & Company, 1966 [1962]); Seroff,

Take, for example, the narrow issue of vocation, since it is here that Stasov's argument lays its foundation, and it is also here that the most vital echo of Stasov's Borodin—that propounded by professional chemists—takes its point of departure. If one is looking for it, one can find many references in Borodin's correspondence that his music and his chemistry were engaged in a zero-sum battle for his time.[94] If one values the music more than the chemistry, therefore, it is easy to claim that the music suffered because Borodin neglected his obligation to it. Borodin himself was aware of this particular narrative, and even at times subscribed to it, writing to an admirer in the year before his death: "I would ask that you not restrict my biography to the musical part alone, since my scientific and teaching activity serves as an explanation why I became a composer late and wrote so little music."[95] For Stasov, as one gleans from his correspondence, the issue of a "vocation" as the fulfillment of one's destiny was quite a serious one, and it explains why he took his program for the autonomy and equality (if not primacy) of Russian art as a mission. He declared to Balakirev as early as 1858, when the constellation of views that he would later mobilize were just beginning to cohere, that one could only be truly happy when following one's true vocation: "I already tried and became convinced that *there is no other happiness* than doing that which each of us is capable of, regardless of whether this will be a grand affair or the tiniest. We are all born only in order to *birth* from ourselves new creations, new thoughts, new life—as women are born in order to birth new people."[96] My goal is not to malign Stasov's views of the history and historical function of art, but to show that they are indeed partisan views. Stasov articulated a world of actors who succeeded or failed based on how deeply they held to their vocation. From today's perspective, where destiny is less of a category, there is no necessity to subscribe to his framework.

And, in fact, most of the brief articles written on Borodin today—the vast majority of them by practicing chemists—do not focus on vocation for the same reasons that Stasov did, although they still use the evidence packaged in his

---

The Mighty Five; M. Montagu-Nathan, *A History of Russian Music* (New York: Charles Scribner's Sons, [1914]); Rosa Newmarch, *The Russian Opera* (New York: E. P. Dutton & Co., [1914]); Alfred Bruneau, *Musiques de Russie et Musiciens de France* (Paris: Bibliothèque Charpentier, 1903); Albert Soubies, *Histoire de la Musique en Russie* (Paris: Société Française d'Éditions d'Art, 1898); and the more recent and thorough Dorothea Redepenning, *Geschichte der russischen und der sowjetischen Musik. Band I. Das 19. Jahrhundert* (Laaber: Laaber-Verlag, 1994). For a friendlier account of Stasov's esthetics, see G. A. Obraztsov, *Estetika V. V. Stasova i razvitie russkogo natsional'no-realisticheskogo iskusstva* (Leningrad: Izd. Leningradskogo universiteta, 1975); and the excellent T. Livanova, *Stasov i russkaia klassicheskaia opera* (Moscow: Gos. muzkyal'noe izd., 1957).

[94] For example: "There is as yet no time to work in the laboratory, to work on my music even less" (Borodin to E. S. Borodina, 17 October 1870 [sic: 1871], *BorP*, I, 307).

[95] Borodin to Ol'ga Akimovna Kochetova, [February–December 1886], *BorP*, IV, 179.

[96] Stasov to Balakirev, 24 June 1858, in Balakirev and Stasov, *Perepiska*, I, 61. Emphasis in original.

original pieces and speak of Borodin in almost identical terms. Many have observed that numerous scientists seem to have strong interests in classical music as opposed to the other arts, although to date there has been little persuasive explanation of it. Yet, despite these possible connections and well-documented instances of historical links between science and music, the case of Borodin has drawn the lion's share of the attention from both within and outside the scientific community.[97] The chemical biographies of Borodin cited at the beginning of this essay generally accept all the terms of Stasov's presentation of Borodin, but instead of viewing Borodin's work in chemistry as a distraction from his true vocation of music, they interpret the fact that Borodin spent so much time on it to argue that the man himself valued his chemistry first and his music second—accepting Stasov's parameters and relabeling the polarities of the terms.

One gets a sense of the underlying moral of the Borodin story from the chemists' perspective in the very first biography of the man in a chemical journal, his obituary by his son-in-law in the *Journal of the Russian Physico-Chemical Society*: "[T]he person of A. P. [Borodin] serves as a most obvious example (among a few others) that in a richly gifted nature analytic, strictly scientific work does not at all exclude the possibility of free, purely artistic creativity and vice versa."[98] Thus, this opposition was already being formed long before C. P. Snow would formulate the classic opposition between the sciences and the humanities (although most prominently literature) in his Rede Lecture of 1959—forever after known as the Two Cultures.[99] As Lionel Trilling has pointed out, however, the common reading of Snow as pointing merely to a failure to communicate between two equal (and equally useful) cultures is quite misleading. Quite the contrary, Snow clearly declared the scientists as the team to bet on, and the literary intellectuals and other humanists as the slow coach that had missed its chance to join the race.[100]

---

[97] These other cases generally display greater propensity to be generalized, while Borodin's case remains highly idiosyncratic. See, for example, Myles W. Jackson, "Harmonious Investigators of Nature: Music and the Persona of the German *Naturforscher* in the Nineteenth Century," *Science in Context* 16 (2003): 121–145; Erwin Hiebert and Elfrieda Hiebert, "Musical Thought and Practice: Links to Helmholtz's Tonempfindungen," in Lorenz Krüger, ed., *Universalgenie Helmholtz: Rückblick nach 100 Jahren* (Berlin: Akademie Verlag, 1994): 295–311; "Chemist-Composers: An Explosive Combination," *Chemical Heritage* 15, no. 1 (Fall 1997): 35; and Martin D. Kamen, "On Creativity of Eye and Ear: A Commentary on the Career of T. W. Engelmann," *Proceedings of the American Philosophical Society* 130 (1986): 232–246.

[98] Dianin, "Aleksandr Porfir'evich Borodin," 376

[99] C. P. Snow, *The Two Cultures* (Cambridge: Cambridge University Press, 1998 [1959]).

[100] Lionel Trilling, "The Leavis-Snow Controversy (1962)," in Lionel Trilling, *The Moral Obligation to Be Intelligent: Selected Essays*, ed. Leon Wieseltier (New York: Farrar, Straus, and Giroux, 2000): 402–426.

This illuminates the underlying current of the chemists' consistent champion-
ing of Borodin: they do not wish to show that it is possible for a chemist to be a
member of the artistic world; they hope to demonstrate that Borodin was *at
root* scientific, and thus it is possible for scientists to belong to both cultures,
while the humanists, parochial and simplistic, are confined to just one. One
might be forgiven in thinking that Borodin would view such an agenda—as he
would Stasov's—as a failure of proper training.

Part V
Geology and Natural Theology

Chapter 10
The Genesis of *Historical* Research on the History of Geology, with Thoughts About Kirwan, de Luc, and Whiggery

David Oldroyd

Few historians of science can be said to have founded a whole new field of study—or at least a wholly new way of studying an important branch of science—but I think that such a claim can fairly be made for Charles Gillispie with his *Genesis and geology* (1951): an "enduring classic" as Nicolaas Rupke (1996, p. xv) has appropriately called it. In the "pre-Gillispie era", if I may label it thus, studies in the history of geology were chiefly undertaken by "scientist-historians". Among English-language texts just three books dominated the field in 1951: Archibald Geikie's *Founders of geology* (1897/1905); Marian Ogilvie-Gordon's excellent (but only partial) translation of Karl von Zittel's "names and dates" (or "kings and queens") history (Zittel, 1901); and Frank D. Adams's *Birth and development of the geological sciences* (1938).

Zittel wrote a somewhat dour and dull book of record, of permanent importance, but he largely treated geology as if it developed in some kind of social and philosophical vacuum. Geikie had early training as a humanist and was a stylish writer who could stand compare with most professional historians for his literary elegance, clarity, and his intimate knowledge of his subject matter. But his book suffered from a distinct anti-German stance—perhaps understandable at a time when tensions were rising between Germany, France, and Britain (and Geikie's wife was French). One can also question his judgement on certain points. His omission of any serious treatment of the work of Charles Lyell was rather extraordinary,[1] but he gave good discussions of the work of Frenchmen such as Guettard and Desmarest and Englishmen like Smith and Murchison. On the other hand, his treatment of Werner and his Freiberg

---

[1] Reasons for this have recently been advanced by Leonard Wilson (2009). According to his view, Geikie's negligible presentation of Lyell in *Founders of Geology* arose from a long-standing bitterness on Geikie's part arising from an error he made when a young man, to which Lyell, perhaps gratuitously, drew attention in his *Principles of geology*.

D. Oldroyd (✉)
Department of History and Philosophy of Science, University of New South Wales, Sydney, NSW 2052, Australia
e-mail: doldroyd@bigpond.com

J.Z. Buchwald (ed.), *A Master of Science History*, Archimedes 30,
DOI 10.1007/978-94-007-2627-7_10, © Springer Science+Business Media B.V. 2012

epigone showed little short of contempt, to the extent that Anglophones have for long failed to recognise the importance of "Freibergian" geognosy, Werner's ill-founded Neptunism notwithstanding. Geikie's historiography was, then, profoundly Whiggish or anachronistic.

Adams' book concentrated on what one might reasonably call the "prehistory" of geoscience. It was somewhat quirky and curiously organised. The book did not "flow" smoothly or stylishly in the manner of the work of a skilled "man of letters" (which criticism certainly cannot be levelled at Geikie) but it *did* open up the whole field of geology's "prehistory" (*i.e.* long before the term "geology" became a commonplace in the early nineteenth century) and provided a starting point for much subsequent work on, for example, early mineralogy or the relation of studies of the Earth to alchemy.

Besides these three main figures, there were a number of "antiquarian" studies in English such as those of John Judd, Leslie Cox, and Victor Eyles and his wife Joan. The Eyles built up a major personal library of early geological books and their contributions yielded many "real historical facts" (but Joan Eyles' hoped for biography of William Smith never materialised).

Such, then, was the situation when Charles Gillispie gave the study of the history of geology a wholly new direction, from which in many ways it has never turned. After studying chemistry and English history as an undergraduate at Wesleyan College (Connecticut) and serving with the American forces in World War II he went to Harvard to undertake postgraduate studies and promptly moved into the field of history of science, but doing so from the perspective of social history. The result was his widely acclaimed *Genesis and geology*. This work started as a doctoral dissertation with the title "Geology and Genesis" (1949). It was published as Harvard Historical Studies, Volume 58 (1951, reprinted 1969) and as a Harper Torchbook paperback edition (1959), through which the book chiefly became known. The book was reprinted in 1996 by Harvard U.P., with an introduction by Gillispie's admirer Nicolaas Rupke and a new foreword by the author. Curiously to my knowledge there has only been one translation—into Chinese. Gillispie's abilities were, however, swiftly recognised in the U. S. and he was appointed to a position at Princeton as early as 1947, before his thesis was completed and published.

Genesis and geology was almost the first book I encountered in history of science (back in 1959) and my Torchbook copy is extremely well thumbed—falling to pieces in fact. I was much taken by the work and it was a significant factor influencing my decision to enrol in the MSc course in history and philosophy of science at University College London. I was in fact delighted with the book! It was highly readable and congenial to my atheistic proclivities (already well developed at that time). But I should acknowledge that it imbued me with certain Whiggish tendencies that have proved hard to cast off.[2]

---

[2] At that early stage of my career I had never heard of Whiggish historiography of course, and there was no mention of such matters at UCL. Indeed I wonder now whether the lecturers had

Gillispie and Rupke have subsequently given us an account of how *Genesis and geology* (1951) was received by reviewers (Gillispie, 1996; Rupke, 1996). Certainly it did not "fall stillborn from the press" in the manner of Hume's *Treatise of human nature*. But Victor Eyles (in *Nature*, 1953) apparently disapproved of the book's author in that he was "not a geologist or even a scientist"; and *Notes and records of the Royal Society* only seemed to like the book because it was "entertaining" (W. N. E[dwards]., 1952), as if that were generally inappropriate for work of serious scholarship. The book was judged to be "not so much concerned with the emergence and growth of scientific ideas as with their social impact"; and Edwards evidently disapproved of this departure from the straight and narrow. On the other hand, Continental historians of science such as Reijer Hooykaas (The Netherlands) and Sten Lindroth (Sweden) thought well of the book. German and French historians, however, seem to have taken rather little notice of it.

The reason for these differences is not hard to discern: Eyles and Edwards were geologists or science historians; Hooykaas and Lindroth were professional historians of science. The old school scientist/historians such as Eyles had almost a fetish about facts, which (one might unkindly say) should be collected and stored in pots (or vessels or jars), almost as envisaged by Mr Gradgrind in *Hard times*; and only a (geo)scientist could assess the worth of putative (geo) scientific facts and order them appropriately.

But, one might ask, what is the *point* of the scientist-historians' activities? An important factor in the 1950s was, I suggest, to assign credit for scientific discoveries correctly; for discoveries and credit are (understandably of course) matters of great concern to practising scientists. For many of them, I think, the two constitute major components of their *raison d'être* (not to mention the development of good theories and successful practical applications of course).

Gillispie, however, as a trained historian, was interested in other questions altogether. He came from a "free-thinking" home background and was—like so many before and since—interested in Darwinism and what is commonly called today the "Darwinian Revolution", which was already a large mark on the horizon for studies in history of science, to the extent that it seemed to Gillispie to be a field that was already well tilled and getting too crowded. So he aimed to uncover some of the intellectual roots of Darwin's work. But to limit his enterprise somewhat he did not concern himself too much with Lamarck(ism) and Humboldt(ianism), which were obviously in the domain of French and German science. We know well from Darwin's autobiography that he was greatly influenced by von Humboldt, but written by a student of English social history Lamarck and Humboldt were largely set aside in Gillispie's study. On the other hand, a great influence specifically identified by Darwin was that splendid exponent of natural theology, Archdeacon William Paley of Carlisle

---

heard of it either! Elsewhere, I have subsequently argued in favour of the inclusion of some Whiggish elements in historiography (Oldroyd, 1989).

Cathedral—whose work delighted, charmed, and instructed Darwin greatly.[3] The study of Paley led Gillispie smoothly and naturally to the *Study in the relations of scientific thought, natural theology, and social opinion in Great Britain, 1790–1850*, as *Genesis and geology* was subtitled, in the style of the PhD mode to which it in fact belonged.

However, the starting point of 1790 was not in fact adhered to. Perhaps Gillispie suffered, as do many PhD students, from what I like to call the "Leibniz syndrome". As a court historiographer Leibniz, was, as the reader will recall, charged with sorting out and substantiating the historical land-claims of the House of Brunswick, to which end he embarked on a monumental history: *Origines Guelficae* (Leibniz, 1750–1780). This was never completed in Leibniz's lifetime but it appears that he tried to trace things back to the point where he wrote a text (titled *Protogaea*) about the origin of the Earth; and this appeared as a preface (1749) to the main work when *Origines Guelficae* was eventually published posthumously by Christian Ludwig Scheidt. Thus one went back to the very origin of the world when trying to clarify the claims of the Guelfs and the Ghibellines![4]

One can see how one can get into such a fix all too easily when working towards an historical PhD.[5] Thus it is that we find that Gillispie's book and thesis took, as he put it, "a running start" at the topic by consideration of natural theology and providentialism as these had developed in Britain from the seventeenth century, a fair way in advance of the Darwinian Revolution. But providentially no ill effect flowed from this precursory leap. For just as Darwin's ideas did not sprout from nowhere or nothing neither did Paley's natural theology.

So, one must suppose, Gillispie began his task of digging into obscure texts (yielding "facts"!). And really for the first time the *intellectual* roots of geology began to be exposed. The tyro historian was thinking about the *social* history of science, where, however, the social framework was deemed to be chiefly theological rather than, for example, industrial or technical. The excavation work must have involved a huge intellectual and physical effort, as he opened up countless printed sources to reveal the conceptual resources deployed in the writings of early British geologists (which, perhaps surprisingly, given the starting point of his enquiries, attracted Gillispie's attention more than did works on plants or animals). The early "geologists"[6] were not "at war" with

---

[3] I too confess to being charmed by Paley's diction. Who could not be when one encounters his captivating suggestion that the exit from the bowel to the outside world is equipped with a divinely designed sphincter muscle that is "vincible when requisite"?

[4] To be fair to Leibniz, his *Protogaea* arose from his interest in Harz Mountain mining operations as much as his studies in political history.

[5] I speak from experience. My own PhD was originally intended to be concerned with the influence of the Chemical Revolution on mineralogy. When submitted, the story started with Paracelsus!

[6] The term is used here anachronistically, for convenience. The science of "geology" didn't exist, as we understand the term, in the seventeenth and eighteenth centuries.

religion as one might suppose from (say) a reading of Draper's *History of the conflict between religion and science* (1874) (a fine book in my opinion![7]). For just about all the protagonists in Gillispie's book were to a greater or lesser degree religious in the conventional theistic sense or were thoughtful deists. So we may say that Darwin's evolutionary biology grew from his geology (which was his initial scientific interest) and from natural theology (along with other things of course), as Gillispie showed in *Genesis and geology*.

To trace the lineage, however, one must inevitably look at the ideas of certain late eighteenth-century naturalists who were specially interested in the Earth and became involved in the great debates between Vulcanists and Neptunists, Huttonians and Wernerians, or Uniformitarians and Catastrophists (all somewhat anachronistic terms for that period). And thus we encounter two interesting figures that appear in *Genesis and geology*, namely the Irishman Richard Kirwan and the Swiss naturalist Jean-André de Luc, who spent much of his life in Britain and thus has a rightful place in Gillispie's book.

Here Gillispie "took sides" and represented these two figures as rather obtuse figures, or at least benighted obscurantists: "Kirwan presents, of course, almost too classic example of the degree to which a perverse conception of natural theology could corrupt a scientific mind" (p. 56). De Luc was presented a little more sympathetically: "[i]t would be a fine distinction, but Deluc might be awarded an intellectual position a cut above Kirwan's and be credited with a somewhat more subtle empirical exegesis" (pp. 56–57). Even so, both Kirwan and de Luc were hammered hard. So should Gillispie be hammered in turn for writing Whiggish history? In fact I have myself made that accusation (Oldroyd, 2009), even while being well aware that Gillispie is cognisant of the problems of Whig historiography.[8]

To appraise (mock?) the thinking of Kirwan and de Luc—whose names had largely been forgotten from the times of their decease to the appearance of *Genesis and geology*—as did Gillispie would surely not be "permitted" by any PhD examiner today. The historian of science is not "allowed" to make derogatory remarks about those about whom he or she is writing. It would likely be ruled out of court as involving *anachronism*. The student of Kirwan today would be expected to take note of the fact that he was President of the Royal

---

[7] I think that, in the last analysis, there is an irreconcilable difference between the findings of science on the one hand and philosophy and religion on the other.

[8] This is apparent from various remarks in his *Essays and reviews in history and history of science* (2007), where, for example, Gillispie more or less accuses (in the politest way, as in all his work) Sambursky's *Physical world of the Greeks* (1956) of being Whiggish in that it selected certain elements from Greek science for favourable attention because they were part of the road forward to modern science; but elements that did not point in that direction were dismissed. Certainly *Essays and reviews* shows that Gillispie regards science as a "progressive" enterprise, so that one can, to a point, "judge" past science by the yardstick of the present, despite the fact that "[c]urrent fashion contemns the merest whiff of Whig history" (p. 316).

Irish Academy, a good analytical chemist, and was appointed to advise the Government on the advancement of mining in Ireland. He should be seen in the context of his times and not denigrated or derided. De Luc's excellent personal characteristics should be remarked and he should not be damned by faint praise. He should not be criticised for interweaving his science and his religion, for that was normal practice in the late eighteenth/early nineteenth century. One "should" look at him with sympathy and note with suitable respect that he was one of the "founders of geology" and was in fact one of those who introduced the very term "geology". He did good (actualistic) work by employing "historical principles" in the study of the Earth. This is the favourable view of de Luc taken by Martin Rudwick, perhaps the leading historian of geology today (2001, 2005).

I imagine that Gillispie would not write about Kirwan and de Luc today in the manner that he did in the 1940s and 1950s. But he did what he did and then "walked away" from the history of geology to other projects, notably the editing of the *Dictionary of scientific biography*. Gillispie acknowledges that he was not trained in geology (a shocking disqualification for an historian of geology in the eyes of Victor Eyles). Yet his pioneering book, though it has no direct followers or emulators, laid one of the major foundation stones for the study of the history of the Earth sciences; and this is so even though by the time others began to write PhDs on the history of geoscience the norms of "polite historiography" had changed significantly and the mocking of the likes of Kirwan and de Luc is no longer acceptable (if you want to get a PhD).

So what can I say, writing in 2010 as a member of a younger generation than Gillispie, but now myself well into retirement? Elsewhere (Oldroyd, 2009), I have compared the treatment of de Luc by Gillispie, Rudwick and Ellenberger and Gohau (1981) and concluded that Gillispie and Rudwick both showed the marks of their religious preferences on religious matters, whereas the Frenchmen (the former a Calvinist and the latter an atheist) provided the most "neutral" account of de Luc. My own mite was thrown on the fire by drawing attention to a manuscript by William Smith's friend, the Reverend Benjamin Richardson, held at Oxford, who spoke of de Luc in disparaging terms, so that even by the judgement of his *own* times de Luc's geology was thought to be out of order by some at least. So in non-anachronistic (non-Whiggish) terms Gillispie might now (in retrospect!) be said to have right on his side. And as I say, de Luc could well have been wholly forgotten were it not for the fact that Gillispie brought him into view for twentieth-century readers.

As aficionados of history of geology will be well aware, Rudwick's great treatise (2005) was designed to show that the emergence of geology as a science was, among other things, linked to the development of *historical* studies of the Earth. I agree that that was of supreme importance, even though it tends to downplay the significance of the role of the mining industry and of Werner's geognostic work (and that of his students) in mapping the Earth's surface in central and eastern Europe and the location and recording of rocks, mineral bodies, and ores of different kinds, without giving primary significance to fossils as was done in Britain and France.

But for Rudwick's thesis to stand, with its downgrading of Hutton as a major figure (being an "ahistorical" thinker) and its representation of William Smith's work as "enriched geognosy" he had to restore the good name of de Luc, that was blackened long before in the minds of many people (including that of the present writer) by Gillispie's *Genesis and geology*. This is possible, up to a point, because de Luc himself realised (I think) that his Genesis-based geology could not work for the whole history of the globe. So he developed what Rudwick has felicitously termed a "binary history". That is, the Earth's history could supposedly be divided into two parts: a pre-Flood era where all sorts of strange things happened, with water disappearing into hypothetical caverns, etc.; and a post-Flood era, where actualistic geological principles and investigative procedures could be applied. For example, de Luc estimated the rate and extent of the infill of Lake Geneva and the rate of accumulation of peat deposits on the North German plains and concluded that they indicated that the processes had been going on since about the time of Noah's Flood, as gauged by the Biblical record. So the "incubus" of de Luc's seemingly erroneous ideas in his *Elementary treatise on geology* (1809) is (at least partly) removed by invoking the notion of a "binary history", and the fruitful union of geology and religion that Rudwick commends, and which he claims can be found in the literature on the history of geology, is maintained. In fact, of course, de Luc's observations on Lake Geneva (for example) were relating to the time since the last Ice Age, so he was "out" by a factor of about two. Whether this was due to his predisposition to find evidence in favour of an empirically datable Noachian Flood or the approximate and inaccurate character of his determinations I am not able to judge. I merely note that for Rudwick de Luc could be placed on the side of the angels, and I suppose by implication Gillispie belonged on the other side! In any case, we should grant that Gillispie *introduced* Kirwan and de Luc to modern readers and in a sense initiated a debate that is still of interest to historians of science today.

If I leave aside my historian's hat here and put on the one that keeps me metaphysically warm I don't mind saying, with Gillispie, that I think that Kirwan and de Luc were, on occasions, talking nonsense. Gillispie could "get away" with such suggestions in the 1940s and 1950s, but today, if I am wearing my historian hat, I can not;[9] and I dare say that Gillispie would hesitate to write what he wrote back in the 1940s if he were submitting a PhD in history of science now. Why should this be so? And how do all these problems lock together?

---

[9] I am not writing a PhD today, and I think I can, with advancing years, afford to be provocative.

The classic Whig historians believed that humanity was making social progress and thus they had a reference point ("the present") to make judgements about events and actions in the past (Butterfield, 1931). Whether or not human societies make progress is, of course, an assumption that is hard to justify or prove; and for myself I don't really know whether it is or is not true. But that science makes progress seems difficult to reject. We do know more about the natural world than we did in the past, even though the history of science is full of examples of false views and faulty reasoning or observation. And one can be almost certain that much of what is thought true today may in fact be false. Even so, speaking very broadly, a "Whiggish tendency" in the historiography of science is acceptable—maybe even commendable? Certainly most scientists are Whiggish in their attitude to earlier works.

I suggest, however, that the "Whig problem" for historiography has more to do with anachronism than the idea of progress. It is obvious (I think) that someone in the past cannot have ideas that have not yet been thought of, or know facts that have not been discovered. Isaac Newton knew nothing of "curved space" and it would be stupid to criticise him for that reason. On the other hand, his "solid, massy, hard, impenetrable, movable particles" (in his *Opticks*) can reasonably be said to have foreshadowed John Dalton's atomism in some sense. So a historian may legitimately make that assertion and seek to trace the historical thread of connections between Newton and Dalton, as was done, for example, in Arnold Thackray's fine book *Atoms and powers* (1970). But one should not think of Newton's small "massy" particles as "proto-Daltonian" atoms. The two ideas (models) arose at different times, in different contexts, and in response to different intellectual problems.

In looking at the history of science one can discern both competent and incompetent work, theories, or arguments. The same is true as regards studies in the historiography of science. Historiography of science (writing about the history of science) is at a "meta-level" with regard to science itself. It is not part of science itself (except on rare occasions, as when a taxonomist searches the literature for synonyms—and that case it is really science that is being done). Or historical work is sometimes used by scientists as part of their "rhetoric". And scientists may preface their scientific work with historical surveys of the literature in order to make the projects in which they are engaged intelligible, both to specialists and to a wider audience too. In doing this they give credit where credit is due; or sometimes the reverse if they wish to draw attention to what they regards as faulty work done in the past and which they hope to rectify. No one gets too excited about this; and we don't hear cries of "Whiggery"!

Much the same occurs amongst historians of science. They like to draw attention to previous work in the field in which they are working and try to correct previous deficiencies or open up new fields (as Gillispie did so successfully in *Genesis and geology*). This can be done with courtesy (as Gillispie does unfailingly in his *Essays and reviews*, 2007).

The issue can be understood more clearly by referring to a hierarchy of "levels", as follows[10]:

Etc., etc.

Level 5   Descriptions, histories, and theories *etc.* of/about histories of the historiography of science.[11]

Level 4   Descriptions, histories, and theories of/about the historiography of science (or philosophy of science, sociology of science, etc.). (Here histories of the historiography of science are most relevant to our present concerns.)

Level 3   Descriptions, histories, and theories about science, or analyses thereof:

    (a)   History (= historiography) of science

    (b)   Philosophy of science

    (c)   Sociology of science

        etc.

Level 2   Descriptions, histories and theories about "the world" (= science).

Level 1   "The world".

Now books and articles do not always stick to any given level all the time so there is occasional fuzziness between the levels, or remarks situated on *quasi-mezzanine* floors! But let us suppose here, for the sake of simplicity, that precise partitioning is possible. It seems to me that problems don't arise if authors stick to the levels where they belong. A scientist can criticise the work of earlier scientists. No problem. Likewise historians (historiographers) of science can legitimately criticise other historians' (historiographers') work in the historiography of science. They do it all the time in book reviews, refereeing, or "re-writing the history books".

The problems arise if we cross levels—if, for example, we are situated or working at Level 3 and criticise or praise earlier scientists' work at Level 2 from the vantage point of the present and from Level 3. Here anachronism can raise its ugly head. If, as an historian of geoscience, I make adverse comments about Kirwan and de Luc in terms of today's geological knowledge then I am guilty of historiographical anachronism. And I should avoid that. It's all right if I criticise or praise the historiography of, say, Archibald Geikie, Leonard Wilson, Martin Rudwick, or whoever, at my own level. (And if alive they may well answer back.) That's legitimate for an historian of science; and from the dialectical exchange of ideas improvements in our understanding of the history of science may ensue.

So that's all right then: stick to your own level and you'll be fine. Don't wander onto the mezzanine floor. But unfortunately it isn't quite so simple.

---

[10] This idea of levels was also given in an earlier paper of mine (Oldroyd, 1980).

[11] Just now I can't think of any examples of texts that unquestionably belong to Level 5; but they could exist in principle, or indeed at Levels 6, 7, 8, *etc.—quelle horreur!*

The trouble is that Kirwan and de Luc were not just scientists: they were physico-theologists. They were *trying* to weave together their religious ideas and their science (which was not therefore strictly "natural science"). So there are two aspects to their life and work that may be considered. But is it the historian of science's task to disentangle the two?

Perhaps, but there is a problem if one tries to do that: it distorts what Kirwan and de Luc were seeking to do. So maybe one should simply describe what they were up to. But that would make for a dreary piece of work and I suspect that *Genesis and geology* had far greater success than its author anticipated just because Gillispie ridiculed the likes of K. and de L. It made it an exciting book! In any case, almost everyone will agree today that there is no philosophical certainty in religious beliefs and principles. Indeed, it's not clear to me that theology does (or can) progress as (I think) its philosophical premise that there is a benevolent creator God (or gods) is false. It follows that if theology doesn't progress we can't have theological Whiggery though there can be theological anachronism. So in a sense Gillispie was justified in pouring scorn on K. and de L., as from *his* point of view their metaphysical views were nonsensical. And that is how he presented them—though perhaps in an "unkindly" way, given that they couldn't "answer back".

I have said that it's all right for a historian of science to praise or blame earlier work in history of science as no change of "level" is involved. So let me pronounce judgement on *Genesis and geology*! I think it was innovative (as said at the outset it virtually founded a whole new area in the historiography of science). I think it was entertaining and informative and revealed an extraordinary erudition in a field that was virtually a new one. (Witness its notable bibliographical essay.) I think the things that Gillispie said about Kirwan and de Luc were correct, but I recognise that people with metaphysical and theological principles different from those that Gillispie and I share will probably not agree about this. The book was anachronistic as regards such principles; but these are not set in stone and undoubtedly and inevitably change over the years. (They don't necessarily progress, but people generally think that *their* views are sounder and better than the views of those who think differently or those in the past who thought differently.) It is easy to conflate Whiggery with historiographical anachronism, but they are not one and the same. Given that K. and de L. mixed theology and science the account of them given in *Genesis and geology* is warranted (according to my way of thinking); but I wouldn't dare to write about them the way that Gillispie did if I were writing a PhD today. I am so glad that things were more liberal in the 1940s than is generally the case today!

Bibliography

Adams, F. D. *The birth and development of the geological sciences*, Baltimore: Williams and Wilkins Co., 1938 (reprinted New York: Dover Publications, 1954).
Butterfield, H. *The Whig interpretation of history*, London: G. Bell and Sons, 1931.

Draper, J. W. *History of the conflict between religion and science*, New York: Appleton, 1874.

Ellenberger, F. and Gohau, G. 'A l'aurore de la stratigraphie paléontologique: Jean-André de Luc, son influence sur Cuvier', in: *Revue d'histoire des sciences* 34, (1981), pp. 217–257.

Eyles, V. A. 'Scientific thought of the early nineteenth century', in: *Nature* 171 (1953), p. 714.

Geikie, A. *Founders of geology*, London: Macmillan & Co. Ltd, 1897 (2nd edn 1905, reprinted Dover Publications, 1962).

Gillispie, C. C., *Genesis and geology: a study in the relations of scientific thought, natural theology, and social opinion in Great Britain, 1790–1850*, Cambridge, MA: Harvard University Press; and London: Oxford University Press, 1951 and 1969 (reprinted New York: Harper & Brothers, Harper Torchbooks, 1959, with original title and also *Genesis and geology: the impact of scientific discoveries upon religious beliefs in the decades before Darwin*). Reprinted with a foreword by Nicolaas A. Rupke and a new preface by the author: Cambridge, MA and London: Harvard University Press, 1996.

Gillispie, C. C., *Essays and reviews in history and history of science*, Philadelphia, PA: The American Philosophical Society, 2007.

Gillispie, C. C. and Holmes, F. L., eds. *Dictionary of scientific biography*, New York: Scribner, 17 vols, 1970–1990.

Leibniz, G. W. edited by C. L. Scheid *et al.*, *Origines Guelficae …*, Hanover: H. E. C. Schleuter, 5 vols, 1750–1780.

Oldroyd, D. R. 'Sir Archibald Geikie (1835–1924), geologist, romantic æsthete, and historian of science: the problem of Whig historiography of science', in: *Annals of Science* 37 (1980), pp. 441–462.

Oldroyd, D. R. 'Why not a Whiggish social studies of science?' in: *Social epistemology* 3 (1989), pp. 355–359.

Oldroyd, D. R. 'Jean-André de Luc (1727–1817): an atheist's comparative view of the historiography', in: Kölbl-Ebert, M., ed., *Geology and religion: a history of harmony and hostility*, London: The Geological Society, 2009, pp. 7–15.

Rudwick, M. J. S. 'Jean-André de Luc and nature's chronology', in: Lewis, C. L. E. and Knell, S. J., eds, *The age of the earth from 4004 BC to AD 2002*, London: The Geological Society, 2001, pp. 51–60.

Rudwick, M. J. S. *Bursting the limits of time: the reconstruction of geological history in the age of revolution*, Chicago and London: The University of Chicago Press, 2005.

Rupke, N. A. 'Foreword', in: Gillispie, C. C., ed., *Genesis and geology: a study in the relations of scientific thought, natural theology, and social opinion in Great Britain, 1790–1850*, Cambridge, MA and London: Harvard University Press, 1996, pp. v–xv.

Thackray, W. W. *Atoms and powers: an essay on Newtonian matter-theory and the development of chemistry*, Cambridge, MA: Harvard University Press; London, Oxford University Press, 1970.

Wilson, L. 'Archibald Geikie and the last elevation of Scotland', in: *Earth sciences history* 28 (2009), pp. 32–56.

WNE [Edwards, W. N.] 'Genesis and geology', in: *Notes and records of the Royal Society* 9 (1952), pp. 378–379.

Zittel, K. A. von. *History of geology and palaeontology to the end of the nineteenth century, translated by Marian M. Ogilvie-Gordon*, London: Walter Scott, 1901 (first German edition 1899).

Chapter 11
Five Discourses of Bible and Science 1750–2000

Nicolaas A. Rupke

11.1 The Gillispie Thesis

Literally hundreds of books and pamphlets and thousands of articles dealing with the combination of Bible and science have appeared during the period covered by this chapter.[1] The reason for this copious output of literature is that different relationships between biblical texts and modern science can be and have been constructed, and that these have proved highly contentious. The issue at stake has been, putting it black-and-white, whether the Bible is a divinely inspired, inerrant and unique source of physical as well as moral truth; or whether it is not, containing unscientific absurdities. Titles of the many publications include such variations on the theme of 'Bible and science' as 'scripture and science,' 'Bible and nature,' 'Bible and astronomy,' 'Genesis and geology,' and 'Moses and Darwin.'

The majority of the authors who write about Bible and science also take sides in the controversy over the nature of the relationship. Scholarship that does not participate in the debates but primarily explores the history of the relationship dates by and large from the post-WW II period and even then has been thin on the ground.[2] Accordingly, the historiographical models canvassed to give structure to the relationship Bible and science have tended to reflect the

---

[1] The literature on the more general combinations of science-religion, science-Christianity, science-theology, or science-church, is more voluminous yet and, for the most part, has not been included in this chapter. Not considered either are esoteric Bible and science relationships such as found in the writings by Mary Baker Eddy (1821–1910) about Bible, science and healing. Also not considered are extra-biblical healing miracles attested to by medical scientists such as Alexis Carrell (1873–1944) or other extra-biblical religion-science *causes célèbres*, of which the Turin Shroud is an example.

[2] For a recent set of contributions see Van der Meer, Jitse M. and S. Mandelbrote (eds.), *Interpreting nature and scripture: History of a dialogue in the Abrahamic religions*, 2 vols. Leiden and Boston: Brill, 2008.

N.A. Rupke (✉)
Georg-August University Göttingen, Göttingen, Germany
e-mail: nrupke@gwdg.de

J.Z. Buchwald (ed.), *A Master of Science History*, Archimedes 30,
DOI 10.1007/978-94-007-2627-7_11, © Springer Science+Business Media B.V. 2012

partisanship of their propounders. Both harmony and warfare models have been put forward, followed in recent years by a so-called complexity thesis which states that Bible-and-science should not be reduced to a common denominator of either a good or a bad relationship but must be seen in the multifariousness of the different times and places of its making. Talk now is more of encounters and engagement between the two, less of conflict.[3]

In broad agreement with, and further development of, the complexity thesis, I here document the history of the relationship Bible and science as a story of multiple discourses that have existed alongside each other during much of the period of this chapter. 'Discourse' is used here in a social theory sense, denoting more than extended speech or writing, namely a coherent set of contentions that generates its own regime of validity inside a particular constituency with distinct sociopolitical values. The multiple discourses approach avoids painting an oversimplified picture of the relationship Bible and science, which relationship, as I document below, cannot be resolved into two polar opposites of cognition – the one of religious conviction, the other of natural knowledge; nor can it be reduced to two competing professions – the one theology, the other science. The history of 'Bible and science, 1750–2000' has been less driven by the encounter between theologians and scientists or their respective beliefs and fields of expertise, than by competing discourses about the Bible-science copula, each one conducted by groups that count theologians and scientists among their own. I propose that we refer to this view as the Gillispie thesis. As Charles Gillispie famously argued in his by now classic doctoral dissertation, *Genesis and geology. A study in the relations of scientific thought, natural theology, and social opinion in Great Britain, 1790–1850*, the 'Genesis and geology' debate of the early nineteenth century was not a matter of religion versus science but of religion within science.[4]

Some five discourses can be recognized, each identifiable by a particular hermeneutic strategy in dealing with Bible and science. (1) The Bible is divinely inspired, literally true and the textual passages that deal with the natural world are imperatively valid for science. (2) Apparent discrepancies between Bible and science disappear if we interpret certain biblical texts the right way, in many places not in a literal but a figurative sense. (3) Bible and science do not clash, because they share no common ground but have separate spheres of validity, the one of moral conduct, the other of physical reality. (4) The Bible and science are fellow travellers who conduct a dialogue in which each informs the other by addressing the same reality but from a different point of view. (5) The Bible is a rag-bag of antiquated stories and in part harmful notions from which science

---

[3] Representative is Ferngren, Gary B. (ed.), *The history of science and religion in the western tradition*, New York, NY [etc.]: Garland, 2000.

[4] Gillispie, Charles C., *Genesis and geology. A study in the relations of scientific thought, natural theology, and social opinion in Great Britain, 1790–1850*, new edn, Cambridge, MA: Harvard University Press, 1996 (1st edn 1951).

sets us free. These five interpretations, in addition to having different religious and sociopolitical anchoring grounds, also have fastened onto characteristic sets of contentious biblical texts and scientific theories, although overlaps between the sets exist.

11.2 Changing Boundaries

The pertinent criticisms of the Bible are many and varied and already by the middle of the eighteenth century added up to a familiar canon of examples. During the past 250 years the number of points of contention in scripture have changed little, few if any new biblical passages having been added to the tally of allegedly reason- and science-defying instances. By 1750, a major encyclopedic apologia started its 32-year publishing history, addressing countless attacks on the Bible and defending its rationality and divinely inspired truth. The 16-volume work, *Die gute Sache der Offenbarung* (1750–1782), was written by the Lutheran professor of theology at the University of Königsberg, Theodor Christoph Lilienthal (1717–1781), upon his death honoured by his colleague Immanuel Kant (1724–1804) with an affectionate lament ('Trauergedicht').

Lilienthal, who was an ardent anti-deist in the Wolffian tradition of Halle University, primarily engaged with the English deists of the seventeenth and early-eighteenth centuries (Charles Blount, Matthew Tindal, Thomas Woolston, Thomas Morgan, Viscount Bolingbroke, John Leland, and also the French Spinozist Simon Tyssot de Patot). Examples of contentious biblical stories that are pertinent to the workings of the natural world – some already discussed at the time of the church fathers, for instance by Augustine – ranged from the creation account and Noah's deluge, via the plagues of Egypt, Joshua's Long Day, and Jonah and the whale, to the New Testament miracles surrounding Christ, in particular His virgin birth and resurrection. Lilienthal insisted that biblical miracles are factual; some had involved supernatural divine action, others could be explained in terms of natural processes.

Even though the number of contentious passages has changed little, the engagement between the discourses has been marked by moving flashpoints and the boundaries between them have repeatedly shifted. In part this has been due to the fact that 'Bible' and 'science' are no static entities but have acquired new meaning and, in the case of science also new content. Biblical and archaeological scholarship have led to novel understandings of the literary character of, for example, the first chapters of Genesis, greatly influencing hermeneutical approaches. Science has undergone – and continues to undergo – more extraordinary changes yet, showing an exponential growth of practitioners and texts, and a never ending rejuvenation of theories and cognitive paradigms. By 1750, the very concept and, with it, the term 'science' was unknown, and many 'natural philosophers' still regarded the Bible as a source of knowledge of the physical world, witness contemporaneous sacred geographies (Willem Albert Bachiene), and the persisting use of sacred chronologies (William Hales). In the process, natural philosophers adopted biblical language and notions of paradise, of Adam and Eve, of Noah and the deluge, and Baconians, Hutchinsonians and Newtonians confidently continued to invest Bible passages with their scientific theories. Subsequent professionalisation of the study of nature caused a breakdown of this 'Baconian compromise',[5] and sacred knowledge was filtered out of mainstream science.

The dynamic interactions are well illustrated by the controversy over pre-adamism.[6] The claim by the French Huguenot Isaac la Peyrère (1596–1696) of pre-adamic humans – that Adam was not the first human being but the first Jew – formed an early chapter in the history of Bible scepticism. Lilienthal and many apologetes since, saw the supposition of an inhabited world before Adam as an assault on the veracity of the Genesis story. The dispute went through different configurations, reflecting changes in both biblical hermeneutics and science. During much of the eighteenth century, language was central to debates about human origins, but through the nineteenth century physical anthropology and paleoanthropology became pivotal. In spite of the heterodox beginnings of pre-adamism, it 'later came to reside among religious conservatives'[7] who attempted to harmonize the Bible with the findings of historical geology and Darwinian evolution. Pre-adamism was not just part of an abstract, intellectual encounter between Bible and science but became implicated in legitimation attempts of racism and slavery (by Samuel George Morton, George Gliddon, Josiah Clark Nott) and thus grounded in the political and social realities of particular constituencies. Let us now look at the different discourses in some detail.

11.3 The Bible as a Guidebook of Science

Lilienthal contributed to a discourse that took the Bible and, in particular, its historical portions, in a literal sense. The Old and New Testament alike are divinely inspired and therefore inerrant. Moses is the author of the Pentateuch and God has spoken through him, revealing in the opening chapters of Genesis the origin and early history of the world. Biblical history is factual, reliable and

---

[5] Moore, James R., 'Geologists and interpreters of Genesis in the nineteenth century,' in Lindberg, David C. and Ronald L. Numbers (eds.), *God and nature. Historical essays on the encounter between Christianity and science*, Berkeley, CA: University of California Press, 1986, 322–350.

[6] Livingstone, David N., *Adam's ancestors: Race, religion, and the politics of human origins*, Baltimore: Johns Hopkins University Press, 2008.

[7] Livingstone, *Adam's ancestors*, 220–221.

in this sense scientific. The Bible has to be taken as a source of natural knowledge that can and must inform science, specifically the scientific theories to which the sacred narrative pertains.

This does not mean that the Bible is a textbook of biology, geology or any other branch of science – as literalists have often remonstrated; but the letter of the historical texts does provide a framework of physical truths about the origin, governance and end of the world. Moreover, literalism need not imply a naïve and simplistic hermeneutic. For example, biblical references to the sun rising or setting have to be given a common sense interpretation as optical reality, they are colloquial speech, just like ours today, and do not mean an endorsement of a geocentric as opposed to an heliocentric solar system. Already John Calvin conceded – as indeed had some of the church fathers – that divine revelation is accommodated to the circumstances of the original Bible writers, a view to which both literalists as well as non-literalists appealed.

Most important to the literalist view have been creation (Genesis 1–2), the flood (Genesis 6–8) and the age of the world as calculated on the basis of the genealogies of the ante- and postdiluvial patriarchs (Genesis 9–11). On the validity of these signposts of the early history of the world depends – they believed and believe – the entire scheme of a divinely guided, eschatological history and Christian soteriology. As the English clergyman-naturalist and convert to Methodism Joseph Townsend (1739–1816) commented: 'The science of geology becomes of infinite importance, when we consider it as connected with our immortal hopes. These depend on the truth of revelation, and the whole system of revealed religion is ultimately connected with the veracity of Moses'.[8]

Thus the deluge became a crucial issue, the more so when during the half century 1780s to 1830s the seventeenth- and eighteenth-century cosmogonical explanations of the earth's crust were superseded by modern stratigraphy and vertebrate paleontology.[9] The outer shell of the globe proved to be made up of an enormously complex sequence of rock formations. Could the formational complexity of the earth's crust be attributed to the deluge and were all these sediments with their organic fossils deposited in the course of approximately 1 year – the duration of Noah's flood? Or had these rocks accumulated over long periods of antediluvian earth history and had the deluge been of little consequence, geologically speaking?

On this question a major bifurcation of opinion took place during the two or three decades following the French Revolution. Some of the Protestant leaders

---

[8] Townsend, Joseph, *The character of Moses established for veracity as an historian, recording events from the creation to the deluge*, London: Longman, Hurst, Rees, Orme and Brown, 1813, 430.

[9] Rudwick, Martin J.S., *Bursting the limits of time: The reconstruction of geohistory in the age of revolution*, Chicago and London: University of Chicago Press, 2005; *Worlds before Adam. The reconstruction of geohistory in the age of reform*, Chicago and London: University of Chicago Press, 2008.

of the new historical geology created space for their science by reinterpreting the creation and deluge stories in a non-literal sense (see next section). They limited the geological impact of the biblical flood by ascribing merely superficial features of sedimentation and landscape erosion to diluvial action. The massive rock formations below the surface had been deposited during earlier periods of geological history. This theory, that the flood had been of limited geological consequence, was referred to as 'diluvialism' (and today as old-earth creationism). By contrast, 'deluge geology' considered most if not all of the sedimentary record as a product of the flood, and simultaneously insisted on a traditional, literal interpretation of Genesis (young-earth creationism). This theory stated that much of the geological column, including the fossils, had accumulated neither before, nor during, but after the six days of creation or, more precisely, after the fall of man, and was nearly in its entirety attributable to the deluge. If death and suffering had come into the world by sin, the fossil record with its evidence of death, extinction and carnivorousness ought to have accumulated in the wake of the fall.

The literalists – also known as Mosaical geologists or scriptural geologists (early representatives were Granville Penn, George Bugg, George Fairholme and George Young) – began taking on a recognizable identity upon the publication, and as critics of, the diluvialism of the Oxford clergyman-geologist William Buckland (1784–1856). More than before, they emphasized the universality and geological effectiveness of Noah's deluge, in that way making it possible to retain for the earth a young age of the traditional order of magnitude as determined by the sacred chronologies of scholars from Scaliger to Ussher.

By now, the literalist discourse was cut loose from its initial establishmentarian moorings at the centre of academe and became located at the provincial periphery. In North America, literalism, although not marginalized to the same extent, was being formulated in reaction to modern geology, too, prominently by the Episcopalian professor of medicine at the University of the City of New York, Martyn Paine (1794–1877). Directing his censure among others at Buckland, Paine insisted on a literal understanding of the stories of creation and deluge. He saw in the coal formations of the Carboniferous proofs of the efficacy of the flood waters to uproot the luxuriant vegetation of the antediluvian period and deposit layers of plant debris intercalated with strata of reworked sediment. Paine's deluge geology went hand in hand with a strong defence of the immateriality of the human soul, and his most substantive essays that argued for a literalist hermeneutic of creation and deluge appeared as major additions to his *Physiology of the soul* (1872), a volume that met with rapturous acclaim across North America.

Paine objected also to the theory of evolution, and deluge geology acquired additional meaning through the second half of the nineteenth century with the emergence of Charles Darwin's theory. Given its utter irreconcilability with the letter of Genesis 1, Darwinism was firmly rejected. Thus young-earth literalism became young-earth creationism which, from the start, used deluge geology to disprove organic evolution by arguing that the geological column with its progressive fossil record, the main pillar of evolution theory, was an artefact of circular reasoning. Young-earth creationism fell on fertile soil among conservatives in the established churches, in particular the Missouri Synod Lutherans and Presbyterians, but also among non-conformist and new religious groups that flourished in North America. A number of the key contributors to literalism were Seventh-day Adventists, prominently among them the Canadian self-taught geologist George McCready Price (1870–1963), who in his *Illogical geology* (1906) argued that the most effective way to defeat the theory of evolution was to deprive it of its framework of geological time. Price's *magnum opus*, entitled *The new geology* (1923), was an audacious and imaginative attempt to revise modern geology in terms of a literal understanding of the first nine chapters of Genesis.

Price's attack strategy against evolution gained popularity with the Creation Science movement in the 1960s, which took off in the wake of the movement's canonical text, *The Genesis flood* (1961), written by the conservative evangelical[10] Old Testament scholar John C. Whitcomb (b. 1924) and the Southern Baptist professor of hydraulic engineering Henry M. Morris (1918–2006). The book significantly enlarged upon Price's young-earth creationism, by addressing such major issues as radioactive dating of the age of the earth.[11] The twentieth-century constituency of 'Bible as a guidebook of science' has been researched in detail.[12] It is part of the fundamentalist movement of orthodox Christianity that has repeatedly clashed with secular science in the courts over the teaching of evolution/creation in schools, and that in recent years has linked hands with creationism in the Muslim world.[13]

11.4 The Adjustable Bible

Already during the second half of the eighteenth century, Jean André Deluc (1727–1817), a Genevan Calvinist, geologist and meteorologist who moved to London to become Reader to the English Queen, helped shape a different discourse from the literalist – one that accepted modern science yet also stuck to a belief in the Bible as God's Word.[14] Apparent discrepancies between scripture and science can be resolved by taking relevant biblical expressions

---

[10] 'Evangelical' is used throughout to denote 'orthodox Christian'.

[11] Whitcomb, John C. and Henry M. Morris, *The Genesis flood. The biblical record and its scientific implications*, Philadelphia, PA: Presbyterian and Reformed Publishing Company, 1962, 331–438. A more recent exposition of the same views is Roth, Ariel A., *Origins: Linking science and scripture*, Haggerstown, MD: Review & Herald Publishing Association, 1998.

[12] The definitive study is by Numbers, Ronald L., *The creationists. From scientific creationism to intelligent design*, 2nd edn, Cambridge, MA: Harvard University Press, 2006.

[13] Numbers, *The Creationists*, 421–427.

[14] Hübner, Marita, *Jean André Deluc (1727–1817): Protestantische Kultur und moderne Naturforschung*, Göttingen: Vandenhoeck & Ruprecht, 2010, *passim*.

and stories in the right sense, in many instances figuratively. The Bible is scientifically accurate, if only we interpret the texts correctly.

As a century later the British Liberal statesman and Prime Minister, Willam Ewart Gladstone (1809–1898), summed up in his *The impregnable rock of Holy Scripture* (1890), the issue is not one of the substance of divine revelation as conveyed to us in the Old Testament but of its literary form.[15] In a crucial concession to geology, the 'adjustable Bible' accepts that the earth is very much older than indicated by sacred chronology and has gone through long periods of time during which successively higher forms of life came into existence. Central is the creation story of Genesis 1, reinterpretations of which open the shutters on the time vistas needed by modern geology. In other words, the sedimentary and fossil record were, by and large, not attributable to the deluge which in turn has been placed on the back burner.

Scientific Christians and theologians alike, concerned that the book of nature not be at variance with scripture, put forward a variety of harmonization schemata.[16] A genre of literature developed that dealt with the congruence of the Bible and science. Not uncommonly, geological textbooks would include a chapter on how to reconcile the new earth history with the biblical accounts of creation and deluge. The Yale University scientist Benjamin Silliman (1779–1864) added a lengthy supplement on 'Consistency of geology with Sacred History' to the American edition of Robert Bakewell, *An introduction to geology* (1833). Edward Hitchcock (1793–1864), president of Amherst College and professor of natural theology and geology included in his frequently reprinted *Elementary Geology* (1840) a chapter on the 'Connection between geology and natural and revealed religion.' James D. Dana (1813–1895), Silliman's student and successor at Yale as professor of natural history and geology, appended to all four editions of his authoritative *Manual of geology* (1863; 1874; 1880; 1895) a Genesis and geology harmonization scheme.

During the early part of the nineteenth century, much of this literature was produced by English-language scientists – experts for the most part in comparative anatomy, paleontology and stratigraphy, such as Buckland at Oxford, whereas during the century's second half the genre was enriched with a number of monographs by theologians, mainly in Germany (notwithstanding Dana's simultaneous writings). The high point of the reconciliation literature was reached with the formidable scholarship of Otto Zöckler (1833–1906), renowned also for his work in the areas of Old and New Testament studies, dogmatics and church history. His *Geschichte der Beziehungen zwischen Theologie und Naturwissenschaft, mit besondrer Rücksicht auf Schöpfungsgeschichte* (2 vols, 1877–1879) is a classic of the genre.

---

[15] Gladstone, William Ewart, *The impregnable rock of Holy scripture*, London: Isbister, 1892, v, 26–91.

[16] Rupke, Nicolaas A., 'Christianity and the sciences, 1815–1914', in Stanley, Brian and Sheridan Gilley (eds.), *The Cambridge history of Christianity, vol. 8: world Christianities, c. 1815–c. 1914*, Cambridge: Cambridge University Press, 2006, 164–180.

Three basic reconciliation exegeses of the hexaemeron existed/exist as part of the 'adjustable Bible', and shifts from one position to another have not been uncommon. The first of the reconciliation attempts – the concordist or 'day-age' interpretation – sees a concordance between the Mosaic days of creation and the stratigraphic succession of major formations and periods of earth history and gives the word 'day' the meaning of 'period' (Deluc, Cuvier, Wiseman, Hugh Miller, Franz Delitzsch). The second schema – the restitution, 'gap' or 'ruin and restoration' exegesis – focuses on the first two verses of Genesis, placing an indefinite and possibly very long time gap between verse 1 and 2 that could accommodate all of geological history, which thus had taken place before the six days of creation (Buckland, John Pye Smith, Andreas Wagner).

The third and least literal exegesis was the idealist which states that the creation days represent ideal 'moments' rather than consecutive periods of time. The creation days were neither actual days nor periods but a logical sequence of aspects of divine creation that functions as a narrative device (Friedrich Michelis, Johann Baptiste Baltzer). A version of this third exegesis was propagated also by Gladstone: 'It seems to me that the "days" of the Mosaist are more properly to be described as CHAPTERS IN THE HISTORY OF CREATION'.[17] In the early-twentieth century, this view was reformulated in Calvinist circles by, for example, the Dutch theologian Arie Noordtzij (1871–1944), becoming known as the framework hypothesis of Genesis 1. To Noordtzij, the six days of creation are the sum of two parallel tridiums, days and nights as well as mornings and evenings being used as a literary framework to lead up to sabbath observance.[18] A yet other version gained a certain popularity during the mid-twentieth century when Air Commodore P.J. Wisemann (1888–1948) (father of Donald, the evangelical archaeologist), published his *Creation revealed in six days: The evidence of scripture confirmed by archaeology* (1948). The days of creation correspond to the sequence in which divine revelation had been recorded on six tablets (a view adhered to also by Robert E.D. Clark and Bernard Ramm); thus 'Babel' in the form of ancient clay tablet accounts was called upon to support 'Bible'.

In the wake of *The origin of species* by Charles Darwin (1809–1882) the discourse of an adjustable Bible was faced with a new challenge which required a further reinterpretation of Genesis 1, this time no longer of its time frame but the nature of God's creative acts. The new adjustment conceded that the aim of scripture was not to supply us with the workings of the creative process but with the glorification of God the Creator. Not all were willing to accommodate the theory of evolution. Zöckler, for one, objected to Darwin's theory;[19] but many others had no great difficulty reinterpreting the process of creation from a

---

[17] Gladstone, *Impregnable rock*, 56.

[18] Noordtzij, Arie, *Gods woord en der eeuwen getuigenis: het Oude Testament in het licht der oostersche opgravingen,* Kampen: Kok, 1924, 111–119.

[19] Zöckler, Otto, *Ueber Schöpfungsgeschichte und Naturwissenschaft,* Gotha: Perthes, 1869, 50.

special, instantaneous and direct divine intervention to a gradual and indirect act, God having used organic evolution to execute His plan of creation. Euphemistically, evolution would be referred to as progressive creation. Already before the publication of *The origin of species*, Darwin's rival, the Anglican theist Richard Owen (1804–1892), redefined creation from a miraculous event to a naturalistic process, naming it 'the ordained becoming of living things'.[20] Yet a particular problem, not always squarely addressed nor readily solved, was that of 'man's place in nature,' which included such issues as 'the antiquity of man,' the fall, and whether or not humans are unique for possessing an immortal soul.

Christian life scientists and biblical scholars alike on both sides of the Atlantic embraced the substance of evolution theory (Asa Gray, James McCosh, Benjamin B. Warfield, Alexander Winchell, St. George Jackson Mivart).[21] Unlike true Darwinians, however, a majority of evangelical evolutionists firmly held to divine providence, and they remained convinced that the pattern of evolution was not random but a goal-directed unfolding towards mankind or, more specifically yet, to a Christ-centred point in the history of the cosmos, as proposed by the Catholic priest and French paleoanthropologist Pierre Teilhard de Chardin (1881–1955).

11.5 The Non-scientific Bible

In the third discourse, the relationship Bible and science is unproblematic for the simple reason that it is non-existent – there is no meaningful connection between the two. 'Hebrew scripture' has no bearing on the modern study of the physical world or the other way around. Bible and science each belong to wholly separate domains of reality. The Bible pertains to the sphere of morality and spirituality; science, by contrast, rules when it comes to physical things. Biblical descriptions of the natural world may have a poetical quality but the belief that these passages should contain divine revelations about the natural world in accord with modern scientific discoveries and theories is absurd. If people wish to see in these poetical passages in the Old and New Testaments some elements of divinity, inspiration and revelation, that is admissible; but inerrancy and literal truth are out of the question, as the Bible bears many imperfections that mark its origins as an historical document.

---

[20] Rupke, Nicolaas A., *Richard Owen: Biology without Darwin*, Chicago and London: University of Chicago Press, 2009, 159.

[21] E.g., Moore, James R., *The post-Darwinian controversies. A study of the Protestant struggle to come to terms with Darwin in Great Britain and America 1870–1900*, Cambridge: Cambridge University Press, 1979; Livingstone, David N., *Darwin's forgotten defenders. The encounter between Evangelical theology and evolutionary thought*, Grand Rapids: William B. Eerdmans; Edinburgh: Scottish Academic Press, 1987.

An eminent representative of this view was the Prussian traveller and scientist Alexander von Humboldt (1769–1859). Little if any reference to the Bible occurs in his voluminous scientific oeuvre, except in his *Cosmos* (German original 1845–1862) where he discussed the Bible, especially the Old Testament, for its poetic descriptions of nature. Damning with faint praise, Humboldt downplayed the importance of the Bible by extolling its qualities as 'Hebrew poetry'.[22]

German Jews appreciated Humboldt's respect for the literary quality of the Old Testament; but orthodox Christians raised the alarm: the great, trend-setting scientist was removing God and religion from the scientific study of nature. Humboldt defended himself by appealing to the example set by 'the very Christian Immanuel Kant,'[23] who in his *Allgemeine Naturgeschichte und Theorie des Himmels* (1755) had separated science from scripture, making no reference in his cosmogonical treatise to the Mosaical cosmogeny and relegating Gold-talk to the sphere of metaphysics. Kant's authority and example were appealed to by other scientific Christian who, while continuing to hold the Bible in respectful esteem, dissociated their scientific work from scripture.

In Great Britain a leading representative of the 'separate spheres' view was Baden Powell (1796–1860), the Anglican clergyman and Savilian professor of geometry at Oxford. Powell addressed the issue of Bible and science in a series of publications, from his *Revelation and science* (1833) to his *The unity of worlds and of nature* (new title to the 2nd ed. of 1856). The Bible is a source of moral and spiritual intimations 'which are, in their essential nature, alien from physical consideration,' Powell insisted.[24]

The third discourse merged with higher criticism of the Bible. Already around 1800, Lutheran theologians in Germany had criticised the Mosaical geologists for taking the Genesis account of creation and flood literally. *Moses und David keine Geologen* (1799), stated the title of a book by the Helmstedt and later Göttingen theologian David Julius Potter (1760–1838). Genesis 1 was to be bracketed with Psalm 104 and represented a 'Schöpfungshymnus,' a creation hymn, the main purpose of which had been the ordination of the sabbath week. In a later critique, too, the original purpose of the creation story was not to give an account of how the world came about but to assert the monotheistic view and fight polytheism.[25] More radical than the hermeneutic revisions that were forced on many Bible believers by science, were those by this critical tradition

---

[22] Humboldt, Alexander von, *Cosmos. A sketch of the physical description of the universe*, 2 vols. Baltimore and London: Johns Hopkins University Press, 1997, vol. 2, 57.

[23] Humboldt, *Cosmos*, vol. 1, xxv.

[24] Powell, Baden, *The unity of worlds and of nature: Three essays on the spirit of the inductive philosophy; The plurality of worlds; and the philosophy of creation*, London: Longman, Brown, Green, Longmans, & Roberts, 1856, 300.

[25] Otto, Eckart, 'Auf welche Fragen antwortet eine antike Schöpfungstheologie im alten Orient und in der Bibel? Die Falle des Kreationismus', in Kraus Otto (ed.), *Evolutionstheorie und Kreationismus – ein Gegensatz*, Stuttgart: Franz Steiner Verlag, 2009, 17–26.

within theology, leaders of which ranged from Potter's older Göttingen colleague Johann Gottfried Eichhorn (1752–1827) to the latter's later successor Julius Wellhausen (1844–1981), both orientalist-theologians. Whereas geology appeared to remove a literal meaning from Genesis, the documentary hypothesis of the historical school in biblical studies went further, reducing the entire Pentateuch from a unitary record of divine revelation to a product of historical change, cobbled together from a variety of pre-existent sources and repeatedly altered in a process of editorial changes. The creation and deluge stories, for example, were traced back from Bible to Babel. Scripture appeared diminished from a divinely revealed, inerrant account of the grand scheme of the world – past, present and future – to a fallible product of human contrivance, in part plagiarised from pagan sources.

Higher criticism did not become a topic of major, public debate in Britain before the appearance of the *Essays and reviews* (1860),[26] questioning the Mosaic authorship and with that the authenticity of the Pentateuch as a divinely inspired account of history. Powell was one of the seven members of the Established Church who authored *Essays and reviews*, one of the 'Septem contra Christum,' as a troubled Cantabrigian called the team of authors. In his contribution 'On the study of the evidences of Christianity' Powell reiterated that the 'region of spiritual things' and the 'domain of physical causation' were two unrelated spheres.[27] Biblical miracles, because they go against the laws of physics, are not believable, and Powell pleaded for a 'Christianity without miracles'. Biblical stories that are claimed to be historical but contradict scientific fact must be 'transformed into truths taught by parables'. A fellow contributor, the lawyer and Egyptologist Charles W. Goodwin (1817–1878), insisted that neither the literal interpretation of Genesis nor the figurative one does justice to the nature of the 'Hebrew cosmogony'. It is not a divine 'but a human utterance, which it has pleased Providence to use in a special way for the education of mankind'.[28]

This discourse, by its very nature, was more practised than written about, and Powell was somewhat of an exception publicly to express his convictions about the separateness of Bible and science. Others did in their work what Powell put in writing. Among these were such Christian giants of Victorian science as John Herschel (1792–1871), Charles Lyell (1797–1875), and Richard Owen (1804–1892). Tellingly, none of these religious men of science consented to having their name put to the so-called Scientists' Declaration, which in response to *Essays and reviews* affirmed the essential harmony between the Holy Writ and physical science. Owen, for one, when asked for his endorsement

---

[26] Important, too, was the bishop of Natal, John William Colenso's, *The Pentateuch and the book of Joshua critically examined*, London: Longman, 1862.

[27] Powell, Baden, 'On the study of the evidences of Christianity', *Essays and Reviews*, 10th ed., London: Longman, Green, Longman, and Roberts, 1862, 112–172, on 152.

[28] Goodwin, Charles W., 'On the Mosaic cosmogony', *Essays and reviews*, 10th ed., London: Longman, Green, Longman, and Roberts, 1862, 249–305, on 305.

of the Declaration, declined by retorting that modern science, not scripture, was the means by which God revealed natural truth.[29]

In the twentieth century, the discourse of a non-scientific Bible increasingly adopted the wider practice that separates religion – rather than just scripture – from science. At the very end of the century, the Harvard evolutionary paleontologist and science popularizer Stephen Jay Gould (1941–2002) revitalised the notion of separate domains, coining the acronym NOMA – nonoverlapping magisteria – to describe the two autonomous realms of religious belief and scientific knowledge.[30]

11.6 The Complementary Bible

In recent decades, a somewhat different discourse from the one that allocates to Bible and science a separate domain of their own has become louder. It recognizes that also science has its limitations, especially in addressing ultimate questions about past and future of the world. The Bible may provide answers to these and thus complement science. The two are fellow travellers, engaged in a dialogue, each contributing its special knowledge in addressing the same reality, but different aspects of this, and each from its own epistemological vantage point.

This belief rides the coattails of post-WW II diminishing expectations of what science can contribute to the public good. The atom bomb and a broad variety of similar derailments of scientific knowledge have demonstrated that science needs to be kept on a leash of moral governance, for the provision of which many are looking anew to religion and its sacred texts. Moreover, the development of atomic physics, particularly of quantum mechanics and, more recently, of the latest strides forward in big bang cosmology, have seemed to legitimise a range of metaphysical and religious speculations in which scientists themselves have indulged, from Pascual Jordan (1902–1980) to, for example, Paul Davies (b. 1946). A new natural theology has emerged, today incorporated in the John Templeton Foundation which promotes the study of the physical world as spiritual capital. Talk is of convergence between religion and science, of synergy and of scientists discovering in their work spirituality if not the divine. The Foundation's preoccupation is with God, religion and spirituality, less with Bible and science, but the relationship between these two does receive some attention.

The focus of interest here is on cosmology and on the question to what extent the biblical and scientific conceptions of the world cohere. Darwinian evolution is thereby taken less into consideration than such issues as the

---

[29] Rupke, *Owen*, 245.

[30] Gould, Stephen Jay, *Rock of ages: Science and religion in the fullness of life*, New York: Ballantine Publishing Group, 1999, passim.

temporal beginning of the universe, the anthropic principle that considers humans and their retrospective understanding of the universe as a prospective outcome of cosmic evolution, and the end times. Eschatology is providing a striking case for dialogue. The Catholic physicist-theologian Stanley Jaki (1924–2009) or the equally Catholic physicist-philosopher Ernan McMullin (1924–2011) have stressed the independent validity of scripture as a source for our understanding of physical reality. An Anglican representative is the British physicist and theologian John Polkinghorne (b.1930), like Jaki a recipient of the Templeton Prize for Progress in Religion. No one view captures the universe, he believes, no single interpretative key will unlock every ontological door. Biblical belief captures reality in a particular way and so does science. As Polkinghorne and his collaborator, the Heidelberg evangelical systematic theologian Michael Welker (b. 1947) state: '... we are not reintroducing the old assumption that science deals with facts and truth while theology handles meaning and value. Our claim is more subtle and demanding: science and theology are both concerned with realities (facts and meanings; truth and value) attentive to the connection between under-standing and what is presented to be understood.'[31]

Yet in this as well as in the view of an adjustable Bible, conceptions of God and religion have for the most part taken the place of scripture references, the latter being relegated to a back seat position. Moses and Genesis have been substituted by references to great theologians and their hermeneutic inventions – e.g., by references to Karl Barth and *Die kirchliche Dogmatik*.

11.7 The Anti-scientific Bible

Gould's NOMA olive branch held out to religion is, in the view of some of his scientific colleagues, just a fig leaf to cover the shame of biblical inconsequenti-ality. Whereas the other discourses were and are conducted by theologians as well as scientists, the anti-scientific Bible discourse has been dominated by philosophers and scientists. Already some of Lilienthal's eighteenth-century deist and atheist adversaries wrote with contempt about the Bible. Of later authors writing disparagingly about scripture, a few examples may suffice. Carl Vogt (1817–1895), known for his materialist philosophy and participation in the Revolution of 1848, systematically denounced and ridiculed the Mosaic hexaemeron, the story of Noah's Ark and especially, too, the unity of human-kind in Adam and Eve.[32]

---

[31] Polkinghorne, John and Michael Welker (eds.), *The end of the world and the ends of God. Science and theology on eschatology*, Harrisburg, PA: Trinity Press International, 2000, 5.

[32] Vogt, Carl, *Köhlerglauben und Wissenschaft: eine Streitschrift gegen Hofrath Rudolph Wagner in Göttingen*, Giessen: Ricker, 1855.

An example from Britain was Joseph McCabe (1867–1955) who, having left the Catholic priesthood, became a freethinker and atheist who not only rejected the literalist and the harmonist interpretations of the Bible but higher criticism as well. In one of his many publication, *The bankruptcy of religion* (1917), he argued that the approach to the Bible by Eichhorn and his followers was merely a ploy by liberal theologians to save what could be saved from the sinking ship of scripture. Higher criticism had been a strategy to preserve religion and theology against the unstoppable march of truth, giving Christianity an opportunity to hold on to the Bible by dealing with it in a more rational, acceptable manner. The Old Testament and its early books were child-like and demonstrably wrong, with 'numerous palpable blunders and inconsistencies'. The churches and the clergy 'imposed upon ignorant Europe a colossal delusion' of the Bible as a supernatural document. Moses not being the inspired author of the Pentateuch, the position of Christ is directly affected, who is reduced to 'a human and fallible person'.[33]

Through the twentieth century, also among these authors a move took place, away from specific references to scripture to general ones about God and religion. The Cambridge philosopher, mathematician and political pacifist Bertrand Russell (1872–1970), in his *Science and religion*, echoed views similar to those who had engaged in the anti-scientific Bible stance, but he rarely referred to the Bible specifically. More recently, in the wake of the resurgence of literalism, a number of Darwinian atheists such as Richard Dawkins (b. 1941) and Daniel Dennett (born 1942) have returned to attacks on the Bible. To Dawkins, for example, the Bible has no valid truth claims, not even in matters of morality. Parts of the Bible 'are odious by any reasonable standards'.[34] Today, biblical morality must strike any civilized person as 'obnoxious'. With respect to ethics, scripture shows the birthmarks of crude and cruel ages. The Bible is to ethics what folk medicine is to scientific medicine: it may contain valuable elements but needs testing by science. Against Gould and his NOMA notion Dawkins and his atheist comrades-in-arms argue that scripture and science are not unrelated magisteria but significantly overlap in the sense that religious faith and its sacred texts such as the Bible have primitive evolutionary origins and are subject to scientific deconstruction.[35]

'To be fair,' Dawkins conciliates, 'much of the Bible is not systematically evil but just plain weird, as you would expect of a chaotically cobbled-together anthology of disjointed documents, composed, revised, translated, distorted and "improved" by hundreds of anonymous authors, editors, copyists, unknown to us and mostly unknown to each other, spanning nine centuries'.[36]

---

[33] McCabe, Joseph, *The bankruptcy of religion*, London: Watts & Co., 1917, 137, 141–142, 145.

[34] Dawkins, Richard, *The God delusion*, London [etc.]: Bantam Press, 2006, 57.

[35] It perhaps should be added that although the anti-scientific Bible writers for the most part were and are atheists, the reverse does not apply: not all atheists are aggressively anti-Bible.

[36] Dawkins, *God delusion*, 237.

If we want to retain the Bible – and Dawkins placates we should – it can be given a place in our literary education: so many expressions in novels and plays would not be understood if we lost our knowledge of biblical stories.[37]

11.8 Conflict Between Discourses

Ever since James Moore's *The post-Darwinian controversies* (1979), the old conflict thesis of Bible and science has been dying a slow but certain death. As pointed out above, among historians of the subject the complexity thesis has taken over, allowing for a multiplicity of relationships between scriptural hermeneutics and the scientific study of the physical world, most of these friendly. The trend of scholarship in recent decades has been away from an epic warfare narrative towards stories of interactions in different places, under different circumstances and, for the most part, to beneficial effect. 'Conflict' and 'warfare' have been declared offside. This has created a problem, however, namely what to do with the historical evidence for private agony, public fights and even warfare that so abundantly does occur in the bibliographical record of the relationship Bible and science. Fights, also fierce ones, frequently have taken place and continue to take place. Where then precisely is conflict located if not between scripture and science?

Some have suggested that tensions between Bible and science have a professional dimension.[38] To a certain extent, 'Bible' and 'science' stand as proxies for 'theologians' and 'scientists,' and the move away from the Bible as a standard of truth has involved a change from theologians to scientists as the experts to whom many of us turn for distinguishing fact from fiction. Professional rivalry can indeed be detected in historical clashes over Bible and science. Yet the Bible-science discourses, for the most part, do not translate into professional theology-science camps. After all, many of the leading figures wanting to retain the authority of scripture have themselves been scientists – the geologists Buckland, Dana or Teilhard de Chardin, the physicists Jaki and Polkinghorne, the hydraulic engineer Morris, and many more. The Gillispie thesis (see above) can be extended: 'religion within science' has taken the form of some five separate discourses and the major, first-order controversies have been located in the interstices between them, each supported by experts from a variety of disciplines. The biblical literalist Paine did not in first instance argue for theology and against science – he himself was a scientist – but took up the cudgels against those who propounded an adjustable Bible such as his fellow scientist Buckland whose harmony schemes he excoriated as 'impudent professions of corroborating Holy Writ' that 'opened the door for a wide-spread

---

[37] Dawkins, *God delusion*, 340–343.

[38] Turner, Frank M., 'The Victorian conflict between science and religion: A professional dimension', *Isis* 69 (1978), 356–376.

infidelity'.[39] The discourses have clashed because they translate into questions of social and political agendas, of life styles, values, and vested interests. Thus far, the study of their sociopolitical locations has been thoroughly carried out only for the literalist one. A desideratum of further research would be the societal anchoring grounds of the others as well.

Acknowledgements I am grateful to Denis Alexander, David Livingstone, Jim Moore and John Riches for their constructive and generous criticism.

---

[39] Paine, Martyn, *A review of theoretical geology* (from the Protestant Episcopal Quarterly Review, 1856), 11.

Chapter 12
"Natural Theology of Industry" in Seventeenth-Century China?: Ideas About the Role of Heaven in Production Techniques in Song Yingxing's *Heaven's Work in Opening Things* (*Tiangong kaiwu*)

Yung Sik Kim

12.1 Introduction

I have chosen the title of my essay intentionally to imitate that of Charles Gillispie's classic paper, "The Natural History of Industry."[1] From that short paper, I, like many other historians of science in my generation, learned to think critically about the relation of science and technology, and about the real nature of written materials on techniques and trades, those in the famous *Encyclopédie* for example. Much of what Gillispie said in showing "natural historical" aspects of many French technical writings of eighteenth century applies to the subject of the present essay, the *Tiangong kaiwu* 天工開物 (Heaven's Work in Opening Things, published in 1637), a famous seventeenth-century Chinese book of production techniques, written by Song Yingxing 宋應星 (1587–1666, ca.), whom Joseph Needham has called "Diderot of China."[2] In this essay, I will reexamine Song Yingxing's attitude to the production techniques and his motivation for writing the book, and call attention to another aspect of the book that has not been noted so far: "natural theology of industry."

Many modern scholars praised the *Tiangong kaiwu* for its discussion of various production techniques of agriculture, mining, and other traditional Chinese industry. Yabuuchi Kiyoshi 藪内清, for example, has pointed out that the book is important as it covers all the important industries, and is

---

[1] Gillispie (1957).

[2] The two major studies on the *Tiangong kaiwu* are: Yabuuchi (1954a); Pan Jixing (1989). In addition, Pan Jixing has written an extensive biography of Song Yingxing: Pan (1990). There have been translations of the *Tiangong kaiwu* into English, modern Chinese, Japanese, and Korean: Sun and Sun (1966); Zhong Guangyan (1978); Yabuuchi (1969); Ch'oe (1997). For various editions and reprints of the *Tiangong kaiwu*, see Pan (1989, pp. 131–171). Needham's reference to Song Yingxing appears in Needham (1954, p. 13).

Y.S. Kim (✉)
Department of Asian History and Program in History and Philosophy of Science,
Seoul National University, Seoul, South Korea
e-mail: kysik@snu.ac.kr

J.Z. Buchwald (ed.), *A Master of Science History*, Archimedes 30,
DOI 10.1007/978-94-007-2627-7_12, © Springer Science+Business Media B.V. 2012

detailed in discussing the production processes of each industry.[3] The most enthusiastic praise has come from Pan Jixing 潘吉星, who has worked on the book most extensively. He referred to the book as a "systematic", "conclusive" treatise of the Chinese agriculture and production techniques up to the Ming times (1368–1643).[4] There have also been comments on the background and the motivation of Song Yingxing's writing of this specialized book devoted to production techniques. Most scholars have pointed out the general intellectual climate of late Ming, which showed a great interest in practical utility.[5] It has also been noted that the publication of the *Tiangong kaiwu* was in line with the widespread interest in late Ming in miscellaneous and strange things that resulted in publication of many miscellaneous books and encyclopedias.[6]

The meaning of the expression "*tiangong*" 天工 (heaven's work) in the title, and thus the role of heaven in the production techniques, has also received some attention. Pan Jixing was of the view that this expression referred to both the natural world and man's work: " '*Tian*' (heaven) of the '*tiangong*' refers to the natural world; '*gong*' (work) refers to man's skill or technique."[7] He then interpreted the whole title, "*tiangong kaiwu*," as referring to cooperation of "heaven's work and man's work in developing the myriad things."[8] Of the roles of heaven and man, however, Pan Jixing put the emphasis on man's role, and suggested a translation of the title: "Exploitation of Products by Artificial Skill from the Nature."[9] Adopted by Joseph Needham in a modified form as "the exploitation of the works of nature," this interpretation has become standard among many scholars.[10] Given this kind of emphasis on man's role, not much

---

[3] Yabuuchi (1954b, p. 2).

[4] Pan (1989, pp. 21, 92). Pan Jixing went as far as saying that at the time there was no book on technology even in the West that could be compared to the *Tiangong kaiwu* in depth and breadth of coverage: Pan (1989, pp. 98–100). He also saw some characteristics of modern science in the book, using such expressions as "experimental science," "mathematization," "critical spirit," and "enlightenment thought": Pan (1989, pp. 76–78).

[5] Yabuuchi (1954b, pp. 15–16); Pan (1989, p. 16). Peter Golas has also pointed out such aspects as "stress on concrete accomplishments whether as an official or in a private capacity," "willingness to accept the legitimacy of profit-taking by merchants," and "interest in scientific and technological learning that had practical application," as well as more immediate motivations for "reputation" and "income": Golas (2007, p. 574).

[6] E.g., Pan (1989, pp. 16–19). Elman (2005, chap. 1). When such interest died out later among the literati, the book went out of sight in China—to appear in Japan.

[7] Pan (1989, p. 72). In this Pan Jixing was following the view of Zheng Wenjiang 鄭文江 (1888–1936), one of the pioneers of the modern studies of the *Tiangong kaiwu*. Dagmar Schäfer's translation, "heaven, work, and the inception of things," is essentially along the same line. Schäfer (2005, pp. 35–60).

[8] Pan (1989, p. 69).

[9] Pan (1989, p. 73).

[10] Peter Golas, for example, has sided with this line of interpretation by emphasizing the importance of "the idea that technology involved human beings employing their skills to make useful the resources provided by nature." See Golas (2007, p. 585).

attention has been given to Song Yingxing's ideas about the role of heaven in the production techniques he discussed in the *Tiangong kaiwu*.[11]

The purpose of this chapter is to throw more light on what Song Yingxing meant with the expression "heaven's work" (*tiangong*) by examining what he thought about the role of heaven in man's acquisition and use of the techniques. His basic idea was that while man plays various roles in producing many things, it is really heaven that lies behind all this. In particular, I will show that heaven's role of taking care of man appears prominently in his discussion of production techniques in the *Tiangong kaiwu*. I believe that such an understanding of heaven's role makes a better sense for the title of the book, and is also in line with the original use of the expression "*tiangong*" in the *Book of Documents* (*Shujing* 書經): "It is the heaven's work; man substitutes it." (*Tiangong, ren qi daizhi* 天工, 人其代之). Examination of Song Yingxing's ideas about "heaven's work" will also shed light on his general ideas and attitudes about production techniques: the nature and the role of production techniques; the place of knowledge about them in the corpus of man's—and scholars'—knowledge; the relation of production techniques and knowledge about them with other human activities and knowledge on the one hand, and the natural world and phenomena in it on the other.

12.2   Content of the Book and Song Yingxing's Motivation

Before looking at Song Yingxing's ideas on the role of heaven, I will reexamine the content of the *Tiangong kaiwu* to find more clues to various aspects of his attitude to the production techniques, and to his motivation for writing the book, which, in turn, will provide a context for my discussion on his views about heaven's role in production techniques. I will begin the reexamination with his preface, a natural place where one would expect to find the author's thoughts about the book and its subject matter.

12.2.1   The Preface: What Song Yingxing Did and Did Not Intend to Do

Song Yingxing's preface to the *Tiangong kaiwu*, a work on which he spent so much effort, is rather brief and casual. He does not seem to have been interested in using the preface to lay out systematically his thoughts about the subjects discussed in the book, or about his own discussions of them. In fact, he did not even explain what he meant by his title, the meaning of which was hardly transparent.

---

[11] Craig Clunas has translated the title as "Heaven's Craft in the Creation of Things", but has not elaborated on it: Clunas (2004, p. 166).

On the other hand, Song Yingxing's remarks about the situation surrounding various commodities like foods and clothes reveal his motivation for writing the book. For example, he pointed out that people were interested only in things that were rare while not knowing well about the things that were common. He then noted that in his time it had become easy for people to see many rare things that they could not see in the past. These remarks reveal an intention on his part to deal with the common things that were frequently seen. Song Yingxing's main interest, however, was not in showing the things themselves, but in showing the processes and instruments used in producing them. He said:

> The royal princes are born and raised deep inside the palace. [Smelling] the good fragrance of the best rice in the royal kitchens, they may wish to look at the plows. [Looking at] the court ladies cutting brocade clothes, they may imagine the looms and the silk. At such times, [if they can] unroll a picture [book] and look at it, it would be like acquiring a heavy jewel.

This echoed the recurring motivation of the Chinese technical books to show the elite readers how the everyday commodities providing them with various comforts had been produced.[12]

In the preface Song Yingxing also made remarks about what he did not intend to do and about the topics he did not discuss. For example, toward the end of the preface comes his famous remark that "scholars with great ambitions" should throw away the book because the book has nothing to do with "fame and advancement." This remark indicates his view that the topics he had chosen for the book fell outside the curriculum for civil service examination, the primary concern of many scholars. Song Yingxing also said that originally he had written chapters on astronomy and music, but eliminated them before printing the book. This shows that originally he did not intend to restrict his topics to production techniques; perhaps the book he had in mind was not a book on techniques only but one covering all concrete subjects (including astronomy and music) lying outside the learning for civil service examination.—full of abstract, abstruse philosophical discourses.[13] In fact, the reason provided by Song Yingxing himself for his decision to exclude the two topics— that "their ways are too exquisite and are not my affairs"—shows his general distaste for abstract and abstruse topics and his inclination to concrete subjects.[14] The preface also shows Song Yingxing's practical bent. His saying that the order of the topics, the "five grains" coming first and the "gold and jade" the

---

[12] For example, Francesca Bray has spoken of "a general interest among the leisured class of the period in how everyday objects were made." Bray et al. (2007, p. 30).

[13] Dagmar Schäfer has seen "a very Confucian idea of presenting an all-encompassing worldview" in Song Yingxing's original plan. See Schäfer (2005, p. 53).

[14] Of course, it is also possible that he did not include them because there had already been many writings on them. Pan Jixing has suggested that the contents of Song Yingxing's other essays, "Tan tian" 談天 (Discourse of Heaven) and "Lun qi" 論氣 (Discussions on *Qi*), correspond to those of these two chapters: Pan (1990, pp. 266, 273).

last reflects his thought about the importance of them: i.e. valuing basic commodities like grains and slighting luxurious things.[15]

Thus, on the whole, this book on production techniques seems to have been born of its author's inclination to the everyday, practical, and productive topics, and of his distaste of the learning geared to civil service examination, which he had failed repeatedly.

12.2.2 Practical Interest of Entrepreneurs

Song Yingxing's selection of the topics in the main part of the *Tiangong kaiwu* was mainly based on the practical utility of the techniques and the products. What he wrote about these topics, however, was usually not the kind that could be used by the actual practitioners of the techniques; rather, his discussions were more frequently about what the entrepreneurs, or owners, of the techniques would be interested in.[16]

Thus, Song Yingxing frequently mentioned the prices and costs of source materials, products, and tools used for producing them. But more typically, he mentioned the ratios of the quantities of various materials involved in the production processes. For example, he mentioned the ratios of different raw materials required in a given process: the ratio of saltpeter to sulphur in making gunpowder is 9 to 1 (III.15.31b)[17]; that of fat and wax to copper in making bells is 1 to 10 (II.8.18b-19a). He spoke of the quantity of products obtained from a given quantity of source material: the weights of oil obtained from a given quantity of different kinds of seeds (II.12.64a-64b); the weights of flour obtained from a given quantity of different kinds of wheat (I.4.56a). He

---

[15] Yabuuchi Kiyoshi saw an emphasis on the importance of agriculture in this: Yabuuchi (1954b, p. 5). Indeed, a large part of the content of the *Tiangong kaiwu* is devoted to agricultural topics: not only the topics like grains and clothes, but chapters on dyeing, sugar, etc. contain discussion of agricultural problems.

[16] The printers of the second edition (published between 1650s and 1680s) seem to have seen this aspect of the book, as can be seen in the content of the advertising sentences on the title page. See Schäfer (2005, p. 42). Peter Golas has given an excellent, if somewhat too modernized, list of the items described in the *Tiangong kaiwu*.

> The topics Song regularly discussed about any given technology included raw materials, areas of production, the technical processes involved, essential operations, equipment used, consumption of raw and processed materials, consumption of energy, products and rate of production, special characteristics of the products and their uses. (Golas, 2007, p. 585.)

[17] Song Yingxing divided the *Tiangong kaiwu* into three parts and 18 chapters, using the same character *juan* 卷 to refer to both the "parts" and the "chapters." In my citation, I put the numbers for the parts (I, II, III) and the chapters (1–18), followed by the page numbers of the original edition (1637), reprinted as volume 1115 of the *Xuxiu Siku quanshu* 續修四庫全書 series (Shanghai: Guji chubanshe 古籍出版社, 1995–1999). "III.15.31b", for example, refers to Part III, Chapter 15, p. 31b.

mentioned different quantities of a product that could be produced by using different tools or methods and from different source materials: the quantity of wheat flour produced by different grinding mills (I.4.56b). He noted the extent to which the quantity of raw materials is reduced after undergoing a given process: three tenths of the wrought iron wasted during firing and hammering (II.10.44b); three tenths of silk yarns lost after boiling, and two tenths if the yarns are good (I.2.40b). He also provided information on the number of men required to carry out a particular work, or number of days required to do a given work—using a particular machine, cows, or human power alone, etc: the ploughing work done by two men is equivalent to the work of one ox, and the work done by rotary harrow pulled by two men is equivalent to the work of three oxen (I.1.4a). Such considerations led him to comment upon the economic aspects of the production techniques: the skin of one sable is less than one square foot, and thus more than sixty sables would be needed for making one fur garment (I.2.45b); a household with 100 sheep will make 100 taels, because each sheep yields wool for three pairs of socks, and each pair produces two lambs per year (I.2.47a); although the price of a good-grade flour is two-tenths higher, it is more economical to use it if the byproducts of gluten and starch are taken into consideration (I.4.56b).[18]

12.2.3 *"Natural History of Industry"?: "Investigation of Things" (gewu) and "Test and Verification" (shiyan)*

Song Yingxing was usually very brief about the actual production methods. Instead, he frequently discussed the origins, histories, and geographical distributions of the production techniques and the products. He also spoke of various differences in the products: kinds (species), parts, sizes, names, methods of production, as well as uses and damages.

In the course of such discussions Song Yingxing sometimes revealed his general views about production techniques. For example, he was interested in the progress and disappearance of the techniques. He pointed out that the firearms techniques progress continuously (III.15.25a): the gunpowder-making technique in current use would no longer be used after ten years (III.15.35a). He noted that the method of producing "Japanese silk" (*woduan* 倭緞) was likely to be lost because it was not used any more (I.2.41a). He seems to have been of the view that knowledge of techniques cannot be kept secret. Discussing the methods of obtaining the grains, he said: "How can its way (*dao* 道) be kept secret forever?" (I.4.53a). These examples reflect not so much entrepreneurs' interest as scholars' intellectual curiosity about the techniques of producing

---

[18] Contents like these made Pan Jixing state that Song Yingxing was dealing with the "quantitative relations." See Pan (1989, p. 77).

various things. Indeed, what we find in the *Tiangong kaiwu* are usually not instructions to do what and how, but records of observations, or of descriptions found in books, on what were done and how they were done.[19] One can discern here an aspect of what Charles C. Gillispie called "the natural history of industry".[20]

It is then possible that Song Yingxing thought of the work he had done to compose the book as work of "investigation of things" (*gewu* 格物), the basis of the Confucian intellectual and moral endeavor.[21] In fact, in the preface, he actually used the term in expressing his respect for his friend and sponsor, Tu Shaokui 涂紹煃 (1582–1645) whose "sincere intention moved heaven, and [who with] his numinous mind 'investigated things'." About people who mistake clam shell powder for oyster lime, he said that it was "because they did not investigate things." (2.11.54b) Also, he frequently used the expression, "principle of things" (*wuli* 物理), thus invoking the word "principle" (*li* 理), the object of the endeavor of "investigation of things," the ultimate aim of which was to attain the "principle of heaven" (*tianli* 天理). For example, speaking about the Sichuan people feeding the Zhejiang-specie silkworms with the leaves of a local tree when mulberry leaves are scanty, he said: "'The principle of things' are one." (I.2.27a) Speaking of the difficult process of using the fragmented jade pieces to fill the cracks in the lutes, he said that it was "nearly impossible to understand 'the principle of things' controlling and obeying each other." (III.18.62a)

Sometimes Song Yingxing gave an impression that what he reported was the result of the examination of the actual procedures.[22] For example, Song Yingxing's saying that he "could not examine in detail" the processes of making the so-called "dragon robes" (*longpao* 龍袍) (II.2.41a) may be taken as implying that he must have examined the detailed processes of the other techniques on which he did not express such a regret. He said a similar thing about the "seventy-two processes necessary for making a porcelain cup": "Its minute details could not be fully exhausted yet." (II.7.14a) Sometimes he seemed to speak of actual observations. For example, after recording various things about the gunpowder, he said: "All of these can be discussed in detail after they have been tested by observation." (III.15.32a)

---

[19] Dagmar Schäfer has noted that Song Yingxing's stance on craft was based on theoretical reflections, not on practical experiences: Schäfer (2005, p. 55).

[20] Gillispie (1957). In fact, in the *Tiangong kaiwu* we frequently find the expression, "*bowu*" 博物 (literally, "broad [study of] things"), which is used in modern Chinese in translating "natural history." For example, in the preface Song Yingxing spoke of "the intelligent '*bowuzhe* 博物者' ('those who broadly [study] things')." He ended Chapter 1 on the grains by saying: "How can a '*bowuzhe* 博物者' neglect them?" (I.1.22a). See also III.16.46b.

[21] On the idea of "investigation of things" (*gewu*), see Kim (2000, chap. 2).

[22] Pan Jixing has noted that some items in the book are based on the examination of the actual practice. See Pan (1989, p. 20).

Song Yingxing even used the expression, "test and verification" (*shiyan* 試 驗). For example, after mentioning different kinds of vegetable oils and different yields of oil per a given quantity of different vegetables, he said that he did not "investigate, test and verify" (*qiongjiu shiyan* 窮究試驗) all of them; for some he did test but did not know completely (II.12.64b). Concerning many writings on gunpowder and firearms, he said that they "have not necessarily gone through 'test and verification.'" (III.15.31b) Again, these assertions seem to suggest that he did perform "test and verification" on the other techniques on which he did not add such comments.

It is difficult to pin down exactly what Song Yingxing meant by the word "test and verification" (*shiyan*). It is not likely, however, that he had "experiment" in mind.[23] Perhaps what he had in mind may be something similar to the idea of "evidential study" (*kaozheng* 考證).[24] Indeed, in his preface, he expressed a regret that he could not verify, by means of books and actual objects, what he wrote from his own observations and hearsays, and that he could not discuss them with other scholars, implying his belief that in order to obtain proper knowledge of the techniques he should have done so. What Song Yingxing said he should have done is exactly the kind of things that were done by the scholars of the evidential studies.[25]

12.2.4 Miscellaneous Casual Knowledge and Critical Spirit

The pages of the *Tiangong kaiwu* also contain a good deal of casual comments, observations, and records of readings and hearsays on all sorts of things, often without any technical significance. In fact, the passages discussed so far in this section are scattered in the book in a complete mixture with this kind of casual miscellanies. To give some examples: Song Yingxing mentioned bans on the sales of saltpetre and sulphur (III.15.32a) and on the private use of the special tiles for palaces or temples (II.7.2b), and noted that the laws had become severe in order to prevent thefts of and fights over silver (III.14.4b). He spoke of merchants coming from Sichuan to sell silk fabric and buy pepper to take back (I.2.41a). He said that wine-makers must choose the yeast sellers who are reliable and have fame (III.17.49a). He mentioned the pearl gatherers of an area offering sacrificial services to a "sea deity" (*haishen* 海神) every third lunar month (III.18.54a), and said that when pearl gatherers come out from water,

---

[23] On the other hand, Christopher Cullen has mentioned the possibility that Song Yingxing actually made measurements: Cullen (1990, p. 312).

[24] Francesca Bray and Georges Métailié have noted that the method of Xu Guangqi 徐光啓 (1562–1633) in compiling his agricultural treatise, the *Nongzheng quanshu* 農政全書, was that of the evidential studies, rather than scientific experimentation: Bray and Métailié (2001, pp. 342–343).

[25] On the scholars of "evidential studies," see Elman (1984).

they must be wrapped by warm blanket, for otherwise they may die of cold (III.18.54b). Speaking of the grinding mills turned by oxen, he said that blinders should be placed on the eyes of the oxen to prevent dizziness, and wooden pots should be suspended under their bellies to collect excretions (I.4.56a). While discussing the carts, he even said that the drivers should stop the carts when they come across pedestrians (II.9.40b).

Among such casual remarks of Song Yingxing are stories of strange things and events that are difficult to believe today. For example, he noted that while nothing happens when people kill one or two bees, an entire group of bees swarm to attack people when more than three bees are killed, calling it the "bee rebellion" (*fengfan* 蜂反), or that when human urine is mixed with the stuff emitted by the bees, the fragrance and sweetness is increased (I.6.79b). He spoke of the story of exquisite red porcelains being produced after a potter jumped into a burning kiln; the story was later enlarged as the kiln producing objects like deer and elephants (II.7.14a). While discussing the damages to rice plants, he spoke of the "ghost fire" (*guihuo* 鬼火) burning rice plants, and provided a long explanation for it (I.1.6a). He also spoke of legendary swords turning into dragons (II.10.44a).

Sometimes Song Yingxing expressed doubts about such reports of strange things and beliefs. For example, about stories of "mercury sea" (*honghai* 澒海) and of mercury produced from certain grass, he said that they were "unfounded and absurd, which gullible people believe." (III.16.42a). He was especially critical of the "recipe books" (*fangshu* 方書), and the "recipe masters" (*fangshi* 方士). For example, he said that some of the things the recipe books spoke about the tin and the arsenic as "absurd words." (III.14.20a) In an appendix on the so-called "cinnabar silver" (*zhusha yin* 朱砂銀), he spoke of the "dishonest recipe masters deceiving people by [the alchemical techniques of] 'furnace and fire,' " and said that he added the appendix because foolish people were greedy and ignorant, and thus were easily deceived by those recipe masters (III.14.7a). His desire to prevent people from being deceived by stories about strange things and events seems to have been part of his motivation to write the *Tiangong kaiwu*.

Song Yingxing frequently criticized the contents of the *materia medica* (*bencao* 本草). For example, after mentioning the methods of obtaining silver, he said that there were no other methods of making silver, and found it "extremely objectionable that the recipe books and the books of *materia medica* contain unfounded conjectures and notes." (III.14.6a) It is possible that the target of Song Yingxing's criticism was the *Systematic Materia Medica* (*Bencao gangmu* 本草綱目) of Li Shizhen 李時珍 (1518–1593), which was widely circulated among scholars of the time.[26] In fact, the *Tiangong kaiwu* made quotations from the *Systematic Materia Medica* in more than 20 places, although Song Yingxing did not explicitly mention the title of the book he was mostly

---

[26] Yabuuchi (1954b, p. 14); Pan (1989, p. 34).

criticizing.[27] It may have been that Song Yingxing wanted to present his own book as containing more accurate and reliable information than this famous book which drew on numerous reports about strange things and events he could not accept.[28]

12.3 Role of Heaven in Production Techniques, and Elevation of the Status of Knowledge about the Techniques

A role of heaven in the production techniques is implied in the rhetorical question at the beginning of the preface of the *Tiangong kaiwu*: "How can [all the numerous things in the world] be [due solely to] man's power?" Song Yingxing made more explicit remarks about heaven's role in the main part of the book. Such remarks can be found scattered throughout various places in the book, but mostly in the introductory paragraphs of chapters, which begin with the phrase, "Master Song said" (Songzi *yue* 宋子曰). For example, in the introductory paragraph of Chapter 3 on the dyes he said "Heaven hung down images; the sages followed them and manifested the five colors [of heaven] into the five colors [of man]." (I.3.49a) The introductory paragraph of Chapter 4 on the preparation of grains begins and ends with remarks suggesting a role of heaven: "Heaven produces the five grains to nourish people. ... Is it possible that the one who made these is not heaven [disguised] in man's shape?" (I.4.53a). Indeed, Song Yingxing's main motivation may have been to show that heaven's role, or "heaven's work" (*tiangong*), is manifest not only in such abstract ideas as human nature (*xing* 性), principle (*li* 理), etc., but in the concrete objects and phenomena also. His remark in the preface that he had originally intended to include the subjects of astronomy and music can be understood along this line as well.

At times Song Yingxing used different expressions in referring to heaven's role. For example, he used the expression, "Creator of things" (*zaowu* 造物) instead of "heaven" (*tian*). After pointing out that "although silk, hemp, fur, and wool all have plain uncolored qualities, [one can] make them precious by dyeing them with different colors," he said: "As for those who say that 'Creator of things' did not care, I do not believe them." (I.3.49a) After noting that silver is not produced within 300 *li* 里 (one *li* being about 400 meters) of a place producing gold, and *vice versa*, he said: "The 'emotion' (*qing* 情) of 'Creator of things' can be seen abundantly [in this] also." (III.14.6a) Sometimes he used the more traditional expression "Creative transformation" (*zaohua* 造化)[29] instead of "Creator of things." While speaking about coals, he said: "Between metal and

---

[27] Pan (1989, pp. 80, 103, 186).

[28] Nappi (2009, chaps. 3–6).

[29] The expression, *zaohua*, usually referred to the subtle, marvelous, mysterious aspect of some natural phenomena. But sometimes the word had a meaning quite close to that of "Creator." See Kim (2000, p. 101).

'earth and stone,' 'Creative transformation' additionally made this specie [of coal] appear." (II.11.55b). After saying that mercury cannot be recovered once it has been converted into vermilion, he said: "The skill of the so-called 'Creative transformation' has already been exhausted." (III.16.42a) After mentioning graphite and saltpetre producing a bright light, he added that this was placed at the end of the book on the heaven's work because "this is what 'Creative transformation' of the *qian-kun* 乾坤, [i.e. hexagrams corresponding to heaven and earth,] conveniently revealed on the earth's surface." (III.18.66a-66b)

Song Yingxing also brought in the expression, "the divine" (*shen*). For example, the introductory paragraph of Chapter 1, in which he spoke of "Shen Nong" 神農 ("The Divine Agriculturalist"), ends with the following remark: "The character '*shen*' is put before the character '*nong*' (agriculture) and is connected to it. How can [agriculture] be the act of man's power alone?" (I.1.1b). In the introductory paragraph of Chapter 9, he said that one who made the cart for the first time should be called a "divine man" (*shenren* 神人) (II.9.29a). After explaining the explosive effect of mixing sulfur and saltpetre in terms of the interaction of the essences of the yin and yang, he said: "This is a 'divine thing' (*shenwu* 神物) that the *qian-kun* [i.e., heaven and earth] worked out fantastically." (II.11.60b) He used the expression, "divine transformation" (*shenhua* 神化), in referring to the disappearance of the coal after burning (II.11.55b).

One thing Song Yingxing aimed at by showing heaven's role in the production techniques seems to have been an elevation of the status of knowledge about them.[30] Such intention can be seen also in his frequent citations of classical passages mentioning the things related with the techniques discussed in the *Tiangong kaiwu*. For example, in Chapter 1 on grains, Song Yingxing spoke of the names of grains appearing in the *Book of Poetry* (*Shijing* 詩經) and the *Book of Documents* (I.1.18b, 19a, etc.). Other ancient texts cited in the *Tiangong kaiwu* include the *Book of Changes* (*Yijing* 易經), the *Rites of Zhou* (*Zhouli* 周禮), the *Record of Rites* (*Liji* 禮記), the *Analects* (*Lunyu* 論語), the *Mencius* (*Mengzi* 孟子), the *Laozi* 老子, the *Zhuangzi* 莊子, the *Xunzi* 荀子, the *Hanfeizi* 韓非子, the *Liezi* 列子, the *Record of the Grand Historian* (*Shiji* 史記), the *Han History* (*Hanshu* 漢書), and the *Classic of Mountains and Seas* (*Shanhaijing* 山海經).[31] In fact, many of the chapter titles themselves came from classical passages. Song Yingxing also mentioned the commentaries of Zhu Xi (II.9.40a, II.10.46a, etc.). By making connections with the classics and with Zhu Xi in this manner, Song Yingxing was obviously trying to enhance the intellectual status of the knowledge about the techniques he was discussing.

---

[30] This was what the fifteenth- and sixteenth-century European writers on technical arts did. See, e.g., Rossi (1970, Chap. 1.) Song Yingxing's contemporary Wang Zheng 王徵 (1571–1644) also showed a similar tendency in the *Yuanxi qiqi tushuo luzui* 遠西奇器圖說錄最 (Selected Records of the Diagrams and Explanations of the Strange Machines from the Far West). See Kim (2010).

[31] Pan Jixing has identified many such citations: Pan (1989, pp. 175–183).

12.4 Heaven and Man: Production Techniques and Natural Phenomena

The actual patterns in which heaven's role is manifest in Song Yingxing's discussion of production techniques reflect his views about the relations between the techniques, natural phenomena, heaven, and man.

12.4.1 Man's Role

Song Yingxing's basic position was that heaven prepares the techniques and transmits them to man; man uses them. For example, speaking of the weaving technique in Chapter 2 he noted that "the looms of the Heaven's Grandchild (*Tiansun* 天孫)[32] transmitted the skill [of weaving] to man. Man saw figured patterns from the original materials, and then embroidered and washed them to get the silk." (I.2.23a) Later in the chapter, he said: "With the looms of the Heaven's Grandchild, the skill of man has become complete." (I.2.39b)

Of course, Song Yingxing did not deny the role of man in production techniques. At times he saw man's role even in preparing the techniques. When heaven does not perform the work necessary for producing things, man does them. For example, when heaven does not rain, man has to draw water to irrigate (I.1.8b). Also, things are not produced by themselves; man has to perform some tasks to get them.[33] For example, Song Yingxing spoke of the mutual dependence of man and crops: "Men cannot live long [by themselves]; the five grains sustain them. The five grains cannot grow themselves; men grow them." (I.1.1a) He also noted that oil inside plants does not flow out by itself, but man has to do something—using the power of water, fire, or tools made of wood and stone—to cause it to flow out (II.12.63a).

In Song Yingxing's view, man's role can be seen mainly in using the techniques. In particular, it is manifest in man's selecting different methods according to different conditions. For example, man chooses different quantities of raw materials. Speaking about the so-called "ten-thousand men killer" (*wanrendi* 萬人敵) bomb, he said that the proportions of mixing "poison gunpowder" (*duhuo* 毒火) and "divine gunpowder" (*shenhuo* 神火) are varied flexibly by man (III.15.35a). In making inkstones, the quantities of precious raw materials to be used are decided by man (III.16.46b). Man chooses the timing of various processes in the production of things. For example, in southern China where

---

[32] Here Song Yingxing is referring to Zhinü 織女, a mythological female figure associated with the weaving, recorded as "Heaven's Grandchild" in the astronomical treatise (*Tian'guan shu* 天官書) of the *Record of Grand Historian* (*Shiji*).

[33] This aspect must have been what made Pan Jixing conclude that the notion that man cannot depend completely on the natural world, but has to do something himself, was "the leading idea" of the *Tiangong kaiwu*: Pan (1989, p. 70). This idea can be traced back to the above mentioned phrase from the *Book of Documents*, "It is the heaven's work; man substitutes it."

there is no frost the time of the harvest of sugarcane is decided by man (I.6.76a). Man also prevents damages that occur while using the techniques. In raising silkworms, for example, the task of preventing the damages of dampness, heat and pressure lies with man (I.2.27b). Also, it is man who chooses among the different methods of preventing birds, rats, and mosquitoes, which spoil silkworms (I.2.29b).

In referring to such roles of man in the use of techniques, Song Yingxing frequently used the expression, "man's work" (*rengong* 人工), in addition to the expression, "man's skill" (*renqiao* 人巧). For example, in speaking of the techniques of producing different ceramic wares suitable for the soils of different localities, he said: "Man's work exhibited differences, and produced good ceramic wares." (II.7.1a) In speaking of the making of tiles in various shapes of birds and animals, he said: "All these are made one by one by man's work." (II.7.2a) And after speaking of a complicated, ingenious method of extracting pure silver from impure one, he said: "Bits of both man's work and heaven's work are seen." (III.14.6b)

12.4.2 Man, the Sages, and Heaven

In the last example, the expression "heaven's work" appears together with "man's work." In saying this, Song Yingxing was obviously thinking of some sort of comparison of man's work with heaven's work. Indeed, he frequently noted the greatness and excellence of man's work, or skill. For example, he said that man's skill of making the rice jelly had a thousand methods (I.6.80b). While speaking of a kind of water mill that can simultaneously perform three different tasks, he said that it was made by "someone whose mind's deliberation leaves nothing." (I.4.55a) He even used the expression, "divine achievement," in referring to the techniques of forging metals (II.10.44a).

It was then natural that Song Yingxing, who saw such great, superior qualities in man's work in the preparation and the use of production techniques, should attribute the origins of some key techniques to the ancient sages (*shengren* 聖人).[34] For example, speaking of dyeing in the introductory paragraph of Chapter 3, he said: "Heaven hung down images; the sages followed them and manifested the five colors [of the heaven] into the five colors [of man]." (I.3.49a) In the introductory paragraph of Chapter 8 he attributed the origin of mining and casting metals to the sages like Yellow Emperor (Huangdi 黃帝) and King Yu 禹 (II.8.17a). In the introductory paragraph of Chapter 17, he pointed out that yeasts needed for making liquors were made by the sages like Yandi 炎帝 and the Yellow Emperor (III.17.48a), implying obviously that ordinary men who are not sages do not reach the level of Yandi and the Yellow

---

[34] In this Song Yingxing was following the tradition of the "Commentary on the Appended Words" (Xici zhuan 繫辭傳, ch. B2) of the *Book of Changes*, which attributed the origins of key institutions and techniques to the ancient sages.

Emperor. In his attribution of the invention of paper making to the high ancient period instead of the Han times (206 B.C.–A.D. 220), one can also see his desire to associate it with the sages (II.11.70a).

It should be noted, however, that Song Yingxing chose to attribute the origin of the techniques to the sages, and not to ordinary men. The sages, although they were human, were endowed with superior qualities that were close to heaven—the "divine," "Creative transformation," and "Creator of things." Indeed, the skill and intelligence involved in various production techniques at such a superior level seem to have made Song Yingxing suspect that they cannot be of solely human quality. In the introductory paragraphs of many chapters, he implied that there must exist some nonhuman agents possessing such superior qualities and performing such marvelous feats involving the techniques discussed in the chapters. For example, after speaking of the great advances in the weapons, he said: "Even though man has a skillful thought, how could he reach such an extreme?" (III.15.25a-25b), implying that perhaps some other agents have intervened. After speaking about different kinds of metals satisfying different needs of man, he said: "As for differentiating good and bad ones, and indicating their weights, who started first to make them depend on each other forever?" (III.14.1a). It is possible that here again he was implying a possibility of a non-human agent. After mentioning the great variety of metal goods produced by the techniques of casting metals, he said: "Who can count them all? In short, man's power cannot reach this [level]." (II.8.17a) After speaking of various methods of extracting oil from plant seeds, he said: "As for such skill and intelligence of man, I do not know from what it was passed and endowed to man." (II.12.63a) After speaking of bees collecting honey from flowers, he asked: "Who was in charge of this and grew and spread them all over the world?" (I.6.74a). After pointing out that every color is available to man, he asked who other than "the extreme divine" (*zhishen* 至神) can provide this (III.16.40a). His saying that even the sages cannot stop wars implies that they are beyond man's will (III.15.25a).

Sometimes Song Yingxing was more explicit in suggesting that it is heaven that is responsible for such supreme abilities and skills. Speaking of those who made such tools as the mortar and the pestle, he said: "Is it possible that the one who made these is not heaven [disguised] in man's shape?" (I.4.54a). After speaking of an exquisite process of obtaining mercury, he said: "This is most sublime and mysterious transformation. Everything is heaven's 'mechanism' (*ji* 機)." (III.16.41b)

Song Yingxing emphasized that heaven's ability manifest in many production techniques is so excellent, mysterious, and abundant that man's skill, however great it may be, cannot reach heaven's work. For example, while discussing the techniques of calcination of stones, he spoke of the examples of changes appearing in nature as the result of heat and fire, such as the colors of alums and the powers of sulphur. He then added that even the great skills of the recipe masters cannot reach even one ten-thousandth of heaven's work (II.11.53a). Thus, in Song Yingxing's view man cannot know heaven's work, or heaven's ability, completely. For example, after noting that everywhere in the world some forms of salts are produced, he wondered: "Who knows the reason why it is so?" (I.5.66a). After speaking of the red fire giving rise to black color and the white mercury changing into deep red, he said: "As for 'Creative transformation' smelting and forging, how can [man's] thought contain them?" (III.16.40a). After mentioning precious gems gathered by man, he asked: "Can the brilliant essences of heaven and earth stop at these few?" (III.18.53a), implying again that what men know are only part of the extreme abundance of heaven and earth.

12.4.3 *Heaven's Role of Taking Care of Man: Natural Theology of Production Techniques?*

One aspect of heaven's great ability Song Yingxing emphasized was that heaven—or "Creator of things" or "Creative transformation"—makes a great variety in creating things. He frequently spoke of the opposite qualities of the things created by heaven and earth. For example, while discussing various metals, he said: "Generally, as heaven and earth produce things, what are bright are the opposite of what are dark and turbid; what are moist are the opposite of what are withered and harsh. If noble things are here, lowly things are there." (III.18.53a) He ended the introductory paragraph of Chapter 2 on clothes with the following remark: "In general, man and things pair each other; the noble and the lowly have distinctions. It is really heaven that does this." (I.2.23b) While speaking of coal, as we have seen, he said that "between metal and 'earth and stone,' 'Creative transformation' additionally made this specie [of coal] appear." (II.11.55b)

Song Yingxing noted that such distinctions in the natural world can be found in the human world also. For example, the main point of introductory paragraph of Chapter 3 on colors was that as there are different colors in nature, there also are distinctions in the colors of clothes people wear (I.3.49a). In this manner the distinctions in nature were analogically extended to the human world. He used the expression, "Its meaning is also like this," to refer to such analogical extension. Sometimes, however, the direction of the analogical extensions was opposite. For example, he noted that as the human order was established owing to the existence of the ten classes among men, "the great earth [also] produced the five metals so that they could be used all under heaven and in the posterity." He then added the same words, "Its meaning is also like this," implying that there is the distinction of the noble and the lowly among metals also (III.14.1a). He even said that "'Creator of things' prepared different clothing materials to make distinctions between the noble and the lowly men, and between men and animals." (I.2.23a).

In such cases where Song Yingxing was showing heaven's care of man, he frequently used the expression, "Creator of things." Song Yingxing actually said that "'Creator of things' cared [about man]," after pointing out, in the passage we have already seen, that clothing materials can be dyed in many different colors (I.3.49a). Speaking of fragrant, colorful, and sweet things that are greatly desired by men, he said: "'Creator of things' had the thought of making it particularly different." (I.6.74a).

One aspect of heaven's care of man that was emphasized by Song Yingxing was that heaven makes sure that whatever is needed by man is produced somewhere and somehow. The best example was salt which is essential for man's life. Song Yingxing noted that while there are places that do not produce grains or vegetables, there is no place where no salt is produced. In the introductory paragraph of Chapter 5 on salt, he first pointed out that while man can go for years without bitter, sour, sweet, or acrid taste, man cannot live without salt even for ten days, and then said that therefore heaven made salt as the source of man's "life *qi*" (*shengqi* 生氣). He added: "It skillfully comes into being [everywhere] and waits [for man's use]." (I.5.66a) Speaking of different forms of salt, he noted that in places far from the sea where the sea salt is rare, other forms of salt are produced for man's use, or transported from other places. In the places where it is difficult to carry salt by boats or carts, "'Creator of things' produces [salt somehow] according to the situation." (I.5.66b) Similarly, he noted that in the coastal area where no limestone is produced, heaven produces oyster shells instead (II.11.54a). In places where there are no coals, plants are abundant, from which "the subtlety of heaven's mind is seen." (II.11.55b) And as we have seen, he said after noting that silver is not produced within 300 *li* of a place producing gold, and *vice versa*, "the emotion of 'Creator of things' can be seen abundantly [in this] also." (III.14.6a) He even noted that the number of honey-bees is small in places producing the sugarcane. The implication is clear that heaven, or "Creator of things," although Song Yingxing did not explicitly mention either word, is behind this to make sweet things available everywhere. (I.6.79a).

The "natural theological" sentiment is unmistakable in these examples showing Song Yingxing's belief in heaven's taking care of man. We might ask what the source of this belief was. Would it have been the Christian idea of God? Given the widespread dissemination of Christian doctrines among the late Ming thinkers, it is not impossible that Song Yingxing picked up such a natural theological sentiment from this intellectual climate.[35] But any answer to this question will remain a speculation until we have come to know a lot more about the details of Song Yingxing's life.

---

[35] Christopher Cullen has noted that Song Yingxing "certainly had some contact with Jesuit teaching." Cullen (1990, p. 316, n. 73). Xu Guangtai 徐光台 (Hsu Kuang-t'ai) has suggested a possible connection between Song Yingxing and Xiong Mingyu 熊明遇 (1579–1649), an influential scholar with a favorable attitude to Catholicism: Xu (2007, pp. 378–379). It is also highly likely that Song Yingxing read Matteo Ricci's (1552–1610) *True Meanings of Heavenly Lord* (*Tianzhu shiyi* 天主實義), which was widely circulated among the Chinese scholars of late Ming, and which contained many remarks of natural theological character. Ricci said, for example: "Considering Heavenly Lord's producing this heaven and earth and these myriad things, there is not a single thing that He did not create for man's use." Zhu (2001, p. 69).

12.5 Concluding Remarks

Song Yingxing's basic idea about the roles of heaven and man in production techniques was that while man does something in producing many things, heaven lies behind all this and does everything. What man does in using the various techniques, for Song Yingxing, was really to use the work of heaven. And it is to be noted that nearly all the remarks on such a role of heaven come up in the introductory paragraphs beginning each chapter where Song Yingxing laid out his general, basic ideas about the techniques discussed in the chapter. Indeed, the importance of heaven's role was reflected in the title of the book itself: "Heaven's Work in Opening Things."

This idea of heaven's role in production techniques, however, was not something entirely new with Song Yingxing. It was essentially in agreement with the traditional Chinese idea of heaven producing everything of the world: man and the ten thousand things. Even the natural theological notion that heaven prepared everything for the sake of man did not go beyond the basically man-centered worldview of the Confucian tradition. What did stand in conflict with Song Yingxing's ideas about the heaven's role in production techniques was the established convention in traditional China to associate the supreme level of human affairs with the human sages and that of natural phenomena with the "divine," "Creative transformation," etc. While traditional Chinese generally accepted the attribution of the beginning of the key institutions and techniques to the ancient sages, they usually used such words as "divine" and "heaven" in referring to some mysterious, marvelous natural phenomena.[36] In view of this convention, Song Yingxing's associating "heaven," "divine", etc. with the human activities of production techniques, was a departure from the tradition.

Song Yingxing seems to have done this because he wanted to show that heaven's role can be seen not only in such abstract and sublime ideas as human nature (*xing*), principle (*li*), and mind (*xin* 心), valued in the learning for civil service examination. Heaven has a role also in such concrete, practical, and lowly things as the everyday commodities and in the techniques of producing them.

Bibliography

Bray, Francesca (2007). "Introduction: The Powers of *Tu*", in Bray, Dorofeeva-Lichtmann, and Métailié (2007), pp. 1–78.

Bray, Francesca and Georges Métailié (2001). "Who Was the Author of the *Nongzheng Quanshu*?", in Catherine Jami, Peter Engelfriet, and Gregory Blue, eds., *Statecraft & Intellectual Renewal in Late Ming China: The Cross-Cultural Synthesis of Xu Guangqi (1562-1633)*. Leiden: Brill, pp. 322–359.

Bray, Francesca, Vera Dorofeeva-Lichtmann, and Georges Métailié, eds. (2007). *Graphics and Text in the Production of Technical Knowledge in China: The Warp and the Weft*. Leiden: Brill.

---

[36] Peterson (1982); Kim (2000, chaps. 6–7).

Ch'oe Chu, tr. 崔炷 註譯 (1997). *Ch'ŏn'gong kaemul* 天工開物. Seoul: Chŏnt'ong Munhuasa 傳統文化社.

Clunas, Craig (2004). *Superfluous Things: Material Culture and Social Status in Early Modern China*. Honolulu: University of Hawai'i Press.

Cullen, Christopher (1990). "The Science/Technology Interface in Seventeenth-Century China: Song Yingxing on *qi* and the *wu xing*," *Bulletin of the School of Oriental and African Studies* 53: 295–318.

Elman, Benjamin A. (1984). *From Philosophy to Philology: Intellectual and Social Aspects of Change in Late Imperial China*. Cambridge, MA: Harvard University Press.

Elman, Benjamin A. (2005). *On Their Own Terms: Science in China, 1550–1900*. Cambridge, MA: Harvard University Press.

Gillispie, Charles C. (1957). "The Natural History of Industry," *Isis* 48: 398–407.

Golas, Peter (2007). "'Like Obtaining a Great Treasure': The Illustrations in Song Yingxing's *The Exploitation of the Works of Nature*," in Bray, Dorofeeva-Lichtmann, and Métailié (2007), pp. 569–614.

Kim, Yung Sik (2000). *The Natural Philosophy of Chu Hsi (1130–1200)*. Philadelphia: American Philosophical Society.

Kim, Yung Sik (2010). "A Philosophy of Machines and Mechanics in Seventeenth-Century China: Wang Zheng's Characterization and Justification of the Study of Machines and Mechanics in the *Qiqi tushuo*," *East Asian Science, Technology, and Medicine* 31: 64–95.

Nappi, Carla (2009). *The Monkey and the Inkpot: Natural History and Its Transformations in Early Modern China*. Cambridge, MA: Harvard University Press.

Needham, Joseph (1954). *Science and Civilisation in China*, vol. 1. Cambridge, MA: Cambridge University Press.

Pan Jixing 潘吉星 (1989). *Tiangong kaiwu jiaozhu ji yanjiu* 天工開物校注及研究 (Critical Annotations and Studies on the *Tiangong kaiwu*). Chengdu 成都: Bashu Shushe 巴蜀書社.

Pan Jixing 潘吉星 (1990). *Song Yingxing pingzhuan* 宋應星評傳 (Critical Biography of Song Yingxing). Nanjing 南京: Nanjing Daxue Chubanshe 南京大學出版社.

Peterson, Willard J. (1982). "Making Connections: The 'Commentary on the Attached Verbalizations' in the *Book of Change*," *Harvard Journal of Asiatic Studies* 42: 67–116.

Rossi, Paolo (1970). *Philosophy, Technology and the Arts in the Early Modern Era*. New York: Harper.

Schäfer, Dagmar (2005). "The Congruence of Knowledge and Action: The *Tiangong kaiwu* and Its Author Song Yingxing," in Hans-Ulrich Vogel, Christine Moll-Murata and Song Jainze, eds., *Chinese Handicrafts Regulations of the Qing Dynasty: Theory and Application*. Munich: Indicum, pp. 35–60.

Sun, E. Z. and Sun, S.-c. (tr.) (1966). Sung Ying-Hsing, *T'ien-kung K'ai-wu: Chinese Technology in the Seventeenth Century*. University Park, PA: State University Press.

Xu Guangtai 徐光台 (2007). "Mingmo xixue yu Bailutong Shuyuan" 明末西學與白鹿洞書院 (The White Deer Grotto and the Western Learning in Late Ming), in Gao Feng 高峰, Hu Qing 胡青, Lai Gongou 賴功歐, and Ye Zunhong 葉存洪, eds., *Zhongguo shuyuan luntan* 中國書院論壇 (Discussion Forum on the Chinese Academies). Beijing: Zuojia chubanshe 作家出版, pp. 370–384.

Yabuuchi Kiyoshi, ed. 藪内清 編 (1954a). *Tenkō kaibutsu no kenkyū* 天工開物の研究 (Studies on the *Tiangong kaiwu*). Tokyo: Kōseisha 恒星社.

Yabuuchi Kiyoshi (1954b). "*Tenkō kaibutsu* ni tsuite" 天工開物について (On the *Tiangong kaiwu*), in Yabuuchi (1954a), pp. 1–24.

Yabuuchi Kiyoshi, tr. 藪内清 譯註 (1969). *Tenkō kaibutsu* 天工開物. Tokyo: Heibonsha 平凡社.

Zhong Guangyan, ed., tr. 鍾廣言 注釋 (1978). *Tiangong kaiwu* 天工開物. Hong Kong: Zhonghua Shuju 中華書局.

Zhu Weijing 朱維靜, ed. (2001). *Li Madou Zhongwen zhuyiji* 利瑪竇中文著譯集 (Collection of Matteo Ricci's Writings and Translations in Chinese). Hong Kong: 香港城市大學出版社.

Part VI
Mathematics

Chapter 13
On the Role of the *Ecole Polytechnique*, 1794–1914, with Especial Reference to Mathematics

In honour of Charles C. Gillispie, Doyen Historian of French Science and Technology

Ivor Grattan-Guinness

13.1 What Kind of New School?

Soon after its founding in 1794 and ever afterwards, the *Ecole Polytechnique* has enjoyed a high reputation as an institution of higher education, especially during its first forty years or so, producing a large number of graduates who went to distinguished careers in science and/or engineering. Yet its teaching programme has always been confined to certain kinds of civil and military engineering and related subjects—especially mathematics, which played a prominent role—and for many decades it had no formal research remit. What happened there, and why did it decline? The bulk of this article summarises the history up to 1914, with special attention paid to the mathematics courses and to the glory decades; at the end some appraisal is offered, including the influence on other countries.

For the engineering profession three dangers arose from the chaos following the French Revolution. Firstly, by 1793 many officers in the army and navy had gone abroad or returned to their families in the countryside; thus the higher ranks needed rapid replenishment (Julia 1995). Secondly, all institutions of higher education were closed in 1793; while the engineering schools were soon open again with few changes, their own roles needed reappraisal. Thirdly, the transport situation of the country had suffered from years of neglect: not only roads but also the new technology of canals, and the coastal harbours and sea travel.

Thus 'the republic needed academics' (Langins 1987a). An emergency council was set up in Paris, and created a new school there: the '*Ecole Centrale des Travaux Publics*' and launched in December 1794. Under supporters such as the chemist François Fourcroy (1755–1809), it seems to have been conceived as the one and only institution to train engineers, both civilian and military. To this

I. Grattan-Guinness (✉)
Middlesex University Business School, London, UK
e-mail: ivor2@MDX.AC.UK

J.Z. Buchwald (ed.), *A Master of Science History*, Archimedes 30,
DOI 10.1007/978-94-007-2627-7_13, © Springer Science+Business Media B.V. 2012

end 400 students were rapidly enrolled, and 'revolutionary courses' in mathematics and chemistry were taught, with the help of dozens of rapidly recruited officers (listed in Fourcy 1828, 389–390). It sounds like a nonsense, similar to a contemporary one that closed after a few months: the *Ecole Normale,* that was intended to provide similar forced feeding for future teachers and administrators (Dupuy 1895).

But the founders realised that most of the military schools were outside Paris, and no naval school could possibly run in the city. Thus the role of the school was changed to that of a *preparatory* institution for the other schools, with a three-year curriculum. The change was reflected in a change of name, made in September 1795, to '*Ecole Polytechnique*'; the adjective first appeared in a pamphlet published that year by an involved politician, Claude Prieur de la Côte d'Or (1763–1832), on 'l'enseignement polytechnique de l'*Ecole Centrale des Travaux Publics*' (Prieur 1795). The adjective conveyed the plurality of techniques.[1]

13.2 The Organisation of the *Ecole Polytechnique*

The *Ecole Polytechnique was* saved from demise by the 'regular courses', which comprised the three-year programme. Since both military and civilian needs were being met, governmental control of the school lay with the Ministries of War and of the Interior. The annual recruitment was lowered to a more realistic level of around 120 students. The relationship was clarified with the other schools; in a move that involved the mathematician and politician Lazare Carnot (1753–1823), these other schools were organised into a collection of '*écoles d'application*' (Carnot 1796). Some of them were civilian, such as the *Ecole des Ponts et Chaussées* and the *Ecole des Mines* in Paris; among the rest was the school for *Génie et Artillerie* created in 1802 at Metz in eastern France as the fusion of two earlier schools at Chalons-sur-Marne and Mézières (Belhoste and Picon 1996). Monge had long taught at the latter school before the Revolution, and so it too exercised considerable influence on the design of the *Ecole Polytechnique*.

The Directorship, a rotating post, had first been given to Elie Lamblardie (1747–1797): director of the *Ecole des Ponts et Chaussées* in Paris, he saw it very much as the father of the new establishment (Langins 1991). The founder *professeurs* of analysis and mechanics were J.L. Lagrange (1736–1813) and of mechanics Gaspard Riche de Prony (1755–1843)—a stark contrast between the leading purish mathematician of the time and a committed engineer who also ran a small school for geographers for a few years at that time, and from 1798 directed the *Ecole des Ponts et Chaussées* for forty years (Bradley 1998).

---

[1] At various times the *Ecole Polytechnique* and all the other schools and institutions carried the adjectives '*Impériale*', '*Royale*' or '*Nationale*' at suitable places in their names; I shall not use them here.

Descriptive and differential geometry was in the hands of Gaspard Monge (1746–1818), who was also much involved in the initial conception of the school and served as Director for two short periods. Each *professeur* had an assisting *répétiteur*, some becoming *professeurs* in due course. A notable early *répétiteur* was Joseph Fourier (1768–1830), recruited after standing out as one of the few good students of the *Ecole Normale*.

One major feature was the distinction made between teaching and examining; so examiners were also appointed. For mechanics and analysis the initial pair were C.J. Bossut (1730–1814) and P.S. Laplace (1749–1827), the latter the leading mathematician after Lagrange (Gillispie 1997), and a great influence on the early development of the school although he never taught there (Langins 1987b). The importance of these subjects was underlined by the decision to call these examinerships 'permanent' while those for all other subjects were 'temporary', though the holders could serve long in post. Until the 1840s the examinations were conducted verbally, with around an hour given to each student.

A *Journal Polytechnique* (its original title) was soon established, initially to fulfil the requirement of publishing the lecture courses. Each volume was composed of a varying number of *cahiers*; the print-run was 1,000 copies, sometimes more. The library was built up quickly (Bradley 1976b), aided by thefts effected during wars (Pepe 1996). The librarian from 1818 was Ambroise Fourcy(-Gaudain) (1778–1842), who produced in 1828 the first history of the school, still a most informative text account enhanced by various lists, including of all the students up to the time of publication (Fourcy 1828).

13.3 Students for and at the *Ecole Polytechnique*

Students were chosen by admissions examiners. The country was divided into three (later four) regions, and also Paris, and an examiner travelled in a region for some weeks during the summer to interview promising young boys nominated by local families or dignitaries. The examiners and some school officers then met together and chose the enrolment for the following autumn (between 75 and 260 students up to 1816, with the mode around 120). These examiners were often mathematicians, occasionally of some note. A few others wrote elementary textbooks, which were often used by candidates for precisely this entrance procedure.

The admission system is a much underrated feature of the success of the school. For under it students were recruited *without* dependence upon family wealth, and it is noticeable that several of the major later figures came from modest backgrounds (Bradley 1976a). At first students were given a small salary to support their living circumstances, although a steep fall in the value of the currency in 1795 made it largely worthless for a time. But during the next decade fees were demanded, so that the pool of potential candidates was reduced.

The school opened in buildings attached to the *Palais Bourbon*, where the parliament met. Students from the provinces usually lived with approved Republican families in Paris.

The students studied for 8.5 days in each *décade* of the revolutionary calendar, then six days per week when the normal calendar was restored in 1806. The academic year covered two terms from late October or early November till the following June, followed by some weeks in the summer devoted to campaigns and exercises. The dominant role of mathematics soon led the students to deploy the catchword 'X' as a name for the school. After graduation they became known as '*polytechniciens*'. Table 13.1 lists the graduates up to 1840 who later held distinguished careers, not necessarily at the school; the number up to the later 1810s is striking, and so is its reduction thereafter.

Table 13.1 *Polytechniciens* with significant later careers in mathematics, 1800–1840

1794	Biot (1795)
	Brisson (1795)
	Cagniard-Latour (1797)
	Francoeur (1797)
	Lancret (1797)
	Malus
	Poinsot (1797)
1795	Duchayla (1796)
1796	Bourdon (1800)
	Reynaud
1797	Français, J.F. (1800)
	Gay-Lussac (1800)
1798	Poisson
	Binet, P. (1801)
1800	Guenyveau
	Plana (1803)
1801	Dulong (1802)
	Dupin
	Terquem (1804)
1802	Navier (1803)
1803	Arago
	Bazaine
	Brianchon (1806)
	Lefebure de Fourcy
	Mathieu
1804	Binet, J.
	Fresnel, A.J.
1805	Cauchy
1807	Burdin (1810)
	Duleau
	Fresnel, L.
	Petit
	Poncelet (1810)
1808	Belanger
	Coriolis
	Lesbros
1809	Raucourt
1811	Olivier (1815)
	Pontécoulant

Table 13.1 (continued)

1812	Babinet (1813)
	Carnot, S.
	Chasles (1815)
	Coste (1813)
	Demonferrand
1813	Enfantin (1814)
	Lechevalier
	Morin (1817)
	Pambour
	Piobert
	Saint-Venant (1817)
1814	Comte
	Duhamel
	Lamé (1817)
	Woisard
1815	Bienaymé (1816)
	Savary
1816	Clapeyron
1817	Bobillier
	Didion
	Elie de Beaumont
1818	Combes
1821	Perdonnet (1822)
1822	Dupuit
	Emy
1823	Reech
	Transon
1825	Gratry
	Liouville
1829	Bravais
	Lalanne
1830	Laurent, P.
	Regnault
1831	Leverrier
1832	Wantzel
1833	Catalan
1834	Delauney
1838	Bonnet
	Marie
	Serret
1839	Bertrand

If a student did not graduate in two years, then the
year of his completion is given

13.4 Reforms, 1799–1804

For six rather disastrous weeks in 1799 Laplace performed as Minster of the
Interior. He proposed that the school have a governing council, the *Conseil de
Perfectionnement*, to supplement the *Conseil d'Instruction* on teaching details
and a *Conseil d'Administration* for management. Lo and behold, Laplace was

one of its founder members, and exercised much influence, in particular reducing the time given to Monge's descriptive geometry (Paul 1980) and transfer-
ring much of it to mechanics. Laplace wanted to confine the programmes at the
Ecole Polytechnique to teaching general theories, which would then be applied
in the more specialist teaching at the other schools (as their collective name
'*écoles d'application*' surely reflects).

A major change of 1799 reduced the course from three to two years, with
appropriate changes in curricula effected in the other schools. A point of
discord was a monopoly under which *polytechniciens* were the *only* recruits
into these schools.

In 1799 the ageing Lagrange was replaced by S.-F. Lacroix (1765–1843), the
leading textbook writer in mathematics of the time; and Monge's *répétiteur* Jean
Hachette (1769–1834) was promoted to *professeur*. Laplace was succeeded as
permanent examiner by A.-M. Legendre (1752–1833).

In December 1804 Bonaparte crowned himself Emperor Napoléon, at a
ceremony attended by Laplace and student Corporal François Arago
(1786–1853; 1803 enrolment) as one of the representatives of the students of
the school. Not by accident Bonaparte had militarised it a month earlier; and
while some of his aims were not fulfilled (Bradley 1975), the atmosphere in the
school changed substantially, from rather free-wheeling to meticulously
bureaucratic. Governmental control still remained with the Ministries of War
and of the Interior, although one may imagine that the balance had changed
somewhat. The rotating Directorship became a permanent Governorship. Students now held official ranks and executed daily drills.

In 1805 the school moved from its buildings near the *Palais Bourbon* to two
former colleges not far from the *Panthéon*. Barracks could now be provided,
and exeats were strictly controlled. The school stayed there until moving out of
town in the mid 1970s.

Among other innovations, from 1806 brief records were made, usually by an
administrator, of the content by topic (though rarely also of method) of each
lecture in a course. These *registres d'instruction* are a most valuable historical
resource, as they supplement the information in the programmes about the content
of the teaching; they are now kept as manuscript volumes in the school's archives.

From the start the syllabi were pretty demanding. The calculus courses went
into aspects of differential equations and the calculus of variations. The
mechanics courses encompassed the basic theory of mass-points, rotating
solid bodies, and fluids, though not much mathematical astronomy: influential
versions were taught by de Prony and then by S.-D. Poisson (1781–1840; 1798
enrolment), who had soon been appointed to the staff after graduation, in effect
as the successor of Fourier. The published versions of all the courses usually
contained more material than was actually taught, and so served as sources for
further study and exercises. In strange contrast to all the meticulous organisation and control, especially after 1804, the staff seem to have been given much
freedom in the choice of version of the subjects programmed, especially in
mathematical analysis and mechanics (Grattan-Guinness 2005, arts. 8–9).

In the early years several courses appeared at first in the *Journal de l'Ecole Polytechnique* (its definitive name from 1796) and then and especially thereafter quite a few appeared as books with Paris publishers. The *Journal* was becoming more and more of a venue for research papers (Lamy 1995); so in 1804 Hachette launched the *Correspondance sur l'Ecole Impériale Polytechnique,* which contained shorter and more educational articles, mathematical problems, and news of the school. Three valuable volumes were produced until 1816, and were reprinted until the mid 1830s.

13.5 Another Revolution, 1815–1816: The Return of the Bourbons

The *Correspondance* ceased to appear because Hachette was dismissed without pension from the school in 1816. His fate, shared by his master Monge, was part of the changes that followed the departures of Napoléon in 1814 and 1816. Indeed, the future existence of the school was then much in doubt, for a major policy of the 'Restoration' was indeed to re-establish the *ancien régime* in as close to its condition of 1789 as possible: the school was a creation of the interim period, and so it should be abolished. On the other hand, if it did disappear, then a similar institution would have to be created in its place.

To consider this dilemma the *Conseil de Perfectionnement* met in March 1816 under the chairmanship of Laplace and prepared a report advocating continuity of the school. Then during the summer Laplace chaired an inter-Ministerial commission, which largely accepted the school's proposals; so rather few modifications were made. The annual enrolment was reduced to around 75 students. The mediocre teaching of physics was improved with the appointment of Laplace's young followers Aléxis Petit (1791–1820; 1807 enrolment) and Pierre Dulong (1785–1838; 1801 enrolment). A course in the theory of machines, which Hachette had launched in 1806, was converted into a new one in geodesy and machines, with a chair initially held by Arago.

Several important changes in personnel occurred, a few for political reasons. Jacques Binet (1786–1856; 1804 enrolment), respectable mathematician and fervent Catholic, became Director of Studies. Permanent examiners Lacroix (Bossut's replacement since 1808) and Legendre in analysis and mechanics were replaced by de Prony and Poisson, whose professorships went to A.M. Ampère (1775–1836) and A.-L. Cauchy (1789–1857; 1805 enrolment).

13.6 Yet Another Revolution, 1830: The Exit of the Bourbons

Cauchy's appointment was to provide the school with some of its main difficulties during the Restoration period. His professorship only a decade after his student time was high reward, though a reflection of his amazing gift for mathematics. But his effectiveness as a teacher, however, is another matter.

During the 1820s the school earnestly asked him to modify his teaching of mathematical analysis rigorously grounded upon a theory of limits, but this fanatic for the King on the Bourbon throne and the Catholic god in heaven paid little attention (Belhoste 1991).

However, political unrest in France increased after the heir to the Bourbon throne was assassinated in 1820 and the regime became severe. Some relaxation came later in the decade; but it was too late, and three days of fighting in the Paris streets during July 1830 saw them out of power (Pinkney 1972). Several *polytechniciens* played a notable part in the revolting forces, and afterwards they were excused examinations and given three months' holiday. Meanwhile *professeur* Cauchy abandoned all his teaching posts and fled into Europe with the Royal Family.

As in 1816, a commission was set up to examine the future of the school, with Arago and de Prony among its members. The school fell entirely under the administration of the Ministry of War. The annual enrolment, which had increased somewhat during the later 1820s, seems to have been reduced to its 1816 level of around 75 students. A period of stability set in, and rather tinged with complacency, such as largely unmodified syllabi. The professorships and examinerships in analysis and mechanics rotated among the three *polytechniciens* Jean Duhamel (1797–1872; 1817 enrolment), Gabriel Lamé (1795–1870; also 1817 enrolment) and Joseph Liouville (1809–1882; 1825 enrolment), together with the Swiss-born mathematician Charles Sturm (1808–1855). Lamé also served as *professeur* of physics for some years. After over twenty years as *répétiteur* to Cauchy and others, in 1838 G.G. Coriolis (1792–1843; 1808 enrolment) succeeded the deceased Dulong as Director of Studies in 1838 until his own death, when Duhamel took over; and Michel Chasles (1793–1880; 1812 enrolment) took the chair in geodesy and mechanics in 1841. New *répétiteurs* included Liouville until his promotion in 1838; and Auguste Comte (1798–1857; 1814 enrolment), who also served as an admission examiner from 1837 to 1844.

13.7 Counter-revolution: The Report of Leverrier, 1850

Laplace's view of the school's curriculum was prevailing over Monge's, but not without disquiet. In 1840 Coriolis wrote an extensive report on policy for the school.[2] He wished to stiffen the admissions curriculum with more geometry and mechanics, and to review the relationship with courses taught at the *écoles d'application*. In particular, he hoped for more teaching of the theory of machines, a subject upon which he was a specialist; and in a particular piece of prescience, he hoped that a course on the theory of electricity could be

---

[2] Parts of Coriolis's report are transcribed in (Grattan-Guinness 1990, 1346–1351); for discussion see pp. 1262–1263.

introduced. He also looked forward to a return to teaching the differential and integral calculus at the school, instead of the (over-)rigorous version introduced by Cauchy.

In 1850 similar criticisms were argued by Urban Leverrier (1811–1877; 1831 enrolment), a distinguished astronomer recently become famous for his role in the discovery of Neptune. In 1850 he chaired a commission to decide the future role of the school. Its members included General J.V. Poncelet (1788–1867; 1807 enrolment), who had just retired from two rather ineffective years as *Commandant* (the first academic in this post); and Duhamel as Director of Studies. In six months of 1850 Leverrier wrote a large and remarkable report on the school, including many aspects of its history drawn from the archives (Leverrier 1850a). He also published a substantial supplement with the *Assemblée Nationale* (Leverrier 1850b). Irritated by the long and recent dominance of the policies of Laplace and Cauchy, Leverrier argued in effect for a return to Monge's aspirations. When the recommendations were approved, reactions at the school were stark: Liouville, Sturm and Chasles resigned immediately in protest at this rejection of the emphasis on teaching purish mathematics that they had inherited and continued from predecessors such as Cauchy.

13.8 Career Opportunities for the *Polytechniciens*

We treat here the double role of the school in serving both civilian and military needs.[3] The normal career path of a *polytechnicien* with military aspirations lay in the army or the navy, where he would hope to rise to high rank. Some specialist organisations provided posts for the academically inclined; for example, the *Dépôt Général de la Guerre* with a *Comité de Fortifications* in Paris, to which Poncelet moved from Metz in 1834 until retiring in 1850. Civilian graduates usually entered the *Corps* corresponding to their specialty, especially *Ponts et Chaussées* and *Mines,* and expected to rise through the ranks of engineers in a *département* and maybe the national administration. New *Ecoles Supérieures* were added as the corresponding technologies advanced; for example, *Télégraphie* in 1878 and *Eléctricité* (especially heavy current) in 1894.

A tiny minority went back into academic life, at the *Ecole Polytechnique* and/ or one of the *écoles d'application,* and/or in the other scientific or engineering institutions of the country such as the civilian *Bureau des Longitudes* for astronomy and navigation or the military *Dépôt Général* just mentioned. A few *polytechniciens,* such as Claude Navier (1786–1836: 1802 enrolment), combined teaching with an active engineering career.

---

[3] Even after militarisation in 1804, the school maintained its double role. This is not always recognised; in particular, Ken Alder misses it in his excellent account of the French armaments industry in the late 18th and early 19th centuries (Alder 1997, ch. 5, esp. p. 306).

This whole system was set up for education and employment in military and civil engineering. Nothing was officially organized for industrial or commercial engineering; so in 1829 the *Ecole Centrale des Arts et Manufactures* was founded in Paris for this purpose, as a private enterprise (Pothier 1887). Among *polytechniciens,* Théodore Olivier (1793–1853; 1811 enrolment) was a founder, and Coriolis and Liouville early professors. Very successful, within a decade or so it was made an *école d'application.*

Teaching posts for *polytechniciens* were also available elsewhere. In 1808 Napoléon, at the special prompting of Fourcroy, replaced the abolished system of universities with a new *Université Impériale de France.* The name is misleading, for the organisation was quite different from its predecessors. The Empire was divided into *arrondissements,* each one (apart from Paris) comprising a few contiguous *départements,* and the primary and secondary schools there were administered under an *académie* (again an absurd use of the word) based in the capital of one of the *départements.* In addition, within some *académies* higher education was available, in *facultés* of science, medicine, law, letters and theology. Hachette, Poisson and Cauchy from the *Ecole Polytechnique* also taught in the Paris *Faculté des Sciences.*

Within the *Université* but outside all the *académies,* a new *Ecole Normale* was set up in Paris, with its own *Facultés* of science and letters. It held no formal link to the earlier short-lived institution of that name (Section 13.1), although training of future teachers and bureaucrats was again a special aim. Even with the existence of *Ecole Normale,* the *Université* was clearly secondary in status to the *écoles d'application* (Grattan-Guinness 1988).

Finally but less important, the old *Collège Royal* in Paris had been quickly relaunched in 1795 as the *Collège de France,* and continued its role of offering free lecture courses without conditions for enrolment or examination of the auditors. Untypically, the courses in mathematics were very elementary, although the physics *professeur,* Jacques Cousin (1739–1810), was competent in differential equations and analysis. Improvements started only when Poisson taught supplementary courses between 1809 and 1812, and especially when Lacroix took the chair in 1815 and held it till his death in 1843. From around 1860 high-level mathematics was taught here (Sédillot 1870).

13.9 On the Period 1850–1914

Historical research on all aspects of this later period is far more limited than on the earlier one, in all respects. So the remarks offered here are rather scattered.[4]

---

[4] Some information may be gleaned from the books written or co-edited by Belhoste in the bibliography, especially (1994a, pt. 2); also (Shinn 1980, chaps. 3–5), (Fox and Weisz 1980, pts. 1 and 2 passim), and various articles in the *Bulletin de la Société des Amis* (*SABIX*) of the school's library, where copies are available.

After the Leverrier changes the school seems to have settled down to another fairly uniform regime, until 1870. The most significant later mathematicians are Joseph Serret (1819–1885: 1838 enrolment), Ossian Bonnet (1819–1892: 1838 enrolment), Joseph Bertrand (1822–1900: 1839 enrolment), Charles Hermite (1822–1901: 1841 enrolment), E.N. Laguerre (1834–1866: 1853 enrolment), Emile Mathieu (1835–1890: 1854 enrolment), Camille Jordan (1835–1922: 1855 enrolment), C.A. Laisant (1841–1908: 1859 enrolment) and G.H. Halphen (1844–1889: 1862 enrolment). The absurd nomenclature of 'permanent' and 'temporary' examiner was dropped in 1845.

The next major discontinuity was literally military: the Franco-Prussian war of 1870–1871, when the French forces manifestly failed to keep the Prussian army at bay. In consequence the school enrolment doubled for a decade or so, from around 120–140 to at least 250; however, no commensurate increase in staff was made, at least at the professorial level. Further, time for academic study was reduced in favour of more compulsory non-academic activities, such as horse-riding. The quality of education must have suffered, thus making the school less attractive for gifted students. It did soon attract two very notable recruits, H. Becquerel (1852–1908: 1872 enrolment) and Henri Poincaré (1854–1912: 1873 enrolment); but the reduction in the number of later eminent figures was maintained. In addition to Poincaré, the major mathematicians between 1870 and 1914 are M.G. Humbert (1859–1921: 1877 enrolment), Maurice d'Ocagne (1862–1938: 1880 enrolment) and Paul Lévy (1886–1977: 1904 enrolment).[5]

The main change in the profession of mathematics in France was a gradual but large shift in balance caused by the rise of the *Ecole Normale Supérieure* (its name from 1845) as a centre for science in general and mathematics (usually pure) in particular (Zwerling 1976). Signs of the change started when candidates such as Gaston Darboux (1842–1917) in 1861 and J. Tannery (1848–1910) five years later chose to become *normaliens* rather than *polytechniciens* after coming top of both lists. Later *normaliens* include Paul Appell (1855–1930), Emile Picard (1856–1941), Emile Goursat (1858–1936), Paul Painlevé (1863–1933), Jacques Hadamard (1865–1963), Jules Drach (1871–1941), Emile Borel (1871–1956), René Baire (1874–1932), Henri Lebesgue (1875–1941), Paul Montel (1876–1975), Maurice Fréchet (1878–1973) and Arnauld Denjoy (1878–1974). By the end of the century the *Ecole Normale Supérieure* was the most prestigious French centre for (mostly pure) mathematics. Apart from Duhem and Baire this galaxy of graduates was to teach at their old school, the Paris *Faculté* and/or the *Collège de France*, so that the latter two institutions also rose in mathematical significance. Their role now came to dominate in the continuing rivalry—or at least mutual concern—between France and the ever rising Germany as mathematical countries.

---

[5] One notes also A. Dreyfus (1859–1935: 1878 enrolment). The full enrolments at the school up to today may be viewed on the website www.bibliotheque.polytechnique.fr, with notes on the careers of some of the more distinguished graduates.

These changes in mathematical balance in France affected the *Ecole Polytechnique* itself. Some of its distinguished graduates continued to occupy its chairs of analysis: Bertrand (1856–1894), Hermite (1869–1876), Jordan (1877–1911) and Humbert (1895–1920). In a new explicit distinction between the examinerships for analysis and for mechanics, examiners for analysis included Hermite (1863–1869), Bonnet (1869–1872) (when he became the *Directeur des Etudes* for six years) and Jordan (1873–1877). But the professor of mechanics at the school from 1905 to 1933 was the *normalien* Painlevé,[6] who was joined in 1912 by Hadamard (succeeding Jordan in analysis and staying until 1937, when Lévy took over). The appointments of both professors and examiners in geometry and applied mathematics are perhaps less notable: one notes Poincaré for astrophysics (1904–1908), where he also taught some statistics, but once again his main appointments were held in the *Faculté*.

In addition to personnel, the old issues about curriculum surfaced on occasion. The annual number of lectures in analysis had been reduced in the 1870s from 80 to 65, though the quantity of mathematics taught remained substantial. In 1895 Bertrand retired from his chair, and the issue of pure versus applied arose again; the school went purish (Gispert 1992) but the chair was offered to (*polytechnicien*) Humbert rather than to the higher-powered *normalien* Picard.[7] Throughout this time the content was adjusted somewhat to accommodate applications; for example, more on solutions of major differential equations such as Laplace's and less on elliptic functions.

As before, curricula in other subjects seem to have been reasonably stable while flexible in content. For example, physics received boosts on occasion, including in Leverrier's (1850) report; after some specialists in optics, Becquerel held the chair from 1895 to 1908. In chemistry Victor Regnault (1810–1878: 1831 enrolment) was appointed from 1841 to 1870.

The number of applicant students to the school remained high, with greater proportions than before drawn from families of private means and from civil servants and less from the professional classes. Still under the Ministry of War, its influence upon the mathematics curricula in French secondary schools gradually decreased. Indeed, it resisted requests made by the Ministry of Public Instruction, especially around 1900, to integrate its courses more into national programmes.

The career options became considerably more complicated. After 1870 the great majority of *polytechniciens* followed the military rather than the civil routes; but stagnation in promotion and the increasing financial attractions of

---

[6] Perhaps Painlevé's appointment was nominal at times, for he held various ministerial offices from the mid 1910s onwards, even briefly serving as Prime Minister.

[7] By chance, the time of this appointment was the centenary of the school, for which they produced a large but superficial celebratory history (Pinet 1894–1897). In 1896 the various *Facultés* in the *Académies* were converted into separate and somewhat autonomous *Universités*.

industry and commerce led to resignations. However, *polytechniciens* were less appropriately trained for those sectors than graduates from the *Ecole Centrale des Arts et Manufactures,* or from the *Facultés,* where the Ministry of Public Instruction had been encouraging industrial support and collaboration for some time (Shinn 1979).

So in an atmosphere of some vacillation and uncertainty the school went on fulfilling its intended role and not adding very much to the academic roster. As the Great War loomed, enrolment rose in 1913 to the levels set following the Franco-Prussian War. An extended visitor to Paris shortly earlier was the historian of mathematics R.C. Archibald, who was extremely unimpressed (1911, 121):

> From being perhaps the leading school of the time with regard to its output of brilliant
> mathematicians it has, then, sunk to a position of wholly inconsiderable importance in
> this respect.

He blamed the 'widening realm' of engineering for his situation; but for once this fine historian had overlooked the reason for the formation and existence of the school, and placed that purpose below the remarkable but unintended generation of major research mathematicians.

13.10 Some Appraisals

13.10.1 On the School

Let us start by recalling that training of students to become major research scientists was never part of the explicit purpose of the school; hence its achievement was unexpected, and search for reasons for its decline after around thirty years should be replaced by wondering about its early rise. I put as the main reason the recruitment of talented students, though even then the quantity of merging talent listed in Table 13.1 is astonishing. Eminent *savants* were appointed to its staff, but a good researcher is not necessarily a good teacher; for example, Lagrange was notoriously awful. Moreover, its status applied only at the level of training; there were no research programmes at the school, and usually the scientific achievements of the *polytechniciens* in later life were not made in connection with it.

A major feature of our summary was the basic disagreement about the aims of the school, and the syllabi appropriate to fulfil them: whether its teaching should be oriented to really practical needs of engineers, or whether more general theories be taught and more advanced theories handled in the appropriate *écoles d'application*. The school and its management resemble a glamorous cruise ship where however there is much disagreement among the management both about the cruises to be offered and the kinds of food to list on the overlong menus.

13.10.2 On France

A curious imbalance in staffing emerged. The *écoles d'application* had distinguished staff of their own for certain periods, such as Navier at the *Ecole des Ponts et Chaussées* from 1819 and Poncelet at the Metz school from 1824; but, much more typically, Navier supplemented luminaries such as Joseph Eisenmann and Louis Bruyère, who were responsible for the teaching of mechanics presumably at a level more advanced than that which had been presented at the *Ecole Polytechnique* by Ampère and Cauchy! This situation suggests that the *écoles d'application* were underrated at the time—as has the histories of most of them since.

The companion story to the 'decline' of the *Ecole Polytechnique* is the extraordinary rise of the *Ecole Normale* to dominance in the French mathematical community after decades of insignificance, to the extent that *normaliens* took over much of the teaching of mathematics at the school. In common with trends in other parts of Europe in the second half of the 19th century, usually they preferred pure or applicable mathematics to applied mathematics on their research work (Gispert 1991), and their various *Cours d'analyse*s suggest that they carried this attitude over in their teaching at the *Ecole Polytechnique*. But the relevance to such teaching to military and civil engineers is dubious, as Leverrier for one—and not the only one—had asked and answered forcefully. The same difference is manifest in the research profile of the French mathematical community all century, where the pure or theoretically applied mathematician or scientist is rather distinct from the *ingénieur savant* (Grattan-Guinness 1995, Chatzis 2010).

Crucial support for the school, and thus for its fame, came from the relatively large number of posts that *polytechniciens* could hold and the vigorous publication industry for books and journals, both far greater than elsewhere until well into the 19th century; and also their own strong sense of bonding to the school and (usually) to each other. As Gillispie has explored in his magisterial study (2004) of French science and technology during and after the Revolution, engineering was connected to science much more closely than applied elsewhere, and the *Ecole Polytechnique* was a key component in this structure.

13.10.3 On Elsewhere

The history of the influence of the *Ecole Polytechnique* outside France has not been well studied, but here are some remarks. From the 1810s onwards mathematical education and research improved in several other countries and states improved; thus the French decline was to some extent an optical illusion. Some foreign improvements were reactions to their achievements, including the school itself, which was emulated in at least Sweden, Turkey and Saxony.

The most striking case was the United States Military Academy at West Point. The key figure was Sylvanus Thayer (1785–1872), who lived in Paris from 1814 to 1817 when the future of the school became in doubt and thus its role debated. After his return home and appointment as Academy Superintendent, Thayer followed the school in various ways, even giving the chair of descriptive geometry to *polytechnicien* Claudius Crozet (1788?-1864; 1805 enrolment) (Rickey 2002). But the differences between the two institutions are considerable, over and above the scientific eminence of France and the still fledgling condition of the U. S. A. The *Ecole Polytechnique* served both civilian and military roles, furnishing a two-year programme followed by further study elsewhere, whereas the Academy provided a full integrated course over its four years, and fulfilled only a military role (Albree and others 2000, ch. 1). So there is no analogue to the civilian side of the school and its connection to the *Ecole des Ponts et Chaussées* and *Ecole des Mines*; the connections that developed between the Academy and the Rensselaer Polytechnic Institute in mid-century might have arisen naturally as an attempt to fill the gap.

Finally, the school exercised some of its international influence through its publications, especially textbooks written both for admission and those based upon the teaching there; several were used outside France, and some were even translated (Grattan-Guinness 2002). But the French institutional structure was not copied; no military college elsewhere achieved the status of the *Ecole Polytechnique*, and universities remained the principal institutions for higher education in all other major countries and states.

Acknowledgements Sections 13.1–13.8 are condensed from an article (Grattan-Guinness 2005) in the *American mathematical monthly*; I am most grateful to the American Mathematical Association, which retains copyright. Section 13.9 was written as an appendix to that paper for an Italian translation; the original is published here for the first time. Section 13.10 is new.

Literature Review

Several histories and celebratory volumes of the *Ecole Polytechnique* have appeared at various times, not always reliable and often derivative from the sources now to be named. The reprint of (Fourcy 1828) in 1987 contains some further information by J. Dhombres. At the time of Fourcy's book the story to date was summarised succinctly in (Hachette 1828). Several relevant documents are contained in an edition of material on educational policy in general for the period 1792–1795 (Guillaume 1891–1907, esp. vol. 5, 627–653). The partial reprint of (Leverrier 1850a) in the *Moniteur universel* continued a tradition there from the start of the school of publishing much information: decrees, opinions and reactions but also data on staff and teaching. (Marielle 1855) is a valuable catalogue of the enrolments up to that time.

Among more recent sources (Grattan-Guinness 1990) contains much information in a detailed study of most of the French mathematical community;

(Dhombres and Dhombres 1989) surveys it from a broader perspective. The bicentenary of the school in 1994 led to several volumes, which are valuable mostly for the account of aspects of developments after 1914, teaching of subjects other than mathematics, and relationships of the school with other countries (Belhoste and others 1994a, 1994b, 1995). Belhoste (2003) reviews various aspects of its role and function up to 1870, including students' background (on which see (Shinn 1980) for more detail), enrolment and behaviour; and the relationship of the school to the applications schools and the associated professional corps.

The school's own archives contain many fine files, especially after its archivists, Natalie Bayle and Claudine Billoux, created remarkable order out of the previous chaos in the mid 1980s (Grattan-Guinness 1985). The website www.bibliotheque.polytechnique.fr contains many sources and lists. Several important files are kept in the archives of the *Ecole des Ponts et Chaussées,* apparently due to borrowings by de Prony.

Bibliography

Albree, J., Arney, D.C. and Rickey, F. 2000. *A station favorable to the pursuits of science: primary materials in the history of mathematics at the United States Military Academy,* Providence (American Mathematical Society) and London (London Mathematical Society).

Alder, K. 1997. *Engineering the revolution. ¸carms and enlightenment in France, 1763–1815,* Princeton (Princeton University Press).

Archibald, R.C. 1911. 'Mathematical instruction in France', *Trans. Royal Soc. Canada,* (1910), 89–152.

Belhoste, B. 1991. *Augustin-Louis Cauchy. A biography,* New York (Springer).

Belhoste, B. 2003. *La formation d'une technocratie. L'Ecole Polytechnique et ses élèves de la Révolution au second Empire,* Paris (Belin).

Belhoste, B. and Picon, A. 1996. (Eds.) *L'Ecole d'application de l'artillerie et du génie de Metz (1802–1870),* Paris (Ministère de la Culture, Direction du Patrimonie, Musée des Plans-Reliefs).

Belhoste, B. et al. 1994a. (Eds.), *La formation polytechnicienne 1794–1994,* Paris (Dunod).

Belhoste, B. et al. 1994b. (Eds.), *Le Paris des polytechniciens. Des ingénieurs dans la ville 1794–1994,* Paris (Délégation à l'Action de la Ville de Paris).

Belhoste, B. et al. 1995. (Eds.), *La France des X: deux siècles,* Paris (Economica).

Bradley, M. 1975. 'Scientific education versus military training: the influence of Napoleon Bonaparte on the *Ecole Polytechnique*', *Ann. Sci., 32,* 415–449.

Bradley, M. 1976a. 'Scientific education for a new society. The Ecole Polytechnique 1795–1830', *Hist. Educ., 5,* 11–24.

Bradley, M. 1976b. 'An early science library and the provision of textbooks: the *Ecole Polytechnique,* 1794–1815', *Libri, 26,* 165–180.

Bradley, M. 1998. *A career biography of Gaspard Clair François Marie Riche de Prony, bridge-builder, educator and scientist,* Lewiston, Queenston and Lampeter (Edwin Mellen).

Carnot, L.N.M. 1796. 'Loi concernant les écoles de services publics', *J. Ecole Polytechique, (1)1, cah.4,* xiii–xxviii.

Chatzis. K. 2010. 'Theory and practice in the education of French engineers from the middle of the 18th century to the present', *Arch. Int. d'Hist. Sci., 60,* 43–79.

Dhombres, J.G. and Dhombres, N. 1989. *Naissance d'un nouveau pouvoir. Sciences et savants en France (1793–1824),* Paris (Payot).

Dupuy, P. 1895. 'L'*Ecole Normale* de l'an III', in *Le centenaire de l'Ecole Normale 1795–1895*, Paris (Hachette), 1–209.

Fourcy(-Gaudain), A.L. 1828. *Histoire de l'Ecole Polytechnique*, Paris (*Ecole Polytechnique*). [Repr. Paris (Belin), 1987 with notes by J.G. Dhombres.]

Fox, R. and Weisz, G. 1980. (Eds.), *The organisation of science and technology in France 1808–1914*, Cambridge (Cambridge University Press).

Gillispie, C.C. 1997. C.C. Gillispie, *Pierre Simon Laplace. A life in exact science*, Princeton (Princeton University Press).

Gillispie, C.C. 2004. *Science and polity in France. The revolutionary and Napoleonic years*, Princeton (Princeton University Press).

Gispert, H. 1991. *La France mathématique. La Société Mathématique de France (1870–1914)*, Paris (Belin).

Gispert, H. 1992. 'L'enseigement de l'analyse à l'Ecole. Un moment du débat théorie/application', *Bull. Soc. des Amis de la Bibl. Ecole Polyt.*, no. 1, 2–14.

Grattan-Guinness, I. 1985. 'On the transformation of the *Ecole Polytechnique* archives', *Br. J. Hist. Sci.*, *19*, 45–50.

Grattan-Guinness, I. 1988. '*Grandes écoles, petite Université*: some puzzled remarks on higher education in mathematics in France, 1795–1840', *Hist. Univ.*, *7*, 197–225.

Grattan-Guinness, I. 1990. *Convolutions in French mathematics, 1800–1840. From the calculus and mechanics to mathematical analysis and mathematical physics*, 3 vols., Basel (Birkhäuser) and Berlin, DDR (Deutscher Verlag der Wissenschaften).

Grattan-Guinness, I. 1995. 'The *ingénieur savant*, 1800–1830: a neglected figure in the history of French mathematics and science', *Sci. Context, 6*, 405–433.

Grattan-Guinness, I. 2002. 'End of dominance: the diffusion of French mathematics elsewhere, 1820–1870', in K.H. Parshall and A. Rice eds., *Mathematics unbound: the evolution of an international mathematical community*, Providence, R.I. (American Mathematical Society), 17–44.

Grattan-Guinness, I. 2005. 'The *Ecole Polytechnique*, 1794–1850: Differences over educational purpose and teaching practice', *Am. Math. Monthly, 112*, 233–250

Guillaume, J. 1891–1907. (Ed.) *Procès-verbaux du Comité d'Instruction Publique de la Convention Nationale*, 6 vols. and undated index in 2 pts., Paris (Hachette).

Hachette, J.N.P. 1828. 'Notice sur la création de l'*Ecole Polytechnique*', in G.E.D. Monnais, ed., *Ephémérides universelles*, vol. 3, Paris (Corby), 251–261; also in *J. Gén. Civil, 2*, 251–263.

Julia, G. 1995. 'Le modèle méritocratique entre Ancien Régime et Révolution', in (Belhoste and others 1995), 33–50.

Lamy, L. 1995. 'Le *Journal de l'Ecole Polytechnique* de 1795 à 1831; journal savant, journal institutionnel', *Sci. Tech. Perspect., 32*, 150 pp.

Langins, J. 1987a. *La République avait besoin de savants*, Paris (Belin).

Langins, J. 1987b. 'Sur l'enseignement et les examens à l'*Ecole Polytechnique* sous le Directoire. A propos d'une lettre inédite de Laplace', *Rev. Hist. Sci., 40*, 145–177.

Langins, J. 1989. 'Histoire de la vie et les fureurs de François Peyrard', *Bull. Soc. des Amis de la Bibl. Ecole Polytechnique*, no. 3, 2–12.

Langins, J. 1991. 'La préhistoire de l'Ecole Polytechnique', *Rev. Hist. Sci., 44*, 61–89.

Leverrier, U.J.J. 1850a. *Rapport sur l'enseignement de l'Ecole Polytechnique …*, Paris. [Part in *Moniteur universel*, (12 January 1851), suppl., i–xxxiv.]

Leverrier, U.J.J. 1850b. 'Rapport supplémentaire … relative aux écoles polytechnique et militaire', *Assemblée Nationale Législative*, no. 865, 44 pp.

Marielle, C.P. (Bacouillard-) 1855. *Repertoire de l'Ecole Polytechnique …*, Paris (Mallet-Bachelier).

Paul, M. 1980. *Gaspard Monges "Géométrie descriptive" und die Ecole Polytechnique …*, Bielefeld (Universität, Institut für Didaktik der Mathematik).

Pepe, L. 1996. 'La formazione della biblioteca dell'*Ecole Polytechnique*: il contributo involontario del Belgio e dell'Italia', *Bull. Stor. Sci. Math., 16*, 155–197.

Picon, A. 1992. *L'invention de l'ingénieur moderne. L'Ecole des Ponts et Chaussées 1747–1851*, Paris (Presses de l'École Nationale des Ponts et Chaussées).

Pinkney, D. 1972. *The French revolution of 1830*, Princeton (Princeton University Press).

Prieur de la Côte d'Or, C. 1795. *Programmes de l'enseignement polytechnique de l'Ecole Centrale des Travaux Publics*, Paris (Imprimerie Nationale); also in (Langins 1987a), 126–198.

Rickey, W.F. 2002. 'The first century of mathematics at West Point', in (Shell-Gellasch 2002), 25–45.

Sédillot, L.P.E.A. 1870. 'Les professeurs de mathématiques et de physique générale au *Collège de France*', pt. 4, *Bull. Bibl. Stor. Sci. Math. Fis., 3*, 107–170.

Shell-Gellasch, A. 2002. (Ed.), *History of undergraduate mathematics in America*, West Point (Military Academy).

Shinn, T. 1979. 'The French science faculty system, 1808–1914: institutional change and research potential in mathematics and the physical sciences', *Hist. Studies Phys. Sci., 10*, 271–332.

Shinn, T. 1980. *Savoir scientifique et pouvoir social: l'Ecole Polytechnique 1794–1914*, Paris (Presse Fondation Nationale des Sciences Politiques).

Zwerling, C.S. 1976. 'The emergence of the *Ecole Normale Supérieure* as a center of scientific education in nineteenth century France', Harvard University Ph.D.

Chapter 14
The Notion of Variation in Leibniz

Eberhard Knobloch

14.1 Introduction

From the very beginning of his scientific studies Leibniz's philosophical and mathematical thinking was strongly influenced by combinatorial ideas. As a young man he temporarily adhered to Lullism but abandoned this theory rather soon because Lullists such as Athanasius Kircher disappointed him. His conception of a combinatorial art was closely connected with an inventive logic and an art of invention. He made a first major effort in this respect in his *Dissertatio de arte combinatoria* that appeared in 1666 (Leibniz 1666; Echeverria/Amunategui 2005).

The underlying key notion was the notion of variation that he defined in the following way (Leibniz 1666, 171f.):

1. Variatio h. l. est mutatio relationis. Mutatio enim alia substantiae est alia quantitatis alia qualitatis; alia nihil in re mutat, sed solum respectum, situm, conjunctionem cum alio aliquo.
2. Variabilitas est ipsa quantitas omnium variationum. . . .
3. Situs est localitas partium.

1. Here variation is a change of a relation. For there is now a change of substance, now one of quantity, now one of quality. Another variation does not change anything of the object but only the connexion, locality, conjunction with something.
2. Variability is the quantity of all variations. . . .
3. 'Situs' is the locality of the parts.

Obviously this classification is reminiscent of Aristotle's hylomorphism. About thirty years later Leibniz used a classification even more reminiscent of

E. Knobloch (✉)
Technische Universität, Berlin, Germany
e-mail: eberhard.knobloch@tu-berlin.de

J.Z. Buchwald (ed.), *A Master of Science History*, Archimedes 30,
DOI 10.1007/978-94-007-2627-7_14, © Springer Science+Business Media B.V. 2012

Aristotle (Leibniz 1976, 40). The manuscript was written between 1690 and 1697 and deals with variations and their numbers saying:

> In variatione aut materiae seu rerum ingredientium, aut formae seu dispositionis ratio habetur. In dispositione cum alia spectari possunt tum maxime ordo. Variationes rerum sunt complexus; variationes ordinis sunt transpositiones. Et complexus sunt vel simplices vel (si eadem res plus semel occurrat) replicati.

> In a variation either matter, that is, the things in question, or form, that is, dispositions are taken into account. In a disposition also other things are considered but mainly the order. Variations of things are combinations; variations of order are transpositions. And combinations are either simple or (if the same thing occurs more often than once) repeated.

Thus we get the following classification:

$$
\begin{array}{c}
\text{variationes} \\
\text{variations}
\end{array}
$$

materiae	formae
of matter	of form
(complexus = combinations)	
simplices replicati	variationes ordinis other aspects
simple repeated	variations of order
	(transpositiones = permutations)

This terminology differs from modern notions insofar as one speaks of combinations (subsets) and permutations (bijective mappings) with or without repetitions. If the order of the elements of a combination matters one considers arrangements with or without repetitions.

In the following paper I would like to clarify where variations played a crucial role in Leibniz's mathematics and then confine myself to their occurrences in his algebraic studies.

14.2 Variations in Leibniz' Mathematics

Where did the combinatorial art play an essential role in Leibniz's mathematics? In his *Dissertatio de arte combinatoria* the most original and most important part concerned the so-called 'caput'-theory. A 'caput' is a given subset of a set. Leibniz asked for the number of variations that contain such a subset ('caput variationis'). Some years ago, Echeverria and Amunategui gave a group-theoretical interpretation of this theory (Echeverria/Amanategui 2005).

Later on, at least four mathematical areas can be identified where Leibniz successfully applied the combinatorial art:

1. Symmetric functions (Leibniz 1976)
2. (Number-theoretical) partitions (Leibniz 1976)
3. Determinant theory (Knobloch 1980)
4. Insurance calculus (Knobloch/Schulenburg 2000).

The following diagram describes the structure of mathematical connections between the concerned areas:

<pre>
                          combinatorial art
         algebra   number theory insurance calculus
   linear   nonlinear   partitions   . . . . .
determinants   finite equations   infinite equations
               symmetric functions
</pre>

14.3 The Solution of Algebraic Equations

In order to understand why symmetric functions played such a crucial role in Leibniz's mathematical writings one has to know how he tried to solve algebraic equations of higher degree. Like his contemporaries he was convinced that these equations can be algorithmically solved. At the end of May or at the beginning of June 1678 he wrote to his friend Ehrenfried Walther von Tschirnhaus (LSB III,2, 428):

> "Observaveram jam olim et fortasse primus radices irrationales altiores exemplo Cardanicarum inveniri posse", "Already long ago I had observed and maybe as the first that the higher irrational roots can be found according to the example of the Cardanic (roots)".

In other words Leibniz generalized the Cardanic approach by using a 'section of the root' (sectio radicis), that is, he represented the root being sought as a sum of terms.

In about 1680/1682 he wrote (Leibniz 1976, 201):

> Assumo jam instar hypotheseos, quod radix quaesita aequationis x, sit polynomium, seu x = 1 + m etc. quae 1, m, etc. sunt quantitates irrationales, posito aequationem carere termino secundo.

> I assume just as an hypothesis, that the sought root x of the equation is a polynomial or x = 1 + m etc. These 1, m etc. are irrational quantities on condition that the second term of the equation is lacking.

If the equation reads $x^n + qx^{n-1} + \ldots + t = 0$, $x = a + b + c + \ldots$, there are n-1 terms of the sum.

Hence Leibniz had to calculate the different powers of x thus dealing with expressions like $y = \overset{..}{ab}$, $z = \overset{..}{abc}$, $\omega = \overset{..}{abcd}$ etc.: These are the elementary symmetric functions or 'formae simplices'.

$f = \overset{..}{a^n}$: These are the power sums or 'potentiae combinatoriae polynomiorum'.

$g = \overset{..}{a^n b^n}$, $h = \overset{..}{a^n b^n c^n}$ etc.: These are the multiform symmetric functions. The third power of

$x = a + b + c$ reads in this notation: $x^3 = 3\overset{..}{a^2 b} + 6\overset{..}{abc} = (a^3 + b^3 + c^3)$

$$+ 3(a^2 b + ab^2 + a^2 c + ac^2 + b^2 c + bc^2) + 6abc.$$

In May 1678 Leibniz was convinced that the calculation of the root can be reduced to a system of linear equations. Exactly for this reason he became interested in such systems of equations. In order to solve them he invented the determinant theory (Knobloch 2000). But first he had to study symmetric functions or 'formae'.

14.4 Symmetric Functions: Tables for the Numbers of Terms in Any Symmetric Function

Around about 1677/1680 he wrote in the treatise 'De formis omnibus ad solas formas simplices reducendis deque aequationum radicibus novissima methodus' ('On the reduction of all symmetric functions to exclusively elementary symmetric functions and a newest method regarding the roots of equations') (Leibniz 1976, 55):

> Usus formarum praeter pulchritudinem et generalitatem contemplationis in eo consistit, ut ope earum inveniamus radicem generalem aequationis affectae cujuscunque gradus.

'Apart from the beauty and generality of the idea, the use of symmetric functions consists in finding the general root of a non-pure equation of an arbitrary degree.' To that end Leibniz had to find out the number of terms of a symmetric polynomial.

Let $l^h m^r n^s \ldots$ be a symmetric polynomial, $p = h + r + s + \ldots$ its degree, k the number of variables of a single term, v the number of admitted variables, d the number of pairwise different exponents.

What is interesting here is the number k of variables that enter a single term of the function. It does not matter whether the exponents are small or great. Only the frequency of their occurrence matters, only the type of their repetition in order to calculate the number of terms of a certain symmetric function.

$l^4 m^4 n^3 o$ or $l^3 m^2 no$ belong to the same type of repetition: one exponent (4 or 1, respectively) occurs exactly twice, two exponents occur once at a time (3, 1 or 3, 2, respectively). Such a type of repetition can be described by means of a number-theoretical partition of k (the number of variables of a single term):

In our case $k = 4$ or $4 = 2 + 1 + 1 = 2.1 + 1.2 = 2r_2 + 1r_1$

Hence there are three steps:

(1) Consider the different exponents.
(2) Determine the frequency of their occurrence.
(3) Describe the type of repetition by a (number-theoretical) partition.

Let $l^3 m^3 n^2 o^2 p$ be given as an example. Hence $k = 5$; let v be 6.

(1) There are $\binom{6}{5} = 6$ possibilities to select a combination of five elements.

(2) Every such combination like lmnop admits $\binom{5}{2} = 10$ terms of the type $l^3 m^3 nop$. In the whole there are $6.10 = 60$ such terms.

(3) Every such term l^3m^3nop admits $\binom{3}{2} = 3$ terms of the type $l^3m^3n^2o^2p$. In the whole there are $60.3 = 180$ such terms.

Between May 1677 and May 1678 Leibniz elaborated a table for the numbers of terms in any symmetric function. At the very beginning he emphasized its usefulness (Knobloch 1973, table after page 248) (Fig. 14.1):

> Numerus terminorum in qualibet forma specimen est artis combinatoriae sed usum praeterea habet maximum ad multiplicationes formarum in se invicem compendio faciendas.

> The number of terms in an arbitrary symmetric function is an example of the combinatorial art, but moreover it is extremely useful in order to multiply symmetric functions by each other in a shortened way.

Step by step the rows refer to the different types of repetition of combinations consisting of one, two, three, four etc. elements, while the columns refer to the number of admitted variables (one, two, three, four etc.). The example $l^3m^3n^2o^2p$ considered above belongs to the same type of repetition as $l^3m^2n^2op$. The sixteenth row ($l^3m^2n^2op$) reveals 180 (our result) in the sixth column ($v = 6$):

The type of repetition in question can be described by:
$k = 5 = 1 + 2 + 2 = 1.1 + 2.2 = 1r_1 + 2r_2$. There are $d = r_1 + r_2 = 3$ pairwise different exponents.

Now the general solution of the general case can be deduced:

$$k = k_1 + \ldots + k_{r_1} + k_{r_1+1} + \ldots + k_{r_1+r_2} + \ldots + k_{r_1+r_2+\ldots+r_k} =$$
$$= 1 + \ldots\ldots + 1 + \quad 2 + \ldots\ldots\ldots + 2 + \ldots \quad + k =$$
$$\qquad r_1 \text{ times} \qquad\qquad\qquad r_2 \text{ times} \qquad\qquad\qquad r_k \text{ times}$$
$$= 1r_1 + 2r_2 + \ldots + kr_k$$

If r_k should occur, it must be equal to 1, all the other r_i must be equal to 0.

Let $r_1 + r_2 + \ldots + r_k = d$ be the number of pairwise different exponents, M the number of terms sought , $k = v$:

$$M = \binom{k}{k_1}\binom{k-k_1}{k_2}\binom{k-k_1-k_2}{k_3}\cdots\binom{k-k_1-k_2-\ldots-k_{d-1}}{k_d} =$$

$$= \frac{k!}{k_1!(k-k_1)!} \cdot \frac{(k-k_1)!}{k_2!(k-k_1-k_2)!} \cdots \frac{(k-k_1-k_2-\ldots-k_{d-1})!}{k_d!(k-k_1-k_2-\ldots-k_{d-1}-k_d)!} =$$

$$= \frac{k!}{k_1!k_2!\ldots k_d!} = \frac{k!}{(1!)^{r_1}(2!)^{r_2}\ldots(k!)^{r_k}}$$ according to the frequencies of $1, 2, \ldots, k$, respectively. If v should be larger than k, M has still to be multiplied by $\binom{v}{k}$. This result admits an important interpretation: M is the number of partitions of k objects into d classes $S_1, \ldots, S_d$ that contain $k_1, \ldots, k_d$ objects.

TABULA PRO NUMERIS TERMINORUM IN QUALIBET FORMA
[Mai 1677 – Mai 1678]

Überlieferung: L Konzept: LH XXXV 14,1 Bl.293. Das Blatt (ehemals 2°) ist an zwei Seiten beschnitten und
  mißt etwa 19,5 x 31 cm; eine trapezförmige Ecke (Höhe 13,5 cm, Grundseiten 7 bzw. 10 cm)
  fehlt. 1 S.

Numerus Terminorum in qualibet forma specimen est artis Combinatoriae sed usum praeterea habet maximum ad
multiplicationes formarum in se invicem compendio faciendas.

Column-head staircase (over columns 1–10):

$$
\begin{array}{cccccc}
 & & & & l & \\
 & & & l & m & \\
 & & l & m & n & \\
 & l & m & n & o & \&c. \\
l & m & n & o & p & q \\
\end{array}
$$

formae	1	2	3	4	5	6	7	8	9	10	
$l.(l^2.l^3)$etc.	1	2	3	4	5	6	7	8	9	10	A
$lm.(l^2m^2.l^3m^3)$&c.	0	1	3	6	10	15	21	28	36	45[1]	B
$l^2m(l^3m.l^3m^2)$	0	2	6	12	20	30	42	56	72	90[1]	2B
$lmn(l^2m^2n^2.l^3m^3n^3)$	0	0	1	4	10	20	35	56	84	120	C
$l^2mn(l^2m^2n.l^3m^3n^2)$	0	0	3	12	30	60	105	168	252	360	3C
$l^3m^2n(l^4m^3n^2.l^4m^2n)$	0	0	6	24	60	120	210				2,3C
$lmno$(etc.)	0	0	0	1	5	15	35	70	126	210	D
l^2mno	0	0	0	4	20	60	140				4D
l^2m^2no	0	0	0	6	30	90	210				$\frac{3,4}{2}$D
$l^3m^2no(l^4m^4n^2o.l^4m^3no)$	0	0	0	12	60	180					3,4D
$l^4m^3n^2o$	0	0	0	24	120	360					2,3,4D
$lmnop$	0	0	0	0	1	6	21	56	126	252	E
l^2mnop	0	0	0	0	5	30					5E
l^2m^2nop	0	0	0	0	10	60		560			$\frac{4,5}{2}$E
l^3m^2nop	0	0	0	0	20	120					4,5E
$l^3m^2n^2op$	0	0	0	0	30	180					$\frac{3,4,5}{2}$E
$l^4m^3n^2op(l^4m^3n^2o^2p)$	0	0	0	0	60	360					3,4,5E
$l^5m^4n^3o^2p$	0	0	0	0	120	720					2,3,4,5E
$lmnopq$	0	0	0	0	0	1	7	28	84	210	F
l^2mnopq	0	0	0	0	0	6	42	168	504	1260	6F
l^2m^2nopq	0	0	0	0	0	15	105	420[2]			$\frac{5,6}{2}$F
l^3m^2nopq	0	0	0	0	0	30	210				5,6F
$l^2m^2n^2opq$	0	0	0	0	0	20	140[3]				$\frac{4,5,6}{2,3}$F
$l^4m^3n^2opq$	0	0	0	0	0	120	840				4,5,6F
$l^3m^2n^2p^2pq$	0	0	0	0	0	60	420				$\frac{4,5,6}{2}$F[5]
$l^3m^3n^2o^2pq$	0	0	0	0	0	90					$\frac{3,4,5,6}{4}$F

Commentary (right column):

A — Numeri literarum seu Naturales

B — Numeri Combinationum seu Triangulares

...... dupli B nam lm dabit l^2m lm^2
ln l^2n ln^2
mn m^2n mn^2

C — Numeri Con3nationum seu Pyramidales

...... tripliC nam lmn dat $l^2mn.lm^2n.lmn^2$
...... Nam varietates l^3m^2n fiunt ex varietatibus
$l^2mn.$ hoc modo quilibet hujus formae terminus,
ut l^2mn dat duos terminos prioris formae, ut l^2mn
dat l^2m^3n, l^2mn^3

D — Numeri Con4nationum seu Triangulo-Triangulares

$l^2mno.lm^2no.lmn^2o.lmno^2$ ex $lmno$
$(l^2)m^2no.(l^2)mn^2o.(l^2)mno^2$ ex l^2mno Ubi patet
$l^2(m^2)no.l(m^2)n^2o.l(m^2)no^2$ ex lm^2no terminum
l^2m^2no bis prodire eodem modo et reliquos omnes.
Ergo ad habendum numerum terminorum formae
l^2m^2no ex variatione unius termini in forma
l^2mno,debent variationes provenientes dimidiari.
$l^3m^2no.l^3mn^2o.l^3mno^2$ ex l^3mno.:Nec metuendum
hic, ut si caeteros formae variandae (l^3mno) terminos, ut lm^3no varies, redeat aliquis terminus produ-
cendae formae in $l^2m^3no.lm^3n^2o.lm^3no^2$ ex lm^3no.
ratio est quia nunquam iterum redibit l^3. Quod secus
erat in producenda forma l^2m^2no, ubi l^2 duobus mo-
dis prodire poterat, uno ex forma varianda, altera ex
variatione.
$l^4m^3n^2o.l^4m^3no^2$ ex l^4m^3no
$l^3m^4n^2o.l^3m^4no^2$ l^3m^4no

E — Numeri Con5nationum seu Trianguli- Pyramidales

$l^2mnop.lm^2nop.lmn^2op.lmno^2p.lmnop^2$ ex $lmnop$
$(l^2)m^2nop.(l^2)mn^2op.(l^2)mno^2p.(l^2)mnop^2$ ex
$l^2mnop.$ $l^2(m^2)nop.l(m^2)n^2op.l(m^2)no^2p.l(m^2)nop^2$
ex $lm^2nop.$Unde patet Numerum formae l^2m^2nop esse
$\frac{4,5}{2}$E. Nam si nulli repeterentur termini variatione pro-
ducti foret numerus 4,5E: idem autem terminus non nisi
bis prodit, quod jam tum praescire possumus, quia 4,5 ,
seu 20 non potest dividi per 3, sed per 2. Nam quater aut
quinquies prodire non posse, manifestum est.
$l^3m^2nop.l^3mn^2op.l^3mno^2p.l^3mnop^2$ ex l^3mnop
Nam numerus terminorum hujus formae fit ex 4,5E nu-
mero terminorum praecedentis formae multiplicando eum
per $\frac{3}{2}$, cujus rei ratio est, quod omisso l^3,quia per variatio-
nem non producitur, perinde est ac si ex m^2nop per varia-
tionem produceremus m^2n^2op, sive si ex l^2mno cujus numerus erat 4D, produceremus l^2m^2no cujus numerus est
$\frac{3,4}{2}$D. quod fit numerum formae variandae multiplicando
per $\frac{3}{2}$.

F — Numeri Con6nationum seu Pyramido-pyramidales

brevius quod sex rerum combinationes 15

$l^2m^2n^2opq.l^2m^2no^2pq.l^2m^2nop^2q.$
$l^2m^2nopq^2$ ex l^2m^2nopq
$l^2m^2n^2opq.l^2mn^2o^2pq.l^2mn^2op^2q.$
$l^2mn^2opq^2$ ex l^2mn^2opq
$l^2m^2n^2opq.lm^2n^2o^2pq.lm^2n^2op^2q.$
$lm^2n^2opq^2$ ex lm^2n^2opq

fit ex l^2m^2nopq quia 4 rerum nopq sex sunt
combinationes quarum duae quaelibet ad cubum
possunt attolli

[1] 54 108 L ändert Hrsg.   [2] 240 L ändert Hrsg.   [3] 280 L ändert Hrsg.   [4] $\frac{4,5,6}{3}$F L ändert Hrsg.   [5] $\frac{3,4,5,6}{3}$F gestr. L

Fig. 14.1 Table for the number of terms in any symmetric function
Source: Knobloch (1973, table after page 248)

Remark 1 The order of the classes matters even if they contain the same number of elements. It matters whether l, m or n, o have the exponent 2 or 3: $l^3m^3n^2o^2p$ is unequal to $l^2m^2n^3o^3p$. Yet, both terms represent the same type of repetition $k = 5 = 1.1 + 2.2$. The different exponents individualize the classes. They are distinguishable so to speak by different colours.

Remark 2 If only the number of classes and their contents matter, M has still to be multiplied by $\frac{1}{r_1!r_2!...r_k!}$, whereby the r_i are the numbers of classes of equal size. The following example might help to clarify the situation. Four objects l, m, n, o have to be put into three classes, no class must remain empty. There is only one type of partition: $4 = 1 + 1 + 2$. We get six possibilities:

l|m|no, l|n|mo, l|o|mn, m|n|lo, m|o|ln, n|o|lm. m|l|no is no new possibility because the order of the classes does not matter. Hence $M = S_k^d = S_4^3 = 6$. Such numbers are called Stirling's numbers of second kind.

But if we seek the number of terms of the symmetric function represented by l^3m^2no ($v = k$ and $k = 1.2 + 2.1$), there are 12 instead of 6 terms. We shall come back to this question in Section 14.8.

14.5 Symmetric Functions: The Multiplication Tables

Leibniz's method of looking for the algorithmic solution of algebraic equations made the multiplication of symmetric functions necessary. As we saw above he used his table for the number of terms in any symmetric function to construct multiplication tables. Hence in 1677 he explicitly said (Leibniz 1976, 180) (Fig. 14.2):

> Continuanda inquisitio de multiplicationibus formarum quia in materia resolutionis aequationum primaria est.

> The investigation of the multiplication of symmetric functions must be continued because it is of highest importance for the solution of equations.

He explains his multiplication method in his letter to Tschirnhaus dating from the end of May or the beginning of June 1678 (LSB III, 2, 435, 441).

Let us assume the we have to multiply l^2m by lm on condition that there are four variables l, m, n, o. There are three steps:

(1) All terms of one symmetric function have to multiplied by one term of the other function. The table for the number of terms of any symmetric function shows which of the two functions has fewer terms than the other in view of the presupposed conditions: l^2m has 12, lm has six terms. Hence it is reasonable to enumerate the six terms of lm which are multiplied by l^2m:

$$\begin{array}{cccccc}
lm & ln & lo & mn & mo & no \\
l^2m \quad l^3m^2 & l^3mn & l^3mo & l^2m^2n & m^2n^2o & l^2mno
\end{array}$$

	l	l^2	l^3	lm	l^2m^2	l^3m^3	lmn	$l^2m^2n^2$	$l^3m^3n^3$	l^2m	l^3m	l^3m^2	l^2mn	l^3mn	l^2m^2n
l	l^2 $2lm$	l^3 l^2m	l^4 l^3m	l^2m $3lmn$	l^3m^2 l^2m^2n	l^4m^3 l^3m^3n	l^2mn $4lmno$	$l^3m^2n^2$ $l^2m^2n^2o$	$l^4m^3n^3$ $l^3m^3n^3o$	l^3m $2l^2m^2$ $2l^2mn$	l^4m l^3m^2 $2l^3mn$	l^4m^2 $2l^3m^3$ l^3m^2n	$[l^3mn]$ $[2l^2m^2n]$ $3l^2mno$	l^4mn l^3m^2n $3l^3mno$	l^3m^2n $3l^2m^2n^2$ $[2l^2m^2no]$
l^2		l^4 $2l^2m^2$	l^5 l^3m^2	l^3m lmn^2	l^4m^2 $3l^2m^2n^2$	l^5m^3 $l^3m^3n^2$	l^3mn $lmno^2$	$l^4m^2n^2$ $4l^2m^2n^2o^2$	$l^5m^3n^3$ $l^3m^3n^3o^2$	l^4m l^3m^2 $2l^2mn^2$	l^5m $2l^3m^3$ l^3mn^2	$[l^5m^2]$ $[l^4m^3]$ $2l^3m^2n^2$	l^4mn l^2m^3n $2l^2mno^2$	$[l^5mn]$ $[2l^3m^3n]$ l^3mno^2	l^4m^2n $l^2m^2n^3$ $3l^2m^2no^2$
l^3			l^6 $2l^3m^3$	l^4m lmn^3	l^5m^2 $l^2m^2n^3$	l^6m^3 $[3l^3m^3n^3]$	l^4mn $lmno^3$	$l^5m^2n^2$ $l^2m^2n^2o^3$	$l^6m^3n^3$ $4l^3m^3n^3o^3$	l^5m $[l^4m^2$ l^2mn^3	l^6m l^3m^4 $2l^3mn^3$	l^6m^2 $l^5m^3]$ $2l^3m^2n^3$	l^5mn l^2m^4n l^2mno^3	l^6mn l^3m^4n l^3mno^3	l^5m^2n $l^2m^2n^4$ $l^2m^2no^3$
lm				l^2m^2 $2l^2mn$ $6lmno$	l^3m^3 l^3m^2n l^2m^2no	l^4m^4 l^4m^3n l^3m^3no	l^2m^2n $3l^2mno$ $10lmnop$	$[l^3m^3n^2\ ]$ $[l^3m^2n^2o\ ]$ $[l^2m^2n^2op]$	$l^4m^4n^3$ $l^4m^3n^3o$ $l^3m^3n^3op$	l^3m^2 $2l^3mn$ $2l^2m^2n$ $3l^2mno$	l^4m^2 $2l^4mn$ l^3m^2n $3l^3mno$	l^4m^3 l^4m^2n $2l^3m^3n$ l^3m^2no	l^3m^2n $3l^3mno$ $3l^2m^2n^2$ $4l^2m^2no$ $6l^2mnop$	l^4m^2n $3l^4mno$ $l^3m^2n^2$ $2l^3m^2no$ $6l^3mnop$	$[l^3m^3n]$ $2l^3m^2no$ $[2l^3m^2n^2]$ $3l^2m^2n^2o$ $3l^2m^2nop$
l^2m^2					l^4m^4 $2l^4m^2n^2$ $6l^2m^2n^2o^2$	l^5m^5 $l^5m^3n^2$ $l^3m^3n^2o^2$	l^3m^3n $[l^3m^2no]$ l^2m^2nop	$l^4m^4n^2$ $3l^4m^2n^2o^2$ $10l^2m^2n^2o^2p^2$	$l^5m^5n^3$ $l^5m^3n^3o^2$ $l^3m^3n^3o^2p^2$	l^4m^3 l^4mn^2 $2l^3m^2n^2$ $3l^2mn^2o^2$	l^5m^3 l^5m^2n $2l^3m^3n^2$ $l^3m^2n^2o$	l^5m^4 $2l^5m^2n^2$ $l^4m^3n^2$ $3l^3m^2n^2o^2$	l^4m^3n $2l^4m^2no$ $l^3m^3n^2$ $[2l^3m^2n^2o\ ]$ $[3l^2mno^2p^2]$	$[l^5m^3n]$ l^5m^2no $3l^3m^3n^3$ $2l^3m^3n^2o$ $l^3m^2n^2op$	l^4m^4n $l^4m^3n^2$ $2l^4m^2n^2o$ $3l^3m^2n^2o^2$ $[6l^2m^2no^2p^2]$

Fig. 14.2 Multiplication table for symmetric functions
Source: Leibniz (1976, 175)

(2) One realizes that there are repetitions. For example the terms 1^3mn, 1^3mo or the terms 1^2m^2n, 1^2m^2o represent the same function, respectively. Yet, only those terms have to be enumerated that have different relations with 1^2m that is

$$lm \quad ln \quad\quad mn \quad\quad no$$

The others are taken into account by a later multiplication of the coefficients.

(3) The products are multiplied by coefficients consisting of apparent fractions. In reality the denominator is always a factor of the numerator. The numerator is the number of terms of the multiplier 1^2m, in our case 12. The denominator is the number of terms of the product, in our case three times 12 and once 4. Hence we get the coefficients:

$$\frac{12}{12} \quad \frac{24}{12} \quad \frac{24}{12} \quad \frac{12}{4}$$

or

$$1^3m^2 + 2\,1^3mn + \quad 2\,1^2m^2n + \quad 3\,1^2mno$$

14.6 Symmetric Functions: The Fundamental Theorem

Within the area of symmetric functions the elementary symmetric functions behave like the prime numbers within the ring of integers. This matter of fact is expressed by the fundamental theorem of the theory of symmetric functions. Already Leibniz has gained this insight. In many manuscripts Leibniz dealt with this subject. Already the title of some manuscripts made this clear as in the case of the manuscript cited above that dated from about 1677/1680: 'On the reduction of all symmetric functions to exclusively elementary symmetric functions etc.' In this manuscript he avows (Leibniz 1976, 54):

> Cum ergo tam paucae sint formae simplices si compositis comparentur, patet magnum calculi circa formas compendium fore, si compositas omnes ad simplices reducere possimus. Id vero succedere posse spes magna est.

> Thus because there are so few elementary symmetric functions compared with composite, it is clear that there will be a considerable abbreviation of calculation regarding the symmetric functions, if we can reduce all composite to elementary. But there is a big hope that this can happen.

Leibniz still expresses a hope. In another manuscript dating from June 4, 1678 he already realizes (Leibniz 1976, 85):

> Fundamentum hujus calculi sumitur ex tabula formarum in se invicem ductarum, modum autem probandi per numeros, sumsimus ex tabula exemplorum cujusque formae, ponendo quodlibet exemplum 1. Hinc patet posse omnes formas resolvi in x, y, z, ω et ex his facta.

> The foundation of this calculation is taken from the table of symmetric functions that are multiplied by each other. But we took the way of proving by numbers from the table of terms of an arbitrary symmetric function putting every term equal to 1. As a consequence it is clear that all symmetric functions can be reduced to x, y, z, ω and to their products.

In other words Leibniz formulated the fundamental theorem of the theory of symmetric functions without proving it. He just expressed a research program. It did not happen by chance that he dated this manuscript. He was well aware of the importance of his discovery. About two years later he elaborated his study *Formarum reductio ad simplices* (The reduction of symmetric functions to elementary symmetric functions). There he systematically and recursively calculated such reductions of certain groups of symmetric functions:

$$a^3b^20 = xyz - 1yxx + 3zx - 0\delta - 3xz + 0yy$$

$$a^3b^21 = yyx - 2zxx + 5\delta x - 5\varepsilon - 3yz + 2zy$$

$$a^3b^2c = zyx - 3\delta xx + 7\varepsilon x - 12\theta - 3zz + 4\delta y$$

$$a^3b^2cd = \delta yx - 4\varepsilon xx + 9\theta x - 21\lambda - 3\delta z + 6\varepsilon y \text{ etc.}$$

The order reveals the rule of formation for such a group of functions. Leibniz ends by saying (Leibniz 1976, 191):

> Perficiendus est hic calculus, mira enim compendia, et omnino totius algebrae clavem continet.

> This calculation has to be perfected because it contains wonderful abbreviations and the key to the whole of algebra.

14.7 Symmetric Functions: Girard's Formula

Leibniz saw a strong parallelism between the powers of polynomials and power sums or combinatorial powers (potentiae combinatoriae polynomiorum) as he called them (Leibniz 1976, 200): If $x = 1 + m + n +$ etc. the power sums $l^i + m^i + n^i +$ etc. are reprensented by means of the elementary symmetric functions of l, m, n etc. In 1629 Albert Girard had given the first three power sums of the second up to the fourth degree in his *Invention nouvelle en l'algèbre* (Girard 1629, f. F 2r°-v°) without mentioning a general law of formation. Newton's recursion formulae were published in 1707, again without any general law of formation (Newton 1707, 251). Only in 1762 Edward Waring published such a rule in his *Miscellanea analytica de aequationibus algebraicis, et curvarum proprietatibus* (Waring 1762, 1–3) and once more in 1770 in his *Meditationes algebraicae* (Waring 1770, 1–4).

Leibniz anticipated Waring's results by about eighty years. Around about 1677/1679 he elaborated a table of the first nine power sums (Leibniz 1976, 195) (Fig. 14.3):

$\int 1$ aequ. x

$\int 11$ aequ. $x^2 - 2y$

$\int 1^3$ aequ. $x^3 - 3xy + 3z$

$\int 1^4$ aequ. $x^4 - 4x^2y + 4xz + 2yy - 4\omega$

$\int 1^5$ aequ. $x^5 - 5x^3y + 5x^2z + 5xyy - 5x\omega - 5yz$

$\int 1^6$ aequ. $x^6 - 6x^4y + 6x^3z + 9x^2y^2 - 6x^2\omega[-12xyz] - 2y^3 + 3zz + 6y\omega$

$\int 1^7$ aequ. $x^7 - 7x^5y + 7x^4z + 14x^3y^2 - 7x^3\omega[-21x^2yz] - 7xy^3 + 7xz^2[+14xy\omega]$
$+ 7y^2z - 7z\omega$

$\int 1^8$ aequ. $x^8 - 8x^6y + 8x^5z + 20x^4y^2 - 8x^4\omega[-32x^3yz] - 16x^2y^3 + 12x^2z^2$
$[+24x^2y\omega][+24xy^2z - 16xz\omega] + 2y^4 - 8y^2\omega - 8yzz + 4\omega^2$

$\int 1^9$ aequ. $x^9 - 9x^7 + 9x^6z + 27x^5y^2 - 9x^5\omega[-45x^4yz] - 30x^3y^3 + 18x^3z^2$
$[+36x^3y\omega][+36x^2y^2z - 27x^2z\omega] + 9xy^4[-27xy^2\omega - 27xyz^2]$
$+ 9x\omega^2 - 9y^3z[-18yz\omega] + 3z^2$

$35x^6y^2$ $50x^4y^3\ 25x^4z^2$

$15x^2\omega^2$ $+25x^2y^4$

Fig. 14.3 Table of the first nine power sums
Source: Leibniz (1976, 195)

He added: 'Nullam unquam tabulam numerorum vidi ex qua plura mysteria pulcherrima duxerim', 'I have never seen any table of numbers from which I drew more most beautiful mysteries.'

Between 1680 and 1682 Leibniz found the general law of formation of an arbitrary term of the representation of a power sum by elementary symmetric functions (Leibniz 1976, 200, 207, 213). The coefficient can be found thanks to a proportionality.

xyz might serve as an example: There are three factors or elementary symmetric functions. Hence the apparent dimension of the term is 3. The elementary symmetric functions have the dimensions one, two, three etc. Hence the hidden dimension of the term is 6. The proportionality reads:

(number of the apparent dimension of the term (3)):(number of the hidden dimension of the term (6))
= (number of the permutations of the elementary symmetric functions of the term
$(3! = 6)$):coefficient sought or: $\frac{6}{3} \cdot 3! = 12$

Let r_i be the frequency of the ith elementary symmetric function, let $x^{r_1} y^{r_2} z^{r_3}$ etc. be the term the coefficient c of which is sought. Then $1r_1 + 2r_2 + 3r_3$ etc. is its hidden dimension, $r_1 + r_2 + r_3$ etc. is its apparent dimension. The number of permutations of the elementary symmetric functions has to be calculated by means of the formula for permutations with repetitions.

$$c = \frac{(1r_1 + 2r_2 + 3r_3 + \ldots)}{(r_1 + r_2 + r_3 + \ldots)} \cdot \frac{(r_1 + r_2 + r_3 + \ldots)!}{r_1! r_2! r_3! \ldots}$$

The sign rule reads: If the apparent and the hidden dimension of a term are odd or even at the same time, its sign is +, otherwise −.

For example (I use Leibniz's notation): $\int 1^6 = x^6 - 6x^4y + 6x^3z + 9x^2y^2 - 6x^2\omega - 12xyz - 2y^3 + 3zz + 6y\omega$

These rules for the calculation of the coefficients are correct though Leibniz did not give any justification.

14.8 Symmetric Functions: The Reduction of Multiform Symmetric Functions to Uniform Symmetric Functions (Power Sums)

Since 1700 Leibniz corresponded with Theobald Overbeck on the reduction of symmetric functions to elementary symmetric functions. This correspondence is especially interesting because it contains new results in this respect. Overbeck summarized the most important of them in an *Algebraic Treatise on polynomials* that was presumably written in 1714 (Knobloch 1973, 146–160). I would like to discuss two problems: the multiplication of power sums or uniform symmetric functions by each other and the reduction of multiform symmetric functions or functions of the type $\Sigma a^m b^n c^p$ etc. to power sums. For technical reasons I shall use this notation instead of Leibniz's point notation (see Figs. 14.4 and 14.6).

14.8.1 The Multiplication of Power Sums By Each Other

Step by step Leibniz multiplied two, three, four etc. power sums by each other and represented the product by symmetric functions. The product of three power sums might serve as an example (*Ascensus tertius* in Fig. 14.4):

15   parum dubitem. Progressum tamen mei calculi breviter conabor explicare: Primo
        quaesivi quid fieret si Formulam a^m multiplicarem per a^n et hoc Productum
        denuo per a^p et hoc per a^q etc: Hunc voco Ascensum discolorem. Ex eo
        pronum erit colligere quid fiat in ascensu concolori si a^m multiplicetur per a^m
        Productum denuo per a^m etc: ad sumendos Quadratos, Cubos etc: τοῦ a^m

20   a^m per a^n fac. $a^{m+n} + a^m b^n$ Ascensus secundus
        Multipl. per a^p

        Fac. $a^{m+n+p} + a^{m+n} b^p$ A s c e n s u s   t e r t i u s
              $+ a^{m+p} b^n$
            $+ a^{n+p} b^m + a^m b^n c^p$

25   Multipl. per a^q

        Fit $a^{m+n+q} + a^{m+n+p} b^q$ A s c e n s u s   Q u a r t u s
            $+ a^{m+n+q} b^p + a^{m+n} b^{p+q} + a^{m+n} b^p c^q$
            $+ a^{m+p+q} b^n + a^{m+p} b^{n+q} + a^{m+p} b^n c^q$
            $+ a^{n+p+q} b^m + a^{n+p} b^{m+q} + a^{n+p} b^m c^q$

Fig. 14.4 Products of power sums
Source: Leibniz (1976, 236)

$$(\Sigma a^m)(\Sigma a^n)(\Sigma a^p) = \Sigma a^{m+n+p} + \Sigma a^{m+n}b^p + \Sigma a^m b^n c^p$$
$$+ \Sigma a^{m+p}b^n$$
$$+ \Sigma a^{n+p}b^m$$

A suitable notation revealed the hidden combinatorial structure of this problem. First Leibniz replaced the powers by the exponents. The exponents of different bases are separated from each other by a vertical line:

$$= mnp + mn|p + m|n|p$$
$$+ mp|n$$
$$+ np|m$$

Secondly he replaced the groups of exponents by the number of their terms. The encircled number is a coefficient that sums partitions of the same type (I write it within brackets) (Fig. 14.5):

$$= 3 + (3)2.1 + 1.1.1$$

Calculus Curtatus per solos Exponentes
m. per n. facit mn+m|n.   A s c e n s u s 2 d u s

35   Multipl. per (p.)
____

      fit mnp+mn|p   A s c e n s u s t e r t i u s
        +mp|n
        +np |m+m|n|p
      Multipl. in q)
____

40   fit mnpq+mnp|q
          +mnq|p+mn|pq+mn|p|q
          +mpq|n+mp|nq+mp|n|q
          +npq|m+np|mq+np |m|q
              +mq|n | p
45   +nq |m| p
              +pq |m|n+m|n|p|q
      Conferendo utramque signaturam sponte se offert methodus calculum
      plenum ex curtato restituendi [.] Ut vero calculum curtatum haberem
      quoque extra seriem id obtinui per C a l c u l u m B i c u r t a t u m.
50   eumque talem
      A s c e n s u s 2 d u s 2.+1| 1
      A s c e n s u s t e r t i u s 3. + ③ 2|1.   + 1|1|1.
                  Nam mnp sunt 3 | hic sunt 2|1. | hic sunt 1|1|1
                     Literae | Literae idque | Literae m n p
55 | 3. modis |

Fig. 14.5 Simplified representation of products of power sums
Source: Leibniz (1976, 237)

In order to find the way of calculating such coefficients we generalize Leibniz's procedure.

Let k be the number of factors. One has to look for all number-theoretical partitions of k. Let

$$k = 1r_1 + 2r_2 + \ldots + kr_k$$

an arbitrary partition. Its factor N reads as follows:

$$N = \frac{k!}{(1!)^{r_1}(2!)^{r_2}\ldots(k!)^{r_k}r_1!r_2!\ldots r_k!}$$

The first k powers of the denominator are necessary because all i! permutations of exponents that are represented by i, $i = 1, 2, 3, \ldots, k$ have to be suppressed. For example let i be 2:

If there is a term like $mn|p|q$ or 2.1.1, a new term $nm|p|q$ cannot occur, because one cannot distinguish between a^{m+n} and a^{n+m}.

The second k factors of the denominator are necessary because all r_i! permutations that are produced by the r_i sections of the same length i of exponents have to be suppressed:

If there is a term like $mn|p|q$ or $a^{m+n}b^p c^q$ the corresponding symmetric function must include the term $a^{m+n}c^p b^q$ because of its symmetry. Hence one cannot distinguish between $mn|p|q$ and $mn|q|p$.

Obviously this problem reminds us of the table for the number of terms of any symmetric function (see Section 14.4, Remark 2). The calculation of the factor N is equivalent to the following problem:

Let X be a set of k objects, let d be the number of classes of a partition so that no class remains empty. What is the number of partitions of the type $k = 1r_1 + 2r_2 + \ldots + kr_k$ (r_i is the number of classes with i objects)?

The number sought is given by N.

14.8.2 The Reduction of Multiform Symmetric Functions to Power Sums

The solution of this problem is based on the preceding result:

$$(\Sigma a^m)(\Sigma a^n) = \Sigma a^{m+n} + \Sigma a^m b^n$$

The last expression of this equation is a multiform symmetric function. Hence one immediately gets:

$$\Sigma a^m b^n = (\Sigma a^m)(\Sigma a^n) - \Sigma a^{m+n} \tag{*}$$

X I I. Sed quia semper Ordo ipsarum m.n.p. etc. idem manet, sufficiet Numeris
335   indicare quot literas quilibet Articulus cujusque Juncturae exhibeat. Coeffi-
cientes autem Circulo includemus ne cum iis confundantur

fit   א $a^m b^n$=1.1.–2.

ב $a^m b^n c^p$=1.1.1.–2.1.+(2)3.

ג $a^m b^n c^p d^q$=1.1.1.1.–2.1.1.+(2)3.1.+2.2.–(6)4.

340   ד $a^m b^n c^p d^q e^r$=1.1.1.1.1.–2.1.1.1.+ 2 3.1.1.+2.2.1.–(6)4.1.–(2)3.2.
+(24)5.

O b s e r v. 1. Hae juncturae sunt ipsae Discerptiones Numeri ultimae Juncturae seu
Numeri literarum Formulae ad Uniliteram reducendam ut si $a^m b^n c^p d^q e^r f^s$ sex
Literarum vellemus ad Uniliteros Valores reducere, hae ordine darentur Junc-
345   turae quae sunt dispersiones Numeri Senarii. 1.1.1.1.1.1. *2.1.1.1.1.* 3.1.1.1
2.2.1.1. 4.1.1. *3.2.1.* 5.1. *3.3.* 4.2. 2.2.2. 6.

O b s. 2. Prospectus Numerorum in Junctura unitate minutorum inter se Multi-
plicati, faciunt Coefficientem Juncturae, ut(6)4.1. quia 4 ÷ 1 facit Prospectum
3.2.1. Sic fiet

350   ה $a^m b^n c^p d^q e^r f^s$=1.1.1.1.1.1.–2.1.1.1.1.+(2) 3.1.1.1.+[2.2.1.1.]–(6)4.1.1.
–(2)3.2.1.+(24)5.1.+(4)3.3.+(6)4.2.–2.2.2.–(120)6.

Verbi gratia (4)3.3. quia hic in junctura bis occurrit 3 cujus Unitate minuti Prospectus bis facient 2.1.2.1. Facit (4) per [Obs. 2.]

O b s. 3. Pro signis + et –. Pares Numeri in junctura Multitudine pari vel nulla
355   dant Signum +. Multitudine Impari dant –[.]

X I I I. P r o b l e m a 5. Ex Numeris Juncturae colligere quot ejus sint juncturae membra[.]

In valore ד s i t   j u n c t u r a 3. 2. denotans mnp | qr

3 literae   2 literae

360   Haec junctura in Abaco A r t i c. X. includit 10. membra. Summa totius juncturae est 5. Ejus Prospectus 5.4.3.2.1. Numeri in junctura data sunt 3.2. Eorum
Prospectus 3.2.1. et 2.1. Fac $\dfrac{5.4.3.2.1.}{3.2.1.2.1.} = 10$.

R e g u l a   p r i o r Prospectus Juncturae summatae dividatur per Prospectus
singulorum in junctura Numerorum s i   t o t a   j u n c t u r a   s i t   d i s c o l o r[.]
365   R e g u l a   p o s t e r i o r Si in juncturam cadat Concoloritas (seu repetitio

Fig. 14.6 Reduction of multiform symmetric functions to power sums
Source: Leibniz (1976, 252)

Leibniz multiplied this equation by Σa^p. Functions of two variables were
eliminated by means of equation (*). In such a way he step by step deduced the
representations of an arbitrary multiform symmetric function (Fig. 14.6). For
example:

$$\Sigma a^m b^n c^p d^q = 1.1.1.1 - 2.1.1 + (2)3.1 + 2.2 - (6)4 \qquad (**)$$

How can one find an arbitrary representation?

Let k be the number of variables. One has to enumerate all possible number-theoretical partitions of k. The example $k = 4$ given above leads to five partitions.

Let the partition of k be represented by $k = 1r_1 + 2r_2 + \ldots . kr_k$

Its coefficient C will be $C = (1!)^{r_2}(2!)^{r_3} \ldots ((k-1)!)^{r_k}$

Let the last term of equation (**) be the example: $4 = 4.1 = kr_k$. Its coefficient $C = (3!)^1 = 6$.

The sign rule reads: If there is an even number of even numbers, the sign is $+$. If there is an odd number of even numbers, the sign is $-$.

These formulae were published by Waring in 1762 and in 1770 for the first time (Waring 1762, 6–8; 1770, 7–10).

14.9 Determinants

In Section 14.3 I mentioned that Leibniz invented determinant theory. He applied it to the solution of linear inhomogeneous equations, to the resultant of two polynomials, and to the elimination of a common variable from algebraic equations (Knobloch 2000). On January 22, 1684, he wrote the most important paper in this respect: *On the elimination of letters from equations or on the reduction of several equations to one equation.*

Determinants are combinatorial aggregates. Leibniz was fully aware of this matter of fact. Here it might suffice to cite his sign rule:

> Terms which emerge from an odd number of transpositions of the left or right subscripts from one another have different signs, and in the case of an even number, they have the same sign.

The modern definition takes, like Gabriel Cramer (1750, 658), the concept of inversion as its basis, while Leibniz used the concept of transposition (in the modern sense of the word). The sign rules are equivalent, nevertheless, since a permutation is an even or odd permutation if and only if it is the result of an (in both cases possibly smaller) even or odd number of transpositions from the original arrangement.

Epilogue

All these examples make clear why Leibniz drew the conclusion that algebra was subordinate to the combinatorial art and not vice versa. In his letter to Tschirnhaus dating from May/June 1678 cited in Section 14.5, he mentions his table of symmetric functions and continues (LSB III, 2, 425):

> Quae alios maximos habet usus, continet enim Algebrae totius arcana; Combinatoriae vero applicationem egregiam. Nam ego Combinatoriae subordinatam puto Algebram quia combinatoriam non habeo pro arte inquirendi numeros possibiles variationum; sed pro arte formarum seu pro scientia generali de Simili et Dissimili.

It provides other benefits of the highest importance because it contains the secrets of the whole of algebra, yet it contains an excellent application of combinatorics. For I believe that algebra is subordinate to combinatorics because for me combinatorics is not the art of investigating the possible numbers of variations but the art of forms or the general science of similar and dissimilar.

Bibliography

Cramer, Gabriel. 1750. Introduction à l'analyse des lignes courbes algébriques. Genf : Frères Cramer and Cl. Philibert.

Echeverria, Alfonso Iommi; Ammategui, Godofredo Iommi. 2005. La *Dissertatio de Arte combinatoria* de Leibniz en seconde lecture. *Studia Leibnitiana* 37, 208–223.

Girard, Albert. 1629. *Invention nouvelle en l'algèbre.* Amsterdam: G. I. Blaeuw.

Knobloch, Eberhard. 1973. *Die mathematischen Studien von G. W. Leibniz zur Kombinatorik,* Auf Grund fast ausschliesslich handschriftlicher Aufzeichnungen dargelegt und kommentiert. Wiesbaden: Steiner. (*Studia Leibnitiana Supplementa* vol. XI).

Knobloch, Eberhard. 1980. *Der Beginn der Determinantentheorie, Leibnizens nachgelassene Studien zum Determinantenkalkül, Textband.* Hildesheim: Gerstenberg. (arbor scientiarum Series B, vol. II).

Knobloch, Eberhard. 2000. First European theory of determinants, in: Karl Popp, Erwin Stein (eds.), *Gottfried Wilhelm Leibniz, The work of the great universal scholar as philosopher, mathematician, physicist, engineer.* Hannover: Schlütersche, pp. 56–64.

Knobloch, Eberhard, Schulenburg, J.-Matthias Graf von der (eds.). 2000. *Gottfried, Wilhelm Leibniz, Hauptschriften zur Versicherungsmathematik.* Berlin: Akademie-Verlag.

Leibniz, Gottfried Wilhelm. 1666. *Dissertatio de arte combinatoria.* Leipzig: Johann Simon Fick and Johann Polycarp Seubold = LSB VI,1, 163–230.

Leibniz, Gottfried Wilhelm. 1976. *Die mathematischen Studien von G. W. Leibniz zur Kombinatorik,* Textband, im Anschluß an den gleichnamigen Abhandlungsband zum ersten Mal nach den Originalhandschriften herausgegeben von Eberhard Knobloch. Wiesbaden: Steiner. (*Studia Leibnitiana Supplementa* vol. XVI).

LSB = Gottfried Wilhelm Leibniz, *Sämtliche Schriften und Briefe,* edited by the Berlin-Brandenburg Academy of Sciences and the Göttingen Academy of Sciences. Berlin: Akademie Verlag, since 1923 (III, 2 means: Third series, second volume).

Newton, Isaac. 1707. *Arithmetica universalis sive de compositione et resolutione arithmetica liber,* ed. by W. Whiston. Cambridge: University of Cambridge.

Waring, Edward. 1762. *Miscellanea analytica de aequationibus algebraicis, et curvarum proprietatibus.* Cambridge: Bentham.

Waring, Edward. 1770. *Meditationes algebraicae.* Cambridge: J. Nicholson.

Chapter 15
Founding Acts and Major Turning-Points
in Arab Mathematics

Roshdi Rashed

Classical mathematics is neither homogeneous nor all of one piece. Some chapters in its development go back as far as Greek mathematics. We have only to think, for example, of plane geometry, the geometry of cones or the geometry of spheres. Others are rooted in Arab mathematics, embracing the algebraic disciplines and work on geometrical transformations. Finally, yet other developments, such as infinitesimal calculus took place in Europe in the seventeenth century. What we can say without fear of contradiction, however, is that the distinctive characteristic of this classical mathematics is that it is 'algebraic and analytical'.

The question that remains is precisely when and how this distinctive characteristic saw the light of day, that is, how this algebraic-analytical reasoning arose and how it developed. The foundations for this new type of rationality came from a number of separate initiatives without which it could neither have emerged nor become established. This article will restrict itself to the initiatives attributable to Arab mathematicians that were what I shall call 'founding acts' in this new rationality. Other such advances followed in Italian mathematics the introduction of imaginary quantities – others again came with Viète and Descartes – the invention of a fully worked out system of symbolic representation – and the same period, which I call the age of the 'liberation of the infinite', brought yet more.

Recent historical studies have now placed beyond doubt, it seems clear, something that people had always suspected was true: that over at least five centuries it was above all in the lands of Islam and in the Arabic language that intensive and fruitful mathematical research took place. Between the eighth and the fourteenth centuries research in mathematics had been conducted by figures such as al-Khwārizmī, Thābit ibn Qurra, Ibn al-Haytham, al-Khayyām, and others, but also by dynasties of scholars who engaged in genuine team-work within established schools. There were the Banū Mūsā, and the Banū Karnīb

R. Rashed (✉)
Univ Paris Diderot, Sorbonne Paris Cité, Laboratoire SPHERE, UMR 7219, CNRS,
F-75013 Paris, France
e-mail: rashed@paris7.jussieu.fr

J.Z. Buchwald (ed.), *A Master of Science History*, Archimedes 30,
DOI 10.1007/978-94-007-2627-7_15, © Springer Science+Business Media B.V. 2012

dynasties, for example; the Niẓāmiyya of Baghdad, the schools of Marāgha, Samarkand among many others. This high-level research was, as one might expect, cumulative, diversified, and sometimes revolutionary. It was certainly cumulative, in that it was constantly enriching the inheritance left by the ancient mathematicians, essentially Greek and Hellenistic, with further progress in all the fields in which they had worked; but it was also cumulative in the sense that it had not taken long for it to become organised into traditions in which each succeeding generation had constantly added its own discoveries to the knowledge acquired by its predecessors. The diversification of this research brought about advances in many areas unknown to the ancients whose development completely redefined both the organisation and the extent of the fields covered by the mathematical sciences. This is observable in algebra, algebraic geometry, combinatorial analysis, integer Diophantine analysis, and trigonometry, among other branches. Finally, it was sometimes revolutionary in breaking ancient taboos and in devising new procedures: treating irrational quantities arithmetically, changing the criteria for admissible geometrical constructions, treating geometrically algebraic algorithms as well as those of quadratic interpolation, explicitly introducing movement in geometry, etc.

A full picture of this scientific activity, or, at the very least, an account of some of the chapters in its development involves writing its history in sufficient detail to show when it began, what conditions made its beginning possible, how it was organised, what obstacles sprang up in its path, and when it came to an end. The only way to answer all these questions and to produce an epistemological history of this kind is, it seems to me, to try to combine the history of the concepts with that of the texts. It is to precisely this kind of history that I have devoted myself over the past four decades. While it is, of course, impossible for me to summarise that work here, I should like in this account to concentrate briefly on the beginnings of this research activity, that is to say, on the ideas and concepts on which it is founded, before going on to examine how these ideas and concepts were renewed, or, in other words, how research of this same kind began to be taken up again.

My aim is not to rewrite the history of this research movement, but to attempt what I might call a phenomenological description of it designed to capture what these beginnings really meant and how far their influence extended. This exercise will help to make clear just where this mathematical research stands in relation to the mathematics of the ancient world and that of the modern era, that is, how it relates to Archimedes, Apollonius, Menelaus, etc., on the one hand, and to Descartes, Fermat, Cavalieri, etc., on the other.

1. A fact that does not receive the emphasis it should is that mathematical research in Arabic began in a manner that can perhaps most appropriately be described as paradoxical. The very process of translating Euclid's *Elements* and Ptolemy's *Almagest* provided the opportunity to make the first complete break with the Hellenistic tradition. Put another way, deviation from the Hellenistic mathematics went hand in hand with the learning of it. It was a colleague in

the Baghdad Academy of al-Ḥajjāj, the translator of Euclid and Ptolemy, al-Khwārizmī who, putting the translation of the *Elements* to good use, was responsible for the first break. This event took place in the course of the first third of the ninth century, when al-Khwārizmī developed a new discipline: algebra. It was a founding act in several respects, new in its assumptions, in the aims it put forward, in the language it forged, and new in the mathematical possibilities it opened up.[1]

It is worth reminding ourselves that in this first book of algebra al-Khwārizmī had a precise end in view that was clearly formulated as such: to develop a theory of equations solvable by radicals, to which arithmetical and geometrical problems alike could be reduced, and which could thus be applied to calculation, trade, inheritances, land-surveying, and so on.

The mathematicians who preceded him as well as his contemporaries in the field, no matter what their languages, formulated their equations in response to the problems they set themselves, whereas al-Khwārizmī, by contrast, started from the equations, that is, from the theory which enabled him to derive and classify them. And there was no limit to the number of problems, whether arithmetical or geometrical, that could be reduced to them. Thus in the very first part of his book, al-Khwārizmī began by defining the primitive terms of his theory, which, by reason of the limitations imposed by the method of resolution by radicals and by his own level of expertise in the field, were necessarily confined to equations of the first two degrees. The primitive terms were: the unknown – the 'thing' (the variable quantity in question) – its square, the positive rational numbers, the rules of elementary arithmetic, equality. The principal concepts that he introduced next were the first degree equation, the second degree equation, the associated binomials and trinomials, normal form, algorithmic solution and the geometrical proof of the algorithm. Al-Khwārizmī was, in effect, concerned to establish each time by means of a geometrical proof that the algorithm was sound and that it was an effective way of reaching the result. In the course of this presentation he takes care to justify the arithmetical treatment of quadratic irrational quantities and to demonstrate geometrically that such treatment is valid. It emerges clearly from this brief account that what was new in al-Khwārizmī's procedure was of a theoretical and not of a technical nature. From a technical standpoint, his book does not in fact reach the level achieved by Diophantus' *Arithmetica*. The theoretical innovation in al-Khwārizmī's algebra lies in the fact that the equation idea is not invoked in the course of the solution of problems, but is a primitive notion stemming from primitive terms whose combination has the potential to produce all possible equations.

But this conceptual break with traditions – Babylonian, Greek, Indian, or whatever – has roots that run still deeper, that draw upon a new mathematical

---

[1] By 'founding act' I do not in any sense mean a symbolic founding gesture or some manifestation of purely subjective significance, but a genuine project whose constituent elements show themselves progressively as the project itself is followed through.

ontology, and also on a new epistemology. The central concept of algebra – the unknown or the 'thing' – does not in fact designate, as according to traditional ontology it had to, a particular existent, but an object that may equally well be numerical or geometrical. In other words, the subject-matter of the new discipline consists neither in geometrical figures nor rational numbers; and the properties that this discipline is deemed to study are not those of measure any more than they are those of position and shape. Its object is something new and is not defined negatively: the mathematical entity finds itself invested with a new meaning, for it is now an entity general enough to admit of several determinations, both geometrical and arithmetical. This original indeterminacy is itself pregnant with logical possibilities, ready to be called into being as and when the means are discovered for studying the object from one viewpoint and/or another. In other words the algebraic object conceived by al-Khwārizmī cannot be obtained by abstraction from particulars; no more can it be arrived at by approximate imitation of a form or idea. There is nothing either Aristotelian or Platonic about the new ontology; it is, as it were, formal and, as such, no doubt the first to be met with in the history of mathematics. Its impact on mathematics, and thereafter on philosophy, was to be considerable, as can be seen from the work of the mathematician al-Karajī and the philosopher al-Fārābī, for example.

It is because algebra is conceived as a science that we can speak of it as an epistemic innovation. Like every mathematical science, it is apodictic, and it has in common with art that its ends lie outside itself, in that it is intended to resolve arithmetical and geometrical problems. Algebra does not fit the Aristotelian-Euclidean pattern.

Finally, the new apodictic discipline is also algorithmic. True enough, the algorithm used in the solution of a problem must itself submit to geometrical proof. If in fact the solution is thought of as no more than a procedure for decision-making, it follows that this procedure must be justified geometrically, i.e. in a different mathematical language. It was in this respect that al-Khwārizmī broke with all earlier traditions and contemporary practice in algorithmic mathematics.

What made the conception of a *sui generis* mathematical science like this possible was, it can be confirmed, the formal and combinatorial choice that led to the establishment of an *a priori* classification of equations. The choice was made in stages, as follows: 1st determine a finite set of discrete elements (the number, the 'thing' (the unknown), and the square of the 'thing'); 2nd using these elements resort to a combinatorial analysis so as to obtain *a priori* all the possible equations; 3rd following the theory, isolate from among the possible cases, those which match the criteria laid down by it. On this principle, out of the eighteen equations he found, al-Khwārizmī retained the six canonical ones, thereby avoiding both redundancy and repetition.

The *a priori* classification of possibilities, which, along with the other features just listed, gives the general shape of the beginning of algebra, also defines an entire area of classical mathematics that would go on expanding. So it was, that by breaking away from the Hellenistic style of mathematics, and indeed from that of all other mathematics that could be known at the time, algebra came into being.

But an authentic beginning can be identified as such on the one hand by the conceptual and textual tradition that it inaugurates, and on the other by the further breaks with existing practice that it goes on to provoke. In this case the tradition is embodied in the names of the mathematicians and in the titles of their works. In the wake of al-Khwārizmī, successors who would build on his work came thick and fast; for one thing, they began to extend the scope of algebraic calculus much further than he had done, for another, to integrate rational Diophantine analysis into algebra, and finally to formulate the proofs of the algorithms more rigorously in the language of Euclidean geometry. Names that may be picked out from among the other mathematicians associated with these advances are those of Ibn Turk, Thābit ibn Qurra, Sinān ibn al-Fath, and above all Abū Kāmil. To this last we owe the first treatise on algebra to include a chapter on rational Diophantine analysis. This book was also well known through its translation into Latin and into Hebrew, and from the fact that Fibonacci borrowed from it.

Other breaks with tradition, or, if preferred, other 'fresh starts' prompted by the new way of looking at mathematical science are already embryonic in the new possibilities opened up by groundbreaking ideas and founding acts. With al-Khwārizmī's algebra it in fact became possible to apply the disciplines to each other: arithmetic to algebra, algebra to arithmetic, algebra to geometry, algebra to trigonometry, and so on. Every one of these applications led to the establishment of new branches of mathematics and, as a result, to redrawing the map of the mathematical continent.

As part of this process, the application of arithmetic to algebra made possible the conception of the algebra of polynomials in the old sense, that is, the algebra of the elements of the ring $Q[x, 1/x]$. This particular 'fresh start' was the work of al-Karajī (end of the tenth century) and his successors, like al-Samaw'al ibn Yahyā, and we have dubbed it the 'arithmetisation' of algebra, because it was this step that made it possible. Arithmetisation led to an unprecedented development in abstract algebraic calculus, which extended to irrational quantities of which there was now an infinite multiplicity of kinds. Combinatorial analysis figured among other means that were forged for this purpose. It was precisely this pattern of polynomial calculus that inspired a new approach to the use of decimals and the invention of decimal fractions. The persistent difficulty with which al-Karajī had to contend was in applying the Euclidean divisibility algorithm to polynomials. But the only invertible elements in the ring $Q[x, 1/x]$ are the monomials. Al-Karajī therefore divided a polynomial by a monomial, not by a polynomial. In order to surmount this difficulty his twelfth-century successor al-Samaw'al came up with the idea of continuous division, which thus entailed approximation (a limited development).

This extension of algebra involved still deeper research on rational Diophantine analysis by introducing a new classification according to forms (linear, quadratic, cubic).

It was, too, in accordance with this algebra, but also in reaction to it, that mathematicians such as al-Khujandī, al-Khāzin, Abū al-Jūd, al-Sijzī, etc. conceived and developed integer Diophantine analysis. They often began by studying numerical right-angled triangles, before raising a lot of other problems, including Fermat's theorem for $n = 3, 4$. Their choice of numerical right-angled triangles and problems of a similar kind can be explained by two considerations: the fact that the domain of solution is restricted to integers, and, above all, the new requirements upon which the mathematicians insisted *viz.* the need to offer Euclidean-style proofs and to justify the algorithms for solving Diophantine equations in Euclidean terms. In this domain there are at the same time signs of an important change of direction to be seen in the search for purely arithmetical proofs, particularly with the help of congruencies.

The application of algebra to number theory enabled new proofs to be offered in areas already worked over, such as the theory of amicable numbers (Kamāl al-Dīn al-Fārisī), but it also allowed mathematicians to conquer new territories: the study of elementary arithmetical functions, sum and number of divisors.

Again, it was in connection with algebra, and more especially with the development of abstract algebraic calculus, that a discipline never before conceived in Hellenistic mathematics was built up: combinatorial analysis. It was in fact entirely for the benefit of algebraic calculus that al-Karajī established the binomial theorem and Pascal's triangle. Explicitly combinatorial interpretation is to be found in the work of a good many mathematicians, in particular that of Naṣīr al-Dīn al-Ṭūsī. Combinatorial analysis, applied to the various branches of linguistics, number theory, proportion theory, as well as to philosophy was explicitly founded on two ideas that, in the majority of subject-areas, distinguished the forms of thought of the period: classifying *a priori* all possible forms, or all the elements of a finite set of discrete possibilities. This is the path followed in fields as diverse as lexicography, prosody, cryptanalysis (codebreaking), and mathematics, among others.

But *pari passu* with the arithmetisation of algebra the foundations were being laid for another programme – its geometrisation. Two advances provided the impetus for this new programme. The first, achieved by several tenth-century mathematicians beginning with al-Māhānī, consisted of translating the problems of solid geometry into cubic equations. The second, which was repeated a number of times in the tenth century, involved the resolution of cubic equations by the intersection of several conic curves (al-Qūhī, Abū al-Jūd, etc.); this latter exercise picked up on what had, furthermore, been a doubly negative situation: no one had yet managed to solve cubic equations by radicals, nor were the means available for justifying the algorithm employed to solve certain forms of cubic and of biquadratic equation, because those particular solutions cannot be constructed with a ruler and compasses. There were consequently several other advances that all tended in the same direction. The first was made by al-Qūhī, who devised a theory for cubic equations equivalent to that of the application of areas for plane equations. Al-Khayyām was the scholar who worked out the first geometrical theory of cubic equations starting from a classification of all possible forms, a method that necessarily excluded equalisation to zero; then a classification according to the curves involved in the solution of equations. Finally, with Sharaf al-Dīn al-Ṭūsī barely half a century after al-Khayyām, came the analytic change of emphasis in the theory, occasioned by the new requirement to prove the existence of positive roots. With these last two, we are now in the presence of the first elementary research in algebraic geometry.

2. Scarcely three decades after al-Khwārizmī, the Baghdad mathematicians launched themselves upon new conquests, this time armed with an ampler knowledge of the heritage left by the Greeks. The research on which they had embarked had in effect given rise to a whole industry dedicated to translating great numbers of Greek books into Arabic, works such as Apollonius' *Conics* and his *Cutting Off of a Ratio*, Archimedes' *Measurement of a Circle* and *On the Sphere and the Cylinder*, the eighth book of Pappus' *Synagoge*, the *Spherics* of Theodosius and of Menelaus .... From the ninth century onwards three mutually linked research traditions are to be identified: infinitesimal geometry, conical geometry and spherical geometry. There still remains the problem of identifying which particular initiatives were the new founding acts that distinguished each of these traditions and led the inheritors of Hellenistic mathematics to develop the areas of study concerned. The advances in question, it seems to me, can be summed up under the following two headings: first of all, 'point-wise transformation' and next 'continuous movement'. Let us now examine them in order.

In the course of certain of their proofs Archimedes and also Apollonius make use of point-wise transformations. In *On Conoids and Spheroids* Archimedes has recourse to orthogonal affinity. Apollonius, particularly in *Plane Loci*, probably makes use of some transformations. This said, Archimedes' book was never translated into Arabic, and all we know about that of Apollonius is what Pappus tells us. Neither Apollonius' book nor Pappus' statements reached the Baghdad mathematicians. The sole exception is the sixth book of the *Conics* in which Apollonius follows a proto-transformational approach in so far as he seeks to determine the conditions for two conic sections to be superposable – that is to say, homothetic or similar – with the help of *symptomata* (properties that characterise each of the three conic sections), without, however, taking any interest in the actual nature of the point-wise transformations themselves. All these reasons point to the conclusion that point-wise transformations did not, as such, form part of the heritage passed on via translation. Now this historical fact finds confirmation in the way the concept of point-wise transformations was treated by the Arab mathematicians: for modern mathematicians, they are not simply something brought into the course of proofs, but loom up more and more as conceptual elements of the geometrical object. From the middle of the ninth century onwards, as advances continued to be made in the subject, study ceased to be confined to figures, but extended to their transformations and the relationships that existed between them. Now it is exactly this that we begin to notice in the work of the three Banū Mūsā brothers, their pupil and collaborator Thābit ibn Qurra, the astronomer-mathematician al-Farghānī (in his book *al-Kāmil*), and many others also. Their successors made a massive investment in this domain, so, at the end of the tenth century al-Sijzī, borrowing a generic name (*al-naql*) from Thābit ibn Qurra, dubbed the subject 'point-wise transformation'. The effective foundation of this subject area came some decades later when Ibn al-Haytham, conceived a whole new geometrical discipline mainly aimed at studying the elements in a figure that remained invariable when all the rest changed; he called this discipline 'the knowns'.

This, then, was the second 'founding act' of Arab mathematics: the introduction of point-wise transformations not just as part of proofs but as a geometrical concept in their own right. This advance, unlike the one that marked the foundation of algebra, was not the result of action by a single individual, but the outcome of several scholars' work in different fields at the same time, these fields being infinitesimal geometry and the 'science of projection', as the ancient mathematicians and bio-bibliographers called it. This founding act goes back in particular to the Banū Mūsā and especially to the youngest of the three brothers, al-Ḥasan, along with their pupil Thābit ibn Qurra, and to the astronomer-mathematician al-Farghānī. The discovery was repeatedly demonstrated, confirmed and developed in greater depth in the following centuries, as can be seen from the work of al-Bīrūnī and Ibn al-Haytham in the eleventh century. The key to our understanding of how it arose lies in the interest in the geometry of cones first generated by the translation of the *Conics* of Apollonius.

Let us turn first to al-Farghānī and the 'science of projection'.

One of the first disciplines born, so to speak, from this founding act was the 'science of projection' (*'ilm al-tasṭīḥ*). This emerged towards the middle of the ninth century. There is nothing surprising in that: it was, after all, in this century that astronomy experienced a rise never equalled since the second century. Astronomers were continuously engaged in the translation of Greek works and some Sanskrit texts as well, but were also submitting the theories and calculations they read in them to critical examination. It was indeed this rise in astronomical research that precipitated the detachment of the field of projections from astronomy to become a branch of geometry, even if its main application still remained in astronomy, or rather in astronomical instruments. Al-Farghānī played an essential part in this transfer by his insistence on placing the procedures employed by the astronomers for the exact representation of a sphere on a sound geometrical footing. As far as we know, he was the first to impose such a condition. These were procedures used in the drawing of geographical maps and in the construction of astronomical instruments such as the astrolabe.

Now providing a firm geometrical basis for procedures entails being able to demonstrate a knowledge of the domains involved. Everything points to the fact that if al-Farghānī was in a position to recognise the need to underpin existing methods, it was because of his very recently acquired knowledge of Apollonius' *Conics*. Though it is true that Apollonius himself does not deal with projections in his treatise, we do however know that propositions 4 and 5 of the first book of the *Conics* answers a particular question concerning the intersection of a conic surface and a plane. Now with the aid of these propositions al-Farghānī was able to offer the following proof:

Let there be a circle of diameter AG, the tangent to this circle at G and any chord BC. The projections from pole A of the points B and C on the tangent are respectively I and K (Fig. 15.1).

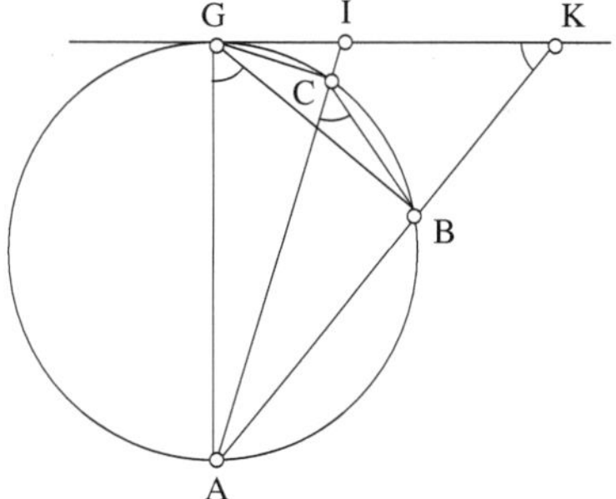

Fig. 15.1

Then $A\hat{G}B = A\hat{C}B$ (inscribed angles) and $A\hat{G}B = A\hat{K}G$ (which have the same complement $B\hat{A}G$); therefore

$$A\hat{C}B = A\hat{K}G \qquad (15.1)$$

and

$$C\hat{B}A = K\hat{I}A. \qquad (15.2)$$

Put a different way, these results may be interpreted in the following manner: Let GB and CG be the respective heights of the triangles GAK and GAI, so that $AG^2 = AB \cdot AK = AC \cdot AI$; thus in the inversion τ of pole A and of power AG^2, we get

$$I = \tau(C) \text{ and } K = \tau(B);$$

And, according to (15.1) and (15.2) the points B, C, I, K belong to a circle that is invariant in inversion τ.

Even if al-Farghānī had not formulated the concept of inversion, it remains true that he had recognised the transformation of a circle into a straight line and that, in this transformation, the extremities of the chord and of the segment are on a circle that is *invariant*.

This lemma in fact comes down to saying: the conic projection from pole A, on to the tangent diametrically opposite, of a chord, is a segment of the tangent such that the extremities of the chord and of the segment are on an invariant circle in the inversion τ, from the same pole A, which transforms the given circle into the tangent straight line. With the help of this lemma, al-Farghānī established that the projection of a sphere, having as its pole the point A on the sphere, on to a plane tangent to the point diametrically opposite, or on to a plane parallel to this plane, is a stereographic projection.

There is no space here to follow the developments that took place in this field from the time of al-Farghānī up to that of Ibn 'Irāq and al-Bīrūnī. To appreciate how much ground had been covered since al-Farghānī, it will be enough to mention only some of the topics studied by al-Qūhī and Ibn Sahl in the course of the second half of the tenth century. These two examined conic projection from a point on the axis or outside the axis of a sphere, cylindrical projection in a direction parallel or not parallel to the axis of a sphere, and obtained a good many new and important results that used to be attributed to later mathematicians. And this was not all; in this field mathematicians like al-Qūhī, Ibn Sahl, Ibn 'Irāq among others thought out new procedures for proofs and forged a new language.

To give an example, when al-Qūhī proves the following property, that, with every circle drawn on to a sphere and whose plane does not contain the pole, the stereographic projection associates a circle in the plane of projection, and the inverse, he uses proposition I.5 of the *Conics* in which Apollonius studies the section of a cone with a circular base, in the case in which the base plane and the secant plane are antiparallel. Al-Qūhī resorts to the technique of rabattement to allow constructions in plane geometry. In this way his proofs are made up of comparisons of ratios, projections and rabattements; in other words, he makes use of traditional techniques, and others that are non-traditional, i.e. projective.

The language employed is likewise mixed: the vocabulary that belongs to the theory of proportions is mingled with terms that from then on are used to refer to projective concepts.

In the tenth century a new field of geometry, for which Ptolemy's *Planisphærium* was no more than a very distant ancestor, was already beginning to open up, one which would go on to be enriched by successive generations.

3. Another application for point-wise transformations that we see developing is in the drawing of conic curves. The problem presented by them probably goes back to the time when geometry was first brought into the study of sundials and burning-mirrors. On the other hand, we can be quite certain that a number of factors contributed to the renewal of interest in the drawing of conic curves.

There was the revival of astronomical research for one thing, the resumption of research on burning mirrors, and, later on, on lenses for another, and finally, from the ninth century, the emergence of the theory of cubic equations. Never, before the ninth century, had the quest for effective procedures for drawing these curves been so intensive, so many-sided, and so uninterrupted. From the tenth century, entire treatises, or chapters within them, were devoted to this question, which consequently took on a variety of separate aspects. Many celebrated mathematicians interested themselves in it, men like Ibrāhīm ibn Sinān, al-Khāzin, al-Qūhī, Ibn Sahl, al-Sijzī, al-Bīrūnī, among others. Methods of drawing using points were put forward, then methods of continuous drawing for which mechanical instruments, like the famous 'perfect compasses', were invented, or other, optical, procedures. Now all these methods, theoretical or technical, rest on one or another affine point-wise transformation, and sometimes even on a projective transformation, as in the work of Ibn Sinān.

This research gave rise to a question no one had previously thought of: would it be possible to obtain conic curves from a circle, that is, by the transformation of a circle, and so use the circle in drawing conic curves? This question, which had been implicit in the writings of Ibn Sinān and Abū al-Waf ā' al-Būzjānī, was spelt out in full by al-Sizjī, a mathematician working in the second half of the tenth century.

Here again everything began with the youngest of the Banū Mūsā, al-Ḥasan, and his pupil Thābit ibn Qurra. Before he had got hold of a translation of Apollonius' *Conics*, al-Ḥasan had made a study of the ellipse and its properties, as a plane section of a cylinder, and also the various types of elliptic sections. Unlike Apollonius, he proceeded by the bifocal method and referred to the ellipse significantly as 'an elongated circular figure'.[2] He shows then that this figure can be obtained from a circle by orthogonal affinity, which is a contraction (or, as the case may be, a dilatation) according as the ratio of the major axis to the minor axis is less than (or, as the case may be, more than) one. His pupil Thābit ibn Qurra, for his part, started from a detailed knowledge of Apollonius' *Conics*. He began by demonstrating the following proposition: The plane sections of two cylinders on circular bases, with the same axis and of the same height are homothetic, the centre of homothesis being their common centre situated on the axis and the ratio of homothesis being the ratio of the diameters of the base circles.[3] Thābit next proved the proposition introduced by al-Ḥasan. It was thus possible to draw an ellipse from a circle with rigorous accuracy, using points. But what of the parabola and the hyperbola? This was a question that Thābit ibn Qurra's grandson, Ibrāhīm ibn Sinān, lost no time in raising and, with the help of a circle, he drew every point of a parabolic section. As for the hyperbola, Ibrāhīm ibn Sinān drew that with the help of a circle and a

---

[2] R. Rashed, *Les Mathématiques infinitésimales du IXe au XIe siècle*. Vol. I: *Fondateurs et commentateurs: Banū Mūsā, Thābit ibn Qurra, Ibn Sinān, al-Khāzin, al-Qūhī, Ibn al-Samḥ, Ibn Hūd*, London, al-Furqān Islamic Heritage Foundation, 1996, chap. I

[3] *Ibid.*, chap. II.

projective transformation designed to transform the circle into a hyperbola whose straight side was equal to the transverse diameter.

For Ibrāhīm ibn Sinān to raise, as a general question, the problem of drawing conic sections from circles by means of points was neither contingent nor circumstantial. The title of his book, *On the Drawing of the Three Conic Sections*, is a programme in itself. On the other hand, a much greater interest than hitherto in the study of geometrical transformations was a necessary precondition for conceiving the project and formulating the question in the first place. Besides, we have only to look at Ibn Sinān's works, such as his *Sundials* or his *Anthology of Problems* to gather how frequently transformations were being used.[4] Now this was an interest that went on growing after Ibn Sinān. In the course of the second half of the century al-Sijzī wrote a treatise whose title perfectly reflects its intention: *All Figures are Based on the Circle*,[5] in which he explicitly goes back to his predecessor's book and as it were generalises it.

But during the course of the tenth century, particularly owing to the requirements to be met in the construction of geometrical problems on the one hand, and the solution of cubic equations by the intersection of conic curves on the other, it was no longer enough to draw these curves by means of points. Henceforward it was necessary to make sure of the continuity of the curves in order to be able to discuss the existence of the points of intersection. The reasons for dissatisfaction were not merely theoretical; there were other, technical reasons too, arising from the making of patterns for parabolic and elliptical burning-mirrors, plane-convex and biconvex lenses, as well as the manufacture of astrolabes and dials. Two contemporaries, Ibn Sahl and al-Qūhī, invented instruments for carrying out continuous drawing, and the entire body of mathematicians working in this area was engaged in the study of these geometrical instruments and in the problems involved in the continuous drawing of curves. Al-Sijzī was among their number, and himself wrote a dissertation on the perfect compasses.[6] Thus everything was now in place for the production of the first treatise entirely devoted to the methods of drawing conic curves: drawing by points and continuous drawing. This was the purpose of al-Sijzī's book bearing the title, *The Description of the Conic Sections*.[7]

---

[4] An edition of these treatises, with (French) translation and commentary is contained in R. Rashed and H. Bellosta, *Ibrāhīm ibn Sinān. Logique et géométrie au Xe siècle*, Leiden, E.J. Brill, 2000.

[5] Edited with (French) translation and commentary in R. Rashed, *Œuvre mathématique d'al-Sijzī*. Vol. I: *Géométrie des coniques et théorie des nombres au Xe siècle*, Les Cahiers du Mideo, 3, Louvain-Paris, Éditions Peeters, 2004.

[6] See R. Rashed, *Geometry and Dioptrics in Classical Islam*, London, al-Furqān, 2005, chap. V.

[7] Edited with (French) translation and commentary in R. Rashed, *Œuvre mathématique d'al-Sijzī*, vol. I.

But the introduction of procedures for continuous drawing brings with it the notion of movement in geometry. Point-wise transformations were, as we have seen, used during proofs, and, this being so, transformations and continuous movement lay at the root of this new phase of the geometry of conics, which mathematicians such as, for example, Kamāl al-Dīn ibn Yūnus (1156–1248) and his pupils, and Athīr al-Dīn al-Abharī (d. 1265) were to go on enriching until the end of the thirteenth century.[8]

4. Point-wise transformations, or continuous movement associated with punctual transformations, alike characterised the founding acts of the new and the revival of the old fields of study in geometry from the middle of the ninth century onwards. This is what we have just seen with 'The science of projection' and 'The drawing of the conic curves'. Advances in three other areas also illustrate the point: geometrical constructions, the theory of parallels, and, lastly, infinitesimal geometry.

The legacy to which the mathematicians of the period had fallen heir – particularly after the translation of the *Burning-mirrors* of Diocles[9] and Eutocius' *Commentary* on Archimedes' *On the Sphere and the Cylinder* – included the construction of certain problems in solid geometry: the two means, the trisection of an angle, Archimedes' straight line, to name only some. To these problems they had added a good many others, and, in particular, they had produced a multiplicity of new constructions. We know that numerous mathematicians had revisited the problem of the trisection of an angle and the construction of a regular heptagon. But they, in contrast to the ancients, had been quick to modify the very criterion for the construction of a solid problem. They had banished transcendent curves as a construction procedure, and kept only the conic curves. This last construction became admissible for the same reasons as construction by means of ruler and compasses for problems in plane geometry. It should be emphasised, however, that the introduction of this new criterion was also due to the need to respond to what the algebraists were doing, since they were beginning to translate problems in solid geometry into cubic equations, as is evident from the work of al-Māhānī at the end of the ninth century. And it was precisely this new criterion that enabled the piecemeal studies of a scattering of particular examples to be brought together under a single heading. But these studies resorted to transformations, and particularly to similarity. Take, for example, the problem of the regular heptagon: you start by constructing one of the triangles whose angles are related according to one or other of the following patterns: (1, 2, 4),

---

[8] See R. Rashed, *Geometry and Dioptrics in Classical Islam*, chap. V.

[9] *Les Catoptriciens grecs.* I : *Les miroirs ardents*, edited with (French) translation and commentary by R. Rashed, Collection des Universités de France, published under the patronage of the Association Guillaume Budé, Paris, Les Belles Lettres, 2000.

(1, 5, 1), (1, 3, 3), (2, 3, 2), before transforming it in order to inscribe it in the circle.[10] In short, it was thanks to conic sections and transformations that this new field of study came to exist as such.

The application of the theory of conic sections by both geometers and algebraists led to a number of developments within the theory itself. The tenth-century mathematicians thus subjected the properties of the harmonic division of conics to closer examination. Ibn Sahl even wrote a dissertation on the subject. His younger contemporary, al Sijzī, embarked on a new topic of study: plane sections and their classification. It was not until the time of Fermat and his successors in the eighteenth century that the matter was taken any further. No less important are the contributions made by mathematicians of the eleventh century: Ibn al-Haytham in Egypt and 'Abd al-Raḥmān ibn Sayyid in Andalusia. Both were engaged in generalising the classic problem: given two magnitudes, find two other magnitudes such that all four are in continuous proportion. This problem is translatable into a cubic equation which algebraists solved using the intersection of two conic curves. Ibn al-Haytham – according at least to al-Khayyām – generalised this problem using four magnitudes between two given magnitudes, which leads to an equation of the fifth degree that is solved by the intersection of a conic and a cubic. Everything points, then, to the fact that Ibn al-Haytham had available to him a method analogous to that used by Fermat in his *Dissertation tripartite*. The true generalisation, however, is that of 'Abd al-Raḥmān ibn Sayyid. According to Ibn Bājja, he wrote a paper on the theory of conics in which he dealt with the intersection of a non-plane surface and a conic surface, that is, in the general case, on skew curves. Ibn Bājja recalls that Ibn Sayyid solved the problem of two means in this way 'for as many straight lines as one wishes between two straight lines, in continuous proportion, and by this route he divided an angle in no matter what numerical proportion'. This problem would not be raised for a second time, we may note, until Jacques Bernouilli (1654–1705) did so. Other developments that should be mentioned include work on the optical properties of conics, a field of study revived by al-Kindī and generalised by Ibn Sahl and Ibn al-Haytham.

There can be no doubt that the theory of parallels constitutes one of the fundamental elements of the geometry of the period. Mathematicians like Thābit ibn Qurra, al-Khāzin, Ibn al-Haytham, al-Khayyām, Naṣīr al-Dīn al-Ṭūsī, among others, had given it their attention. In two successive monographs, which formed the basis for future research, Thābit ibn Qurra intentionally introduced the notion of continuous movement in defining the concept of equidistance between parallels.[11]

---

[10] *Les Mathématiques infinitésimales du IXe au XIe siècle*, vol. III: *Ibn al-Haytham. Théorie des coniques, constructions géométriques et géométrie pratique*, London, al-Furqān, 2000.

[11] R. Rashed and Ch. Houzel, 'Thābit ibn Qurra et la théorie des parallèles', *Arabic Sciences and Philosophy* 15. 1, (2005), pp. 9–55, reprinted in R. Rashed (ed.), *Thābit ibn Qurra. Science and Philosophy in Ninth-Century Baghdad*, Scientia Graeco-Arabica, vol. 4, Berlin/New York, Walter de Gruyter, 2009.

Finally, transformations were employed on a massive scale in a vast domain that may be called 'infinitesimal geometry', or 'infinitesimal mathematics', and that includes the measurement of the areas of curved surfaces and the volumes of curved solids, isoperimetric and isepiphanic problems, the solid angle, the study of the variations of functional expressions such as trigonometric functions.

Let us direct our attention to the most time-honoured of the examples that have just been cited, the measurement of areas and volumes. Research in this domain had effectively died out after Archimedes, and the first steps in its revival had to wait until the ninth century with al-Kindī and the Banū Mūsā, the effect, no doubt, of the first meeting of its kind between the Archimedean tradition and that of Apollonius. This meeting did not take place in an intellectual vacuum, but in a milieu well-informed about the algebra of al-Khwārizmī and his successors. Now the Banū Mūsā and their pupil Thābit ibn Qurra, who had been the first to engineer this encounter between the two traditions, directed their research along two channels which constantly branched out in new directions and developed in increasing depth: arithmetisation that was much more substantial and systematic than before; more deliberate and more frequent use of point-wise transformations. To find a rapid illustration of the approaches involved, we need look no further than Thābit ibn Qurra's method of determining the area of a parabolic segment. He began by setting out twenty-one lemmas of which eleven were arithmetic. These arithmetic lemmas concerned the summation of the numerous arithmetic progressions. He next proved four lemmas on the sequences of segments, using the arithmetic lemmas. It was those sequences of segments that he used to work out the necessary majoration. From these lemmas Thābit ibn Qurra undertook the calculation of the area of a parabolic portion.

He would go on to employ this arithmetically based approach in the calculation of the volume of the paraboloid of revolution. Later on, Ibn al-Haytham would use it to determine the volume of the paraboloid generated by the rotation of a parabola round its ordinate.[12] He also began with arithmetic lemmas in which he calculated the progressions of the powers of integers [numbers], i.e. the sums $\sum_{i=1}^{n} k^i$ for $i = 1, 2, 3, 4$; and arrived at a general rule using the somewhat archaic procedure of complete induction. He then went on to prove the following double inequality

$$\sum_{k=1}^{n}\left[(n+1)^2 - k^2\right]^2 \le \frac{8}{15}(n+1)^5 \le \sum_{k=0}^{n}\left[(n+1)^2 - k^2\right]^2.$$

But al-Ḥasan ibn Mūsā and Thābit ibn Qurra called on transformations in demonstrating other areas of different kinds of plane sections of a right cylinder and of an oblique cylinder, the area of an ellipse and the area of elliptical segments.

---

[12] *Les Mathématiques infinitésimales du IXe au XIe siècle*, vol. II: *Ibn al-Haytham*, London, al-Furqān, 1993.

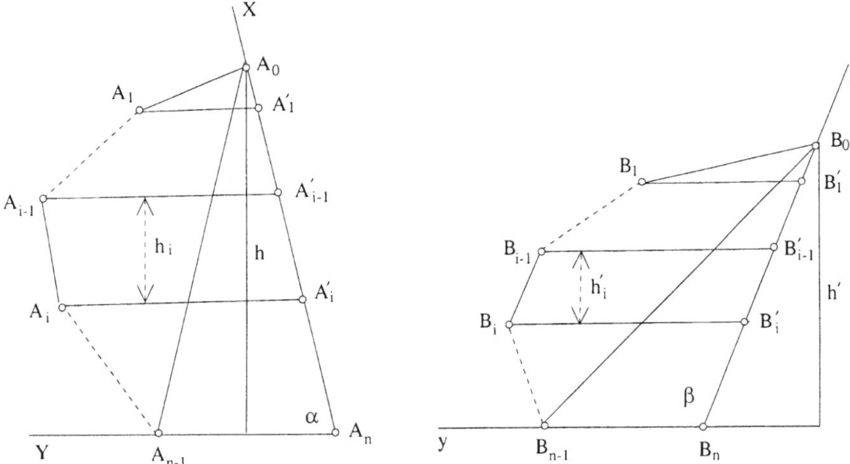

Fig. 15.2

In this exercise, the mathematicians' principal recourse was to orthogonal affinities, to homothesis, and to the composition of these transformations; and they showed that such composition preserved the areas.

This approach, founded on point-wise transformations, was to be adopted by the successors of the Banū Mūsā and Thābit ibn Qurra, who sought to reduce the number of lemmas. In this spirit, Ibrāhīm ibn Sinān (909/946), wrote a short treatise re-examining the measurement of the parabola.[13] Ibn Sinān's central idea, which he was anxious to prove first of all, was this: that the proportionality of areas remains invariant in affine transformation. Then all he needed were two lemmas of a single proposition to complete the study (Fig. 15.2).

Proposition 1 *Let there be two convex polygons* $A = (A_0, A_1, \ldots, A_n)$ *and* $B = (B_0, B_1, \ldots, B_n)$. *Project points* $A_1, A_2, \ldots, A_{n-1}$ *on to* $A_0 A_n$ *parallel to* $A_{n-1} A_n$ *at points* $A'_1, A'_2, \ldots, A'_{n-1} = A_n$ *and points* $B_1, B_2, \ldots, B_{n-1}$ *on to* $B_0 B_n$ *parallel to* $B_{n-1} B_n$ *at points* $B'_1, B'_2, \ldots, B'_{n-1} = B_n$. *If*

$$\frac{A_0 A'_1}{B_0 B'_1} = \cdots = \frac{A'_{n-2} A_n}{B'_{n-2} B_n} = \lambda,$$

and

$$\frac{A_1 A'_1}{B_1 B'_1} = \cdots = \frac{A_{n-1} A_n}{B_{n-1} B_n} = \mu,$$

---

[13] An edition of this work with (French) translation and commentary is to be found in *Les Mathématiques infinitésimales du IXe au XIe siècle,* vol. I.

then

$$\frac{\text{tr. } (A_0, A_{n-1}, A_n)}{\text{p. } (A_0, A_1, \ldots, A_n)} = \frac{\text{tr. } (B_0, B_{n-1}, B_n)}{\text{p. } (B_0, B_1, \ldots, B_n)}.$$

In this proposition, Ibn Sinān proceeds with the aid of transformation T as defined in the terms of the proposition. This is an affine transformation. Ibn Sinān shows that it preserves the ratios of the areas in the case of the triangles and the polygons.

The second proposition is expressed as follows:

Proposition 2 *The ratio of the areas of two portions of a parabola is equal to the ratio of the areas of the two triangles associated with them.*

In this proposition, Ibn Sinān shows that the affine transformation preserves the ratio of the area of a portion of a parabola to that of an associated triangle, and the ratio of their homologues. The underlying property is in fact the preservation of the area ratios (even curvilinear areas) by affine transformation. For that, Ibn Sinān used Archimedes' axiom to show that it is possible to inscribe in a parabolic portion a polygon whose area differs as little as may be wished from that of the parabola.

This being established, the calculation of the ratio of the area of a parabolic portion to that of the associated triangle no longer requires an infinitesimal approach, but only the fact that the ratio does not depend on the portion under consideration; and this was what Ibn Sinān established in the third proposition.

Proposition 3 *The area of a parabolic portion is four thirds of the area of the triangle associated with it.*

Thus Ibn Sinān's strategy for improving his grandfather's proof and reducing the number of propositions from twenty-one to just three, is based on a combination of affine transformations and infinitesimal methods.

Point-wise transformations and continuous movement were, from the middle of the ninth century and throughout the tenth, among the main founding elements of the different chapters of geometry.

In order to understand how geometry developed in this period, it is essential to appreciate the part played by these two elements in the advances that led to the establishment of the separate subject areas. The presence of these notions, which more and more often intervened in their work could not help raising new questions for the mathematicians, and confronting them with new tasks. How could the role of either one of them be made legitimate? How could the notion of movement be admitted in the terms of propositions and in proofs, when it had never been defined? These two questions, which are clearly connected, raised another just as important: if from now on the business of mathematicians was to be the relationships between figures, their transformations and their movements, was it not time to rethink the notion of 'place' and space? The question of place could no longer be allowed to remain in the shadows; and it was quite equally impossible to continue to preserve the Aristotelian *topos*, the place-envelope. These questions began to emerge at the end of the tenth century, as certain of al-Sijzī's writings attest, before becoming, with Ibn al-Haytham, food for reflexion and invention. Let us, very briefly, turn our attention just to the question of continuous movement.

The attitude of Platonist geometers, dictated as it was by the Theory of Ideas, resulted in their outlawing *de jure* any consideration of movement in the elements of geometry; the attitude of Aristotelian geometers was the same, because of their doctrine of abstraction. The truth, however, may be that their stance was due less to their commitment to a particular ontology than to the fact that the kind of geometry that essentially concerned the study of figures had no great need for the notion of movement.

Even when the need for movement made itself felt, albeit very slightly, it is not unusual to find it being avoided *de jure*, only to be brought in surreptitiously, or unintentionally. Is not this Euclid's position in the *Elements*? He avoids movement, but admits it in disguise by resorting to superposition.

Superposition in fact necessarily involves displacement, even if the change of place is only in the mind's eye. And when he defines a sphere, we are conscious that, so to speak despite himself, he is opening the door to movement. Outlawing movement, however, remained the order of the day for a long time to come. We may call to mind al-Khayyām's criticism of the use Ibn al-Haytham made of movement in his attempt to prove the fifth postulate.[14]

It is not the same thing to resort to movement *de facto*, without concerning oneself about the legitimacy of its use. By saying nothing at all on the question of legitimacy, one avoids contradicting received opinion – hence the success achieved alike by the ancient geometers and those of the ninth and tenth centuries through what we might call their practical, not to say pragmatic use of movement and transformations. In any case, this was the dominant position among the geometers in antiquity who were concerned with curves, be they transcendent or algebraic; just as it was later that of Archimedes in both his *Conoids and Spheroids* and his *Spirals*, and of Apollonius in his *Conics*, etc. During the ninth and tenth centuries movement and transformations came to be used more and more.

It is something else again to include movement among the primitive terms of geometry. That would be to take a positive attitude in regard to movement and its role in definitions and proofs. But such an approach requires a reconfiguration of the concepts of geometry or at the very least a certain number of them. They need to be rethought in terms of movement, and the notion of geometrical 'locus' must be looked at afresh. Now, for an adaptation of this kind, there obviously needed to be new bases, a new and distinct discipline and a different

---

[14] R. Rashed and B. Vahabzadeh, *Al-Khayyām mathématicien,* Paris, Librairie Blanchard, 1999. English version : *Omar Khayyam. The Mathematician*, Persian Heritage Series n° 40, New York, Bibliotheca Persica Press, 2000 (without the Arabic texts).

methodology. Ibn al-Haytham was the first to my knowledge to have attempted this reorganisation by conceiving the discipline of the 'knowns', by working out an *ars analytica* and reformulating the notion of 'geometrical locus'.[15] After him further attempts of this kind would have to wait until the second half of the seventeenth century, and in particular for Leibniz's *analysis situs*.

There are other founding acts in other geometrical disciplines: the combination of spherical geometry and trigonometry to do away with Menelaus' theorem, for example. Later on there would be many others: Descartes' invention of symbolism, the introduction of imaginary quantities, exact representation. The same is true of other branches of mathematics: decimal arithmetic, the classification of mathematical propositions (Ibn Sinān, al-Samaw'al), the technique of analysis and synthesis, and, more generally, the philosophy of mathematics. In a word, in all these fields the principal founding acts of classical mathematics are to be discovered. It would not be until the Italian algebraic school in the sixteenth century, Descartes' *Géométrie*, Fermat's Diophantine analysis, and also the infinitesimal geometry of the eighteenth century, that other founding acts would take place, which, along with those we have just picked out, shaped the beginnings of modern mathematics, and, in consequence, the formation of the new rationality.

---

[15] See *Les Mathématiques infinitésimales du IXe au XIe siècle*, vol. IV: *Méthodes géométriques, transformations ponctuelles et philosophie des mathématiques*, London, 2002.

Part VII
Medicine and Health

Chapter 16
Chemotherapy by Design

John E. Lesch

In 1988 the Nobel Prize in Physiology or Medicine was awarded jointly to Sir James W. Black, Gertrude B. Elion, and George H. Hitchings "for their discoveries of important principles for drug treatment." The abstraction of the award citation subsumed two distinct lines of research. Black's work introduced the first members of what became new classes of drugs: beta-blockers used in treatment of cardiovascular and other conditions, and H_2-blockers, used to treat acid-peptic disorders. Hitchings' and Elion's collaboration had yielded effective drugs for use in a remarkable variety of conditions, including cancer, gout, organ transplantation, malaria, and bacterial and viral infections.[1]

This chapter is a first approach to a description and analysis of what may be called the Hitchings-Elion research program, which spanned more than four decades by the time that the investigators received their Nobel award. Three aspects of the Hitchings-Elion program deserve particular emphasis: its coherence and unity across almost half a century of work that engaged a variety of collaborators; its embodiment of both rational and empirical elements; and its character as industrialized research. The close interdependence of these characteristics of the program may best be appreciated by tracing its course from its beginnings in the early 1940s through its various embodiments up to the eve of its recognition by the Caroline Institute in 1988.

---

[1] Folke Sjöqvist, Presentation speech, in *Nobel Lectures in Physiology or Medicine 1981–1990*, Tore Frängsmyr, Editor-In Charge, and Jan Lindsten, Editor (Singapore: World Scientific Publishing Co., 1993), 409–411. Parts of this paper appeared in different form in John E. Lesch, *The First Miracle Drugs: How the Sulfa Drugs Transformed Medicine* (New York: Oxford University Press, 2007), 256–268; and idem., "Dreams of reason: historical perspective on rational drug design," 2008 Kremers Award Lecture, *Pharmacy in History* 50 (2008): 131–139.

J.E. Lesch (✉)
University of California, Berkeley, CA, USA
e-mail: jlesch@calmail.berkeley.edu

J.Z. Buchwald (ed.), *A Master of Science History*, Archimedes 30,
DOI 10.1007/978-94-007-2627-7_16, © Springer Science+Business Media B.V. 2012

16.1 Formation of a Research Program

When George Hitchings joined the Wellcome Research Laboratories in Tuck-
ahoe, New York in 1942 as "head and sole member of the Biochemistry
Department," he had already been working in the field of biochemistry for
over a decade. Following bachelor's and master's degrees in chemistry at the
University of Washington, he began graduate school at Harvard in 1928.
Working in Cyrus Fiske's laboratory in the Medical School's Department of
Biological Chemistry, Hitchings was assigned to develop analytical methods for
the purine bases, a project that became his dissertation and yielded several early
publications. Taking his doctorate in 1932 in the midst of the Depression, he
was able to continue working for several years with temporary appointments at
Harvard in cancer and nutritional research, and at Western Reserve University
in electrolyte research.[2]

At Wellcome Hitchings was given modest resources, but also a free hand to
develop his own program. He later recalled that by that time he had been
interested in chemotherapy for several years, but that

> academia stood, rather disdainfully, apart from all this activity, and stated that it was
> premature to attempt chemotherapy because there was not sufficient knowledge of
> biochemistry, physiology, and pharmacology to sustain any kind of meaningful opera-
> tion. ... But when we came on the scene in 1942, there was a bright, shining star on the
> horizon, which had arisen from the work on Prontosil and its active principle sulfani-
> lamide, and from the recognition by Woods and Fildes that this was a case of meta-
> bolite antagonism. Thus, the antimetabolite theory was born.[3]

Hitchings referred here to the work of British medical bacteriologist Paul
Fildes and biochemist Donald Woods. In publications that appeared in 1940
Woods and Fildes asserted that sulfanilamide, and by extension, other sulfa
drugs, acted on bacteria by interfering with an enzyme that helped to synthesize
a nutrient the bacteria needed for growth and reproduction. Sulfanilamide
closely resembled a compound (substrate) acted upon by the enzyme to produce
the needed nutrient, so sulfanilamide was able to compete with the substrate
and displace it. In this way sulfanilamide prevented the formation of the
nutrient and thereby blocked the growth and reproduction of the bacteria.
Unable to increase in numbers, the invading bacteria were then destroyed by
the defenses of the human or animal host. Woods and Fildes identified the
substrate as p-aminobenzoic acid.[4]

---

[2] George H. Hitchings, Autobiography, in *Nobel Lectures in Physiology or Medicine 1981–
1990*, Tore Frängsmyr, Editor-in-Charge, and Jan Lindsten, Editor (Singapore: World Scientific Publishing Co., 1993), 471–475.

[3] George H. Hitchings, "A biochemical approach to chemotherapy," *Drug Intelligence and
Clinical Pharmacy* 16 (November 1982): 843–848 (on 843).

[4] D. D. Woods, "The relation of p-aminobenzoic acid to the mechanism of the action of
sulphanilamide," *British Journal of Experimental Pathology* 21 (1940): 74–90; D. D. Woods
and P. Fildes, "The anti-sulphanilamide activity (in vitro) of p-aminobenzoic acid and related

Fildes went on to generalize these findings into a program for the discovery of new antibacterial drugs. In 1940 he published "A rational approach to research in chemotherapy" in *The Lancet*. His argument was that antibacterial substances as a group function by interfering with an essential metabolite in the bacterial cell. The kind of inhibitions produced by sulfanilamide required "an inhibitor so closely related in formula to the essential metabolite that it can fit the same enzyme, and sufficiently unrelated to be devoid of essential metabolic activity." For Fildes, this involved the further conclusion that "chemotherapeutic research might reasonably be directed to modification of the structure of known essential metabolites to form products which can block the enzyme without exhibiting the specific action of the metabolite." With this statement, Fildes had converted a particular, if spectacular, result with a known antibacterial agent into a proposal for a research program that might identify many others yet unknown.[5]

In the 1940s and 1950s other researchers put Fildes' program into practice in the search for new antibacterial drugs. More important, other researchers were inspired to take a second step of generalization that opened up a still wider research horizon. This involved the definition of the concept of antimetabolite as a substance that interfered with the action of an essential metabolite in a living cell. This could mean bacteria (as it did for Fildes), but it could also mean other kinds of infectious microorganisms, eventually including viruses, or neoplastic (cancerous) cells that appeared within an organism.

The beginnings of a transition to the broader concept can be seen as early as 1941 in a paper by John Lockwood, and American surgeon and bacteriologist at the University of Pennsylvania. Lockwood saw reason for optimism in the Woods-Fildes theory, and said that

> it is perhaps pardonable to suggest that we may be provided with a new method of approach to the treatment of cancer, a disease in which unrestrained proliferation of tissue cells is similar in some respects to the proliferation of bacteria in invasive infections. If the difference between malignant cells and normal cells should be found to be due to the local activity of some chemical growth factor, a compound of similar chemical configuration might be administered to cancer patients which would block the activity of the proliferative factor without exhibiting its physiological effects.[6]

After he joined Wellcome Hitchings saw an opportunity to use the expanded concept of antimetabolite to bring together in a novel way his interest in the biosynthesis of nucleic acids and a search for new chemotherapeutic agents.

---

compounds" (abstract), *Chemistry and Industry* 18 (February 24, 1940): 133–134. On the background and formulation of the Woods-Fildes theory, see Lesch, *The First Miracle Drugs* (ref. 1), 251–262.

[5] Paul Fildes, "A rational approach to research in chemotherapy," *Lancet* 238 (1940): 955–957.

[6] John S. Lockwood, "Progress toward an understanding of the mode of chemotherapeutic action of sulfonamide compounds," in *Chemotherapy*, University of Pennsylvania Bicentennial Conference (Philadelphia: University of Pennsylvania Press, 1941), 9–28 (on 26).

Research following the Woods-Fildes theory had shown that sulfanilamide was antagonized not only by p-aminobenzoic acid but also by the bases of the nucleic acids and by some amino acids, in certain combinations. Growth factors (later called folic acid) involved in the synthesis of purine and pyrimidine bases had also been identified. Hitchings reasoned that preparation of synthetic analogs of the purine and pyrimidine bases might provide antimetabolites that would serve at the same time as tools for the biochemical study of nucleic acid synthesis and as potential chemotherapeutic compounds. "It seemed that this was a fertile field to explore," he later recalled, "and that one might use the antimetabolite principle to explore folic acid's enzymes and metabolic pathways. We felt that it was highly probable that, in the course of these explorations, we would discover exploitable information that could be used in chemotherapy."[7]

To implement this project Hitchings little by little assembled a small group of collaborators. His first recruit was Elvira Falco, then an assistant in Wellcome's Bacteriology Department. Hitchings and Falco together designed a system to screen purine and pyrimidine compounds for biological activity, using the bacterium *Lactobacillus casei*. Gertrude Elion, a chemist, joined the group in 1944, and concentrated mostly on synthesis of purine analogs. In 1947 Peter B. Russell arrived from Cambridge University, bringing expertise in organic chemistry and some familiarity with medicinal chemistry.[8]

Hitchings later recalled that when this project began, "none of the enzymes and metabolic pathways toward the nucleic acids were known." Nevertheless the black box screening system devised by himself and Falco using *L. casei* quickly yielded promising results. *L. casei* would grow either on a growth factor (folic acid) or on a mixture of purine and the pyrimidine thymine. The system was set up so that it could show either stimulation effects or antagonistic effects of analogs of bases of the nucleic acids. Early screening revealed that analogs could be found that had a marked inhibitory effect not only on *L. casei*, but also on some pathogenic bacteria. Encouraged by these results, Hitchings and his colleagues expanded the biological screening procedures, and added toxicity testing on growing rats.[9]

A few others joined the Hitchings research group in the mid-1940s, but the number remained small, and all shared a single large laboratory. Fortunately,

---

[7] Hitchings, "A biochemical approach to chemotherapy" (ref. 3), 843 (quote); George H. Hitchings, Gertrude B. Elion, Elvira A. Falco, Peter B. Russell, and Henry VanderWerff, "Studies on analogs of purines and pyrimidines," *Annals of the New York Academy of Sciences* 52, Art. 8 (July 7, 1950): 1318–1335; George H. Hitchings, "Selective inhibitors of dihydrofolate reductase," Nobel Lecture, December 8, 1988, in *Nobel Lectures in Physiology or Medicine 1981–1990*, Tore Frängsmyr, Editor-in-Charge, and Jan Lindsten, Editor (Singapore: World Publishing Co., 1993), 476–493 (on 476).

[8] Hitchings, "Selective inhibitors" (ref. 7), 476; Hitchings, Autobiography (ref. 2). On Elion, see below.

[9] E. A. Falco, G. H. Hitchings, and M. B. Sherwood, "The effects of pyrimidines on the growth of *Lactobacillus casei*," *Science* 102 (1945): 251–254; Hitchings, "A biochemical approach to chemotherapy" (ref. 3), 843; and Hitchings, "Selective inhibitors" (ref. 7), 476.

collegial relations were friendly. "Under the leadership of Falco," Hitchings later recalled, "a constant flow of banter developed covering a wide range of subjects and degrees of seriousness. We never had any obstacles to interpersonal communication."[10]

Encouraged by the results of expanded biological screening using the *L. casei* system, Hitchings in 1947 entered into arrangements with two outside entities for expanded testing of the purine and pyrimidine analogs being prepared in his laboratory. One of these was with the Sloan Kettering Institute in New York, which would test compounds for antitumor activity using the sarcoma 180 model in mice. The other was with laboratories that would conduct expanded antibacterial and antimalarial testing.[11]

In addition to making possible increased numbers of tests for antitumor activity, the connection with Sloan Kettering benefited Hitchings' research group in other ways. Impressed with the potential of the compounds and associated biological information coming from the Wellcome team's work, Cornelius P. Rhoads, the Sloan Kettering director, offered the group increased financial support. This assistance, which continued into the early 1950s when it was replaced by internal money from Burroughs Wellcome, allowed for a doubling of the number of members of Hitchings' group, to a total of around fifteen people. The link with Sloan Kettering also led to contacts with researchers and clinicians that proved valuable as the research proceeded.[12]

One of the first compounds sent by the Hitchings group to Sloan Kettering for testing in 1948 was 2, 6-diaminopurine, synthesized by Gertrude Elion. Sloan Kettering researchers found it to be active in sarcoma 180 tests in mice, and clinical trials conducted by Joseph H. Burchenal at Memorial Hospital gave promising results in treatment of patients with leukemia. Hitchings later recalled that these early results were "sufficient to establish cancer chemotherapy as a continuing primary goal of our group."[13]

The early findings on 2, 6-diaminopurine were also one of the first visible results of Gertrude Elion's concentration on the chemistry and metabolism of purines, an assignment she had taken on not long after joining the Hitchings group. The daughter of immigrant parents, Elion had followed education in New York City public schools with four years at Hunter College, where she graduated in 1937 with a major in chemistry. Unable to afford graduate school, she found jobs scarce, and as she later recalled, "the few positions that existed in laboratories were not available to women." After working in a temporary

---

[10] Hitchings, Autobiography (ref. 2).

[11] Hitchings, "Selective inhibitors" (ref. 7), 476–477; and Hitchings, Autobiography (ref. 2).

[12] Hitchings, Autobiography (ref. 2).

[13] Hitchings, "Selective inhibitors" (ref. 7), 477; Joseph H. Burchenal, David A. Karnovsky, Elizabeth M. Kingsley-Pillers, Chester M. Southam, W. P Laird Meyers, George C. Escher, Lloyd F. Craver, Harold W. Dargeon, and Cornelius P. Rhoads, "The effects of the folic acid antagonists and 2, 6-diaminopurine on neoplastic disease, with special reference to acute leukemia," *Cancer* 4 (1951): 549–569.

teaching position and as an unpaid laboratory assistant, she began graduate studies in chemistry at New York University in 1939, supporting herself by teaching in New York City secondary schools. With her masters degree in hand in 1941, she spent a year and a half doing routine quality control work for a food company, then six months in a laboratory at Johnson and Johnson in New Jersey. When the latter position ended, she found herself with multiple job offers from research laboratories. Among these was an invitation to join the Hitchings group, which she accepted in 1944.[14]

The decision to join the Wellcome Research Laboratories proved decisive for Elion's career. In Hitchings' group she found a work environment that gave full scope to her drive and intellectual ambition. Encouraged to learn and to take on increasing responsibility, she found her opportunities quickly expanding. "From being solely an organic chemist, I soon became very much involved in microbiology and in the biological activities of the compounds I was synthesizing," she later recalled. "I never felt constrained to remain strictly in chemistry, but was able to broaden my horizons into biochemistry, pharmacology, immunology, and eventually virology."[15]

That this was the case was no doubt due in part to Hitchings' own qualities as colleague and research manager. The small size of his research group, especially in the early years before it acquired support from Sloan Kettering, was also a factor, since it mitigated against a highly specialized division of work. Equally or more important were the specifically industrial goals of the research, which aimed not simply at new biochemical knowledge, but also at the development of effective chemotherapies. Implementation of such goals called for use or creation of whatever kinds of knowledge, skills, or instruments could be brought to bear on the problems, regardless of their provenance in specialized academic fields.

The modest but unmistakable success of 2, 6-diaminopurine brought Elion and the Hitchings group squarely into the emerging field of cancer chemotherapy. In 1948, the same year that the Wellcome Research Laboratories sent 2, 6-diaminopurine to Sloan Kettering for testing. Sidney Farber and his colleagues at the Children's Medical Center in Boston published a paper reporting promising results in treatment of acute leukemia in children, using aminopterin, a folic acid antagonist. Farber was careful in his conclusions, stressing the small number of patients in the study, the temporary character of the remissions obtained, and the toxicity of the compound. With these reservations, he nevertheless saw in his results "a promising direction for further research concerning the nature and treatment of acute leukemia in children."[16]

---

[14] Autobiography of Gertrude B. Elion, "The Nobel Prize in Physiology or Medicine 1988," *The Oncologist* 11 (2006): 966–968.

[15] Elion, "Autobiography" (ref. 14), 967.

[16] Sidney Farber, Louis K. Diamond, Robert D. Mercer, Robert F. Sylvester, and James A. Wolff, "Temporary remissions in acute leukemia in children produced by folic acid antagonist, 4-aminopteroyl-glutamic acid (aminopterin)," *New England Journal of Medicine* 238 (1948): 787–793.

In his article Farber credited the contributions of researchers in the Lederle Laboratories and the Calco Chemical Division, both components of the American Cyanamid Company, "who are responsible for the chemical research that made possible these studies on children." Behind this acknowledgment lay several years of a collaboration between industrial and clinical researchers that was distinct from, but that in some respects paralleled, the collaboration that had begun to develop between the Hitchings group and Sloan Kettering.[17]

The involvement of Lederle Laboratories in cancer chemotherapy appears to have been prompted in the first instance by a collaboration that it began in 1944 with another medical researcher, Richard Lewisohn. In 1937 Lewisohn, a surgeon at Mount Sinai Hospital in New York City, had begun investigating the antitumor effects of spleen extracts, and in 1939 he had set up a screening program to identify other chemical agents that might cause regression of tumors. By 1941 he was focusing on a search for B group vitamins in yeast, and then barley, extracts. Lewisohn reported promising results in treatment of breast cancer, but an independent investigation conducted in 1943 at Memorial Hospital in New York City at the instigation of Lewisohn's sponsor, the International Cancer Research Foundation, failed to confirm his findings. When folic acid was isolated by Lederle researchers led by Yellepragada SubbaRow in 1944, Lewisohn surmised that this compound might be the active substance in his yeast and barley extracts.[18]

In 1944 Lederle researchers supplied Lewisohn with a growth factor isolated from *Lactobacillus casei*, presumed to be pteroylglutamic acid (folic acid). With this substance Lewisohn obtained inhibition of cancers in mice. Further investigation, however, showed that the substance supplied was a related but distinct compound, pteroyltriglutamic acid, and that pteroylglutamic acid itself was ineffective in treatment of mouse cancer.[19]

Prompted by this finding, Lederle chemists synthesized both pteroyldiglutamic acid and pteroyltriglutamic acid, naming them diopterin and teropterin, respectively. By 1947 the Lederle researchers had begun a collaboration with Farber, and the compounds were passed on to him for clinical testing. In a preliminary clinical report published in late 1947, Farber called for further investigation of teropterin in clinical trials. He also noted in this report and in his 1948 paper that treatment with either diopterin or teropterin accelerated the leukemic process in patients, in comparison to patients not so treated. Based on this finding, Farber suggested two distinct therapeutic approaches. One of these would make use of the acceleration phenomenon by following administration

---

[17] Farber, "Temporary remissions in acute leukemia in children" (ref. 16), 787, 793.

[18] Walter Sneader, *Drug Discovery: A History* (Chichester, UK: Wiley, 2005), 248–249. On the Lederle Laboratories work on folic acid, see Y. SubbaRow et al., "Folic acid," *Annals of the New York Academy of Sciences* 48 (1946): 255–349; and M.E. Hultquist, et al., "Folic acid (supplement). Synthesis of pteroylglutamic acid (Liver *L. casei* factor) and pteroic acid—Part II," *Annals of the New York Academy of Sciences* 48, Art. 5 (supplement) (1947): i–vi.

[19] Sneader, *Drug Discovery* (ref. 18), 249.

of diopterin or teropterin with radiation or nitrogen mustard therapy. The other would employ treatment with folic acid antagonists supplied by the chemists. Among these antagonists was aminopterin, the subject of Farber's 1948 paper.[20]

Behind the synthesis of aminopterin was an effort on the part of chemists at Lederle Laboratories, and also at laboratories in Bound Brook, New Jersey that were part of the Calco Chemical Division of American Cyanamid, to prepare other folic acid analogs as possible antagonists of folic acid. The beginnings of this program remain to be clarified. Part of the background lies in the expansion of pharmaceutical research and production by American Cyanamid beginning in 1936, when Calco set up a new pharmaceutical division to conduct research on sulfonamides and built the first American pilot plant for production of sulfanilamide. American Cyanamid became a leader in the sulfa drugs field, manufacturing not only sulfanilamide, but also sulfapyridine (under license from the British firm May & Baker), sulfathiazole, and sulfaguanidine, and in 1940 introducing sulfadiazine, all of which were heavily used during World War II. In 1937 American Cyanamid set up new general research laboratories in Stamford, Connecticut, and from this time on the company's pharmaceutical research involved collaborations of Bound Brook with either Lederle (at Pearl River, New York) or Stamford.[21]

One glimpse of the evolving interest in the antimetabolite concept within American Cyanamid by the mid-1940s may be found in a paper published by Richard O. Roblin, Jr. in 1946. A chemist in the Chemotherapy Division of the Stamford Research Laboratories, Roblin set out to survey current literature on what he called metabolite antagonists, remarking that "the concept that substances chemically related to a metabolite may interfere with the normal function of that metabolite in living cells is attracting widespread interest among chemists and biologists." In an article of 122 pages that included 471 references, Roblin summarized work to date (the paper was received for publication in December, 1945) on antagonists of vitamins, hormones, and cell metabolites, crediting the Woods-Fildes theory as the stimulus to many of these

---

[20] Sneader, *Drug Discovery* (ref. 18), 249–250; Sidney Farber, Elliott C. Cutler, James W. Hawkins, J. Hartwell Harrson, E. Converse Peirce, 2nd, and Gilbert G. Lenz, "The action of pteroylglutamic conjugates on man," *Science* 106 (1947): 619–621; Farber et al., "Temporary remissions in acute leukemia in children" (ref. 16), 787. On nitrogen mustard as a cancer chemotherapy, see Alfred Gilman and Frank S. Philips, "The biological actions and therapeutic applications of the B-chloroethyl amines and sulfides," *Science* 103 (1946): 409–436; and Rose J. Papac, "Origins of cancer therapy," *Yale Journal of Biology and Medicine* 74 (2001): 391–398.

[21] James M. Smith, Jr., Donna B. Cosulich, Martin E. Hultquist, and Doris R. Seeger, "The chemistry of certain pteroylglutamic acid antagonists," *Transactions of the New York Academy of Sciences*, Series II, 10 (1948): 82–83; Anthony S. Travis, "From color science to polymers and sulfa drugs: Calco Chemical Company and American Cyanamid between two World Wars," *Chemical Heritage* 23, 3 (Fall 2005): 8–13.

investigations. He concluded with remarks that closely paralleled the views that the Hitchings group was putting into practice:

> Since in many respects it is a relatively new and rapidly developing field, it is not possible to assess all the implications inherent in the broad concept of metabolite antagonists. However, as an approach to the mechanism of action of a number of drugs, as a guide in the synthesis of new therapeutic agents, and as a means of evaluating the normal mode of synthesis and function of metabolites in living cells, the concept appears to offer many possibilities as yet unexplored.[22]

With established expertise in the biochemistry of folic acid and in pharmaceutical and organic chemistry, the researchers at Lederle and Bound Brook were well positioned to supply folic acid antagonists to clinical researchers. The first such compound supplied to Farber in 1947 was pteroylaspartic acid, which he found to have some effect in reducing numbers of leukemic cells. Exploring the effects of molecular modifications, the chemists found that replacement of the hydroxl substituent on the 4-position of the pteridine ring by an amino group increased the potency of folic acid antagonists. One of the compounds that emerged from this discovery was aminopterin, which reached Farber in November 1947. Another was the compound synthesized at Bound Brook in the summer of 1947 and at first called amethopterin. Found to be effective but less toxic, amethopterin replaced aminopterin in leukemia chemotherapy by the early 1950s, under the new name methotrexate.[23]

Galvanized by the promise of 2,6-diaminopurine and other findings, and with an expanded research staff made possible by support from Sloan Kettering, the Hitchings group in 1948 began to divide responsibilities for the different components of the laboratory's research. Henceforth Falco, Russell, and Hitchings himself concentrated on pyrimidine analogs. Elion, who had already developed special expertise on purines and purine metabolism, would continue to focus her attention on purine analogs.[24]

In each case the research would continue with two aims. First, it would try to elucidate the roles of purine and pyrimidine bases in nucleic acid synthesis and thus in growth, and the part played by folic acid in the synthesis of these bases. Second, it would try to identify among the analogs new chemotherapeutic agents. The theory was that tissues that depended for survival on rapid

---

[22] Richard O. Roblin, Jr., "Metabolite antagonists," *Chemical Reviews* 38 (1946): 255–377 (on 255 and 366).

[23] A. L. Franklin, E. L. R. Stockstad, M. Belt, and T. H. Jukes, "Biochemical experiments with a synthetic preparation having an action antagonistic to that of pteroylglutamic acid," *Journal of Biological Chemistry* 169 (1947): 427–435; B. L. Hutchings, J. H. Mowat, J. J. Oleson, E. L. R. Stockstad, J. H. Boothe, C. W. Waller, R. B. Angier, J. Semb, and Y. SubbaRow, "Pteroylaspartic acid, an antagonist for pteroylglutamic acid," *Journal of Biological Chemistry* 170 (1947): 323–328; Sneader, *Drug Discovery* (ref. 18), 250–251. For an early clinical report on amethopterin, see Leo M. Meyer, Franklin R. Miller, Manuel J. Rowen, George Bock, and Julius Rutzky, "Treatment of acute leukemia with amethopterin (4-amino, 10-methyl pteroylglutamic acid)," *Acta Haematologica* 4, 3 (1950): 157–167.

[24] Hitchings, Autobiography (ref. 2).

growth—parasitic microbial or cancer cells, for example—should be especially sensitive to compounds antagonistic to substances needed for growth. The researchers expected that specific differences in the biochemistry of different kinds of cells would allow for identification of compounds with selective action, although the identity of these compounds could not be predicted in advance of biological and clinical screening. As Hitchings later put it in a revealing passage of his Nobel lecture, "by 1947, six or seven of us were pursuing this work, and the feeling in the group was, 'Now we have the chemotherapeutic agents; we need only to find the diseases in which they will be active.'"[25]

Although the purine and pyrimidine lines of research undertaken by the Hitchings group were closely related both temporally and conceptually, the sequence of developments within each line may best be understood by considering them in turn. In what follows we will look first at the work related to purines and purine analogs, then at research on pyrimidines and their analogs, in each case focusing on the development of new chemotherapeutic agents. Finally we will consider what may be seen as a related but distinct line of research undertaken primarily by Elion beginning in 1968, namely the search for antiviral drugs and study of their mechanisms of action.

16.2 Purines and Purine Analogs

By 1951 the group led by Elion had synthesized over one hundred purines, all of which were then screened for activity in *L. casei*. In the course of this work they had found that the substitution of oxygen by sulfur at the 6 position of the molecule in the natural purines guanine and hypoxanthine produced purine analogs that were inhibitors of purine metabolism. Two of these compounds were 6-mercaptopurine and 6-thioguanine. The *L. casei* screen showed that the inhibitory effect of 6-mercaptopurine could be reversed by hypoxanthine, a compound described by Hitchings as "more or less the core of purine metabolism." Animal tests at Sloan Kettering showed that 6-mercaptopurine was active against a number of rodent tumors and leukemias, and in 1952 Cornelius Rhoads organized a cooperative clinical trial with around a dozen investigators, including notably Joseph Burchenal at Memorial Hospital.[26]

Clinical results soon revealed the activity of 6-mercaptopurine in acute leukemia in children. Excited by the preliminary findings, Rhoads passed the news on to journalist Walter Winchell, and soon there were press reports that a

---

[25] Hitchings, "Selective inhibitors" (ref. 7), 476.

[26] Gertrude B. Elion, "The purine path to chemotherapy," Nobel Lecture, December 8, 1988, *Nobel Lectures in Physiology or Medicine 1981–1990*, Tore Frängsmyr, Editor-in-Charge, and Jan Lindsten, Editor (Singapore: World Publishing Co., 1993), 447–468 (on 449); Hitchings, "A biochemical approach to chemotherapy" (ref. 3), 843–844.

new leukemia treatment had been found, and that Hitchings had supplied it to Sloan Kettering. Hitchings later recalled what followed:

> As you may imagine, the roof fell in on me. Within two days I had 600 letters on my desk and phone calls from all over the world; we were in one terrible bind. We had limited supplies of the drug, no idea what it would cost, and no mechanism for distribution or for dealing with the many pathetic appeals that we received.

No immediate response to this demand was possible. The company did promptly file a New Drug Application with the Food and Drug Administration. Following personal visits to the sites of the clinical trials by the FDA official in charge of reviewing new applications, 6-mercaptopurine was approved for commercial release in September, 1953. Even then, production problems remained, and the clinical studies were not made public until the end of April, 1954, when the New York Academy of Sciences held a symposium on the new drug.[27]

When 6-mercaptopurine entered medical practice in 1953, standard drug treatment for acute childhood leukemia consisted of methotrexate and steroids. Median life expectancy for children so afflicted was three to four months, and only about thirty percent of patients lived for a year. In some individuals the disease was entirely resistant to chemotherapy. Treatment with 6-mercaptopurine raised median survival time to twelve months, and some patients treated with 6-mercaptopurine and steroids were able to remain in remission for years.[28]

6-mercaptopurine did not solve the problem of childhood leukemia, but did indicate a way forward. Elion, Hitchings, and their co-workers were encouraged to continue, and other cancer researchers joined the search for antimetabolites of nucleic acid bases. With the development of other drugs and of combination chemotherapy, physicians were eventually able to cure around eighty percent of patients with acute childhood leukemia.[29]

As the position of 6-mercaptopurine in the clinic was consolidated, Elion and her co-workers carried on with metabolic studies of the compound, hoping to find ways to improve its therapeutic properties in cancer treatment. Meanwhile, from the mid-1950s other researchers were elucidating pathways of

---

[27] Hitchings, "A biochemical approach to chemotherapy" (ref. 3), 844; Roy Waldo Miner, editor, *6-Mercaptopurine, Annals of the New York Academy of Sciences* 60 (December 6, 1954): 183–508.

[28] Elion, "The purine path to chemotherapy" (ref. 26), 449; Hitchings, "A biochemical approach to chemotherapy" (ref. 3), 844.

[29] Elion, "The purine path to chemotherapy" (ref. 26), 449; Cornelius P. Rhoads, editor, *Antimetabolites and Cancer.* A symposium presented on December 28–29, 1953 at the Boston meeting of the American Association for the Advancement of Science (Washington, DC: American Association for the Advancement of Science, 1955); Sneader, *Drug Discovery* (ref. 18), 253. On childhood leukemia see also John Laszlo, *The Cure of Childhood Leukemia: Into the Age of Miracles* (New Brunswick: Rutgers University Press, 1996). On the status of antimetabolite research by the early 1950s, see D. W. Woolley, *A Study of Antimetabolites* (New York: Wiley, 1952).

purine biosynthesis. Elion and Hitchings realized early on that much of the 6-mercaptopurine administered to a patient was metabolized *in vivo*, so that very little of the compound was excreted unchanged. Especially important was the breakdown of the medicine by the enzyme, xanthine oxydase, yielding 6-thiouric acid. Other reactions affected the sulfur on the molecule. In an effort to modify the metabolism of 6-mercaptopurine so that it would not be readily converted into other compounds *in vivo*, Elion's group first introduced various substituents on the purine ring. With the exception of thioguanine, a derivative they already knew, the resulting compounds lacked antitumor activity. So they tried a different approach, adding removable "blocking groups" to the molecule's sulfur atom, in the hope that these groups might protect the sulfur from oxidation and hydrolysis. The idea was that once inside cells the blocking group might be removed, releasing 6-mercaptopurine, and that ideally this would be effected by an enzyme specific to tumor cells.[30]

The most promising compound to come out of this approach was azathioprine, synthesized in 1957. Able to act as a pro-drug for 6-mercaptopurine, azathioprine also proved to have a better chemotherapeutic index than its parent compound in a mouse cancer, adenocarcinoma 755. Unfortunately its chemotherapeutic index for human leukemia was not significantly better than that of 6-mercaptopurine, ending its prospects as an improved replacement for the latter in cancer chemotherapy.[31]

Azathioprine might have been shelved, had it not been for the intervention of clinicians interested in a different kind of chemotherapy. At Tufts University in Boston, William Dameshek and Robert Schwartz were seeking drugs that might enable human bone marrow transplantation as a means of treating aplastic anemia, leukemia, or radiation damage, but none of the compounds they tried had succeeded in suppressing the immune response. It occurred to Schwartz that the immunoblastic lymphocytes formed in an immune response were very similar to leukemic lymphocytes. If this was so, he reasoned, might not proliferation of the cells formed in the immune response be suppressed by the same agent that suppressed proliferation of leukemic cells, that is, by an antimetabolite? With this idea in mind, he wrote to Hitchings to obtain 6-mercaptopurine, and to Lederle Laboratories to obtain methotrexate. Hitchings replied immediately with a supply of 6-mercaptopurine, but Schwartz's letter to Lederle did not reach its destination. Hitchings later reflected that if the circumstances had been reversed, Schwartz's experiment might have ended, since methotrexate was not active in the system he was using. Instead, a new opening appeared for Dameshek's and Schwartz's investigations.[32]

---

[30] Elion, "The purine path to chemotherapy" (ref. 26), 449–451; Hitchings, "A biochemical approach to chemotherapy" (ref. 3), 844.

[31] Elion, "The purine path to chemotherapy" (ref. 26), 451–452; Hitchings, "A biochemical approach to chemotherapy" (ref. 3), 844; Sneader, *Drug Discovery* (ref. 18), 253.

[32] Elion, "The purine path to chemotherapy" (ref. 26), 452; Hitchings, "A biochemical approach to chemotherapy" (ref. 3), 844–845; Sneader, *Drug Discovery* (ref. 18), 253.

In a series of experiments Schwartz showed that in rabbits injected with a foreign antigen, such as bovine serum albumin, immune response was suppressed by 6-mercaptopurine, and that the suppression effect was strongest when 6-mercaptopurine was given at the same time as the antigen. In the latter case the lack of response could persist for weeks. He also showed that with the right combination of drug and procedures the response could be made specific, with rabbits becoming tolerant to one antigen while mounting an immune response to others. Prompted by Schwartz, the Hitchings-Elion group set up an immunological screening test in which they measured the immune response of mice to sheep red cells, a test that enabled them to identify new drugs and drug combinations, and to extend the investigation in other ways. As Hitchings later pointed out, Schwartz's demonstration of a chemically induced immune tolerance using the antimetabolite, 6-mercaptopurine, represented "an extremely important breakthrough in the field of immunology."[33]

The work of Schwartz and Dameshek drew the attention of Roy Calne, a young British surgeon investigating kidney transplantation in dogs. Collaborating with Hitchings' and Elion's Burroughs Wellcome colleagues in Beckenham, England, Calne used 6-mercaptopurine to suppress immune response, and succeeded in extending the life of a transplanted kidney from the usual 8–10 to 44 days, a new record.[34]

Calne subsequently came to the United States on a Commonwealth Fund Fellowship, with the plan of continuing his work at Peter Bent Brigham Hospital in Boston. Peter Bent Brigham was then a major center for transplantation research, but with the exception of a donation between identical twins, all transplanted kidneys had been rejected. On the advice of the Burroughs Wellcome group in Beckenham, Calne made a stop in Tuckahoe on his way to Boston, and came away with several compounds, including what the Hitchings-Elion group then called 57-322, or azathioprine. Soon Calne reported to Hitchings that azathioprine was superior to 6-mercaptopurine in suppressing immune response, and that one dog had already carried a transplanted kidney for several months. Similar successes led to the first human kidney transplantation with azathioprine as the only immunosuppressive agent. The recipient had been near death, but recovered and lived more than two years after the surgery.[35]

Under the trade name Imuran, azathioprine was joined with prednisone in a standard immunosuppression regimen in the early 1960s. Between 1965 and 1972 some 25,000 kidney transplantations were done in the United States, and

---

[33] Elion, "The purine path to chemotherapy" (ref. 26), 452; Hitchings, "A biochemical approach to chemotherapy" (ref. 3), 845.

[34] Elion, "The purine path to chemotherapy" (ref. 26), 452; Hitchings, "A biochemical approach to chemotherapy" (ref. 3), 845.

[35] Elion, "The purine path to chemotherapy" (ref. 26), 452; Hitchings, "A biochemical approach to chemotherapy" (ref. 3), 845.

numbers increased thereafter. Improvements came from new drugs, and from antigen typing and matching, and transplantation of other kinds of organs became possible. Subsequent investigations showed that azathioprine, 6-mercaptopurine, and thioguanine were also useful in treatment of autoimmune disease, including systemic lupus and rheumatoid arthritis. The antimetabolite concept, embedded in the Hitchings-Elion program, had helped to open another field of medicine.[36]

Elion and her co-workers were not done with 6-mercaptopurine. They knew from metabolic studies that 6-mercaptopurine was broken down in the organism, and that the enzyme responsible for its oxidation was xanthine oxidase. They reasoned that they should be able to potentiate 6-mercaptopurine in treatment of leukemia by inhibiting xanthine oxidase with an antimetabolite. Since xanthine oxydase had been a test enzyme in the Hitchings group's early search for substrates and inhibitors of the natural purines, there were a number of inhibitors at hand. The one they chose was allopurinol, an analog of the natural purine hypoxanthine that early screening had shown to have no inhibitory effect on bacteria or tumors, and to be non-toxic. Mouse studies showed that allopurinol did potentiate the antitumor and immunosuppressive effects of 6-mercaptopurine. Similar results emerged from studies of use of the compound in treatment of human granulocytic leukemia, undertaken in collaboration with a physician, Wayne Rundles, at the Duke University School of Medicine. Later clinical studies showed, however, that the potentiation was accompanied by a proportional increase in toxicity, so that the chemotherapeutic index of 6-mercaptopurine remained unchanged.[37]

With their attention focused on xanthine oxidase, Elion and Hitchings realized that the enzyme was responsible not only for the oxidation of 6-mercaptopurine, but also for formation of uric acid from the natural purines hypoxanthine and xanthine. Since the painful condition of gout is due to deposits of uric acid crystals in joints or kidneys as a result of excess uric acid in the blood or urine, treatment with allopurinol to inhibit the formation of uric acid opened the way to a new and effective treatment for this disease. Several problems had to be confronted in animal and human studies, including the potential long-term effects of a drug that would need to be taken for the patient's lifetime. One especially significant finding was that in the organism allopurinol not only acted as an inhibitor of xanthine oxidase, but also as a substrate of the same enzyme, which converted allopurinol by oxidation into

---

[36] Elion, "The purine path to chemotherapy" (ref. 26), 452–453; Hitchings, "A biochemical approach to chemotherapy" (ref. 3), 845; George H. Hitchings and Gertrude B. Elion, "Chemical suppression of the immune response," *Pharmacological Reviews* 15 (1963): 365–405; G. Wolberg, "Antipurines and purine metabolism," in M. A. Bray and J. Morley, editors, The Pharmacology of Lymphocytes, *Handbook of Experimental Pharmacology* 85 (1988): 517–533.

[37] Elion, "The purine path to chemotherapy" (ref. 26), 453; Hitchings, "A biochemical approach to chemotherapy" (ref. 3), 845–846; Sneader, *Drug Discovery* (ref. 18), 254.

the xanthine analog, oxypurinol. This result was clinically significant, since oxypurinol was found to bind to and inactivate the enzyme. It was also found to have a longer half-life in the organism, enabling steady-state levels of the drug to be more readily achieved in patients. Since allopurinol was completely absorbed in oral administration while oxypurinol was not, Elion and her co-workers concluded that allopurinol was the ideal pro-drug for oxypurinol. Allopurinol went on the market in 1966, and by the 1970s was among the standard drugs used in treatment of gout.[38]

16.3 Pyrimidines and Pyrimidine Analogs

In 1948, the same year that Elion launched the development of purine analogs as drugs with synthesis of 2, 6-diaminopurine, her colleague Elvira Falco initiated a second line of research with synthesis of a pyrimidine analog, p-chlorophenoxy-2,4-diaminopyrimidine. The Hitchings group had begun work with pyrimidines and their analogs as early as 1945. Now the compound prepared by Falco indicated that pyrimidines with the 2, 4-diamino structure could be of special interest for the antimetabolite research program. The lead was intensively pursued by Falco and Peter Russell. They found that not only did compounds of this group strongly inhibit *L. casei*, but that molecular modification also yielded compounds that were markedly selective in their inhibitory action on different species of organism. The practical implications were clear. As Hitchings later recalled, "it appeared probable that we would be able to tailor such compounds for specific actions against pathogenic species of many kinds."[39]

One notable compound to emerge from this line of research in the 1950s was pyrimethamine, an antimalarial. Peter Russell had pointed out earlier the resemblance of a particular compound of the 2, 4-diaminopyrimidine group to a hypothetical structure of a known antimalarial, proguanil, and it was on the basis of this insight that the Hitchings group made an arrangement in 1947 for an outside laboratory to conduct the testing of compounds as antimalarials. The first commercial product to come out of this testing, marketed with the trade name Daraprim, pyrimethamine was a potent and highly selective anti-malarial. Another important compound that came out of the Hitchings group's 2, 4-diaminopyrimidine program was trimethoprim, an equally potent and highly selective antibacterial.[40]

---

[38] Elion, "The purine path to chemotherapy" (ref. 26), 453–456; Hitchings, "A biochemical approach to chemotherapy" (ref. 3), 845–846; Sneader, *Drug Discovery* (ref. 18), 254.

[39] Falco, Hitchings, and Sherwood, "The effects of pyrimidines on the growth of *Lactobacillus casei*" (ref. 9); Hitchings, Autobiography (ref. 2); Hitchings, "A biochemical approach to chemotherapy" (ref. 3), 846.

[40] Hitchings, "Selective inhibitors" (ref. 7), 476–477; Hitchings, "A biochemical approach to chemotherapy" (ref. 3), 846.

How to account for the remarkable specificity of these compounds? While the Hitchings group pursued its investigations in the 1940s and early 1950s, other researchers were elucidating the biochemistry of folic acid and its metabolism. To Hitchings and his colleagues the action of pyrimidine analogs on *L. casei* suggested that they were in some way antagonistic to folic acid, probably by inhibiting an enzyme that reduced folic acid to folinic acid. By 1950 they had concluded that their compounds were indeed acting as selective inhibitors of this enzyme. Continuing biochemical investigation led to the enzyme's isolation and the specification of its action as the reduction of dihydrofolate to the biologically active tetrahydrofolate, and thus its name, dihydrofolate reductase.[41]

Hitchings and his colleagues conjectured early on that the fine structure of dihydrofolate reductase varied from species to species. They reasoned that an analog that closely resembled the substrate, dihydrofolate, in structure, such as methotrexate, would fit most of the binding sites of the enzyme, regardless of its variations, and thus would not be selective in its activity. Smaller molecules, in contrast, would bind to only some of the sites of the enzyme, and might at the same time bind to sites that were distinct in each species. If so, this would account for the high specificity of action of compounds such as pyrimethamine and trimethoprim. This view of a structural basis for selectivity of action was later confirmed by further investigations, including amino acid sequencing and x-ray crystallographic studies of purified enzymes from various species.[42]

For bacteria, at least, another form of selectivity was available in addition to the inhibition of dihydrofolate reductase by trimethoprim. Pathogenic bacteria, unlike humans, are able to synthesize their own dihydrofolate. Research following the Woods-Fildes theory had shown that it is this synthesis that sulfonamides inhibit by competing with an essential substrate, para-aminobenzoic acid. This opened the possibility of what Hitchings called a "sequential blockade," in which the combination of a sulfonamide and trimethoprim would inhibit the same metabolic pathway at two distinct stages, producing a stronger effect on the bacterium than either drug alone. From this reasoning came the major antibacterial co-trimoxazole, a combination of trimethoprim and sulfamethoxazole, approved by the FDA in 1973 and marketed under different trade names including Septra and Bactrim.[43]

---

[41] Hitchings, "A biochemical approach to chemotherapy" (ref. 3), 846; Hitchings, "Selective inhibitors" (ref. 7), 477–478.

[42] George H. Hitchings, "The utilisation of biochemical differences between host and parasite as a basis for chemotherapy," in L. G. Goodwin and R. H. Nimmo-smith, editors, *Drugs, Parasites, and Hosts* (Boston: Little, Brown and Company, 1962), 196–210; George H. Hitchings and Sheila L. Smith, "Dihydrofolate reductases as targets for inhibitors," *Advances in Enzyme Regulation* 18 (1980): 349–371; Hitchings, "Selective inhibitors" (ref. 7), 477–481; Hitchings, "A biochemical approach to chemotherapy" (ref. 3), 486–487.

[43] Hitchings, "Selective inhibitors" (ref. 7), 482; Hitchings, "A biochemical approach to chemotherapy" (ref. 3), 847. On trimethoprim, including doubts about the utility of the trimethoprim-sulfamethoxazole combination, and the eventual marketing of trimethoprim

16.4 Antivirals

In 1967 Burroughs Wellcome appointed Hitchings Vice President in Charge of Research. At the same time Elion became head of the company's Department of Experimental Therapy, a position she was to hold until her retirement in 1983. Elion later remarked that colleagues sometimes described her department as a "mini institute," since it included sections of chemistry, enzymology, pharmacology, immunology and, eventually, virology. By whatever name, she found that the interdisciplinary arrangement "made it possible to coordinate our work and cooperate in a manner that was extremely useful for development of new drugs."[44]

Within a year Elion's department began to turn its attention to antivirals. Looking back from a later vantage point, Elion advanced three reasons for this change of direction. Twenty years of work on purine analogs and the new drugs it had yielded, including 6-mercaptopurine, thioguanine, azathioprine, and allopurinol, had accomplished much, and opened the way for a fresh start. The compound that had initiated all of this, 2, 6-diaminopurine, had already shown intriguing antiviral activity in 1948, although Elion and her co-workers had not followed up on that lead. Finally, a recent publication by Frank Schabel, Jr., of Parke, Davis & Company and the Southern Research Institute in Birmingham, Alabama, had reported that a purine nucleoside, adenine arabinoside (ara-A) inhibited growth of both DNA and RNA viruses.[45]

As she reflected on Schabel's findings, it occurred to Elion that the arabinoside of 2, 6-diaminopurine might be as active against viral DNA and RNA as adenine arabinoside, given known biochemical similarities of diaminopurine and adenine. An organic chemist colleague, Janet Rideout, synthesized diaminopurine arabinoside, and since at the time Elion's department lacked a virus laboratory, Elion sent the compound on to John Bauer at Wellcome Research Laboratories in Britain for antiviral screening. Soon Bauer reported that diaminopurine arabinoside was very active against both herpes simplex virus and vaccinia virus, with less toxicity to mammalian cells than adenine arabinoside. Elion later recalled that this promising result "began our antiviral odyssey," and initiated several years of work in her department on purine arabinosides.[46]

In 1970 Elion's department moved with the rest of Wellcome Research Laboratories from Tuckahoe, New York to North Carolina. At the same

---

as a stand-alone drug, see also David Greenwood, *Antimicrobial Drugs: Chronicle of a Twentieth Century Medical Triumph* (Oxford and New York: Oxford University Press, 2008), 254–256.

[44] Hitchings, Autobiography (ref. 2); Elion, "Autobiography" (ref. 14), 967.

[45] Elion, "The purine path to chemotherapy" (ref. 26), 456–457; Sneader, *Drug Discovery* (ref. 18), 259; F. M. Schabel, Jr., "The antiviral activity of 9-b-D-arabinofuranosyladenine (ara-A)," *Chemotherapy* 13 (1968): 321–338.

[46] Elion, "The purine path to chemotherapy" (ref. 26), 457–458.

time Howard Schaeffer joined the group as head of the Organic Chemistry Department, bringing with him a lead into a new approach to the antiviral research. Schaeffer's work had shown that acyclic nucleosides, and not only nucleosides in which the sugar ring was intact, could be acted on by enzymes. This finding opened the possibility of use of acyclic nucleoside analogs as antimetabolites.[47]

Promising early results of antiviral screening led the Wellcome researchers to focus on acyclic nucleoside analogs, with the labor divided among three groups. The chemists, Schaeffer and Lilia Beauchamp, synthesized the compounds, the U. K. Wellcome unit that included Bauer and P. Collins conducted the antiviral screening in animals, and Elion's department studied mechanisms of action, enzymology, and *in vivo* metabolism. The researchers were not surprised to find that, in a parallel with earlier work on purine arabinosides, the 2, 6-diaminopurine analog proved highly active against herpes simplex virus. They were surprised to find that the guanine analog, acycloguanosine or acyclovir, was more than one hundred times as active as the 2, 6-diaminopurine analog. Elion and her colleagues published the early results with acyclovir in 1977 and 1978.[48]

Acyclovir was impressive in its selectivity as well as its potency. Highly active against herpes simplex viruses and the varicella zoster virus that causes chicken pox, and with some activity against other herpes viruses, it lacked activity against other kinds of virus, and was not toxic to the mammalian cells in which the herpes viruses grew. Convinced that understanding the biochemical basis of this selectivity would yield valuable insights into the herpes viruses, Elion and her colleagues dedicated resources to this project, including the establishment of an in-house virus laboratory that expanded the capabilities of Elion's department. They found the basis of acyclovir's selectivity in an enzyme specific to herpes viruses, viral thymidine kinase, which in the infected cell begins a process that leads to the incorporation of acyclovir triphosphate into the viral DNA and termination of the DNA chain. These studies opened the way for further research on enzyme differences in normal and virus-infected cells, and on other enzymes specific to viruses, investigations that would contribute to the search for other antiviral drugs.[49]

---

[47] Elion, "The purine path to chemotherapy" (ref. 26), 458.

[48] Elion, "The purine path to chemotherapy" (ref. 26), 458; Sneader, *Drug Discovery* (ref. 18), 259; Gertrude B. Elion, Phillip A. Furman, James A. Fyfe, Paulo de Miranda, Lilia Beauchamp, and Howard J. Schaeffer, "Selectivity of action of an antiherpetic agent, 9-(hydroxyethoxymethyl) guanine," *Proceedings of the National Academy of Sciences* 74 (December 1977): 5716–5720; H. J. Schaeffer, Lilia Beauchamp, P. de Miranda and Gertrude B. Elion, "9-(2-hydroxyethoxymethyl) guanine activity against viruses of the herpes group," *Nature* 272 (April 1978): 583–585. On acyclovir see also Greenwood, *Antimicrobial Drugs* (ref. 43), 375–378.

[49] Gertrude B. Elion, "The chemotherapeutic exploitation of virus-specified enzymes," *Advances in Enzyme Regulation* 18 (1980): 53–66; Elion, "The purine path to chemotherapy" (ref. 26), 458–463; Sneader, *Drug Discovery* (ref. 18), 259.

Entering medical practice in the 1980s, acyclovir had a major impact on treatment of herpes virus infections. In genital herpes it alleviated symptoms and reduced time to healing in first infections and, used prophylactically, reduced the frequency of recurrences. It reduced the period of acute pain in shingles (herpes zoster). In immunosuppressed individuals, such as those undergoing bone marrow transplantation, acyclovir could prevent activation of herpes simplex infections during the period of greatest vulnerability to infection. It could save the lives of people with herpes encephalitis, if given in time. Acyclovir could also be an effective treatment for cold sores, caused by herpes simplex infection.[50]

16.5 Conclusion

By the time that Hitchings and Elion delivered their Nobel lectures in December, 1988, the AIDS epidemic had emerged as a major health crisis. Scarcely mentioned in their talks is that Burroughs Wellcome researchers were largely responsible for the first antiretroviral drug used to treat HIV, zidovudine (azidothymidine, AZT), which had received a product license from the FDA just one year earlier and which the company marketed under the trade name Retrovir.[51]

That the Burroughs Wellcome researchers were able to respond so quickly in the wake of identification of the retrovirus (LAV, later HIV) in 1983 was due to the prior existence and the characteristics of the research program examined here.

In June 1984 the Burroughs Wellcome researchers set up a program to identify compounds that might act against HIV. A nucleoside chemist, Janet Rideout, was put in charge of selecting compounds for testing. One of those she selected was AZT, a compound originally synthesized twenty years earlier by Jerome Horwitz at the Michigan Cancer Foundation as a possible chemotherapeutic agent in leukemia. It was probably chosen in part because of its known activity against animal retroviruses. But it helped that the Burroughs Wellcome researchers had already tested it for antibacterial action, and had it at hand. By December 1984 they had positive results for AZT against two types of animal retroviruses, Friend leukemia virus and Harvey sarcoma virus. They then sent samples of AZT to the National Cancer Institute, where researchers had developed a method of testing compounds for activity against HIV, growing the virus in immortalized human T4 cells.[52]

---

[50] Elion. "The purine path to chemotherapy" (ref. 26), 462–463; Sneader, *Drug Discovery* (ref. 18), 259.

[51] Sneader, *Drug Discovery* (ref. 18), 260–261.

[52] Sneader, *Drug Discovery* (ref. 18), 260–261.

Within two weeks of receiving AZT, the NCI researchers had concluded that it was highly effective against HIV. By June 1985 the findings of the NCI investigators had been confirmed by others at Duke University. The FDA gave approval for a Phase I clinical trial in July 1985, and January 1986 a randomized, double-blind clinical trial in 282 patients had begun. The trial was interrupted after sixteen weeks because of distinctly lower mortality among patients receiving AZT.[53]

Even in brief outline, the AZT story exemplifies three fundamental characteristics of the Hitchings-Elion program. One of these is the continuity and coherence of the antimetabolite research program over more than four decades and many changes in the research environment. The Burroughs Wellcome researchers had long experience in synthesizing and testing compounds, including nucleosides and nucleoside analogs, as potential antimetabolites against a variety of cells (microbial pathogens, cancer cells) and viruses. The prior testing of AZT as an antibacterial by them and the accumulated information about it that was already available in consequence were results of this program. Their understanding of AZT as an antimetabolite that inhibited an enzyme, later identified as reverse transcriptase, specific to the pathogen, was a natural extension of the antimetabolite concept.

The work on AZT also exemplifies the joining of rational and empirical elements in the Hitchings-Elion program and its extension. Rational, because AZT was a member of a defined class of compounds, nucleosides and nucleoside analogs, considered to be potential antimetabolites with selective activity on different kinds of pathogens, but especially investigated by Elion and her colleagues as antiviral agents. Empirical, because only screening could determine which members of this class had selective action on specific pathogens. Within such an approach, specific results could not be predicted, and in this sense AZT is an instance of the group's strategy as defined by Hitchings in one compact formulation: "choose a promising field of work and remain untargeted but opportunistic so that the accumulated knowledge dictates the target."[54]

Finally, the effort from which AZT emerges embodies the character of the Burroughs Wellcome program as industrialized research. This means not simply that the research is located in and paid for by industry, but also that the work is done in a research organization in which the primary goal is to produce viable chemotherapeutic agents, that is entities that are at the same time medical technologies and commercial products. To this end the research is organized not to advance knowledge within a particular discipline, although new knowledge is produced, but in a collaborative and interdisciplinary process

---

[53] Sneader, *Drug Discovery* (ref. 18), 261.

[54] George H. Hitchings, "The Bertner Foundation Memorial Award Lecture—salmon, butterflies, and cancer chemotherapy," *Pharmacological Basis of Cancer Chemotherapy* (Baltimore: The Williams & Wilkins Company, 1975), 25–43 (on 30).

that selects and coordinates the knowledge and techniques needed to advance the goal of producing new medicines. In the course of this work outside individuals and institutions, including clinicians and research laboratories, are enlisted as needed.

The beginnings of this organization are already visible in the small group that Hitchings assembled in the mid-1940s. Although larger and more articulated by the 1980s, its essential features were still in place. By the time they went to Stockholm both Hitchings and Elion had retired. The research group and program they had created, and that had enjoyed so many successes, would continue.

Chapter 17
American Health Reformers and the Social Sciences in the Twentieth Century

Evan M. Melhado

17.1 Introduction

This essay situates American health reformers, during the first three-quarters of the twentieth century, in the broader context of reform in American social policy and in the tissue of relationships that connected reformist impulses with the social sciences. Reformers came in many stripes, and the tenor of reform ideas changed over the course of the century from the Progressive Era, through the Hoover years, the New Deal, and the post-war liberal consensus and the conservative era that followed. Nevertheless, despite the diversity and the changes, until the late 1960s a broad similarity and continuity marked thinking among reformers about American public policy, about health policy in particular, and about the role in it of social-scientific expertise. In particular most reformers held that maintaining the security of citizens (individuals and families) was a broad, public responsibility; that programs to serve this end typically operated by removing some spheres of life from the dominance of markets or offering shelter from or recourse to the effects of markets; that programs should lie primarily on the national level (or, if below, conform to national parameters); and that programs required technical knowledge, wielded disinterestedly by experts trained especially in the social sciences and working in public, non-profit, and non-partisan institutions. From the 1970s, however, this consensus eroded and market-based approaches came to the fore. Devised by experts from a greater diversity of settings, these policies aimed to shape constraints on individual decision making within local or regional markets, to optimize market functioning, to improve efficiency, constrain resource use, and foster for-profit enterprise. To characterize especially the pre-market phase of policy, primarily in health care, this essay opens with a short account of the long complex relations between social science and social reform, and it then explores three contexts in which those relations found expression. Thereafter, it exploits

E.M. Melhado (✉)
Department of History and Medical Humanities and Social Sciences Program,
University of Illinois at Urbana-Champaign, Urbana, IL 61801, USA
e-mail: melhado@uiuc.edu

J.Z. Buchwald (ed.), *A Master of Science History*, Archimedes 30,
DOI 10.1007/978-94-007-2627-7_17, © Springer Science+Business Media B.V. 2012

both primary and secondary materials to develop an alternative to one prominent interpretation of health reformers before the shift to markets. The essay also succinctly characterizes later shifts in the application of social-scientific research to health reform, and it concludes with summary comment on the politics of the most recent reforms in the light of the prior history.

Reform impulses in social policy both reflected and influenced the emergence of the social sciences in America (especially sociology, economics, and political science) from the early-twentieth century (Fisher 1993; Jordan 1994; Ross 1991; Smith 1994). Reformers pursued careers as social scientists, usually in new academic social-science departments, but also in independent research institutions; others, often trained in the social sciences but operating in diverse other settings (industry, business, labor, charities, philanthropies, government), made or sponsored practical applications of social-scientific knowledge; others, having obtained perhaps little formal training in the social sciences but having acquired a knowledge of them from practical experience, served as journalists, cultural and political analysts, critics, and publicists; and others, possessing the characteristics of any of the last three categories, operated as practical agents of political functionaries or intellectual brokers within governments or between them and academic and research institutions (Berkowitz 1995 and 2003; Goodwin 1995; Karl 1969; Pells 1973; Recchiuti 2007).

The diversity of reformers extended beyond their forms of activity and employment to their politics, but the poles of the political spectrum they spanned emerge more clearly from an alternative to the usual antipodes of "left" and "right." At one end lay those possessing a strong sense of social solidarity with the working class (and, increasingly after World War II, the middle class), a conviction that capitalistic development had deprived workers of a fair share of economic benefits, and an interest in collective measures to improve distribution. These reformers included frank but non-revolutionary socialists (Pells 1973; Ross 1991: esp. ch. 4)—such as Isaac Max Rubinow (1875–1936), a prominent expert on social insurance and a major figure among health reformers (Kreader 1976)—as well as institutional economists, discussed below, and their fellow travellers. They aimed to improve, not replace capitalism; they were meliorists. At the opposing pole were those who had benefitted most from capitalism but regarded its social problems as threats to its stability and therefore to their economic and social wellbeing; they were conservators. They aimed to rationalize capitalist institutions while managing the social disruptions resultant from large-scale business and economic activity (Brinkley 1995; Haber 1964; Jordan 1994; Murphy 1988; Tripp 1987; Wiebe 1967). Among them were agents of the Rockefeller philanthropies that supported social-scientific research (Fisher 1993), and such figures as Henry Dennison (1877–1952), i.e., businessmen or those in the employ of business who were mindful of new, scientific practices in engineering and management (Jordan 1994: 156, 211; McQuaid 1977; Reagan 1999: esp. ch. 5).

In the social policies that reformers championed, however, these distinctions often dimmed, for they shared several characteristics that fostered both an interest in collaboration across political divides and a confidence that reformers could actually succeed in crafting social policy and putting it into effect. One was a commitment to science as the foundation for social betterment. It was through the social sciences that reformers could study, characterize, evaluate, and ultimately ameliorate (e.g., Smith and White 1929) the social problems of capitalism, a conviction captured in the widespread, variously (and often vaguely) defined term, "social control." Although for some it meant a scientific understanding of the forces underlying the cohesiveness and functionality of an individualistic society, for most it signaled the construction, on the basis of social-scientific knowledge, of levers for influencing social evolution—i.e., history—often through the exercise of planning. In some cases "social control" fused both meanings, in that, as Ross put it (1991: 249): "the means and purposes" of the social scientists "were but the socializing mechanisms and social purposes of society itself" (Alchon 1985; Clark 1936; Hawley 1997; Simons and Sinai 1932). Social control, for diverse reformers, could mitigate the threats to capitalism—its own destructive outcomes and the disruptive forces it unleashed—and improve its capacity to provide for the wellbeing of citizens.

Another factor that facilitated collaboration and optimism was that, during the early-twentieth century, social-scientific reformers moderated the intensity and directness of their activism (Bannister 1987; Fisher 1993; Fox 1979; Furner 1975; Morgan and Rutherford 1998; Ross 1991; Rutherford 1997). Around the turn of the century, social-scientific intellectuals often took leading roles in reform, agitating directly or educating the public. By World War I, however, reformers began to withdraw from activism, instead analyzing social phenomena more disinterestedly (Fink 1997; Furner 1975). One restraint on activism lay in the emergent departments of social-science within universities that drew support from philanthropists in business. Outspoken reformers among social-science departments had elicited negative reactions from the leaders of universities and their patrons. Famous academic-freedom cases that in several instances resulted in disciplining or dismissal of faculty established limits on advocacy (Bannister 1987: ch. 2; Fink 1997: ch. 2; Furner 1975; Hofstadter and Metzger 1955; Ross 1991: 116–118, 309–310). Moreover, new career paths in the university elevated scholarship over agitation (Laslett 1990; Ross 1991: 305, 392–393). Yet another restraint was that social-scientific agitators often found themselves without followers, for they either failed to appeal to the major interests such as labor or business; or else public officials refrained from confronting business or other interests. The reformers, moreover, had little power of their own; politics often proved impervious to the outcomes of research, the aspirations of planners, and the rationales for the substantive reforms that activists had supported. The alternative to activism was playing the expert, using research to address fellow experts rather than the broad public or to advise either leaders of contending interests or government in its role as mediator (Miller 1984). John R. Commons (1862–1945), for example, of the Department of Political Economy at the University of Wisconsin, after having served in mediating institutions like the National Civic Federation, learned that the "place of the economist was that of adviser to the [trade union] leaders, if they wanted him, and not that of propagandist to the masses" (quoted by Fink 1997: 66–69; Gruchy 1947: 172; Rogers 2009: 41; Ross 1991: 194–195, 202; Tyrrell 1986: 52).

Shorn of patent political commitments, politically diverse reformers could more easily ally with others across the spectrum of opinion. They could educate one another, arm each other to educate the leaders of their respective interest groups—who, as figures scientifically informed or perhaps even formally trained, seemed likely susceptible to arguments about social progress and capable of educating their rank-and-file—and help each other bring the interests they represented into agreement. Through such collaboration, diverse reformers revealed another theme that lent unity to their efforts: that they could vouchsafe rational social development by replacing politics with science (Jordan 1994; Ross 1991: 249–253, 392–393; Smith 1994: 87–90). Theirs was a faith in what might be called a "Republic of Professionalism," reminiscent of the shared commitments and sociability of the *philosophes* in the Enlightenment captured by the "Republic of Letters" (Daston 1991; Diner 1980; Goodman 1994; Jordan 1994: 155–165; Karl 1974: 50; Recchiuti 2007). Reformers thus hoped to collaborate in substituting enlightened for unmitigated self-interest through interaction with leaders of the interests, whom they understood to share their commitments to science, rationality, and the public interest. As argued below, this orientation profoundly affected the interactions of health reformers with the medical profession and their expectations about the responsiveness of the profession to their message.

That meliorists could join conservators in serving a broad public interest was another consequence of their growing moderation (Bannister 1987: ch. 2; Kunitz 1981; Lubove 1968). Although motivated by concern for the deleterious impact of capitalism on workers, the meliorists increasingly presented themselves as champions less of labor *per se* or as servants of any specific, class-based interests than of society itself. If social problems resulted from the behavior of particular interests or the clashes among them, then the public interest lay in remedying the problems through their disinterested characterization and analysis and through expert construction of harmonizing reforms. Reformers who adopted this self-conception tended to find employment in the academy, the non-profit sector, non-profit organizations for social or scientific advance, philanthropic institutions, and governmental agencies; there they claimed status as disinterested experts, fashioning the sort of legitimacy that Brint (1994) calls "social-trustee professionalism." In health care, this orientation marked reformers until the rise of markets and the growing dominance of another model of professionalism, Brint's "expert professionalism," which increasingly found legitimacy in serving interested parties that shared a commitment to markets.

The desire to uphold capitalism possessed links to the traditions of American exceptionalism, the conviction that the virtuous republic founded in the New World would not decay or exhibit the class distinctions, social problems, and social decline that marked European history (Ross 1991: ch. 2; cf. Rodgers 1998a: 34–44). As industrial development increasingly threatened America with the same kinds of social decay plaguing Europe, reformers aimed to reconstruct the exceptionalist promise. As shown below, they did so in two ways. Some turned to sub-historical forces that, with little explicit human intervention, would resolve the social problems of industrial capitalism; others, including the health reformers, in harnessing disinterested expertise to social policy, aimed to craft socially progressive measures that would preserve capitalism while protecting citizens from its risks. Paradoxically, the latter group, reflecting a moderation in the exceptionalism that in the nineteenth century had prevented social critics from looking toward Europe, now, benefitting from a transatlantic consciousness of common vulnerability to the problems of industrial capitalism, looked to Europe (and would continue to do so through the New Deal) with admiration for the social policies it created in response to capitalism (Rodgers 1998a). By trying to adapt European experience to peculiarly American conditions, they reconciled their interest in Europe with their persistent commitment to exceptionalism.

Social-scientific reformers of diverse politics shared to varying degrees many goals, sensibilities, expectations: To collaborate despite political differences; to invoke science over politics; to educate interests that, when led by scientifically knowledgeable elites, could be induced defer to the public interest; to save capitalism from its own destructive outcomes; to invoke and modify foreign precedents in social politics to enhance social solidarity so as to restore the promise of American exceptionalism. These themes find illustration in three contexts in which the relations of the social sciences to social reform loomed large. One took form at the University of Chicago, especially from the late 1910s, in a distinctive and prominent school of sociology, as well as other important departments in the social sciences. These departments served as a principal locus of connections between the emergent social sciences to the activities of the Social Science Research Council and to Rockefeller philanthropy, which supported the council and through it, the social-scientific disciplines. A second context, arising roughly simultaneously, was the institutionalist movement in economics; the third, overlapping temporally and substantively with the second, was the activities of social scientists located at or heavily influenced by the University of Wisconsin and by the "Wisconsin Idea" for the achievement of social reform. To serve the discussion of health reformers in its concluding section, here this essay takes up these contexts sufficiently to characterize expressions of the tension between the social sciences and reform and to provide context for the subsequent analysis of health reformers.

17.2 Reformist Traditions and the Social Sciences

17.2.1 The Social Survey, Chicago Sociology, and Ogburn

At Chicago, social scientists, especially in sociology, struggled to concentrate on, as the title of Fisher's book (1993) puts it, "the fundamental development of the social sciences" (Bannister 1987; Bulmer 1984, 1996; Faris 1967; O'Connor 2001; Turner 1991). To that end, they espoused a scientism that assisted them in separating from the principal embodiment of social reformism, the social survey movement, emergent around the turn of the century in the context of social welfare. The Pittsburgh Survey of 1907–1908 is only its most prominent early instance (Bulmer 1984; Bulmer et al. 1991; Greenwald and Anderson 1996; Harrison 1930; Kellogg 1909–1914; O'Connor 2001: ch. 1). Inspired by the Charity Organization Society, the institutionalization of its methods in the Russell Sage Foundation (which helped organize and finance surveys), and the practices of social work, the movement aimed to characterize and exhibit antecedent convictions about the proper targets of reform, to illustrate how the consequences of particular social problems propagated through diverse contexts, and to suggest how experts (especially social workers) could craft meliorative measures. Instancing a familiar Progressive-era faith that a social problem, once clearly characterized and exposed to enlightened public opinion, would elicit reform activity, the survey movement emphasized publicity, informing of public opinion, and engagement of local notables. Chicago, however, aimed at the causes that underlay social problems.

An early product of Chicago sociology that elevated science over reform was *The Polish Peasant in Europe and America*, by William I. Thomas and Florian Znaniecki (1918–1920). It stressed empirical study in scientific sociology; it focused Chicago sociology on exploiting its own city as a laboratory (Park et al. 1925; Ross 1991: 306–311; cf. Karl 1974: 31); and it introduced the conception of social disorganization, seen as characteristic of especially immigrant communities in the city (Carey 1975; Matza 1969: 44–47; Ross 1991: 347–357). That conception helped direct attention away from social problems to their underlying social forces and thereby encouraged more disinterested analysis. Another factor that favored science over the survey was statistics (Bulmer 1984: esp. chaps. 9–10; Camic and Xie 1994; Ross 1991: 369–371, 385–385, 429–433; Smith and White 1929). Its chief promoter in sociology was Franklin H. Giddings (1855–1931) at Columbia, where his students and his colleague, institutional economist Wesley Clair Mitchell (1874–1948), pursued it. It established an important beachhead in Chicago after 1927, when Giddings's student, William F. Ogburn (1886–1959), arrived there. Unlike surveys, which sampled both the opinions of experts and notables and the experiences of individual workers and their families, social scientists aimed to collect and analyze empirical data in the light of theory.

Reform impulses nevertheless lay close to hand (Fisher 1993; see also, Bulmer 1984; Bulmer et al. 1991; Carey 1975; Faris 1967). One reason was that philanthropic funders of social-scientific research, like the Rockefeller philanthropies, hoped to preserve capitalism by keeping potentially destructive problems from fulminating. Moreover, academic researchers needed to collaborate with civic leaders and social-service agencies, which possessed direct and practical knowledge of social problems, channels to local authorities and sources of funding, and a desire to exploit social knowledge in the interest of social melioration, i.e., broadly, the world of social welfare (Carey 1975; Karl 1974: 50–60; Matza 1969; Smith and White 1929). Social scientists also often served as advisors or even administrators of governmental agencies and thus could scarcely ignore reform. To resist these pressures, Ogburn (1930) and others (Ross 1991: ch. 10) called for an intellectual austerity remote from reform, but could not shake off dependence on reform-minded sources of funds or the pull of governmental service.

Ogburn's scientism and its tensions with reformism loomed in his conception of "cultural lag," a theme important in the discussion below of health reformers. Ogburn had "declared himself as an advocate of 'the historical method in the analysis of social phenomena,' and defined historical sociology as 'the history of society, the development of culture, and the evolution of social institutions'" (Ross 1991: 442–443, quoting Ogburn 1921: 71). Although Ogburn saw himself as advocating historical method, in his statistical approach, he pursued not the Rankean descent of the historians into factual details (Iggers 1962), but a search for deep-seated historical laws. His model was the institutionalism (treated below) of Mitchell, and in particular Mitchell's influential statistical analysis of business cycles (1913). For Ogburn, Mitchell's data suggested that institutions, especially those that Thorstein Veblen (1857–1929) had called the "pecuniary" ones important to business cycles, standardized mass behavior, creating recurrent patterns of economic decline and recovery (Mitchell 1924; Rutherford 2000: 293–294). For Ogburn, this notion of standardization fortified his conviction that at work beneath historical diversity were simple processes leading in a particular, progressive direction. Such a process was a sequence of cultural lag and readjustment (Ogburn 1922).

Changes in the material culture (typically innovations in science and technology) created problems in the non-material culture; a consequent "maladjustment" arose of culture to technology, and the lag in adjustment of non-material to material culture (the most common form of cultural lag) perpetuated social problems until events caused a readjustment (Bannister 1987: ch. 11; Ross 1991: 442–445; Volti 2004). An extended discussion of workmen's compensation served as Ogburn's principal example (1922: 213–236) of an emergent problem (accumulating industrial accidents and their toll on workers) and adjustment (workmen's compensation, emergent in the 1910s, the first American example of social insurance). However, reform held a minor place in Ogburn's analysis. In discussing compensation, Ogburn, revealing a common postwar pessimism about reform (Bannister 1987: ch. 11), did nothing to portray the reformist activities that had generated legislation, to identify opposing interests, or to show how reformers overcame opposition. Regarding psychological maladjustments to modern conditions, Ogburn offered few and restricted suggestions for reform, claiming that consciously undertaken changes in culture were just as difficult to achieve as the changes some hoped for in the biological nature of man; both were "almost impossible" (Ogburn 1922: 346). "[I]t is futile to plan any wholesale and powerful control of the course of social evolution" (364). Serving science (generic identification of the kinds of forces at work beneath of the flux of history) seemed to trump the reform impulse (concrete identification of the tasks facing reformers); the fundamental, simple processes (lag and readjustment) outweighed the historical details and offered little room for intervention.

Ogburn's cultural lag had widespread appeal, appearing not only in the disciplines of sociology (Becker 1928: see title; Burgess 1929: 139–40; Groves 1932: inter alia 339–341), history, and social-science generally (Higham 1965: 147; Karl 1974: 191; Volti 2004), but also in public philosophy (e.g., Ratner 1939: 4–5), reformist thought and cultural criticism (Pells 1973), and health policy (Fox 1983; Hirsh 1939; Moore 1927; Rorem and Sigmond 1965: 80; Somers and Somers 1961: 161). So commonplace did cultural lag become that it often appeared without attribution, as some of these examples show. Like others of his generation (Ross 1991: 428–448), Ogburn dealt with a bewildering pace of change by invoking the subhistorical processes that would bring progressive social readjustment Upholders of exceptionalism could take comfort in his implicit commitment to progress through science, technology, and invention, without concern for organized reform movements. Reform-minded scientists, on the other hand, and lay social critics found in cultural lag a basis for understanding rapid social dislocations, a platform for criticizing outmoded practices, and a rationale seeking controls over—and, in the event of the failure of reform, blaming—"selfish interests."

17.2.2 Institutional Economics

Institutional economics showed no ambivalence about social reform (Rutherford 2000), which it saw as the fundamental motive of social-scientific research (e.g., Clark 1924). More than other reformers, institutionalists emphasized "social control," the creation of mechanisms to remedy the failure of markets to serve the public interest. Pure science was not merely foreign to institutionalists; it was the enemy: orthodox economics lacked proper scientific foundations because, inter alia, it traded in abstractions that were not realized on the ground and proceeded from an unjustifiable preference for competitive markets and their supposed power to serve the public interest.

Taking much inspiration from Veblen as moderated by John Dewey (1859–1952), institutionalists portrayed "pecuniary" institutions as having crystallized and preserved practices that eventually fell into conflict with the public interest. They accounted many outcomes of economic activity as unfair (imposing costs on labor and fostering other harmful externalities) and sometimes chaotic. Corporate behavior was therefore "affected with the public interest" as John M. Clark (1884–1963) and several other institutionalists put it (Rutherford 2000: 298–299). Institutionalists also studied labor institutions, especially unions, an interest indebted in part to the intellectual ancestry of institutionalism in the German historical school of economics (Ross 1991: 102–106, 411–412, Rutherford 2000: 302–303). Sympathy for the union movement pervaded much institutionalist work, prompting the creation of labor economics (Kaufman 1993: esp. 84–91; McNulty 1980: chaps. 6 and 7) and generating among institutionalists an acute sense of social solidarity with the working class. However, the institutionalists were not unequivocal allies of labor; in representing the pubic interest, they sought to harmonize diverse interests, not favor one or another (Kaufman 2003; Recchiuti 2007: 148; Stromquist 2006: 91–93), and this harmonization, they anticipated, would extend the reach of social solidarity beyond their own ranks (Gruchy 1947: 172–180). Social policy could help labor improve its place in the industrial economy, without giving labor all that it wanted; while capital, although required to make concessions, would pursue economic activity under stable and harmonious social and political conditions; the two sides would each approach and acknowledge the legitimacy of the other.

17.2.3 The Wisconsin Idea

It was especially one branch of institutionalism, associated with John R. Commons and University of Wisconsin, that exercised an enduring influence on social policy after World War II. Its approach to reform eschewed major structural changes and aimed to harmonize interests by invoking the power of government and the knowledge of experts. Responding much less to Veblen than to his own mentor, Richard T. Ely (1854–1943), Commons focused on the labor movement and its history (Miller 1984; Rutherford 2000). Labor history revealed for Commons the dominance of the "banker capitalists" and the degradation of workers' rights. The Wisconsin Idea was a "vision of an activist state promoting the general welfare through nonpartisan commissions, composed of the most talented citizens. Serving as the public's 'expert' in efficiency, such commissions would regulate business in the 'public,' as opposed to the 'private,' interest" (MacLean 2007: 53). Central to the Wisconsin Idea was the role of the university; the approach had its origins in Ely's thinking, and it took wing with assistance of Charles van Hise (1857–1918), president of the University of Wisconsin (1903–1918). The idea encouraged academic social scientists to serve not as advocates of the interests but as impartial, expert investigators into social, economic, and industrial conditions; to expose the information to public scrutiny; to work as consultants in the drafting of legislation; and, from inside new governmental institutions, to administer new laws dispassionately. Over his career, Commons and those he inspired pursued remedies that included labor legislation, support of unions, social insurance, and creation of new public institutions that would constitute a "fourth branch" of government to harmonize interests and secure workers' rights (Chasse 1991; Gruchy 1947: ch. 3; Kaufman 2003; Rutherford 1983).

In the Progressive Era, the American Association for Labor Legislation (AALL, founded 1906), which drew heavily on Wisconsin economists and operated under the leadership of Commons's protégé John B. Andrews (1880–1943), inaugurated the American movement to create programs of social insurance (then, on the state level) (Chasse 1991, 1994; Lubove 1968; Murphy 1988). The association began with the successful campaign for workmen's compensation and proceeded to the "next great step in social legislation," the drive—which failed miserably—during the second decade of the twentieth century for compulsory health insurance for the working class (Walker 1969). As is well known, that failure was only the first in a series of fruitless efforts, later pursued on the federal level, to achieve some form of what gradually came to be seen as a national health program, i.e., national health insurance (NHI) (Starr 1982b).

However, the Wisconsin tradition in social insurance was hardly barren, for many of those who built, sustained, and expanded Social Security, such as Arthur J. Altmeyer (1891–1972), Edmund Witte (1887–1960), Robert Ball (1914–2008), and Wilbur J. Cohen (1913–1987), had Wisconsin connections and understood themselves, at least in part, as perpetuating the Wisconsin Idea (Berkowitz 1995, 2003). Under this leadership, the Social Security program gradually grew in its range of benefits—some, such as disability insurance (1956) and eventually Medicare and Medicaid (1965), bearing on health—and other, similarly motivated health reforms, such as the Hill-Burton hospital program (1946), lay the foundations of the American health system (Fox 1986; Starr 1982a). A disconnect therefore seems to exist between the perpetuation of the Wisconsin approach in the traditions of Social Security and other programs, and the persistent failure of reform efforts, also rooted in the Wisconsin Idea, to achieve NHI. The disconnect, Fox shows (1983), is in part a creation of traditional historiography, which emphasized the failed efforts to establish health insurance, while obscuring other achievements of activists in health care. However, he also argues that the failures of those pushing for NHI were in part the result of their own perceptions and behavior. Although his argument is persuasive, the traditions of social science here outlined provide an opportunity to portray health reformers differently.

17.3 Interpreting Health Reformers

17.3.1 Reformers, Historicism, and Ogburn

His analysis (Fox 1983, 1986; see also his 1997) rests on two chief pillars: he locates in the appeal of Ogburn's "cultural lag" the self-understanding of reformers who pursued NHI despite recurrent defeats; and he locates in one version of historicism—that the industrial economies of Europe shared a path of historical development which could be expected to manifest itself in America—the reformers' tendency to invoke cultural lag in blaming the medical profession for obstructing progress. Reformers thus saw their goals as imminently attainable, but, having met stubborn opposition and vicious denunciation from medical interests, accused the profession of perpetuating a cultural lag that blocked the progress Europe enjoyed. Moreover, relying on Richard Pells's study (1973) of Depression-era intellectuals, Fox sees in Ogburn's doctrine the notion of a vanguard dedicated to furthering social progress by reducing cultural lag, and he supposes that reformers understood themselves as constituting just such a vanguard, which, because of its many failures, viewed itself as beleaguered. Even on matters where the profession was flexible, reformers rejected compromise, apparently of fear of complicity in blocking progress; they took an elitist stance and in effect removed themselves from the fray of social policy.

This argument possesses an element of truth, but it has weaknesses, and it eclipses other perspectives emergent from the history of twentieth-century reform impulses. One difficulty lies with Fox's conception of historicism. Such claims do find support in the work of American reformers, but a broader picture suggests reformers possessed a more nuanced view of the utility to America of European precedents in social policy. Rodgers (1998a) convincingly places diverse American reformers within the context of a broad, north-Atlantic culture of social politics, in which shared conceptions of social vulnerability to the transformations wrought by industrial capitalism inspired a cluster of convictions about social policy: that industrializing nations needed broadly similar policies, less to achieve specific, shared goals or a common form of polity (e.g., a welfare state or a social-insurance state) than to shelter some features of social and communal life from the reign of the market; that some countries had moved farther or faster in that direction than other, lagging ones (especially America); and that experiences in one country could be studied for their utility to others and perhaps imported with modifications. Such importations, or attempts to achieve them, inevitably suffered refraction through both the perceptual lenses of their analysts and the politics and culture of the recipient countries. Thus, Americans, seeing policies abroad that were already in place, may have been less than fully informed about their histories and the politics at work in their creation. Moreover, their efforts at transplantation of social policies scarcely proceeded unhindered, for while the causes of problems in the exemplary country may have resembled those in the receiving country, the array of interests, their relative strengths, and their modes of interaction usually differed. Nevertheless, as shown below, American health reformers, mindful of the diversity of European experience, of the mistakes they believed Europeans committed on the path to workable policies, and of at least some peculiarities of American experience, hoped to revise European precedents and import them to preserve the promise of American exceptionalism. If American reformers were historicists, their historicist faith amounted to more than the simple anticipation that in European experience could or should be read the American future.

Another problem with Fox's interpretation lies in its deployment of Ogburn's "cultural lag." Fox notes its wide currency but does not then pause at the idea that doctrine so commonplace constituted a cultural resource too handy and too fashionable to ignore; it would have been surprising had health reformers, given proper motivations, failed to exploit it. Because it was hardly peculiar to health reformers, in according it a major role in their self-conception and behavior, Fox may have put more weight on it than it can bear. Moreover, Fox's conception of the doctrine finds less support than he claims for it in Pells's study (1973). Pells examines a highly specific context, the response of diverse intellectuals to the crisis of the Depression. Seeking a remedy in collectivism, they identified as the principal obstacle to its achievement the persistent indivi- dualism of American society, which persistence they labeled an example of cultural lag. Pells's group also maintained that adjustment would result from a transvaluation of values that they hoped to spur, initially through their own efforts as cultural critics and then, from about 1935, through a hoped-for coalition with workers, farmers, and professionals. Because their focus lay more on culture than on politics, and, to the extent it covered politics, aimed more at a broadly conceived reform of society than at specific measures, these intellectuals, Pells notes, often failed to ask the hard questions about what workers and their own group wanted. Although increasingly stricken with a sense of crisis, they expected history to turn the cultural tide and open a path to a socialist transformation. Although unlike Ogburn these intellectuals did not invoke simple, subhistorical processes, they did long for a likely distant histor- ical turn that would rescue them from the stagnation—economic, moral, and political—of the Depression. In sum, their goals were broadly cultural, belat- edly political, passively expectant, and generally vague. Health reformers, as suggested shortly, scarcely fit that bill.

Did Pells's intellectuals see themselves as a beleaguered vanguard? If so, it was because most of them, as Pells makes clear, lacked credentials as experts and could only nurse from the political periphery their hopes for a collectivized, planned society. Their alienation reflected a history not of defeat in pressing concrete reforms, but of powerlessness to affect political development beyond arguing the virtues of collectivism and anticipating salvation through historical change over which they could expect to exert little social control. By contrast, those few of Pells's group who were knowledgeable about social science belonged to the same intellectual circles as the social-scientific reformers and thus could envision themselves in roles of advisers, managers, and technocrats, especially since the "radical" aspirations of Pells's group to achieve a society disinterestedly managed differed little, as Pells points out, from those of the New Deal itself. Whether nursed from a distance as politically marginalized (and thus beleaguered) radical critics who lacked expertise, or sustained directly by expectations of participating as experts in the politics or administration of the New Deal, the dreams of Pells's intellectuals anticipated nothing more than the progressive notion that science, wielded disinterestedly by experts, was superior to politics. If this was elitism, it had long been a standard refrain among social scientists, journalists, and social critics. As suggested below, moreover, most health reformers responded to defeats less by adopting the alienation of those of Pells group who could wield only cultural criticism (and not expertise) than by finding other opportunities to exploit their knowledge and skills.

In pressing for health insurance, American reformers expected no gradual transvaluation of values toward collectivism, but anticipated imminent attainment of specific, collectivist measures, like workmen's compensation and health insurance. These views created one context in which their studies of European policy loomed large, seeming to suggest Fox's historicism. Reformers' confidence in the proximity of their goals rested partly on the similarity they documented between contemporary American developments and those emergent in Europe on the eve of the innovations there in social policy. On both sides of the Atlantic they saw an inchoate proliferation of especially private insurance mechanisms. These similarities seemed to authorize the conclusions that the new scientific medicine, practiced under a dynamic capitalism, generated problems that were becoming as evident here as they had been in Europe , that existing social forces were generating responses, and that insurance was the most familiar and efficient choice (Brown 1937: esp. 170–173; Davis 1916; Falk 1934: 121–123, 1936: ch. 3; Rorem 1931; Rubinow 1913: chaps. 1–2, 14, 18, esp. 283, 1916b: chaps. 1, 15, 1934: ch. 14; Simons and Sinai 1932: 20–21, 176–180; Warren 1915: 81). As Warren and Sydenstricker put it (1916: 50–52) about America after commenting on Europe, "worth emphasis here is the unmistakable tendency upon the part of all concerned—worker as well as employer, the individual as well as the State—toward the adoption of the insurance principle as the most practicable and the most efficient method of attacking the problem of sickness" (cf. Commons 1915b: 304; Rubinow 1913, 1916a: 341–342). Reformers thus aimed to bolster the forces leading to change and to elicit from European experience the principles and standards for health insurance (Davis 1916; Rubinow 1916b; Falk 1936).

17.3.2 Foreign Experience and American Conditions

It is this interest of American reformers in European developments that could be taken as evidence of Fox's historicism. However, the exceptionalist traditions in America and a keen sense for American conditions combined to suggest to reformers in America that, whatever Europe offered would have to be modified here, for America should anticipate a distinctive voyage toward novel social policies. It was in the exceptionalist mode that the founders of institutionalism departed from Veblen (Ross 1991: 371–386) and that other institutionalists, in opposing mainstream economics, argued that not theory but on-the-ground institutions and conditions, the products of specific historical experience, explained economic phenomena. Only thus could they find scientific foundations for social control that would bend the American economy to the service of social welfare (Gruchy 1947: e.g., 405–410 on Tugwell; Ross 1991: 410–414). Similarly, Commons found in his historical studies of labor that American circumstances created social problems and mandated social reforms specific to American conditions (Chasse 2004; Church 1974: 587, 601–602; Commons 1909; Moss 1996: ch. 7; Ross 1991: 202–204; Rutherford 1997: 183–184). The Wisconsin Industrial Commission, of which Commons was a major architect, did embody European precedents, but it was a creation specific to its intellectual, political, and legal context (Rogers 2009: ch. 2). Many other American social-scientific reformers in the early-twentieth century devised similar notions about distinctively American routes for social development. The New History, which the rapidity of social change had elicited from historians, led figures like Charles A. Beard (1874–1948) and Frederic Jackson Turner (1861–1932) to examine distinctively American circumstances for explaining events (Tyrrell 1986: ch.1; Ross 1991: 270–274, 345–346). American reformers, although perhaps in retrospect naïve in their expectations about the ease of transplanting to America European cultivars in social policy, nevertheless understood that their nurturance in American soil and climate entailed alternative structures and practices.

If American students of European social policy need not be taken to have found across the ocean the clear path for American social development, what did study of Europe offer them? One answer is data and experience to which they could apply their expertise in crafting a peculiarly American product, improve on European practice, or at least guide experimentation to determine what could be made to work. For reformers, science, not politics, would guide social evolution and provide social control. Europe had evinced, to use Mitchell's term (Jordan 1994: 88), the "jerky" course of social politics. By taking account of European steps and missteps in health care, America could avoid many of the pitfalls, evident to both American reformers and physicians, that hindered the creation of smoothly functioning programs responsive to the problems of an emergent industrialized economy (Pumphrey 1972: 34–35, quoting Davis 1916; CCMC 1932: 106; Rubinow 1913: 21, 244–249, 1916b: ch. 1). In compensation insurance, America indeed had repeated the errors of Europe and exposed the failures of American lawmaking; the point now was to avoid them both in health insurance (Rubinow 1916b: 8). Thus Rubinow's *Social Insurance* bore the subtitle, *With Special Reference to American Conditions*; and the CCMC as the Committee on the Costs of Medical Care (1932: 127f), Davis (1934), I.S. Falk (1936: ch. 13, 357–358), and labor economist Harry A Millis (1873–1948; 1937: 135, 146–157) recognized differences between American and European conditions and suggested the corresponding differences in the design of insurance arrangements. Commons's example has already been noted. If this was historicism, its utility lay at least as much in allowing the substitution of expertise for politics so as to create a well-functioning and peculiarly American set of social policies as in determining the probable path of social policy. Indeed, in regard to the organization of medical practice, some reformers believed they could apply American experience to improve European social policy.

In particular, two novelties, which paradoxically intensified the opposition of medical interests, seemed to superior to European practice. Both amount to what would now be called supply-side reforms: replacing solo practice with group clinics featuring specialty practice coordinated by generalists and linked with hospitals; and lowering the distinctions between medical practice and public health, by emphasizing prevention within medicine and by coordinating and even integrating clinics and health departments. Both typically entailed hierarchical deployment of local and regional health resources. In regard to group clinics, some early reformers commented only circumspectly (e.g., Rubinow 1916b: ch. 15). However, Michael M. Davis (1879–1971) early pressed for group organization that went well beyond Europe (Davis 1916; Pumphrey 1972), and others gradually grew convinced that organized practice were far superior to what seemed to them increasingly outmoded solo practice (e.g., [CCMC] 1932; Falk 1934; Moore 1927: pt. 3, esp. ch. 12; but cf. Simons and Sinai 1932: 102–105). Some reformers, moreover, explicitly called for lowering the barriers between medicine and public health and coordination of medicine with prevention and other public-health purposes ([CCMC] 1932: ch. 3; Davis 1934, 1941; Moore 1927: chaps. 9, 17–19, app. 10). It was the perceived deficiencies of Europe, not its novelties, that encouraged such proposals.

17.3.3 The Educational Project

Is there an alternative to Fox's emphasis on historicism that will account for the criticisms, invoking cultural lag, that reformers launched at medicine? A likely answer emerges from exploring one major task reformers set themselves, educating the medical profession, their frustration in failing to achieve that end, and their experience of vicious denunciation at the hands of the profession. In this effort, reformers' goals were two: to reduce medical opposition by showing that novel social policies would not prove significantly injurious to the profession; and to suggest that, precisely to preserve its interests in the impending creation of new programs and to assure their efficacy, medicine needed to take a leading role in their development. Reformers argued that American physicians' dissatisfaction with domestic programs for workmen's compensation, as well as European physicians' dissatisfaction early forms of health insurance, resulted from the failure of the profession to assert itself at the outset (Commons 1915b: 305; Rubinow 1916b; Warren and Sydenstricker 1919: 783–784; Falk 1934: 118, 121–123; Simons and Sinai 1932: 42, 79–82, 177–178). Moreover, European experience also taught reformers that any arrangements for the supply of scientific medical services had to accord physicians significant discretion, if exercised in cooperation with the public ([CCMC] 1932: 53–55, 134–137, 149; Davis 1931b: 256; Falk 1934; Simons and Sinai 1932: 114; Warren and Sydenstricker 1919: 784). For American reformers, educating the medical profession was urgent.

Reformers' faith in the efficacy of education rested chiefly on three ideas. One was the conviction that the groups in opposition were not each univocal; that some within each shared something like reformers' own perceptions or were at worst indifferent; that the great majority of each group, while perhaps skeptical, was subject to convincing; and that only an intransigent minority was beyond reach (Davis 1931b: 253–254; Julius Rosenwald Fund 1937: 30–31; Murphy 1988: 3; Pumphrey 1972; Rubinow 1915: 383, 385). If one could enroll the leaders of opposing groups, they in turn could enroll the rank-and-file or at least reduce its opposition. Second, at least initially the educational effort reflected not elitism (i.e., the notion that reformers, educated in the social sciences, were more able to guide policy than ignorant physicians; or the idea that as cultural critics, reformers could understand what physicians could not even perceive), but on the contrary an early faith in the goodwill of physicians as members of the same social stratum to which reformers belonged, those who were active in the science-based professions and who, reformers expected, would serve in the capacity as enlightened leaders of the interests. Industrial reformers like Dennison had emerged from business (Julius Rosenwald Fund 1937: 25); business interests could be expected to be susceptible to demonstrations of their deleterious impact on the working class and assume social responsibility (Clark 1916); so too, health reformers believed, could reformers emerge from medicine. Third, health reformers' approach to the medical profession reflected the orientation widespread among reformers in many fields toward harmonizing the interests, disseminating a sense of social solidarity, and inducing particular interests to bow to the public interest. The earliest American form of social insurance, worker's compensation, had responded precisely to this concern; other forms of social insurance would surely follow (Davis 1916; Kerby 1917; Rubinow 1915: 384, 1916a: 342, 1916b: ch. 1; Walker 1969). In the "Republic of Professionalism," reformers believed, reasonable people could find a common basis for socially beneficial action.

Hence figures like Rubinow (Kreader 1976; Rubinow 1915), Commons (1915a and b, 1918), Davis (1931a; Pumphrey 1972: 35–36), E. H. L. Corwin (1931), Warren (1915), and Sydenstricker (Warren and Sydenstricker 1919), dutifully trekked to meetings of the organized medical interests, made presentations, and published papers, and offered commentary in their books ([CCMC] 1932: 22–24; Leven 1932; Rubinow 1916b: 240, 245–246; Falk 1936: 341–342, 358–360) to argue that doctors' concerns for their own interests were exaggerated, their pejorative labeling of health insurance as "state medicine" was inappropriate, and their skepticism of proposed reforms was unjustified. Reformers' agendas, they claimed, did not in fact conflict with physicians' true interests, and, insofar as reforms aimed for better and more reliable compensation for physicians and better training for the general practitioners (increasingly beleaguered by specialism; Stevens 1998: ch. 7), could in fact advance medical interests while improving social wellbeing (e.g., [CCMC] 1932: 22–24, 29–33; Davis 1931b: 253; Falk 1936: 341–342; Rubinow 1916b: 245–248; Warren and Sydenstricker 1919: 783–786). At first, this approach seemed to be working;

early in the Progressive-era campaign for health insurance, the leadership of the American Medical Association (AMA), drawn from elite physicians little engaged in medical practice, did get on board and seemed to be taking the profession along (Numbers 1978).

In retrospect, however, this educational project was clearly doomed. Historians have shown that an emergent scientific medicine had induced the bulk of the profession to turn away from the broad social questions that continued to occupy parts of its elite and toward institutionalizing technologically advanced, individualized, acute care (Brandt and Gardner 2000; Fox 1995; Reiser 1978; Rosenberg 1979, 1987: ch. 13; Rothman 1997; Stevens 1989, 1998: ch. 7). The rank-and-file of the profession did possess, as reformers expected, a sense of fiduciary responsibility, but its embrace of the individual patient was too narrow to encompass the social contract between the profession and its polity. Physicians therefore eventually did elicit one of the reformers' standard criticisms: the profession was ignorant of the social side of medicine, and thus could not succeed in asserting its interests while serving the public interest (Davis 1931b: 253–255; Rubinow 1934: 198, 205; Simons and Sinai 1932: 42, 167; Stern 1941: 215–216). This judgment reflected elitism far less than frustration. Physicians did have clear interests in preserving both their autonomy and their profitability; to explain to the members of medical societies that the AALL standard bill was not "state medicine" was beside the point. Practicing physicians conformed far less well to Brint's (1994) social-trustee professionalism, which did accord more with the elite of the profession, than to his "expert professionalism," in which practitioners served specific clients and controlled the conditions and rewards of practice. Reformers were not unaware of this orientation among physicians (Davis 1931b: 251 252; Simons and Sinai 1932: 89–90, 109–110), but they had hoped that, by their educational efforts and those of the leadership of the profession, physicians could have been turned around in both their own and the public interest.

In Ogburn's broadly disseminated doctrine of cultural lag, health reformers, unlike Ogburn himself, had found not cause to suppose that reform was "almost impossible," but, like many reformers in diverse spheres, a means to characterize and explain a social problem (e.g., Moore 1927). However, again unlike Ogburn and some of those who exploited his doctrine, health reformers also had supposed that reform was possible because inter alia the interests were educable. The failure of their expectations about education therefore likely proved profoundly dismaying: how could a distinguished professional, scientific elite fail, in the face of reformers' efforts, to acknowledge its social responsibilities? how could the commitment of the profession to the very same scientific progress that underlay reformers' enthusiasm for medicine induce its principal bearers to turn relentlessly inward? how could the profession refuse to study the social side of medicine and help guide change? how could the profession remain impervious, given its scientific and humanitarian commitments, to the appeal of social solidarity in the public interest? how could it fail to see that its own interests lay in supporting reform? Worse, physicians had reacted to reformers by issuing vicious characterizations of their movement as a nefarious effort to deprive them of autonomy and social standing and by invoking the traditions of voluntarism as a red line that the reformers had illegitimately crossed. Faced with this resistance, health reformers, again as in diverse other spheres, made alternative use of cultural lag, as a handy idiom that could account not only for the problems of health care, but also for their own failures to remedy them. Behavior that violated their expectations about members of the "Republic of Professionals" and that obstructed progressive change could be explained by suggesting that social progress was the victim of physicians' willful ignorance of social concerns and the philistinism they revealed in pursuit of status and profit. However, the implications of this experience, that professional solidarity and notions of the public interest and social solidarity could not by themselves sustain reform, dawned only slowly on reformers.

Perhaps Fox's interpretation of reformers as historicists is more fully applicable to their historians Indeed, it was in a review of a collection of studies on the history of American health policy that Fox's views took form, and there (1983) he argues that the historiography amounts to an extrapolation of reformers' own views. In the light of the preceding discussion, it might be more accurate to suggest that the historians, having taken a historicist view, perhaps in the light postwar, social-scientific accounts of the welfare state (1998a: 22n23 [on 515–516] and accompanying text; 1998b: 30), easily read into their sources their own convictions about health care in America and its deficiencies in comparison with Europe. They thus ignored the complex phenomena of transatlantic exchange and the traditions of exceptionalism that animated much thinking among those trained in the social sciences about the character and fate of American social policy.

17.3.4 *Reformers Following Defeats of NHI*

How else did frustrated reformers react to their failure to achieve NHI? Perhaps the most enduring and characteristic response was to persist in research, education, and (with philanthropic or other private support) demonstration projects. Illustrative is the career of Michael M. Davis, who held a Ph.D. in sociology from Columbia. Early in his career he encountered and grew interested in a variety of social problems. He cultivated connections with reformers in social welfare, used foundation support to pursue his interest in supply-side reform (emphasizing group practice and pay clinics for the lower middle class), and engaged in extensive literary activity, consulting work, and (within the context of the American Hospital Association) organizational efforts (Julius Rosenwald Fund 1937: passim and esp. 37; Pumphrey 1972). He continued to call for physician leadership in health reform (Davis 1931b: 253–254, 256; Julius Rosenwald Fund 1937). He used his post at the Julius Rosenwald Fund *inter alia* to help organize the next major landmark in the history of American health reform, the Committee on the Costs of Medical Care (CCMC), active 1927–1932, which, through extensive studies of American health care, took up the case for both insurance and supply-side reforms, as well as arguing for greater emphasis on prevention and coordination of medicine and public health. Its work, although ill received by medicine, provided an example for the subsequent history of health surveys and analysis of the distribution and need, as reformers understood it, for medical resources and medical care (Fox 1986: 45–51; Starr 1982a: 261–297, 1982b). Similarly, I.S. Falk (1889–1984), a prominent and enduring figure among health reformers, for decades advocated public health insurance, but he also pursued a diverse, influential career in public health and health policy (Roemer 1985; Stevens 1985). A multifaceted activism—in education, research, publication, demonstration projects, and novel programs—whether in public or private settings—was typical of health reformers in the twentieth century.

Eventually, two broad approaches to health-care research and education emerged, which Fox (1990; Mechanic 1990, 1993) characterizes with the ideal types "collective welfare" and "social conflict." Particularly after the establishment of Medicare and Medicaid, which increased the magnitude of the health economy and placed much of its escalating cost in the public purse, "health-services research," as it came to be called, took form in response to the emergent demand for information and analysis to guide policymaking (Anderson 1991; Flook and Sanazaro 1973; Ginzberg 1991). As politicians increasingly took notice of health issues, research changed its character, at first seeking to attach itself more firmly than in the past to prevailing, centrist values. Odin W. Anderson (1914–2003), a sociologist early prominent in health services research, argued (1966, 1991) that, under American conditions of liberal democracy and interest-group politics and in the face of the limitations of social-scientific knowledge, health research tended to support the status quo. Moreover, he found that researchers were most likely to be heard when their studies comported with prevailing values and acknowledged the deficiencies of science. In the emergent postwar liberalism, this point of view animated much health reform: as interest in intrusive planning declined, reform looked increasingly like a matter of incremental changes attached to and modestly extrapolating from existing practices. Thus did experts devise and inaugurate many postwar health policies, and architects of social security nurture and expand their program (Berkowitz 1995, 2003; Fox 1986).

17.4 Recent Reforms

Anderson's position implies more than acquiescence to incrementalism; rather, in incrementalism many reformers have found virtues. One is its capacity to reduce the problem of choice among policy goals by fostering opportunism among those in a position to act (Brown 2005; Jost 2004; Peterson 2005). Given lack of consensus among reformers on priorities; the diversity of affected interests; the need of technical expertise in application to issues that the public is little able to grasp; and the rarity of the social, political, and economic conditions that permit major changes of public policy, opportunistic, incremental activity on the part of experts, especially those in governmental employ, have been able to make a significant difference over time without arousing opposition. Incrementalism also has entailed acknowledging something that earlier reformers, even after the New Deal, sometimes failed adequately to perceive, not only that a multiplicity of interests hinder reforms, but also that a federal system offers significant structural obstacles to reforms on the national level (Morone 1998; Rodwin 1987); by taking on smaller, more politically self-contained tasks, reformers have often been able to find a path through the thicket of structures, institutions, and stakeholders.

Moreover, incrementalism, in the sense of close attachment of prevailing principles and practices, has been able to facilitate such major reforms as the occasional rare realignment of political and social forces may permit. Medicare illustrates the point: it was conceived and sold as extending the very successful private Blue Cross hospital plans to the largely uninsured elderly, and, although it rested on Social Security for its financing mechanism, it exploited largely existing Blue Cross plans for its administration. (Medicaid, by contrast, at first attracted little attention, for when first proposed its transformative potential eluded virtually all parties; Mann and Westmoreland 2004). The Obama reforms, although more complex, also illustrate the point. The Patient Protection and Affordable Care Act (PL 111–148, as amended by the Health Care and Education Reconciliation Act, P.L. 111–152; henceforth, ACA), passed at the end of March, 2010, was clearly a major policy innovation. Its passage resulted from a combination of widespread discontent with the health system—owing to the erosion of employer-based insurance, the increasing risk among the middle class of becoming uninsured, the escalation of costs that (especially in cases of major illness) could bankrupt even the insured, and an array of quality problems in the provision of care (Cohn 2007; Fronstin 2009; Glied 2009; Gould 2010; [IOM] 2004; Moore 2007; Shen and Long 2006)—with a transient change in the political complexion of the country. However, its elements were hardly unfamiliar.

The ACA was a bundle of measures that (like Medicare) had been under discussion for years; that in several instances had been proposed by conservative figures; that had been proposed to reform, expand, or constrain prevailing arrangements, public and private. Moreover, its constituent measures—like many of the diverse health reforms that had failed when bundled into President Clinton's plan but gradually achieved passage piecemeal (Rich and Merrick 2009)—might have been taken incrementally. Only had the ACA sought fundamental change, whether by returning to the older, less frankly market-based policies prevalent until the 1970s; or by taking cost escalation rather than entitlement as its major target; or by reconstructing financing (e.g., through a single-payer system) would it have possessed a transformative character. It was much more the integration of its familiar components into a large-scale set of reforms, little understood by the public and many politicians, that made the ACA so controversial and that aroused conservatives to reassert vehemently the traditional themes of antistatism and distrust of government and governmentally wielded expertise. The political survival of the ACA depended on the incremental character of its elements and its having been considered under a temporarily favorable constellation of political forces.

That constellation permitted reform because it sustained measures, even some with conservative pedigrees, that otherwise would likely have been blocked by conservatives. That reform (if incremental in its components) required a political breach in the power of conservative opinion reflects changes in both polity and policy. Especially from the mid-1970s, at the end of the long period of postwar expansion, the prevailing liberal dispensation disintegrated, and a greater diversity of political views, especially conservative ones, grew prominent, forcing changes in policy thinking. Anderson's cautious point of view marked a growing realization that traditional estimates of social solidarity (too high) and antistatism (two low) were inaccurate (Oakman et al. 2010; Zelizer 2003; generally, Jost 2004), and it presaged a major shift within health-oriented research that would accompany the turn, a concomitant of the rise of conservatism, toward markets in health care. A third ideal-type in Fox's classification (1990) of health research emerged under the impact of health economics and epidemiology, the economizing model. As it arose, social-trustee professionalism suffered eclipse by the rise of expert professionalism, to which health-care researchers increasingly conformed. In doing so, they claimed that the strictures of economic theory were morally and politically neutral (Rice 2003), but they adopted values that comported with sponsors of research of diverse but far more often than before conservative ideological stamp, thus dramatically increasing their power to affect public policy (Critchlow 1993; Fox 1990; Melhado 1998, 2000).

Market-based health care made patent the persistent, conflicting images in America of what health benefits and entitlements society owes its citizens; of what values should guide health policy; and of what roles government, the private sector, and especially markets should take in financing, organizing, and distributing health benefits. Viewed in this light, the Obama reforms, at least in their high visibility as an integrated collection of measures, seem to reflect only part of the spectrum of opinion, that on the center-left. Political circumstances safeguarded a package that took inspiration largely from older views of social solidarity and the utility of government in bringing interests together and assisting citizens in need. Yet the first two years of the Obama administration afforded those views only a brief return to center stage. Health reformers, once more or less agreed about the importance of governmental efforts to secure the distribution of health benefits in pursuit of social justice, have come to reside on all parts of a wide political spectrum. To the extent that they have departed from collective-welfare and social-conflict models, they have proven more effective in shaping public policy; and they have reflected and fostered a greater diversity in the voices and forces that make American health policy.

References

Alchon, Guy. 1985. *The Invisible Hand of Planning: Capitalism, Social Science, and the State in the 1920s*. Princeton, NJ: Princeton University Press.

Anderson, Odin W. 1966. Influence of Social and Economic Research on Public Policy in the Health Field: A Review in *Health Services Research I*. Ed. Donald Mainland. pp. 11–51. *Milbank Memorial Fund Quarterly/Health and Society* 44(3, pt. 2).

——. 1991. *The Evolution of Health Services Research: Personal Reflections on Applied Social Science*. San Francisco: Jossey-Bass.

Bannister, Robert C. 1987. *Sociology and Scientism: The Quest for Objectivity, 1880–1940*. Chapel Hill and London: University of North Carolina Press.

Becker, Howard. 1928. Sargasso Iceberg: A Study in Cultural Lag and Institutional Disintegration. *American Journal of Sociology* 34(3) 492–506.

Berkowitz, Edward D. 1995. Foreword by Joseph A. Califano, Jr. *Mr. Social Security: The Life of Wilbur J. Cohen*. Lawrence, KS: University Press of Kansas.

——. 2003. *Robert Ball and the Politics of Social Security*. Madison, WI: University of Wisconsin Press.

Brandt, Allan M., and Martha Gardner. 2000. Antagonism and Accommodation: Interpreting the Relationship between Public Health and Medicine in the United States during the Twentieth Century. *American Journal of Public Health* 90(5) 707–715.

Brinkley, Alan. 1995. *The End of Reform: New Deal Liberalism in Recession and War*. New York: Alfred A. Knopf.

Brint, Steven G. 1994. *In an Age of Experts: The Changing Role of Professionals in Politics and Public Life*. Princeton, NJ: Princeton University Press.

Brown, Esther Lucile. 1937. *Physicians and Medical Care*. New York: Russell Sage Foundation.

Brown, Lawrence D. 2005. Incrementalism Adds Up? Ch. 11 in *Healthy, Wealthy, and Fair: Health Care and the Good Society*. Eds. James A. Morone and Lawrence R. Jacobs. pp. 315–335. Oxford and New York: Oxford University Press.

Bulmer, Martin. 1984. *The Chicago School of Sociology: Institutionalization, Diversity, and the Rise of Sociological Research*. Chicago and London: University of Chicago Press.

——. 1996. The Social Survey Movement and Early Twentieth-Century Sociological Methodology. Ch. 2 in *Pittsburgh Surveyed: Social Science and Social Reform in the Early Twentieth Century*. Eds. Maurine W. Greenwald and Margo Anderson. pp. 15–34, 248–251. Pittsburgh: University of Pittsburgh Press.

Bulmer, Martin, Kevin Bales, and Kathryn Sklar, Eds. 1991. *The Social Survey in Historical Perspective, 1880–1940*. Cambridge, UK: Cambridge University Press.

Burgess, Ernest W. 1929. Basic Social Data. Ch. 4 in *Chicago, an Experiment in Social Science Research*. Eds. T[homas] V[ernor] Smith and Leonard D[upee] White. pp. 139–176. University of Chicago Studies in Social Science. No. 17. Ed. by a Committee of the Social Science Departments. Chicago: University of Chicago Press.

Camic, Charles, and Yu Xie. 1994. The Statistical Turn in American Social Science: Columbia University, 1890–1915. *American Sociological Review* 59(5) 773–805.

Carey, James T. 1975. Foreword by James F. Short, Jr. *Sociology and Public Affairs: The Chicago School*. Sage Library of Social Research. Vol. 16. Beverly Hills and London: Sage.

[CCMC] Committee on the Costs of Medical Care. 1932. *Medical Care for the American People*. Committee on the Costs of Medical Care. Publication No. 28. Chicago: University of Chicago Press.

Chasse, John Dennis. 1991. The American Association for Labor Legislation: An Episode in Institutionalist Policy Analysis. *Journal of Economic Issues* 25(3) 799–828.

——. 1994. The American Association for Labor Legislation and the Institutionalist Tradition in National Health Insurance. *Journal of Economic Issues* 28(4) 1063–1090.

——. 2004. John R. Commons and His Students: The View from the End of the Twentieth Century. Ch. 4 in *The Institutionalist Tradition in Labor Economics*. Eds. Dell P. Champlin and Janet T. Knoedler. pp. 50–74. Armonk, NY, and London: M.E. Sharpe.

Church, Robert L. 1974. Economists as Experts: The Rise of an Academic Profession in the United States, 1870–1920. Ch. 12 in *The University in Society*. 2 vols. Ed. Lawrence Stone. Vol. 2. pp. 571–609. Princeton, NJ: Princeton University Press. Written under the auspices of the Shelby Cullom Davis Center for Historical Studies, Princeton University.

Clark, John Maurice. 1916. The Changing Basis of Economic Responsibility. *Journal of Political Economy* 24(3) 209–229 (Reprinted as ch. 3 in Clark 1936).

——. 1924. The Socializing of Theoretical Economics. Ch. 3 in *The Trend of Economics*. Ed. and with an introduction by Rexford Guy Tugwell. pp. 73–102. New York: Alfred A. Knopf (Reprinted as Ch. 1 in Clark 1936).

——. 1936. *Preface to Social Economics: Essays on Economic Theory and Social Problems*. Eds. Moses Abramovits and Eli Ginzberg. New York: Farrar and Rinehart.

Cohn, Jonathan. 2007. *Sick: The Untold Story of America's Health Care Crisis—And the People Who Pay the Price*. New York: HarperCollins.

Commons, John R. 1909. American Shoemakers, 1648–1895: A Sketch of Industrial Evolution. *Quarterly Journal of Economics* 24(1) 39–84.

——. 1915a. Health Insurance. *Wisconsin Medical Journal* 17(1) 218–223.

——. 1915b. Social Insurance and the Medical Profession. *Wisconsin Medical Journal* 13(7) 301–306.

——. 1918. Health Insurance. *Wisconsin Medical Journal* 17(1) 218–223.

Corwin, E[dward] H[enry] L[ewinski]. 1931. State Medicine in Europe. *Transactions of the American Hospital Association* 33 573–579.

Critchlow, Donald T. 1993. Think Tanks, Antistatism, and Democracy: The Nonpartisan Ideal and Policy Research in the United States. Ch. 7 in *The State and Social Investigation in Britain and the United States*. Eds. Michael J. Lacey and Mary O. Furner. pp. 279–322. Woodrow Wilson Center Series. Washington, DC: The Woodrow Wilson Center Press, and Cambridge, UK: Cambridge University Press.

Daston, Lorraine. 1991. The Ideal and Reality of the Republic of Letters in the Enlightenment. *Science in Context* 4(2) 367–386.

Davis, Michael M. 1916. The Medical Organization of Sickness Insurance. *Medical Record* 89 (2) 54–58.

——. 1931a. Effects of Health Insurance on Hospitals Abroad. *Transactions of the American Hospital Association* 33 579–585.

——. 1931b. *Paying Your Sickness Bills*. Chicago: University of Chicago Press.

——. 1934. The American Approach to Health Insurance. *Milbank Memorial Fund Quarterly* 12(3) 203–217. Michael M. Davis: America Organizes Medicine. 1941. New York and London: Harper and Brothers.

——. 1941. *America Organizes Medicine*. New York and London: Harper and Brothers.

Diner, Steven J. 1980. *A City and Its Universities: Public Policy in Chicago, 1892–1919*. Chapel Hill: University of North Carolina Press.

Falk, I[sidore] S[ydney]. 1934. The Present and Future Organization of Medicine. *Milbank Memorial Fund Quarterly* 12(2) 115–125.

——. 1936. *Security Against Sickness: A Study of Health Insurance*. Garden City, NY: Doubleday, Doran.

Faris, Robert E. L. 1967. *Chicago Sociology, 1920–1932*. Chandler Publications in Anthropology and Sociology. Ed. Leonard Broom. San Francisco: Chandler Publishing Company.

Fink, Leon. 1997. *Progressive Intellectuals and the Dilemmas of Democratic Commitment*. Cambridge, MA, and London: Harvard University Press.

Fisher, Donald. 1993. *Fundamental Development of the Social Sciences: Rockefeller Philanthropy and the United States Social Science Research Council*. Ann Arbor, MI: University of Michigan Press.

Flook, Emma Evelyn, and Paul J. Sanazaro, eds. 1973. *Health Services Research and R & D in Perspective*. Ann Arbor, MI: Health Administration Press.

Fox, Daniel M. 1979. From Reform to Relativism: A History of Economists and Health Care. *Milbank Memorial Fund Quarterly/Health and Society* 57(3) 297–336.

——. 1983. The Decline of Historicism: The Case of Compulsory Health Insurance in the United States. *Bulletin of the History of Medicine* 57(4) 596–610.

——. 1986. *Health Policies, Health Politics: The British and American Experience, 1911–1965*. Princeton, NJ: Princeton University Press.

——. 1990. Health Policy and the Politics of Research in the United States. *Journal of Health Politics, Policy, and Law* 15(3) 481–499.

——. 1995. *Power and Illness: The Failure and Future of American Health Policy*. Paperback ed. (Orig. cloth, 1993). Berkeley, CA: University of California Press.

——. 1997. Policy and Vulnerability: Foundations in Twentieth-Century Health Affairs. *Minerva* 35(3) 311–319.

Fronstin, Paul. 2009. Sources of Health Insurance and Characteristics of the Uninsured: Analysis of the March 2009 Current Population Survey. *EBRI Issue Brief* 334.

Furner, Mary O. 1975. *Advocacy and Objectivity: A Crisis in the Professionalization of American Social Science, 1865–1905*. Lexington, KY: University Press of Kentucky for the Organization of American Historians.

Ginzberg, Eli, ed. 1991. *Health Services Research: Key to Health Policy*. Cambridge, MA: Harvard University Press.

Glied, Sherry. 2009. Covering the Uninsured as a Quality Improvement Strategy [Editorial]. *Health Services Research* 44(2 Pt. 1) 323–326.

Goodman, Dena. 1994. *The Republic Letters: A Cultural History of the French Enlightenment*. Ithaca, NY, and London: Cornell University Press.

Goodwin, Craufurd D. 1995. The Promise of Expertise: Walter Lippman and the Policy Sciences. *Policy Sciences* 28(4) 317–345.

Gould, Elise. 2010. Employer-sponsored Health Insurance Erosion Continues in 2008 and Is Expected to Worsen. *International Journal of Health Services* 40(4) 743–776.

Greenwald, Maurine W., and Margo Anderson, Eds. 1996. *Pittsburgh Surveyed: Social Science and Social Reform in the Early Twentieth Century*. Pittsburgh, PA: University of Pittsburgh Press.

Groves, Ernest R. 1932. *An Introduction to Sociology*. New rev. ed. Longmans' Social Science Series. General Ed. Ernest R. Groves. New York: Longmans, Green.

Gruchy, Allan G. 1947. *Modern Economic Thought: The American Contribution*. New York: Prentice-Hall.

Haber, Samuel. 1964. *Efficiency and Uplift: Scientific Management in the Progressive Era, 1890–1920*. Chicago and London: University of Chicago Press.

Harrison, Shelby M. 1930. Introduction: Development and Spread of Social Surveys. In *A Bibliography of Social Surveys: Reports of Fact-finding Studies Made as a Basis for Social Action; Arranged by Subjects and Localities. Reports to January 1, 1928*. By Allen Eaton. In collaboration with Shelby M. Harrison. pp. xi–xvii. New York: Russell Sage Foundation.

Hawley, Ellis W. 1997. *The Great War and the Search for a Modern Order: A History of the American People and their Institutions, 1917–1933*. Reissued 2nd ed. of 1992. Prospect Heights, IL: Waveland Press.

Higham, John. 1965. With Leonard Krieger, and Felix Gilbert. *History: The Development of Historical Scholarship in the United States*. Englewood Cliffs, NJ: Prentice-Hall.

Hirsh, Joseph. 1939. The Compulsory Health Insurance Movement in the United States. *Social Forces* 18(1) 102–114.

Hofstadter, Richard, and Walter P. Metzger. 1955. *The Development of Academic Freedom in the United States*. New York: Columbia University Press.

Iggers, Georg G. 1962. The Image of Ranke in American and German Historical Thought. *History and Theory* 2(1) 17–40.

[IOM] Institute of Medicine. 2004. Committee on the Consequences of Uninsurance. *Insuring America's Health: Principles and Recommendations*. Washington, DC: National Academy Press.

Jordan, John M. 1994. *Machine-Age Ideology: Social Engineering and American Liberalism, 1911–1939*. Chapel Hill, NC, and London: University of North Carolina Press.

Jost, Timothy Stoltzfus. 2004. Why Can't We Do What They Do? National Health Reform Abroad. *Journal of Law, Medicine, and Ethics* 32(3) 433–441.

Julius Rosenwald Fund. 1937. *Eight Years' Work in Medical Economics, 1929–1936; Recent Trends and Next Moves in Medical Care*. Chicago: The Fund.

Karl, Barry Dean. 1969. Presidential Planning and Social Science Research: Mr. Hoover's Experts. *Perspectives in American History* 3 347–409.

——. 1974. *Charles E. Merriam and the Study of Politics*. Chicago: University of Chicago Press.

Kaufman, Bruce E. 1993. *The Origins and Evolution of the Field of Industrial Relations in the United States*. Cornell Studies in Industrial and Labor Relations. No. 25. Ithaca, NY: ILR Press.

——. 2003. John R. Commons and the Wisconsin School on Industrial Relations Strategy and Policy. *Industrial and Labor Relations Review* 57(1) 3–30.

Kellogg, Paul Underwood, ed. 1909–1914. *The Pittsburgh Survey: Findings in Six Volumes*. 6 v.: fronts., plates.; 24 cm. Russell Sage Foundation Publications. New York: Charities Publication Committee.

Kerby, William J. 1917. United States Department of Labor. Bureau of Labor Statistics. [Address by Chairman, Conference Sessions on Sickness (Health) Benefits and Insurance. I. Existing Agencies]. In *Proceedings of the Conference on Social Insurance. Called by the International Association of Industrial Accident Boards and Commissions. Washington, D.C., December 5 to 9, 1916*. Ed. Commissioner of Labor Statistics Royal Meeker. pp. 419–420. *Bulletin of the United States Bureau of Labor Statistics*. Whole number 212. Workmen's Insurance and Compensation Series. No. 10. Washington, DC: USGPO.

Kreader, J. Lee. 1976. Isaac Max Rubinow: Pioneering Specialist in Social Insurance. *The Social Service Review* 50(3) 402–425.

Kunitz, Stephen J. 1981. Efficiency and Reform in the Financing and Organization of American Medicine in the Progressive Era. *Bulletin of the History of Medicine* 55(4) 497–515.

Laslett, Barbara. 1990. Unfeeling Knowledge: Emotion and Objectivity in the History of Sociology. *Sociological Forum* 5(3) 413–433.

Leven, Maurice. 1932. *The Incomes of Physicians: An Economic and Statistical Analysis*. Committee on the Costs of Medical Care. Publication. No. 24. Chicago: University of Chicago Press.

Lubove, Roy. 1968. *The Struggle for Social Security, 1900–1935*. Cambridge, MA: Harvard University Press.

MacLean, Elizabeth Kimball. 2007. Joseph E. Davies: The Wisconsin Idea and the Origins of the Federal Trade Commission. *Journal of the Gilded Age and Progressive Era* 6(3) 248–284.

Mann, Cindy, and Tim Westmoreland. 2004. Attending to Medicaid. *The Journal of Law, Medicine, and Ethics* 32(3) 465–473.

Matza, David. 1969. *Becoming Deviant*. Englewood Cliffs, NJ: Prentice-Hall.

McNulty, Paul J. 1980. *The Origins and Development of Labor Economics: A Chapter in the History of Social Thought*. Cambridge, MA: MIT Press.

McQuaid, Kim. 1977. Henry S. Dennison and the "Science" of Industrial Reform, 1900–1950. *American Journal of Economics and Sociology* 36(1) 79–98.

Mechanic, David. 1990. Commentary: The Role of Sociology in Health Affairs. *Health Affairs* 9(1) 85–97.

——. 1993. Social Research in Health and the American Sociopolitical Context: The Changing Fortunes of Medical Sociology. *Social Science and Medicine* 36(2) 95–102.

Melhado, Evan M. 1998. Economists, Public Provision, and the Market: Changing Values in Policy Debate. *Journal of Health Politics, Policy, and Law* 23(2) 215–263.

——. 2000. Economic Theory, Economists, and the Formulation of Public Policy. *Journal of Health Politics, Policy, and Law* 25(1) 233–256.

Miller, Harold L. 1984. The American Bureau of Industrial Research and the Origins of the 'Wisconsin School' of Labor History. *Labor History* 25(2) 165–188.

Millis, Harry Alvin. 1937. *Sickness and Insurance: A Study of the Sickness Problem and Health Insurance*. Chicago: University of Chicago Press.

Mitchell, Wesley Clair. 1913. *Business Cycles*. Memoirs of the University of California. 3. Berkeley, CA: University of California Press.

——. 1924. The Prospects of Economics. Ch. 1 in *The Trend of Economics*. Ed. with an Introduction by Rexford G. Tugwell. pp. 3–34. New York: Alfred A. Knopf.

Moore, Harry H. 1927. With an Introduction by The Committee of Five of the Washington Conference on the Economic Factors Affecting the Organization of Medicine. *American Medicine and the People's Health: An Outline with Statistical Data on the Organization of Medicine in the United States with Special Reference to the Adjustment of Medical Service to Social and Economic Change*. New York and London: D. Appleton.

Moore, Michael. Writer, producer, and director. 2007. *Sicko* [film]: Dog Eat Dog Films.

Morgan, Mary S., and Malcolm Rutherford. 1998. American Economics: The Character of the Transformation. Introduction in *From Interwar Pluralism to Postwar Neoclassicism*. Eds. Mary S. Morgan and Malcolm Rutherford. pp. 1–26. Annual Supplement to Volume 30, History of Political Economy. Durham, NC, and London: Duke University Press.

Morone, James A. 1998 (orig. 1990). *The Democratic Wish: Popular Participation and the Limits of American Government*. Rev. ed. New Haven and London: Yale University Press.

Moss, David A. 1996. *Socializing Security: Progressive-era Economists and the Origins of American Social Policy*. Cambridge, MA, and London: Harvard University Press.

Murphy, Donald J. 1988. John B. Andrews, the American Association for Labor Legislation, and Unemployment Reform, 1914–1929. Ch. 1 in *Voluntarism, Planning, and the State: The American Planning Experience, 1914–1946*. Eds. Jerold E. Brown and Patrick D. Reagan. Foreword by Ellis W. Hawley. pp. 1–23. Contributions in American History. No. 130. Series Ed. Jon L. Wakelyn. New York: Greenwood.

Numbers, Ronald L. 1978. *Almost Persuaded: American Physicians and Compulsory Health Insurance, 1912–1920. The Henry E. Sigerist Supplements to the Bulletin of the History of Medicine*. NS, no. 1. Series ed. Lloyd G. Stevenson. Baltimore: Johns Hopkins University Press.

Oakman, Tara Sussman, Robert J. Blendon, Andrea L. Campbell, Alan M. Zaslavsky, and John M. Benson. 2010. A Partisan Divide on the Uninsured. *Health Affairs* 29(4) 706–711.

O'Connor, Alice. 2001. Poverty Knowledge: Social Science, Social Policy, and the Poor in Twentieth-century U.S. History. *Politics and Society in Twentieth-century America*. Eds. William Chafe *et al*. Princeton and Oxford: Princeton University Press.

Ogburn, William Fielding. 1921. The Historical Method in the Analysis of Social Phenomena. *Papers and Proceedings of the American Sociological Association* 16 70–83.

——. 1922. *Social Change with Respect to Culture and Original Nature*. New York: B.W. Huebsch.

——. 1930. Presidential Address: The Folkways of a Scientific Sociology. *Papers and Proceedings of the American Sociological Association* 24 1–11.

Park, Robert E., Ernest W. Burgess, and Roderick D. McKenzie. 1925. With a bibliography by Louis Wirth. *The City*. Chicago: University of Chicago Press.

Pells, Richard H. 1973. *Radical Visions and American Dreams: Culture and Social Thought in the Depression Years*. New York: Harper and Row.

Peterson, Mark A. 2005. The Congressional Graveyard for Health Care Reform. Ch. 7 in *Healthy, Wealthy, and Fair: Health Care and the Good Society*. Eds. James A. Morone and Lawrence R. Jacobs. pp. 205–233. Oxford, New York, NY: Oxford University Press.

Pumphrey, Ralph E. 1972. Michael M. Davis and the Development of the Health Care Movement. *Societas—A Review of Social History* 2(1) 27–41.

Ratner, Joseph. 1939. Introduction to John Dewey's Philosophy. In *Intelligence in the Modern World: John Dewey's Philosophy*. John Dewey. Modern Library of the World's Best Books. Ed. and with an introduction by Joseph Ratner. New York: The Modern Library (published by Random House). pp. 3–241.

Reagan, Patrick D. 1999. Designing a New America: The Origins of New Deal Planning, 1890–1943. In *Political Development of the American Nation: Studies in Politics and History*. Eds. Sidney M. Milkis and Jerome M. Mileur. pp. 3–241. Amherst: University of Massachusetts Press.

Recchiuti, John Louis. 2007. *Civic Engagement: Social Science and Progressive-era Reform in New York City*. Philadelphia, PA: University of Pennsylvania Press.

Reiser, Stanley Joel. 1978. *Medicine and the Reign of Technology*. Cambridge, UK: Cambridge University Press.

Rice, Thomas. 2003. Foreword by Uwe E. Reinhardt. *The Economics of Health Reconsidered*. 2nd ed. Chicago: Health Administration Press.

Rich and Merrick. 2009. The Health Care Reforms of the 1990s: Failure or Ultimate Triumph? Unpublished MS. Courtesy of the first author.

Rodgers, Daniel T. 1998a. *Atlantic Crossings: Social Politics in a Progressive Age*. Cambridge, MA, and London: Belknap Press of Harvard University Press.

———. 1998b. Exceptionalism. Ch. 1 in *Imagined Histories: American Historians Interpret the Past*. Eds. Anthony Molho and Gordon S. Wood. pp. 21–40. Princeton, NJ: Princeton University Press.

Rodwin, Victor G. 1987. American Exceptionalism in the Health Sector: The Advantages of "Backwardness" in Learning from Abroad. *Medical Care Review* 44(1) 119–154.

Roemer, Milton I. 1985. I.S. Falk, the Committee on the Costs of Medical Care, and the Drive for National Health Insurance. *American Journal of Public Health* 75(8) 841–848.

Rogers, Donald W. 2009. *Making Capitalism Safe: Work Safety and Health Regulation in America, 1880–1940*. The Working Class in American History. Editorial advisors James R. Barrett *et al.* Urbana and Chicago: University of Illinois Press.

Rorem, C. Rufus. 1931. *Private Group Clinics: The Administrative and Economic Aspects of Group Medical Practice, as Represented in the Policies and Procedures of 55 Private Associations of Medical Practitioners*. Publication. No. 8. Washington, DC: Committee on the Costs of Medical Care.

Rorem, C. Rufus, and Robert M. Sigmond. 1965. Ch. 3. The Organization of Medical and Health Services. D. Public Policy and Financing for Health Services. Medical Education and Practice: Relationships and Responsibilities in a Changing Society. In *Medical Education and Practice: Relationships and Responsibilities in a Changing Society*. Eds. Stewart G. Wolf, Jr., and Ward Darley. pp. 79–91. Report of the Tenth Teaching Institute, Association of American Medical Colleges, Colorado Springs, Colorado, December 9–12, 1962. Evanston, IL: Association of American Medical Colleges.

Rosenberg, Charles E. 1979. Inward Vision and Outward Glance: The Shaping of the American Hospital, 1880–1914. *Bulletin of the History of Medicine* 53(3) 346–391.

———. 1987. *The Care of Strangers: The Rise of America's Hospital System*. New York: Basic Books.

Ross, Dorothy. 1991. *The Origins of American Social Science*. New York: Cambridge University Press.

Rothman, David J. 1997. *Beginnings Count: The Technological Imperative in American Health Care*. David J. Rothman. A Twentieth Century Fund Book. New York and Oxford: Oxford University Press.

Rubinow, I[saac] M[ax]. 1913. *Social Insurance with Special Reference to American Conditions*. New York: Henry Holt.

——. 1915. Social Insurance and the Medical Profession. *Journal of the American Hospital Association* 44(5) 381–386.

——. 1916a. Address of the President, I.M. Rubinow: "The Relations between Private and Social Insurance". *Proceedings of the Casualty, Actuarial, and Statistical Society of America* 2, Pt. 3(6) 335–346.

——. 1916b. *Standards of Health Insurance*. New York: Henry Holt.

——. 1934. *The Quest for Security*. New York: Henry Holt.

Rutherford, Malcolm. 1983. J.R. Commons's Institutional Economics. *Journal of Economic Issues* 17(3) 721–744.

——. 1997. American Institutionalism and the History of Economics. *Journal of the History of Economic Thought* 19(2) 178–195.

——. 2000. Understanding Institutional Economics, 1918–1929. *Journal of the History of Economic Thought* 22(3) 277–308.

Shen, Yu-Chu, and Sharon K. Long. 2006. What's Driving the Downward Trend in Employer-Sponsored Health Insurance? *Health Services Research* 41(6) 2074–2096.

Simons, A[lgie] M[artin], and Nathan Sinai. 1932. *The Way of Health Insurance*. Publications of the Committee on the Study of Dental Practice of the American Dental Association. No. 6. Chicago: University of Chicago Press.

Smith, Mark C. 1994. *Social Science in the Crucible: The American Debate over Objectivity and Purpose, 1918–1941*. Durham and London: Duke University Press.

Smith, T[homas] V[ernor], and Leonard D[upee] White, Eds. 1929. *Chicago, an Experiment in Social Science Research*. University of Chicago Studies in Social Science. No. 17. Ed. by a Committee of the Social Science Departments. Chicago: University of Chicago Press.

Somers, Herman M., and Anne R. Somers. 1961. *Doctors, Patients, and Health Insurance*. Washington, DC: Brookings Institution.

Starr, Paul. 1982a. *The Social Transformation of American Medicine: The Rise of a Sovereign Profession and the Making of a Vast Industry*. New York: Basic.

——. 1982b. Transformation in Defeat: The Changing Objectives of National Health Insurance, 1915–1980. *American Journal of Public Health* 72(1) 78–88.

Stern, Bernhard J. 1941. *Society and Medical Progress*. Princeton, NJ: Princeton University Press.

Stevens, Rosemary A. 1985. I.S. Falk and the Challenge of Facts [Editorial]. *American Journal of Public Health* 75(8) 827–828.

——. 1989. *In Sickness and In Wealth: American Hospitals in the Twentieth Century*. New York: Basic.

——. 1998 [orig. 1971]. *American Medicine and the Public Interest*. Updated ed. New Haven, CT: Yale University Press.

Stromquist, Shelton. 2006. *Reinventing 'The People:' The Progressive Movement, the Class Problem, and the Origins of Modern Liberalism*. The Working Class in American History. Editorial advisors James R. Barrett *et al.* Urbana and Chicago: University of Illinois Press.

Thomas, William Isaac, and Florian Znaniecki. 1918–1920. *The Polish Peasant in Europe and America: Monograph of an Immigrant Group*. Chicago: University of Chicago Press (vols. 1–2); Boston: Richard G. Badger (vols. 3–5). Reprints (earliest and best-known): Boston: Richard G. Badger, 5 vols., 1920; unabridged and slightly revised: New York: Alfred A. Knopf, 2 vols., 1927; New York: Dover (rpt. of Knopf ed., 1927), 1958; see Bulmer 1984: 45n1 on 238. Diverse later editions exist.

Tripp, Joseph F. 1987. Law and Social Control: Historians' Views of Progressive-era Labor Legislation. *Labor History* 28(4) 447–483.

Turner, Stephen P. 1991. The World of the Academic Quantifiers: The Columbia University Family and Its Connections. Ch. 10 in *The Social Survey in Historical Perspective, 1880–1940*. Eds. Martin Bulmer *et al.* pp. 269–290. Cambridge, UK: Cambridge University Press.

Tyrrell, Ian. 1986. *The Absent Marx: Class Analysis and Liberal History in Twentieth-century America*. Contributions in American History. Number 115. New York: Greenwood Press.

Volti, Rudi. 2004. Classics Revisited: William F. Ogburn, Social Change with Respect to Culture and Original Nature. *Technology and Culture* 45(2) 396–405.

Walker, Forrest A. 1969. Compulsory Health Insurance: "The Next Great Step in Social Legislation." *Journal of American History* 56(2) 290–304.

Warren, Benjamin S. 1915. Sickness Insurance: A Preventive of Charity Practice. *Transactions of the Section on Preventive Medicine and Public Health of the American Medical Association at the Sixty-Sixth Annual Session, Held at San Francisco, June 22 to 25, 1915*, pp. 74–94.

Warren, Benjamin S., and Edgar Sydenstricker. 1916. *Health Insurance: Its Relation to the Public Health*. 2nd ed. Public Health Service Bulletin. No. 76. Washington, DC: USGPO.

——. 1919. Health Insurance, the Medical Profession, and the Public Health. *Public Health Reports* 34(16) 775–789.

Wiebe, Robert H. 1967. *The Search for Order, 1877–1920*. The Making of America. General ed. David Herbert Donald. New York: Hill and Wang.

Zelizer, Julian E. 2003. The Uneasy Relationship: Democracy, Taxation, and State Building. Ch. 11 in *The Democratic Experiment: New Directions in American Political History*. Eds. Meg Jacobs, William J. Novak, and Julian E. Zelizer. pp. 276–300.

Chapter 18
Quantity and Polity: Asylum Statistics and the Drive for Medical Evidence

Theodore M. Porter

The revolutions in France, so disturbing to mental balance, were to have a long career as imputed etiological factors for insanity. For the statistics of lunacy, however, the more fateful date in the year of revolution was not July 14 but January 7, 1789, when a Committee of the House of Commons assembled to inquire into George III' mental state. Disagreement about the king's condition and the proper treatment had caused a furor among his physicians, spreading to the Parliament and the nation. Who, if anyone, was qualified to treat the royal patient, and who could say whether he would recover in time to forestall the need to appoint a regent? What confidence could be placed in the irregular regime of Reverend Dr. Francis Willis? The committee interrogated its experts:

> Whether if Nine Persons out of Ten, placed under the Care of a Person who had made this Branch of Medicine his particular Study, had recovered, if they were placed under his Care within Three Months after they had begun to be afflicted with the Disorder, Doctor Warren would not deem such Person, either very skilful or very successful? (Committee appointed 1789, p. 20)

Richard Warren, Physician to the King, said he would, accenting the conditional.

> Whether, in order to induce Doctor Warren to believe, that, for Twenty-seven years, Nine persons out of Ten had been cured, he would not require some other Evidence than the Assertions of the Man pretending to have performed such Cures?

Warren replied now forthrightly: "I certainly should." (25) One needed proper records.

18.1 The King Is Mad: Long Live Statistics

The therapeutic claims of Dr. Willis, a man so bold as to treat the madness of the king, and the insistence of the committee, supported by more established physicians, that he should have been able to supply written evidence, typify an

T.M. Porter (✉)
Department of History, University of California, Los Angeles, CA, USA
e-mail: tporter@history.ucla.edu

J.Z. Buchwald (ed.), *A Master of Science History*, Archimedes 30,
DOI 10.1007/978-94-007-2627-7_18, © Springer Science+Business Media B.V. 2012

emerging ethic of accountability, a new statistical spirit. Scholars have failed so far to notice that the mania for numbers was particularly suited to the administration of madness. Historians have chronicled an explosion of statistical activity across a range of scientific, bureaucratic, and professional projects in the early nineteenth century. The Danish statistician Harald Westergaard (1932) wrote of an "era of enthusiasm" in statistics beginning about 1830. Charles Gillispie's research (1963, 1972) on the applications of probability theory, first to political and social questions and afterwards to molecular physics, identified the link of mathematics and natural science to rationalizing efforts in administration and politics, making precise an intellectual trajectory sketched out in a chapter by John Theodore Merz (1903) on "the statistical view of nature." The historical and social investigation of tallies, tables, and measurement systems, along with the technologies and typologies that make them possible, has emerged since about 1980 as a major area of social science and humanistic scholarship (e.g. Gillispie 1980, 2004, Krüger et al. 1987a, b).

There is no scholarly consensus on the turning point in the historical surge of quantification. The decade of the 1820s, or perhaps its terminus in 1830, offers a plausible choice (Porter 1986, 2003, Hacking 1989), but a case can also be made for 1790 or 1800 (in round numbers), or 1789, the launching in Europe and North America of the era of censuses. Yet we are reminded of land surveys and geodetic measurements, population estimates, weather records, tables of diseases and therapeutic outcomes, and quantitative forms of experimental physics involving heat and electricity that seem to take off about 1730 (Frängsmyr et al. 1990, Rusnock 2002). This logic draws us ever backwards, from John Graunt's and William Petty's population counts and calculations in the 1660s to the Renaissance fascination with double-entry bookkeeping, merchant arithmetic, and astrological tables. A popular work of big history announces a turning point of quantitative thinking in 1250 (Crosby 1997), yet still we are nowhere near a *fons et origo*, not even in the European West.

We can be impressed by the sweep of the quantifying bustle that accompanied the transformative political, economic, and scientific developments of the early nineteenth century without pretending to have located a clean historical rupture. While the reactions to George III's mental breakdown in 1789 mark a convenient beginning for a historical study of asylum statistics, the story remains halting and episodic until the 1830s. By then, the cascades of numbers that engulfed so many aspects of social, governmental, and scientific life were clearly recognizable as a historic movement. Statistics had become and would remain a key template for knowledge and an irrepressible instrument of administration. The quantitative sensibility did not quite sweep all before it, but for two centuries at least there has never been a down market in numbers. King Canute could not command, but might have secured a reputation in social science by prognosticating, that the statistical tide should never ebb.

The exponential growth of quantification was also qualitative, the creation of new methods and meanings. The increasing prevalence of academic appointments and university degrees in the human and economic sciences, allied to professional training and meritocratic career patterns in government and business, accommodated the appearance of a specifically mathematical form of statistics, which now has the attention of science. Yet the labor of gathering and ordering numbers has remained fundamental in statistics, and before 1890 there was little else. "Statists" in the nineteenth century were very much conscious of their primary role as providers of information and agents of accountability. In this liberal era, their reporting was directed not only to states, but to citizens, including even the "masses," who were urged to use numbers to better their lives and to comprehend the limits as well as the possibilities of public action. Statistics, in this form, typified the scientific spirit of an age that hoped to make knowledge reliable and accessible. (Porter 2009)

18.2 Accountability and the Statistics of Lunacy

Physicians were less committed to an ideal of public science. Principles there were, but individuals differed, and expert judgment would be needed to apply general rules to particular cases. Treatment by numbers, applied indiscriminately, would undermine the professional standing of medicine, and doctors construed popular knowledge as mere quackery. Notwithstanding the interests and intellectual investments at stake, the "numerical method" gained prominence during the 1830s and 1840s, an age of statistical hope and therapeutic nihilism (Cassedy 1984, Matthews 1995). Ordinary medicine, however, was affected less enduringly and far less profoundly by statistics than was the treatment of the insane. This may appear surprising if we think of psychiatry as a humane enterprise involving close observation of a patient's actions and speech. But "alienism," the treatment of the insane, was not fully captured by medicine until the 1830s, and its incongruities in a profession devoted to bodily cures remained a problem for much longer, even to our own day (Scull 1993). Asylum medicine, like hospital medicine, flourished largely as an area of public health, and its ethic of individualism was correspondingly weak. While ordinary physicians earned their income directly from the patients they visited in their houses, asylums were akin to, and competed with, poorhouses and prisons. Some of their inmates were confined for the sake of public safety, and most, lacking the means to pay the costs of their treatment, were designated paupers. Also, as public or charitable institutions, asylums were subject to normal standards of accountability, which meant providing numbers not only of revenues and expenditures, but also of patients admitted and discharged.

Asylum statistics, in fact, originated as a form of bookkeeping. The prototype of the patient reckoning–doubling as evidence of therapeutic effectiveness–was a balance sheet. John Stype's (1720) edition of John Stow's survey of London includes some of these tables for Bethlem Hospital (Bedlam) beginning in 1704. The "Disturbed Men and Women then brought in" that year numbered 64, while 50 were "Cured of their Lunacy and Discharged," and 20 more were "Buried," leaving 130 patients under cure. The table for 1705 shows 72 admissions, 34 cures,

29 deaths, and 137 patients remaining–arithmetic that seems to leave two patients unaccounted for. The next reporting period, "as it was Published," is "1705 to 1706," evidently a single year, since the number of "Distracted Men and Women brought in," 72, as well as the 52 cures and the 13 buried are about right for a twelve-month period. This time the 148 patients remaining are more numerous by four than the arithmetic implies, if indeed the 137 remaining at the end of the last period were the ones present at the beginning of this one. Stype gave another table extending from Easter 1706 to Easter 1707, followed by three years with no information, a table for 1711, five years without numbers, and tables for 1717 and 1718.

If the accounting seems desultory, the evidence of therapeutic effectiveness was still more so. Jonathan Andrews (1991) remarks that the annual reports of Bethlem from 1680 to 1705 claimed cure rates between 57 and 82%. Yet the patients discharged as cured were not distinguished in the records from those sent away still insane. Certainly some did not regain their sanity, since Bedlam had an announced policy to limit the residence of patients to about a year. (The internal evidence of the numbers, with annual admissions about half the figure for patients remaining under care, implies an average stay of two years.) We do not know how the cure rates were calculated, and Stype's proclamation on the subject is consistent with no calculation at all: "So that by God's Blessing for Twenty Years past, ending 1703, there have been above two Patients in three cured, as the Physician hath told me..." And yet the higher powers at Bedlam were not so lax as to leave these numbers to happenstance. Andrews remarks that they kept the death rate down by proactively discharging weak or debilitated patients.

Officials often treated their records as proprietary, even in the late eighteenth century, but lunacy accounts, like weather records, had begun to benefit from a sprit of private initiative. The physician William Black (1781, p. 229), whose dedication to political and medical arithmetic was nicely compatible with his promotion of smallpox inoculation, complained in 1781 that the "relieved, cured and discharged, are jumbled into one list" so that none but the "eminent physician" there, Dr. Monro, could resolve what proportion are cured. This eminence, thus flattered, introduced Black to his son, through whom Black made the acquaintance of another praiseworthy individual, the resident apothecary at Bethlem, John Gozna, "whose learning and curiosity induced him to keep a *private* register of all the patients, upon which, as incontrovertible data, I have founded and collected all the following tables and propositions." (Black 1789, pp. 129–130) He referred in 1788 (p. 235) to lunacy outcomes (or to the record book) as an "untrodden wilderness," through which he would proceed, in the manner of early astronomy, by first establishing secure facts. In the crucial year of George III's madness (1789, p. 130) he claimed a solution to the pressing problem of prognostication. "I may with safety assert, that mine are the only numerical and certain data that ever have been published in any age or country, by which to calculate the probabilities of recovery, of death, and of relapse in every species and stage of insanity, and in every age." (Macalpine and Hunter 1969, pp. 297–299)

In the wake of this investigation, the officers at Bethlem began to express pride in their recordkeeping, which solidified their status as a well-conducted institution. Black did not openly challenge Willis's cure rate with his calculations, but his successor as Bethlem apothecary, John Haslam, emphatically did. Haslam was all the more suspicious of Willis's claims because the cures at Bethlem were nowhere near 90%. His table based on patients discharged from 1784 to 1794 indicated a total of 574 cured and 1090 sent away uncured. Contrary to Willis's assertion that age did not matter, this table showed highly favorable prospects for new patients aged 10 to 20 and increasingly dire outcomes for older ones. The odds against recovery for a man of age 50, such as King George himself, were 4:1. And among patients admitted (against normal policy) to Bethlem after more than a year of illness, there was not a single lasting recovery.

> When the reader contrasts the preceding statement with the account recorded in the report of the Committee who have attended His Majesty, &c. he will either be inclined to deplore the unskilfulness or mismanagement which has prevailed among those medical persons who have directed the treatment of mania in the largest public institution in this kingdom, of its kind, compared with the success which has attended the private practice of an individual; *or to require some other evidence, than the bare assertion of the man pretending to have performed such cures.**

Haslam even supplied documentation historical in the form of a footnote: "*Vide Report, Part II. p. 25," referring to Willis's testimony nine years earlier (Haslam 1798, pp. 112–114).

18.3 A Statistical Specialty

Could a royal bout of lunacy have elicited such numerical documentation a decade or half a century earlier? Betting on a king's life had been known for centuries, and the setting of odds was impeded by the threat of execution for *lèse majesté* rather than by any conceptual obstacles. By 1750, the mathematics of chance and political arithmetic were familiar to the educated. The novelty of this episode has more to do with information practices: first, the expectation that an institution like Bethlem should keep basic records on all its patients, and second, that it should gather them up and make them accessible to the public. We can find other examples. The County Hospital for the Sick and Lame at Winchester began printing an annual report (now accessible at *Eighteenth-Century Collections Online*), with double-entry financial records as well as a rudimentary breakdown of patient outcomes, in October 1737, just under a year after its founding on Michaelmas, 1736. This ethic of statistics and publicity did not spring up all at once, but evolved gradually (Higgs 2004, Crook and O'Hara 2011), and the investigation of 1789 belongs still to the early phases of its rise. In 1760, it might not yet have been possible to come up with numbers for patient recoveries at Bethlem even if a compelling reason to do so had somehow arisen.

The opening up of institutional secrets emerged as crucial in British parliamentary investigations of madhouses in the early nineteenth century, which culminated in an 1815 report. Along with eyewitness accounts of female patients kept naked in bare, squalid cells behind a gate whose keeper claimed no access to a key, the committee called attention to the obstruction of their access to patient books. The magistrate Godfrey Higgins, describing his efforts to inspect the Yorkshire Asylum, replied when asked about records of patient registration: "There was a set of books regularly kept by the apothecary, and also another set by the steward, both of which purported to be a correct account of admissions of patients, and how they were disposed of, but I have reason to believe that those accounts were false, and that they were kept falsely on purpose." He pointed to discrepancies between closely-held books and the official ones that formed the basis for newspaper accounts. A steward refused to deliver up the books he kept, claiming they were his own property, and later testified that he had destroyed them. Just after the court of governors ordered an investigation of the York Asylum, its buildings caught fire, immolating several patients along with the books. (Committee on Madhouses 1815, pp. 4–6)

These 1815 hearings signaled the start of a monumental expansion of asylum systems that was to continue for a century and a half. During much of the nineteenth century, care for the mentally ill ranked among the most costly social programs in Europe and North America. While asylums typically fell to the charge of regional governments such as the American states or English counties, and often depended on philanthropic initiative, they have a notable place in the early history of the welfare state, first as curative institutions showcasing the growth of public investment in a healthy citizenry, and later as custodial ones to protect the population from degeneracy. (Tomes 1984, p. 294) In both guises, asylums and hospitals helped give shape to ideals of public accountability.

This holds not only for financial accounts, but especially for patient accounts, i.e. statistics. These were cultivated for internal as well as external reasons. The English physician Thomas Percival, author of an early treatise on *Medical Ethics* (1803), emphasized record keeping as an element of sound medical practice. The physician or surgeon, he wrote, should draw up an account of every case that is "rare, curious, or instructive." Hospital registers should include "three tables: the first specifying the number of patients admitted, cured, relieved, discharged, or dead; the second the several diseases of the patients, with their events; the third the sexes, ages, and occupations of the patients." This would advance knowledge of healthy and unfavorable "situations, climates, and seasons," of the effects of particular trades and manufactures, and of the attack or cessation of epidemics. Finally, "physicians and surgeons would obtain a clearer insight into the comparative success of their hospital and private practice; and would be incited to a diligent investigation of the causes of such difference." (pp. 15–17)

The question of cure rates, which rose to the surface in regard to George III, became centrally important in the nineteenth century. Especially in Britain, the United States, and Germany, tables proliferated in the public and bureaucratic reports. None were so universal, though, as the patient table or table of population movement, which resembled the first of the accounts recommended by Percival, supplying, in columns, the number of patients at the beginning of the year, new admissions, patients released cured, improved, unimproved, and dead, and the patient number at the end of the year.

TABLE 2.

Whole No.		No. of each sex.	Recov'd.	Improv'd.	Not imprv'd.	Want of room.	Died.
Patients disch'd,	106						
Males, .		62	30	10	2	14	6
Females, .		44	27	5	.	10	2
Recent cases:							
Discharged, .	57						
Males, .		32	26	5	.	.	2
Females, .		25	22	1	.	.	1
Old cases:							
Discharged, .	49						
Males, .		30	5	5	2	14	4
Females, .		19	4	4	.	10	1
Remains, .	138						
Males, .		80	.	20	60	.	.
Females, .		58	.	12	46	.	.

Per cent. of Recoveries of those discharged in the course of
the year, ...   53 3-4 per cent.
Per cent. of Recoveries of cases discharged of less than one
year's duration, ...   84 1-5 per cent.
Per cent. of Recoveries of all other cases discharged,   .   18 2-5 per cent.

Tables such as this one from the Worcester asylum in Massachusetts (1837, p. 148) attested to the proper functioning of the institution. Every patient entered here as cured or improved gave donors and legislators another reason to invest money in specialized institutions for the mentally ill, rather than leaving them to rot in prisons and poorhouses. "It is evident," wrote the American Theodric Romeyn Beck (1830) in conclusion to a survey of asylum statistics, "that the most humane, the most efficient, as well as the most economical plan, would be, for the state to erect in its various great divisions, extensive Lunatic Asylums, provided with proper medical attendance, and all the safeguards so essential both to the patients and the public. Let these be increased, if the increase of the malady demands it. The burden of their support will fall equally upon all; the success of their treatment which, we might reasonably anticipate, would leave vacancies for new cases." (82)

Beck also used the statistical reports to compare American institutions with each other, and with foreign ones. His tables from 1830, still in a very early phase of the asylum movement, showed cure rates in America comparable to the most famous European asylums. American asylum directors often showed conspicuous satisfaction in the superiority of their calculated results to those of celebrated Old World alienists such as Esquirol at Bicêtre or Samuel Tuke at the Retreat in York (79–80).

PROPORTION OF CURED.

	Admitted.	Cured.	Centesimal proportion, or No. cured in every 100.
New-York Lunatic Asylum, from 1795 to 1821,	1584	700	44.19
Bloomingdale Asylum, 7½ years,	1043	436	41.80
Pennsylvania Hospital, from 1752 to 1828,	3487	1254	35.96
Friends' Asylum near Philadelphia, 8 years,	158	53	33.54
Connecticut Asylum, 5 years,	196	100	51.01
Mean,			41.30

	Admissions.	Cured.	Per cent.
The *Cork Lunatic Asylum*, (1798 to 1818,)†	1431	751	52.49
Salpetriere and Bicetre, Paris, (1801 to 1821,)‡	12,592	4968	nearly 30
Aversa near Naples, (1814 to 1823,)‖			29.70
Senevra Hospital, Milan, (1802 to 1826,)‡			58
Charenton, Paris, (1826—7—8,)§			33
Bethlem, London, (1817 to 1820,)‡			54
St. Luke's, London, (1800 to 1819,)‡			46

Yet an absolute cure rate could not capture the achievement, still less the potential, of lunatic asylums dispensing the new moral treatment. These figures, it was understood, included many cases that had become hopeless through neglect or ill treatment. The proper measure of what asylums could contribute to the welfare and prosperity of a people was the cure rate for new cases, and this was much higher.

Proportion of Cured, in Recent and Old Cases.

Bloomingdale Asylum.	Admitted.	Cured.	Per cent.
Recent cases,	581	341	58.69
Old cases,	422	76	18.00

Connecticut Asylum.	Admitted.	Cured.	Per cent.
Recent cases,	97	86	88.66
Old cases,	99	14	14.14

The pressure of competition, as in any free market, inspired vigorous emulation and improvement, pushing these numbers still higher. Especially in the United States, where the results of this early version of evidence-based medicine were widely publicized, cure rates for fresh patients rose to a wondrous level that had seemed merely boastful when claimed by Dr. Willis.

18.4 The Basis of Asylum Science

Tables of population movement and patient outcomes circulated readily for much of the nineteenth century, appearing as news items in the national journals for asylum doctors that sprung up in 1843 or 1844 in France, the United States, and Germany, and a decade later in Britain. The *Allgemeine Zeitschrift für Psychiatrie*, for example, published asylum reports alongside its medical-scientific papers, and included a section of miscellany composed mainly of patient and case outcome numbers in tabular or paragraph form snipped from reports originating in Switzerland or Silesia, Ohio or the Faroe Islands. Cure rates were announced, reprinted, compared, and discussed from Paris and Berlin to the farthest reaches of European commerce. At the launch of this journal, the editors claimed that such numbers could lead this hopeful science of *psychiatry*, whose innovative Greek name was well-suited to methodological pronouncements, into a brighter future. Heinrich Damerow's editorial introduction dismissed, however, any idea that progress would come from indiscriminate heaps of numbers. Alas, "the statistical relationships of insanity in Germany are so far mere fragments lacking completeness or unity." One of the tasks of the journal was to remedy this problem, so that, before long, "at one time and following one method a general census of the insane in Germany can be carried out, and the results laid down in our journal." (Damerow 1844, p. xiv).

Practical *alienists* (the usual word in French and English) or *Irrenärzte* (in German) were well aware of the ambiguities concealed by such numbers. The first paper in the new journal following Damerow's introduction was a triennial report by Ernst Albert Zeller ([1843] 1844), on the institution in Winnenthal that was gaining renown under his direction. He explained there that his cure rates had declined from previous reports because he now waited longer to discharge patients whenever a relapse was feared. Later the same year, coeditor Carl Friedrich Flemming (1844, pp. 430–1) explained that the service of the journal as an archive of asylum reports was of self-evident worth. Every science, he announced, passes through two epochs, a first when carefully-gathered facts are the greatest need, and a second when the mass of facts becomes overpowering and the problem of ordering becomes paramount. Some homogeneity of form, particularly in the classification of disease types, is needed to connect the tables with general principles. Two years later, Damerow (1846, p. 17) called for a more uniform statistics, extending also to the rest of Europe and to America. Complete censuses of the insane, carried out according to a uniform principle, can be most valuable, but meanwhile he requested asylum reports from near and far, which could help to validate a true, anthropological conception of psychiatry and advance the effectiveness of treatment that provides the true test of knowledge. The theoretical discussions at the Paris Academy of Medicine, by contrast, were like shouting "Allah ist gross!" (p. 29). To be sure, Damerow was vexed specifically by barren French theory, not theory as such. But he also was intimidated by what he perceived as a Statistics Gap, for the elaborate measures of insanity in the 1843 census volume (Statistique de la France 1843, 305–70) had moved him to lament that the "insanity statistics of Germany are fractions without a whole." (Damerow 1845) Harmonization, the key to progress in asylum medicine, was a nationalistic project in uneasy alliance with scientific internationalism.

Almost all the crusaders for better statistics of madness, and not alone German ones, avowed their opposition to the mere heaping up of undigested numbers. There were several proposals, including a particularly ambitious one by the French (Lunier 1869), to standardize the statistics of insanity as the basis of a true and accurate science that could incorporate the many factors bearing on the frequency of insanity and the prospects for a cure. One thing they almost never investigated statistically, but only through cases, was the effectiveness of specific medical remedies. Here, medical individualism remained supreme. But asylums were not, in general, sites of careful, individualized treatment, and as they grew and grew beyond all bounds, they became still less so. The statistical way of knowing, though adapted to the requirements of external authorities, was never merely imposed from outside. Even at the apogee of the moral treatment, statistics was in a way intrinsic to institutions whose very existence, after all, advertised the bounds of individual reason. The sense of a seething mob requiring to be fixed into tables was reinforced by the deluge of chronic, often pauper, patients, whose Malthusian increase was becoming an undeniable reality. The call for censuses of the insane was motivated partly by a desire to gauge the need for asylum beds, but mainly to determine if the shocking expansion of insanity was real or only apparent. Most alienists thought the latter. While the pressures of modern life or the epidemic of drunkenness might well bear some responsibility for mental illness, demand seemed to expand to meet every increase in the supply of asylum spaces. It was, perhaps, a tribute to these benevolent institutions, to which so many families were now willing to entrust their stricken husbands, wives, children, and parents.

18.5 Statistics Outlasts the Critiques

Intellectually, the extraordinary explosion of insanity constituted a double challenge to the claims of statistics. One of the most cherished numbers in the asylum reports was the ratio of insane to healthy in the population. Even after the alienists learned to make the category clear, every estimate of the true ratio quickly proved too low. Eighteenth-century figures show insanity as extremely rare, while early nineteenth-century counts converged on a ratio of 1:1000. By the 1840s, 1:500 seemed more defensible, with a few insanity-prone places like Norway and Scotland showing 1:300. Two decades later, alienists acknowledged that a more complete count rather than a surplus of the insane was responsible for this high number (Dahl 1868). Soon, the proportion of mentally ill surpassed this limit, too, as it had overflowed every previous barrier.

Still more dispiriting than errors of enumeration were the failures of outcome measurement, a simultaneous failure of statistics and of treatment. Despite the laudable patient results, especially for new cases, asylums seemed to have no impact on the overall prevalence of insanity. By the 1870s, the idea of the asylum as a curative institution was losing its hold (Tomes 1984, pp. 293–4). Although mental hospitals could be defended as all the more needful amidst the teeming lunatics, plummeting cure rates undermined such advocacy. The most discouraging aspect of the mature asylum, the heaving multitude of hopeless cases, still permitted the argument that the failure was more a result of delayed treatment than of ineffective remedies. Asylum advocates put ever more emphasis on the financial savings that would be possible if only there were sufficient space for new patients and if only their relatives could be taught to seek help without delay. But this did not happen. Instead, the numbers chronicled a melancholy trajectory of increasingly ineffective institutions, with percentage cures sinking toward single digits.

At least these numbers did not lie. But how could doctors reconcile such dismal figures with their steadfast insistence on a narrative of scientific progress lifting alienism, with the rest of medicine, ever upwards? Pliny Earle (1887), formerly superintendent of the private Bloomingdale Asylum in New York and for many years a great champion of asylum statistics, sallied forth in the 1877 report of the Northampton Asylum in Massachusetts with a critique of these calculations. His argument, which gained quick international notoriety and remains well-know to historians, emphasized failures of statistics, and particularly the problem of readmissions. A patient who was discharged as cured and then relapsed could be cured again, in which case he would contribute his unit to the numerator of the cure-rate fraction not just once, but twice, and sometimes more than twenty times. The old doctrine of the curability of insanity, Earle now contended, had been an illusion from the outset, and there was little prospect that anything so commonplace as statistics of the circumstances and causes of insanity could supply the key to effective treatment. (Grob 1983)

While Earle's loss of faith in treatment belonged to the era in which he wrote it, his critique of statistics was part of a decades-long, transnational tradition. Isaac Ray (1849, p. 24), asylum superintendent in Rhode Island, wrote in the *American Journal of Insanity* that statistics could never be simply a "general expression of the facts themselves," never merely "a process in arithmetic. It is, rather, a profound, philosophical analysis of materials carefully and copiously collected…" A few years later, Damerow (1855, p. 440) announced a change of heart on statistics, to which he had previously looked as foundation of an effective psychiatry. Not causal relationships (*ursächliche Verhältnisse*) but numerical ratios (*Zahlenverhältnisse*) are all we can expect from statistics. It only turns tables of numbers into more numbers, on the dubious assumption of thoroughly homogeneous cases. And what a buzz of activity, of troop movements and official visitors and overcrowded rooms and crushing poverty, are omitted from the numbers, as he recalled himself from the experience of the asylum in Halle. A decision to classify a patient as curable or chronic or to discharge her as recovered will often depend on such mundane institutional considerations. Even death, as a statistical outcome, was institutionally constructed. "It is not a rare occurrence," complained the trustees of the Worcester Lunatic Hospital (1851), "that subjects, *not for cure*, but for care and nursing only, reach the Hospital in the last stages of existence, and a few short days, or weeks it may be, add their names to the lists of mortality." In this case, state law left the institution no choice, though in other cases, as we have seen, the decision to admit of reject a patient was made for the sake of the statistics.

It is precarious to generalize over the hundreds of insane asylums set up in so many jurisdictions in the nineteenth century, but my sense is that practical faith in statistics was declining already by the 1850s. Yet institutional momentum kept the reports coming, and as states assumed a more active regulatory role or appointed boards of oversight, the statistical tables became still more closely bound to the legal and budgetary requirements of asylum medicine. And at the intersection of science and polity, asylum statistics was reinvigorated in the 1870s and 1880s, when it did not so much follow as generate the systematic study of human heredity to which Francis Galton would attach the name *eugenics*. The statistical approach, still anchored in the results of inquiries carried out in social institutions, did not ebb but swelled in the era of Mendelism, and became still more integral to the study of human heredity in those more recent periods identified with DNA and genomics.

Statistics was very far from dominating medical knowledge even among nineteenth-century alienists, who also presented case narratives in the lunacy reports and whose medical writings were devoted mostly to sciences of the body. The language of quantity, however, had privileged status as evidence, especially in accounts produced for political leaders, administrators, philanthropists, and the general public. Inside and outside of science are not easily separated, and the relation of social institutions and of numbers to medical science has never been simply external. Medical knowledge and practice were shaped by such engagements, and the dialogue of quantity and polity in asylums and mental hospitals is part of an enduring dynamic of evidence-based practice.

References

Andrews, J., *Bedlam Revisited: A History of Bethlem Hospital, c1634–d1770*, Ph.D. dissertation, London University, 1991.
Beck, T., "Statistical Notices of some of the Lunatic Asylums in the United States," *Transactions of the Albany Institute*, vol. I, part I, 1830, 60–83.

Black, W., *Observations Medical and Political, on the Small-pox, and the Advantages and Disadvantages of General Inoculation, especially in Cities, and on the Mortality of Mankind at every Age in City and Country* ..., London: J. Johnson, 1781.

Black, W., *A Comparative View of the Mortality of the Human Species at all Ages; and of the Diseases and Casualties by which they are destroyed or annoyed*, London: C. Dilly, 1788.

Black, W., *An Arithmetical and Medical Analysis of the Diseases and Mortality of the Human Species*, second edition, corrected and improved, London: C. Dilly, 1789.

Cassedy, J., *American Medicine and Statistical Thinking*, Cambridge: Harvard University Press, 1984.

Committee appointed to Examine the Physicians who have attended His Majesty During His Illness, Touching the present State of His Majesty's Health, *Report*, Ordered to be printed 13th January 1789.

Committee on Madhouses, *Report from the Committee on Madhouses in England*, printed 11 July 1815.

Crook, T. and G. O'Hara, *Statistics and the Public Sphere: Numbers and the People in Modern Britain c. 1800 to 2000*, London: Routledge, 2011

Crosby, A., *The Measure of Reality: Quantification and Western Society, 1250–1600*, Cambridge: Cambridge University Press, 1997.

Dahl, L., "Ueber einige Resultate der Zählung der Geisteskranken in Norwegen, den 31 December 1865," *Allgemeine Zeitschrift für Psychiatrie*, 25, 1868, 839–846.

Damerow, H., "Einleitung," *Allgemeine Zeitschrift für Psychiatrie*, 1, 1844, i–xviii.

Damerow, H., "Statistique de la France. Section III. Aliénés," *Allgemeine Zeitschrift für Psychiatrie*, 2, 1845, 723–730.

Damerow, H., "Die Zeitschrift. Ein Blick rückwärts und vorwärts," *Allgemeine Zeitschrift für Psychiatrie*, 3, 1846, 1–32.

Damerow, H., "Kritisches zur Irrenstatistik aus der Anstalt bei Halle," *Allgemeine Zeitschrift für Psychiatrie*, 12, 1855, 440–467.

Earle, P., *The Curability of Insanity: A Series of Studies*, Philadelphia, PA: J. B. Lippincott Company, 1887.

Flemming, C., "Einladung an die Irrenanstalten-Directoren zur Benutzung gemeinschaftlicher Schemata zu den tabellarischen Uebersichten," *Allgemeine Zeitschrift für Psychiatrie*, 1, 1844, 430–440.

Frängsmyr, T., Heilbron, J. and Rider, R., *The Quantifying Spirit in the Eighteenth Century*, Berkeley, CA: University of California Press, 1990.

Gillispie, C., 'Intellectual Factors in the Background of Analysis by Probabilities', in: Crombie, A., ed., *Scientific Change*, New York: Basic Books, 1963, pp. 431–453.

Gillispie, C., 'Probability and Politics: Laplace, Condorcet, and Turgot,' *Proceedings of the American Philosophical Society*, 116, 1972, 1–20.

Gillispie, C., *Science and Polity in France at the End of the Old Regime*, Princeton: Princeton University Press, 1980.

Gillispie, C., *Science and Polity in France: The Revolutionary and Napoleonic Years*, Princeton: Princeton University Press, 2004.

Grob, G., *Mental Illness and American Society, 1875–1940*, Princeton: Princeton University Press, 1983.

Hacking, I., *The Taming of Chance*, Cambridge: Cambridge University Press, 1989.

Haslam, J., *Observations on Insanity: with Practical Remarks on the Disease, and an Account of the Morbid Appearances on Dissection*, London: F. and C. Rivington, 1798.

Haslam, J., *Observations on Madness and Melancholy: including Practical Remarks on those Diseases; together with Cases: and an Account of the Morbid Appearances on Dissection*, Second edition, London: J. Callow, 1809.

Higgs, E., *The Information State in England: The Collection of Information on Citizens, 1500–2000*, London: Palgrave Macmillan, 2004.

Krüger, L., L. Daston, and M. Heidelberger, *The Probabilistic Revolution, Volume 1: Ideas in History*, Cambridge, MA: MIT Press, 1987a.

Krüger, L., G. Gigerenzer, and M. Morgan, eds., *The Probabilistic Revolution, Volume 2: Ideas in the Sciences*, Cambridge, MA: MIT Press, 1987b.

Lunier, L., "Projet de statistique applicable à l'étude des maladies mentales arrêté par le Congrès Aliéniste International de 1867. Rapport et exposé des motifs," *Annales médico-psychologiques* [5] 1, 1869, 32–59.

Macalpine, I and R. Hunter, *George III and the Mad Business*, New York: Pantheon Books, 1969.

Matthews, J. *Quantification and the Quest for Medical Certainty*, Princeton: Princeton University Press, 1995.

Merz, J., *A History of European Thought in the Nineteenth Century*, vol. 2, Edinburgh: Blackwood, 1903, chap. 12.

Percival, T., *Medical Ethics, or a Code of Institutes and Precepts, adapted to the Professional Conduct of Physicians and Surgeons*, Manchester: S. Russell, 1803.

Porter, T., *The Rise of Statistical Thinking, 1820–1900*, Princeton: Princeton University Press, 1986.

Porter, T., 'Genres and Objects of Social Inquiry, from the Enlightenment to 1890', in: Porter, T. and Ross, D., eds., *The Cambridge History of Science, Volume VII: Modern Social Sciences*, Cambridge: Cambridge University Press, 2003, chap. 2.

Porter, T., "How Science Became Technical," *Isis*, 100, 2009, 292–309.

Ray, I., "The Statistics of Insane Hospitals," *American Journal of Insanity*, 6, 1849, 23–52.

Rusnock, A., *Vital Accounts: Quantifying Health and Population in Eighteenth-Century England and France*, Cambridge: Cambridge University Press, 2002.

Scull, A., *The Most Solitary of Afflictions: Madness and Society in Britain, 1700–1900*, Cambridge: Cambridge University Press, 1993.

Statistique de la France. Administration Publique, publiée par le Ministre de l'Agriculture et du Commerce, Paris: Imprimerie Royale, 1843, 305–370.

Stype, J. and Stow J., *A Survey of the Cities of London and Westminster, Containing The Original, Antiquity, Increase, Modern Estate and Government of those Cities, Written at first in the Year MCXCVIII by John Stow . . . Corrected, Improved and very much Enlarged . . . by John Stype*, 2 vols., London: Printed for A. Churchill, 1720.

Tomes, N., *A Generous Confidence: Thomas Story Kirkbride and the Art of Asylum-Keeping, 1840–1883*, Cambridge: Cambridge University Press, 1984.

Westergaard, H., *Contributions to the History of Statistics*, London: P. W. King and Son, 1932.

Worcester State Hospital, *Reports and other Documents relating to the State Lunatic Asylum at Worcester* (includes first to fourth annual reports) Boston: Dutton and Wentworth, State Printers, 1837.

Worcester State Hospital, *Nineteenth Annual Report of the Trustees of the State Lunatic Hospital at Worcester, Dec. 1851*, Boston: Dutton and Wentworth, State Printers, 1851.

Zeller, E., "Bericht über die Wirksamkeit der Heilanstalt Winnenthals vom 1. März 1840 bis 28 Febr. 1843, *Allgemeine Zeitschrift für Psychiatrie*, 1, 1844, 1–79 (reprinted from *Medizinisches Correspondenzblatt des Württembergischen Ärztlichen Vereins*, 13, no. 38, 1843, 297–322).

Chapter 19
The Question of Efficacy in the History of Medicine

Nathan Sivin

People who write about early medicine take a great deal of trouble to avoid addressing the question of efficacy. Some trustfully accept physicians' claims—Western or Eastern, ancient or modern.[1] Others dismiss reports of cures except for the few instances, mostly of drugs, in which biomedicine has validated the therapy. Physicians generally agree with the skeptics, shrugging off the rich literature of medical case records as mere anecdotes, by which they mean that none of it is worth thinking about. The utility of religious or other popular curing does not come up.

Surely one has to understand differently therapy that powerfully affects suffering and treatments that cannot cause significant change. Personal preference, whether that of the true believer or the inveterate skeptic, is not a productive basis for thinking critically. Clarifying what makes therapy effective can correct some common misunderstandings about health care before modern times—or, for that matter, at any time.

Yes, any time. If we want to find a reasonable way to approach the problem of efficacy, we have to set aside the double standard that puts the highest level of contemporary biomedical knowledge in one category and everything else in another. Anthropologists—when not restricting themselves to ethnology—sensibly demand that any general principle, if it is true of one culture, be true for all. If it is true of every culture but the one that invented academic anthropology, it can't be general. Concentrating on principles that hold for not only popular but scientific medicine, for not only ancient but modern medicine, seems to me a good way to avoid self-delusion. In exploring the meaning of

---

[1] This essay is part of a book in progress about the spectrum of health care in ancient China. I have been working on various avatars of it since the early 1970s, when I had the pleasure of working with Charles Gillispie on coverage of China for the *Dictionary of Scientific Biography*. I offer this in thanks for all I have learned from his scholarship.

N. Sivin (✉)
History and Sociology of Science, University of Pennsylvania, Philadelphia,
PA 19104-6304, USA
e-mail: nsivin@sas.upenn.edu

J.Z. Buchwald (ed.), *A Master of Science History*, Archimedes 30,
DOI 10.1007/978-94-007-2627-7_19, © Springer Science+Business Media B.V. 2012

efficacy, therefore, I will mention examples from both the modern United States and ancient China. My concern here is not to determine whether the therapeutic means of a given time and place are generally effective or not—a question too vague to be answerable—but to see how the issue of efficacy can be productive in studying the past, recent or distant.

19.1 What Does Efficacy Mean?

The sophisticated analysis in Daniel Moerman's *The Meaning Response* (2002, pp. 16–21), based on his field experience and his survey of a large literature in medicine and the social sciences, is an excellent starting point. Moerman, a medical anthropologist and authority on ethnopharmacy, makes it clear that efficacy in health care—the response of the patient to therapy—includes several distinct things.

What Moerman calls *the autonomous response* is the patient's ability to recover spontaneously. As Moerman puts it, abnormal conditions of the body tend to regress to the mean (see also Lock 2000). The distress due to a peak in blood pressure will ordinarily, before long, subside as the pressure becomes lower. He notes as well the general tendency of organic systems toward homeostasis. This response, for all of the primary-care physicians with whom I have discussed their work, makes patience (on their part and the patient's) the most effective resource for the self-limiting majority of everyday problems—colds, flus, muscular aches and pains, and common digestive problems.

Second is the patient's *specific response* to biological, chemical, or physical intervention. It is this sort of response on which public spokesmen for medicine base its claim to be an applied science. The centrality of the specific response in modern medical research is responsible for the primacy of the controlled, randomized double-blind test, the object of which is to exclude the effects of everything but specific agents. Clinicians often insist that if a drug is not specific, any therapeutic success can be due only to the placebo effect (e.g., Shapiro & Shapiro 1997, pp. 13–19; see also Csordas & Kleinman 1996).

Finally, the *meaning response* encompasses "the biological consequences of knowledge, symbol, and meaning"; that is, everything else that can affect the body's healing, from the physician's persuading the patient that the problem is curable, to the moral support of people around the patient, to the color of the prescribed pills (Moerman 2002, p. 4). It challenges the quasi-religious faith of medical hardliners that such matters may affect the patient's subjective illness, but can have no effect whatever on biological realities. This unwillingness to confront anything but the physical aspects of disease traps narrowly trained physicians in what Ann Harrington has called an "existential deficiency." It leaves them powerless to help patients with the non-biological components of suffering (Harrington 2008, p. 17).

The analytic power of medical anthropology and sociology comes from their systematic examination of ailment and cure in a very wide variety of cultures distributed in space and—to a smaller extent—in time. This broad perspective invalidates much of the conventional wisdom. Studies of these kinds find greatly varied patterns of belief among practitioners and patients. Patients' medical experience is not uniquely medical, but is part of, and one with, their experience of life as a whole. They respond to therapists, whether cardiologists or possessed mediums, in ways that are part of their responses to other people in general, and in particular to people with powerful knowledge or people whose status differs notably from theirs. These are social dimensions of the meaning response.

This implies that physicians engaged in therapy, however they may present themselves, are not scientists manipulating laboratory animals, but human beings interacting in complex ways with other human beings who sought them out because they were suffering and needed help. For common self-limiting ailments, their medicaments often matter less than their social, moral, or spiritual influence (Waldram 2000). In any system of health care, a practitioner who knows how to facilitate more than one of the three kinds of response is more likely to help patients than one who cannot.

As a corollary, those trained narrowly in such medical fads as evidence-based practice, and reliance on tests *rather than* examination of the patient's signs and symptoms and empathetic listening, are likely to be least successful except in acute emergencies for which specific therapy is known. As Kleinman & Sung put it, "…the physician is trained to systematically ignore illness [i.e., the patient's experience]. This represents a profound distortion of clinical work which is built into the training of physicians. It pays off on the application of biomedical technology to the control of disease, a less common but crucial clinical function, while it founders on the psychosocial and cultural treatment of illness, which is a much more common clinical function. Failure to heal illness is not articulated in the health professional's system of evaluating the efficacy of healing, but it is articulated in patient non-compliance and dissatisfaction, use of alternative health care facilities, poor and inadequate care. and medical-legal suits" (Kleinman & Sung 1979, p. 24). Once physicians enter practice, most of them cultivate their ability to listen attentively to their patients; but increasingly medical-school curricula ingrain attitudes that many are unable to overcome, above all reductive mindsets that are useful in laboratory research but obstacles to effective clinical care.

19.2 What Role Does the Placebo Effect Play?

Before proceeding further, let us look at efficacy in the long-term evolution of biomedicine. As a historian of medical professionalism has put it, "the peculiar history of physicians in Eurocentric societies provided a focus for understanding the role of the healer who was found in all cultures of which we have knowledge" (Burnham 1996, p. 23.). Peculiar though that history was, in most instances this has been the only focus that non-social-scientists have used. But the history of diseases, and of the placebo effect, is not so simple.[2] Let us first consider the use of placebo in studies of efficacy, which is by no means a recent innovation.

Blind assessment combined with placebo (medicaments known to be chemically inert) or other sham controls go back to 1784, T. J. Kaptchuk has shown, and were used frequently from then on. Out of this development, about 1955 randomized controlled trials evaluated by statistical tests of significance became the hallmark of clinical research that claimed scientific objectivity. Such trials have assumed that all therapy, no matter whether with effective drugs or inactive pills, incorporated "a monolithic effect which was present to the same degree and same direction in both the treatment and dummy arms … one could simply subtract the amount of the placebo effect to determine the presence (or absence) of specific drug effect." But no one has ever rigorously tested this curious assumption. Few clinical experiments have investigated the placebo effect itself. For obvious ethical reasons, almost no trials have incorporated a group of patients who receive no treatment at all to determine how the results differ from those of placebo.

In other words, what most researchers without further thought call the placebo effect is (in Kaptchuk's words) a hodge-podge of "nature taking its course; regression to the mean; routine medical and nursing care; regimens such as rest, diet, exercise, and relaxation; easing of anxiety by diagnosis and treatment; the patient-doctor relationship; classic conditioning and learnt behaviours; the expectation of relief and the imagination; and the will and belief of both patient and practitioner." It also includes the circumstances of research itself: "the method of recruiting patients, manner of giving informed consent, procedures for blinding, vehicle of delivery (colour of pills, pills *vs* injection), provider characteristics, provider verbal attitudes, and physical setting of the environment" (Kaptchuk 1998a, b, quotations from 1998b, pp. 1724–1725).

It is not hard to see why Moerman's "meaning response" is a better label than "placebo effect" for what results from this promiscuous and poorly-studied mixture. In fact, endless confusion has arisen because the placebo itself, an inert substance, cannot—at least from the pharmacodynamic point of view—be what causes the physiological changes (Moerman 2002, p. 94). They cannot be its specific effects. What causes them, in Kaptchuk's first few instances in the last paragraph, is the body's autonomous processes, and in the others, the personal meaning of the circumstances to the patient.

The thoughtless usage of "placebo effect" results from the collective irrationality of highly rational medical researchers. As David B. Morris has put it, "for over two hundred years the best scientific minds have steadfastly denied the bond between what we think and how we feel." That has changed gradually as

---

[2] "Placebo effect" is a woefully confused term, which some users restrict to the use of inert placebo as a drug and most use for the sum of verified effects that medicine cannot explain. See H. Spiro (1997, pp. 40–41).

the evidence against the complete segregation of conscious and unconscious mental processes has become undeniable. But many are still committed to the strict dichotomies of the eighteenth century. Some consider placebo "a kind of gratifying fraud, pleasant but useless"; some assume it does not work for any purpose except those already proven one by one in clinical tests; for others less narrow-minded, the question is not "whether placebos work but rather . . . *how* they work" (Morris 1997, pp. 188, 195).[3]

Some of the scanty clinical studies of the meaning response have separated patients who improve after treatment with inert drugs from those who do not, and have tested the two groups for individual differences in personalities. They find no significant correlation between personality traits and response. Moerman reports that the response to placebo has "no significant impact on the character and quality of meaning effects." Other tests reveal strong effects due to the *therapist's* "nature, character, personality, behavior, and style," especially intellectual and emotional interest in the patient's needs, and ability to persuade patients "that things will turn out well." In some studies, enthusiastic physicians have registered considerably higher therapeutic effectiveness than disinterested ones.

Studies of placebo, although few, have shown variations in effectiveness depending on the color of the inert pills, their size, their reputed price, the number taken per day, and whether the medicine was taken by mouth or injected. Sham surgery works even better than injection, and "high-powered machines with snappy names . . . may be at the top of the heap."

The reasons postulated for these patterns have consistently had to do with meaning. For instance, blue placebo pills generally have been more effective as sedatives than as stimulants. But this was not true in studies of Italy in the 1970s. Blue substitutes for sleeping pills worked nicely for women, who tended to associate blue with the robes of the protective Virgin Mary, but not for men, who connected the color with the uniforms of the national soccer team, Azzurri, and thus with powerful movement and excitement (Moerman 2002, pp. 35–49).

Because ways of experiencing the world differ from one culture to another, the effect of the meaning response on accepted patterns of medical disorder varies greatly. For instance, although U.S. medical manuals assume that menopause is common to all women, purely biological in origin, and needs treatment, in Japan and other countries its symptoms are infrequent or absent (Lock 1994).

The very frequent diagnosis of Attention Deficit Hyperactivity Disorder (ADHD) in the school systems of many American states—and the resulting use of psychotropic drugs, commonly without examination by licensed physicians—divides the United States from the rest of the world. The most pertinent difference is the expectation of many U. S. schools that young children sit still for long periods and otherwise exhibit considerable self-discipline. This demand is often at odds with the early training of the same children by their parents.

---

[3] Morris represents the third stance. For examples of the first and second in the same volume see, respectively, Shapiro & Shapiro and Price & Fields.

Some couples encourage freedom and creativity (whatever that may mean to them) over predictable behavior, and others simply make little effort to teach self-control. Circa 1980, U.S. medical manuals attributed ADHD to minimal brain dysfunction, but neurological examination so rarely played a role in diagnosis and dosing that this reductive etiology is now rarely mentioned.[4]

Moerman's survey of drug experiments for ulcers and hypertension show great national differences between the U.S., Germany, and Brazil, for both drug and placebo. The variation in placebo effectiveness, as always, "carries over to the drug healing rate" (Moerman 2002, pp. 72–83).

These are only examples of what little we can infer so far about the meaning response in disease and therapy. The biomedical view of the body as a complex of purely biological, chemical, and physical systems has discouraged inquiry into its non-material aspects. Few clinical researchers challenge the conventional wisdom that dichotomizes mind and body, the psychological and the somatic, pain sensation and pain affect, and so on.[5]

Psychosomatic medicine, the enterprise that aimed to overcome this either-or approach to mind and body, is now only a distant memory. Research in it flourished from ca. 1930 to the 1950s, and it even gained a modest role in educating medical students, but from that decade on it disappeared from curricula—and with it much of the flexibility that might have integrated Moerman's varieties of response. It was a casualty of the drive to recast medicine into the mold of physical science. This successful campaign transformed what it discarded into mere psychological factors or cultural peculiarities, and relegated them to "illness but not disease, to therapy but not etiology, to symptoms but not pathology, and finally to the course of the disease but not to its cause."[6]

The picture I have just sketched indicates that an adequate account of therapeutic efficacy depends on three kinds of bodily response to medical disorder: the body's autonomous ability to recover normal functioning, its specific response to biological, chemical, and physical agents, and the response of the individual to interaction with other people, which depends on the quality and meaning of that interaction. The combination of these three responses is useful in understanding any therapy in any time and place.

---

[4] The classic study is R. A. Rubinstein & R. T. Brown (1984). Lock & Scheper-Hughes (1996, p. 67), discuss ADHD as an instance of recasting social frictions and miseries as "individual pathologies rather than as socially significant signs." On current National Institutes of Health doctrine re ADHD, see Medline Plus at http://www.nlm.nih.gov/medlineplus/attentiondeficithyperactivitydisorder.html (accessed 2011.10.11). For other examples of important discrepancies in diagnosis and therapy in the U.S. and Western European countries see Payer (1988). This meritorious book is out of date, but no one has done a comparable survey since.

[5] See, for instance, Price & Fields (1997), which even takes the last dichotomy seriously (pp. 133–134).

[6] Aronowitz (1998), p. 51, in a historical analysis of ulcerative colitis. On p. 52 Aronowitz lists several reasons for the demise of psychosomatic medicine.

19.3 Efficacy for Which Disorders?

It is natural to assume that what curers in other times and places contended with were the same diseases that physicians diagnose today. If we affirm that biomedicine provides infallible knowledge of every time and place, that is an obvious corollary. Nevertheless, medical therapy treats human beings in societies. Thus comprehending it depends on our understanding—still all too fallible—of thought and feeling, of culture, of the ways people interact, and of how the varieties of human suffering are related to one another.

It is no news that diseases vary according to place, change over time, and come into being and pass away. Typhus and typhoid are now rare enough that American clinicians are seldom able to diagnose them. Paralytic polio is no longer a constituent of U.S. health statistics. I have already mentioned the precipitous rise of Attention Deficit Hyperactivity Disorder (ADHD) in those statistics. Drunkenness, a "despised although tolerated moral condition" (formerly a crime in many localities), became the medical disorder alcoholism within living memory (Stein 1973, p. 367). Even more recently, senile dementia has metamorphosed into Alzheimer's disease, a term that doctors outside the U.S. still follow Alzheimer in using it for the rare pre-senile version. These instances are all the more remarkable because medicine cannot explain the origin of any of these diseases, prevent them, or cure them. New organic diseases, such as Ebola hemorrhagic fever, Lime disease, and HIV AIDS, appear with some regularity.

When we look some distance into the past, the picture becomes even more obviously discrepant. Fever today is elevated body temperature, a mere symptom that the patient may or may not feel, but that the physician measures to a tenth of a degree. In the U.S. in the late nineteenth century fever was a disease with many symptoms, including cold sensations and elevated pulse rate. In a diagnostic handbook very widely used in the early 20th century, it included even more symptoms. Although the well-trained, experienced biomedical physician of 1930 likely felt that his knowledge and habits of reasoning held for all time, as we look back from the present day his limitations are all too obvious (Dunglison 1874, p. 416; Savill 1930, p. 498; Sivin 1987, pp. 107–109).

The same is true when we look at other cultures. The definition of *re* 熱 in dictionaries of modern Chinese is "fever." Indeed that is what it means to Chinese M.D.'s. Most practitioners of Traditional Chinese Medicine, trained from the 1950s on using textbooks that aimed to impose science on old ideas, also now understand it as fever. But in classical medicine its meaning was quite different. Authors paired *re* with *han* 寒 as antonyms. *Han* meant cold feelings within the body—a symptom that the patient reported, not a physical sign—and *re* meant hot sensations of exactly the same kind; the combination *hanre* 寒熱 meant alternating hot and cold sensations. In some sources before modern times a different word, *fare* 發熱, meant the dynamic manifestation of heat at the periphery of the body—which is not the same thing as elevated temperature there or within the body. The ancient roster of diseases, and even of symptoms, not only differed greatly from the constituents of modern nosology, but changed regularly.

Let me give an example. Hilary Smith (2008) has studied in considerable detail the repeated metamorphoses of the medical disorder *jiaoqi* 腳氣 through Chinese history. Dictionaries of the modern language list its meaning unambiguously as "beriberi," and that is how most historians of medicine have understood it. But that definition that held true only for a limited period in the late nineteenth and twentieth centuries. *Jiaoqi* began as an often fatal disease and in time became a group of diseases, repeatedly changing its great range of symptoms. In modern times, for biomedical physicians, it became beriberi, but it is now increasingly athlete's foot! To practitioners of classical medicine over the centuries, even as its manifestations proliferated and changed, it remained—like fever in early modern Europe—one disease.

Competent modern-day physicians are sensibly reluctant to diagnose diseases in people they have not examined. Andrew Cunningham, in a recent study that questions "the continuous identity of past diseases with modern diseases," has looked closely at the pitfalls. In England, as he demonstrates for several points over the past four hundred years, the categories used for reporting causes of death differed fundamentally from those that physicians today are qualified to choose between. The law for much of this time designated bystanders with no medical education to decide for the record what disease killed someone. In 1672 the requirement was "two ancient Matrons" appointed as Searchers; in the act of 1836 it was "some person present at death, or in attendance at the last illness." It was the sum of such reports that entered official records and statistics. Looking closely at "how diagnosis happens," Cunningham argues that "the identity of any disease is made up of a compound of elements, of which the biological or medical is only one, and sometimes the least important." With due attention to the social dimension, he demonstrates that human decisions regularly change disease entities. The notion that reality lies elsewhere than in how witnesses identified these entities in the language of a given time and place may be unimpeachable as mystical conviction, but has no value for historical study (Cunningham 2002, pp. 13, 16).[7]

To take the example just given, someone familiar with modern nosology may well notice that in the sixteenth century several symptoms of *jiaoqi* resemble those of modern gout, but to say that *jiaoqi* at the time *was* gout is an elementary error. It is easily corrected by citing its other symptoms that had nothing to do with gout (Smith 2008, pp. 183–186). To understand how the two differ is, as Cunningham puts it, a matter of studying historically the *social process of diagnosing them*. His point extends perfectly to eleventh-century China, where the diseases were those that a wide variety of curers recognized, and where the

---

[7] For significant national differences in diagnosing causes of death in the late twentieth century see Payer (1988, p. 25).

recorded cause of death was determined less often by a therapist than by the family of the deceased.

Instead of beginning with a biomedical disorder, let us take the case of someone ca. 1000 whom a classical physician diagnoses with Pulmonary Abscess Disorder (*feiyong* 肺癰). The standard reference for nosology was still a handbook completed in 610. The widely cited descriptions in that book provide a common denominator for cases where, as usual, we know nothing about a doctor's own classification. According to *Zhu bing yuan hou lun*, this ailment is due to damage to the lung functions caused by wind and cold factors when the *qi* 氣 vitalities are depleted. The cold factors, contending with the yin vitalities (*xue* 血), congeal them to form a hidden abscess. If hot factors are also involved, the result is likely to be purulent, bloody sputum often accompanied by trembling (Chao Yuanfang 巢元方 610, *juan* 33, pp. 177b-178a).

The M.D. without a body to examine but willing to diagnose words will perhaps conclude that the patient is ill with pulmonary tuberculosis or another ailment characterized by hemoptysis. But it is not that simple. Instead of merely conning a list of symptoms, as Cunningham reminds us, we need to ask about the process of diagnosis. Were there other options, who chose, and how?

There were in fact a great many alternative diagnoses. There is a very large class of Depletion Exhaustion Disorders (*xulao* 虛勞) that any doctor would also have considered. Among the seventy-five disorders that it comprises are several obvious prospects, for instance Depletion Exhaustion with Hot Sensations (*xulao re* 虛勞熱), Depletion Exhaustion with Alternating Cold and Hot Sensations (*xulao hanre* 虛勞寒熱), Depletion Exhaustion with Bone Steaming (*xulao guzheng* 虛勞骨蒸), Depletion Exhaustion with Upset and Inner Tension (*xulao fanmen* 虛勞煩悶), Depletion Exhaustion with Vomiting due to Reversed Flow of Yin *Qi* and with Spitting Blood (*xulao ouni tuoxue* 虛勞嘔逆唾血), and Depletion Exhaustion with Vomiting Blood (*xulao ouxue* 虛勞嘔血), which vary not only by symptoms but by the etiologies their names reflect.

Other likely disorders occur among the thirty-four consumptive disorders in the class of Possession Disorders (*zhu* 注, 疰), also due to *qi* depletion and exhaustion, as well as to possession by the *qi* of ghosts, which can be transmitted from a dying patient to a bystander.[8]

Every one of these more than a hundred distinct pathological entities has its own list of symptoms and indications, often as long as that of Pulmonary Abscess Disorder. An M.D. willing to accept some symptoms and blankly ignore others may decide that all are simply names for pulmonary tuberculosis or some other modern entity, but that is too high-handed a procedure and too inattentive a method for anything but a parlor game. A physician who insists on

---

[8] For *xulao*, see *Zhu bing yuan hou lun, juan* 3, p. 17a–*juan* 4, p. 27b; *xulao re, juan* 3, p. 19b; *xulao hanre, juan* 3, p. 22b; *xulao guzheng, juan* 4, pp. 23a–23b; *xulao fanmen, juan* 4, p. 23b; *xulao ouni tuoxue* and *xulao ouxue, juan* 4, p. 24a; possession disorders, *juan* 24, pp. 130a–134b.

understanding past experience is likely to concur with the historian in taking all of the sources—and their nosology—seriously.

In fact, modern medical knowledge, if combined with careful attention to exactly what the original records say rather than substituted for the recorded concepts, can be indispensable in study of distant times and places. An excellent instance is the long-debated death of the nineteen-year-old Tongzhi Emperor on 12 January 1875. Generations of scholars have debated whether it was due to smallpox or syphilis, convinced that they could ignore the cause of death in his medical records, *zouma yagan* 走馬牙疳 (literally, "galloping gum ulcerations"). A recent study has reconstructed from the extremely detailed accounts of the emperor's medical care a remarkable denouement.

On 9 December 1874, the emperor was indeed diagnosed with *douzhen* 痘疹, a term that at the time coincided fairly closely with what English now calls smallpox. Within a week his condition had stabilized, and in less than two his doctors announced that recovery was well under way. He decided unilaterally to stay in bed for a hundred days to ensure complete recovery. But by 23 December he had developed large purulent sores about his waist, and these spread over his back and arms until by 7 January 1875 his face and gums swelled and he developed grave infections in his mouth and gums. As they worsened, he died on 12 January.

The historians never settled their long-running debates, although with publication of the palace medical archives it became clear that there had been no symptoms of syphilis, and that the emperor recovered quickly from smallpox. Finally in 1998 Chang Che-chia plausibly explained for the first time what the archives and other important sources revealed. "Galloping gum ulcerations" were indeed what killed the young ruler. What led to them were symptoms well known in Europe, but that Chinese physicians at the time did not include among their disorders, namely bedsores. It took the whim of a youthful emperor to stay in bed for six weeks, a circumstance that so far as is known did not otherwise occur under the eyes of an imperial physician. That allowed the infection to become systemic. His several outstanding physicians, with no experience of bedsores, were flummoxed—and he died (Chang Che-chia 1998, pp. 85–120). Modern knowledge can be most helpful in sorting out the ambiguity of ancient records, but only after their content is well understood.

19.4 Conclusions

Any investigation of curing, no matter where or when, may benefit if the inquirer knows aspects of biology, chemistry, and physics, and of clinical practice. But translating traditional accounts of disease directly into biomedical knowledge leads to grossly inadequate understanding. We have seen that therapy depends as greatly on the body's tendency toward homeostasis, and on the patient's response to social and personal interaction, as it does on technology—and frequently more so. This implies that the present-day physician's humanistic cultivation may play as large a role as her mastery of pathology.

We have also noted that there is no consensus within the medical profession about the ill-assorted gaggle of phenomena usually lumped together as "the placebo effect." Some medical spokesmen use that term to dismiss all scientifically unexplainable curing; others take it seriously but are mystified by it; a very few have thrown odd rays of light on it. Practically no physician is inclined to undertake research directed toward understanding it. In any case the notion is too vague to be useful in historical investigations. A more productive method is to follow Moerman in pondering the role of the meaning response—of the curative power of meaning—in every sort of therapy.

We can deduce from the discussions above several aids useful in exploring the varieties of curing in all times and places:

- Although every kind of health care is unique, we can take as a working hypothesis that, in a given time and place, all practitioners are likely to share some notions of the causes and character of disease, even as they disagree about the rest. To assume that, say, religious curing has nothing in common with the classical medicine of the scholarly doctors is counterproductive. What in a given instance they have in common is an empirical question worth asking.
- The way curers and patients evaluated the strengths and weaknesses of medical care in their time and place can vary greatly, and always calls for attention. The criteria of modern medicine (for instance, the ability to deal with acute emergencies and serious traumata) are, in historical studies, likely to be irrelevant. What criteria are pertinent is also an empirical question.
- Every culture slices up the spectrum of suffering in its own way. It can be seriously misleading to assume that the diseases, or even the symptoms, that people experienced long ago or in a different culture are the same as those of the U.S. *or China* today.

What we learn from open-minded but critical study can throw light not only on the medical experience of ancient times and faraway places, but on current assumptions about how to improve health care and promote wellness.

Bibliography

Aronowitz, R. *Making Sense of Illness. Science, Society, and Disease.* Cambridge History of Medicine. Cambridge: Cambridge University Press, 1998.
Burnham, J. "Garrison Lecture: How the Concept of Profession Evolved in the Work of Historians of Medicine," in: *Bulletin of the History of Medicine,* 70 (1996), pp. 1–24.
Chang, Che-chia. "The Therapeutic Tug of War. The Imperial Physician-Patient Relationship in the Era of Empress Dowager Cixi (1874–1908)." Ph.D. dissertation, Asian and Middle Eastern Studies, University of Pennsylvania, 1998.

Chao Yuanfang 巢元方. *Zhu bing yuan hou lun* 諸病源候論 (Origins and symptoms of medical disorders), 610. Reprint, Beijing: Renmin Weisheng Chubanshe, 1955.
Csordas, T., & Kleinman, A. "The Therapeutic Process," in: Sargent, C., & Johnson, T., editors. *Medical Anthropology: Contemporary Theory and Method.* Rev. ed. Westport, CT: Praeger, 1996, pp. 3–20.
Cunningham, A. "Identifying Disease in the Past: Cutting the Gordian Knot," in: *Asclepio,* 54 (2002), pp. 13–34.
Dunglison, R. *Medical Lexicon: A Dictionary of Medical Science: Containing a Concise Explanation of the Various Subjects and Terms.* New edition. Philadelphia: Henry C. Lea, 1874. First published 1839.
Harrington, A., editor. *The Placebo Effect. An Interdisciplinary Exploration.* Cambridge, MA: Harvard University Press, 1997.
Harrington, A. *The Cure Within. A History of Mind-Body Medicine.* New York: W. W. Norton, 2008.
Kaptchuk, T. "Intentional Ignorance: A History of Blind Assessment and Placebo Controls in Medicine," in: *Bulletin of the History of Medicine,* 72 (1998a), pp. 389–433.
Kaptchuk, T. "Powerful Placebo: The Dark Side of the Randomised Controlled Trial," in: *The Lancet,* 351 (1998b), pp. 1722–25.
Kleinman, A., & Sung, L. "Why Do Indigenous Practitioners Successfully Heal?," in: *Social Science and Medicine,* 13B (1979), pp. 7–26.
Lock, M. *Encounters with Aging. Mythologies of Menopause in Japan and North America.* Berkeley: University of California Press, 1994.
Lock, M. "Accounting for Disease and Distress: Morals of the Normal and Abnormal," in: Albrecht, G., Fitzpatrick, R., & Scrimshaw, S., editors. *Handbook of Social Studies in Health and Medicine.* London: Sage, 2000, pp. 259–76.
Lock, M, & Scheper-Hughes, N. "A Critical-Interpretive Approach in Medical Anthropology: Rituals and Routines of Discipline and Dissent," in: Sargent, C., & Johnson, T., editors. *Medical Anthropology: Contemporary Theory and Method.* Rev. ed. Westport, CT: Praeger, 1996, pp. 41–70.
Medline Plus. http://www.nlm.nih.gov/medlineplus/attentiondeficithyperactivitydisorder. html (accessed 2010.10.21).
Moerman, D. *Meaning, Medicine and the "Placebo Effect."* Cambridge Studies in Medical Anthropology. Cambridge: Cambridge University Press, 2002.
Morris, D. "Placebo, Pain, and Belief: A Biocultural Model," in: Harrington 1997, pp. 187–207.
Payer, L. *Medicine & Culture: Varieties of Treatment in the United States, England, West Germany, and France.* New York: Henry Holt, 1988.
Price, D., & Fields, H. "The Contribution of Desire and Expectation to Placebo Analgesia: Implications for New Research Strategies," in: Harrington 1997, pp. 116–37.
Rubinstein, R., & Brown, R. "An Evaluation of the Validity of the Diagnostic Category of Attention Deficit Disorder," in: *American Journal of Orthopsychiatry,* 543 (1984), pp. 398–414.
Savill, T. *A System of Clinical Medicine: Dealing with the Diagnosis, Prognosis, and Treatment of Disease: For Students and Practitioners.* 8th edition. New York: William Wood, 1930. First published 1903–1905.
Shapiro, A., & Shapiro, E. "The Placebo: Is It Much Ado about Nothing?" in: Harrington 1997, pp. 12–36.
Sivin, N. *Traditional Medicine in Contemporary China.* Science, Medicine, and Technology in East Asia, 2. Ann Arbor: Center for Chinese Studies, University of Michigan, 1987.
Smith, H. "Foot *Qi:* History of a Chinese Medical Disorder." Ph.D. dissertation, History and Sociology of Science, University of Pennsylvania, 2008.

Spiro, H. "Clinical Reflections on the Placebo Phenomenon," in: Harrington 1997, pp. 37–55.
Stein, H. "Ethanol and Its Discontents: Paradoxes of Inebriation and Sobriety in American Culture," in: *Journal of Psychoanalytic Anthropology,* 5 (1973), pp. 355–77.
Waldram, J. "The Efficacy of Traditional Medicine: Current Theoretical and Methodological Issues," in: *Medical Anthropology Quarterly,* 14 (2000), pp. 603–25.

Part VIII
Science and Industry in France

Chapter 20
Secrecy, Industry and Science. French Glassmaking in the Eighteenth Century

Marco Beretta

> *Far more significant for civilization and culture than progress in the metallurgical arts up to the eighteenth century was the great advance in glass making.*
> Lewis Mumford, Technics and Civilization (1934)

Glass has always played a crucial role in directing the investigation of natural phenomena onto revolutionary and innovative paths. Since antiquity philosophers of nature perceived the peculiar quality of this material and their increasing attention to the extraordinarily rich and varied productions made by craftsmen after the introduction of glassblowing techniques reveal the special status glass had in the economy and culture of the Roman Empire. Because of its chemical nature, large scale production of glass was a difficult technological process which required, in addition to dexterity, a broad knowledge of different operations and devices. The fusion of glass requires high temperatures that can only be obtained with the construction of special furnaces and crucibles, combined with the accurate use of salts, namely soda and potash, which enables to decrease the degree of glass' fusion. The amorphous nature of glass, which partly explains its frailty, requires the addition of stabilizing ingredients, such as lime, as well as special treatments that make the whole process of glass production a lengthy and costly technology. With the addition of metal oxides glass may assume any color and through the expert control of temperature it may imitate all sorts of gems and minerals. Glass furnaces must be kept active, day and night, most of the year and following the introduction of glassblowing in the first century BCE, the large capital that had to be invested in a workshop, necessarily turned glassmaking into an industry rather than into a craft. The

An earlier version of this chapter has been presented at the workshop "Artisanal-Scientific Experts in Eighteenth-Century France and Germany", Max Plance Institute for the History of Science, Berlin, October 22–23, 2010. I wish to thank Bruno and Jean-François Belhoste and Ursula Klein for their remarks and suggestions.

M. Beretta (✉)
University of Bologna, Bologna, Italy
e-mail: marco.beretta@unibo.it

J.Z. Buchwald (ed.), *A Master of Science History*, Archimedes 30,
DOI 10.1007/978-94-007-2627-7_20, © Springer Science+Business Media B.V. 2012

complex structure of glassmaking is evidenced by the fact that during the Middle Ages, when the supply of alkali, silica, and combustibles were no longer easily available, only a few European centres of production survived and up to the twelve century glass became a commodity as rare as in pre-roman time.

In order to cover the initial investment, glassmakers were particularly keen to protect their techniques and innovations with strict secrecy codes. As I have argued elsewhere, Ancient glassmakers contributed to the emergence of alchemy (Beretta, 2009). The chemical ambiguity of glass was well known to ancient alchemists and rather than seeing it as puzzling problem, they envisaged the possibility of producing unlimited chemical transformations. What now appears as an unstable and yet solid arrangement of molecules, in ancient times evoked the hidden virtues of matter and, at the same time, the marvelous potentials of chemical manipulations of ingredients that, brought to a glassy state, could take nearly any natural form. The chemical nature of glass did indeed fit the ancient philosophies of matter and the belief in the possibility of transmuting one substance into another. At the same time, the extraordinary possibility of producing ever new commodities forced glassmakers to protect their inventiveness with secrecy. This tendency was institutionalized with the creation of guilds and it is interesting to note that the municipal authority of Venice, the most important European centre of glass production of the Middle Ages, was extremely severe when it came to cases where the secrecy of the guild were exposed. Thus, innovation was kept under the rigid control of the guilds and the circulation of reliable recipes for glass production was limited to the production of ordinary glassware. Unlike other metallurgical arts, no textbook on glassmaking circulated before the appearance of Antonio Neri's *L'arte vetraria*, published in 1612. Neri's work was not in fact, as it is often credited, a technical textbook, but a revised and updated collection of recipes designed to connect glassmaking with alchemy.

The aura of secrecy surrounding glassmaking was partly due to its close connections with the alchemical tradition, partly due to the guild organizations but it also had to do with the complexities of the chemical operations, the devices involved and the costs of innovation. Also because of these reasons, the transfer of knowledge in glassmaking was an incredibly difficult procedure and I do not know of any other craft in which the know-how of ancient techniques has been so frequently irreversibly lost.

In France glassmaking arrived relatively late and in Paris the first mention of a guild of glassmakers came as late as 1292 (Lespinasse, 1892, pp. 745 ff.) with the first statutes regulating its activities issued for the first time only in 1467. By 1292, out of a total population of ca. 250,000, the approximately 130 Parisian guilds employed some 5000 craftsmen, of which only 17 were glassmakers (Saint-Léon, 1922, pp. 219 ff.).

Things changed rapidly. By the middle of the sixteenth century, glass was not exclusively produced by the guilds of the glassmakers (*vitriers* and *faenciers*) but also by a new guild which included the makers of mirrors, spectacles and the opticians (*mirotiers, lunetiers, opticiens*) who played an important role in the specialization of French glassmaking. Interestingly already the statutes of 1467 refers to glassmaking both as a "mestier" as well as a "science" (Lespinasse, 1892, p. 747) and it worth to note that, a few decades later, Bernard Palissy made his career in science and alchemy as a glassmaker. The ennoblement of glassmaking is further evidence in the fact that during the seventeenth century, the aristocracy was granted the exclusive privilege to create and direct a glass manufacture.

Despite its social prestige, French glassmaking struggled for several centuries and its was only by the end of the eighteenth century, when the guilds were abolished in 1791, that it managed to compete with the more flourishing Italian, English and German manufactures. Yet, the history of French glassmaking is of great interest because it illustrates in a very clear way the efforts to overcome the rigid structure of the guilds system and the traditional separation between the artisans and academic scientists (Scoville, 1942).

During the second half of the seventeenth century the guild system, which favoured the activities of small workshop, was progressively challenged by the creation of ever larger manufactures which, thanks to the large funds granted by bankers and financiers, attracted the attention of the Crown. Colbert's economic and cultural reforms are at the origin of these efforts (Mahoney, 2010). The foundation of the Royal manufacture of Saint Gobain (Frémy, 1909; Hamon and Mathieu, 2007), which eventually became the model of Sèvres and the Gobelins, was inspired by Colbert's effort to challenge Venice's technological and economic marked superiority in the production of large glass mirrors. Once the capital was ensured, technological expertise was the first need for a successful enterprise and in the autumn 1664, Colbert, through an aggressive strategy, managed to employ, not without risks, a community of glassmakers from Murano and Altare. By a necessity or by caprice, the guild system was now openly challenged and the impossible became possible: the secret knowledge of the experts was for sale and could be successfully transferred to countries which lacked distinguished tradition for glassmaking.

The Manufacture Royale des Glaces was founded in Paris in 1665, one year before the foundation of the Académie Royale des Sciences, and in 1693 its premises were transferred to Saint-Gobain. The most evident sign of the progress achieved in French glassmaking, was the furnishing, during the early 1680s, of the large mirrors adorning the Chateau de Versailles. The successful substitution of glassblowing with the innovative technique of casting glass to produce ever larger glass mirrors was a typical example of the advantages provided by the scale of French manufactures: their unprecedented dimensions made possible technological innovations which were economically too risky for small workshops. Significantly, this particular innovation was transferred during the first half of the eighteenth century in the production of laminated lead. However important these new techniques were, the privileges accorded to Saint-Gobain and Colbert's proverbial centralism hindered competition and the subsequent circulation of knowledge.

Despite this shortcoming, Colbert's policy of reforms was characterized by his constant efforts to improve the role of expert knowledge in the development of French arts and manufactures. In the case of glassmaking the crucial role played by a deep and up to date knowledge of chemistry was progressively coming to the forefront. The interaction between crafts, manufactures and theoretical science was in fact on Colbert's agenda of economic reforms. To fulfil his ambitious plan, the Ministry of finance commissioned to the members of the Académie des Sciences in Paris the realization of a *Description des arts et métiers.*

Apart from the late and posthumous publication (1774) by Pierre Le Vieil's *L'art de la peinture sur verre et de la vitrerie* (Gillispie, 1980, pp. 350 ff.), it is rather surprising that the art of glassmaking was not included in the project and it was only at the end of the eighteenth century, when both the project and the Académie were close to an end, that Antoine Laurent Lavoisier asked the engineer Pierre Loysel to prepare an up-to-date treatise on the subject (Goupil, 1993). Before Lavoisier, however, several academicians had been conscious that glassmaking was an extremely important chemical art and that much of its progress depended on the improvement of the techniques, the chemical operations and their theoretical understanding. The first to pave way in this direction was René-Antoine Ferchault de Réaumur who, as early as 1709, systematically began to collect material for the *Description* and engaged with an extraordinarily rich succession of inventions and discoveries connected with the chemical arts. Réaumur's work devoted to the preliminary production of the *Description des arts et métiers,* emphasized a utilitarian vision of science which made him an exceptional figure in the Académie des Sciences. In a draft devoted to the usefulness of applied sciences presented to the Académie between 1716 and 1727, Réaumur pointed out the central role played by chemistry in the growth of several arts in the following terms: "Chemistry, whose investigations seems rather frivolous to those who do not know its true purpose, could become on the of the most useful parts of the Academy. [...] The conversion of iron into steel, the method of plating or whitening iron to make tin-plate, the conversion of copper into brass, three great industries which the Kingdom lacks, are in province of chemistry. It is the business of chemistry, also, to investigate the mineral substances used in dyeing and ores and minerals. Glass-works, pottery works, faïencers, porcelain – industries which all need to be improved – also concern it." (Réaumur, pp. 104–105).

In spite of his utilitarian inclination, Réaumur did show a remarkable interest for the theoretical implications entailed in the improvement of technology and it was this perceptive attitude towards applied science that made him appreciate the role of craftsmen.

The fact that one of the most important inventions made by Réaumur is related to glassmaking has often passed unnoticed. In fact, the intersection of industrial entrepreneurship, experimental skills and pure research are highlighted in Réaumur's successful imitation of Chinese porcelain with white opaque glass in 1740. The extraordinary impact of this invention stimulated further chemical research on the properties of glass at the Royal manufacture of Sèvres and within the Académie des sciences. It is significant in this respect that an academic chemist, Jean Hellot, was called in 1751 to act as a technical adviser and that Pierre Joseph Macquer substitute him in this position after his death in 1766 (Guerlac, 1959; Gillispie, 1980, pp. 390–406). It is in this context that professional chemical expertise began to be an essential component of French glassmaking. If at the beginning of the eighteenth century the production of a glass manufacture was confined to ordinary glassware, such as bottles and glasses, already around 1750 the broadening of the chemical expertise became increasingly a necessity. The extensive use of glassware in many scientific disciplines such astronomy, meteorology, experimental physics and, not the least, in chemistry pushed for the production of more sophisticated products.

A typical example of this evolution is the introduction of flint glass in France. Although the chemical role of lead used for making glass similar to crystal and to other transparent stones had been known since the Renaissance and several recipes had already been published in Neri's treatise, George Ravencroft understood that by using lead oxide and by multiplying the quantitative tests, the quality of the artifacts could be considerably improved. Several decades of work were necessary to perfect a relatively standardized and economic method for producing flint glass, but Ravencroft's efforts paved the way for an approach to glassmaking which emphasized the central role of chemical analysis. It is indeed not surprising that between the end of the seventeenth and the first half of the eighteenth century, professional chemists were involved in systematic research which had a direct connection to glassmaking. Ravencroft's discovery of flint-glass made it clear that ingredients were to be examined more systematically. The French centralized efforts, promoted by the Académie and the most distinguished chemists of the eighteenth century, to create a manufacture of flint glass were not successful but they were nevertheless conductive to an increasingly closer relations between academic scientists and the artisans. During the visit to the newly constructed Manufacture de cristaux de la Reine at Moncenis, a delegation of the Académie, composed by Lavoisier, Fourcroy, Vandermonde and Monge, was impressed by the size of the furnaces and first attempts to use coke as a combustible, but their correspondence with the director of the manufacture, Ignace De Wendel, remained without any fruitful results (Lavoisier, 1993, pp. 86–91).

Despite these difficulties, the collaborations between instrument makers, academic chemists and glass manufacturers became tighter and at times successful. This is the case, for instance, of the production of large burning lenses and the solution of major technological challenges, such as the casting of a large pane of thick and perfectly transparent glass with a given convexity. While the use of the burning lenses became common among chemists already during the second half of the seventeenth century, until 1774 the members of the Académie des Sciences in Paris used the lens (89 cm. of diam.) acquired by the Philippe d'Orléans in 1699 from the German alchemist Ehrenfried-Walther Tschirnhaus (Smeaton, 1987). After the resistance in the mid 1740s of Saint-Gobain's manufacturers to engage in a costly and apparently impossible production of a large burning lens, Buffon lamented that nobody in France was able to make thick glass as transparent as that produced in Bohemia (Buffon, 1748, p. 309). Only in 1774 Claude de Bernières, Controleur Général des Ponts et Chaussées, made at the manufacture of Saint Gobain a large lens which, following Buffon's instructions, could be filled with water or alcohol. Bernières' lens was filled the lens with alcohol. It and consisted of two segments of a glass sphere of a 2,60 m. radius, sealed together to leave a space 1,30 m. in diameter and 17 cm of thick space in-between filled with alcohol. The lens, assembled by the technician of the Louvre, Charpentier, was acquired by Jean-Charles-Philibert Trudaine de Montigny and used by Lavoisier, Macquer, Cadet and Brisson in the Jardin du Louvre during the month of October. The results, partly hindered by the unfavorable season, were limited to the fusion of a few metallic specimens. Yet the laborious construction of this apparatus showed that the glassmakers of Saint Gobain, in collaboration with chemical experts and academicians, could finally produce what little more that twenty years earlier was regarded as an impossible technological achievement.

During the second half of the eighteenth century, the intersections of different kinds of chemical expertise were at the origin of another crucial progress in glassmaking. The chemical role of soda and potash as fondants was either poorly understood or kept secret. In his *Specimen Beccherianum* (1703), Georg Ernst Stahl identified the chemical differences between soda and potash and established that only the base of the marine salt was a natural alkali (Stahl, 1738, p. 139). A few years later, Andreas Sigismund Marggraff found a method for distinguishing between soda and potash that he named respectively "fixed mineral alkali" and "fixed vegetal alkali." (Véron, 2009, pp. 14–15). At the end of the century, chemists throughout Europe discovered at least eight methods for producing soda artificially from common salt (Gillispie, 1957) and the Académie Royale des Sciences de Paris actively engaged in promoting innovative research in this field, and more generally in glassmaking, by dedicating four prizes (in 1758, 1766, 1781, 1786 and 1788) to it. Thus, the success of Nicolas Leblanc's process at the end of the century was the culmination of decades of intense research involving chemists and experts. The dialogue was difficult. While chemists always appreciated the value of the arts in their theoretical works, experts and craftsmen were still rather suspicious of the academic world.

One final example will clarify my point. During the eighteenth century Paul Bosc d'Antic was among the most successful researchers in French glassmaking. The rise and fall of his career offers a good example of the radical transformation of glassmaking during the Enlightenment and of its marked separation from the alchemical and guild tradition.

Trained as a physician in Montpellier, d'Antic became interested in chemical technology in Paris while attending Réaumur's and Nollet's courses. Whilst he was well acquainted with the academic science, Bosc d'Antic soon became a specialised glass technician and it is possible, but as far as I know not documented, that this specialization was guided by Réaumur's interest in the subject.

Sent by the Académie Royale des Sciences to direct the glass manufacture of Saint Gobain in 1755, Bosc-d'Antic successfully solved the difficulties of the royal manufacture by introducing new larger crucibles and by testing new chemical combinations to imitate the crystal glass of Bohemia (Bosc d'Antic, 1780). These breakthroughs made transparent the advantages of chemical expertise. However, due to his peculiar background, Bosc-d'Antic represented a new kind of expert and, as soon as he demonstrated his abilities at Saint-Gobain, he hoped for public recognition from the academic world. He wished to be both a technical consultant and an academic scientist. That such an ambition seemed to be at hand became clear quite soon. Back in Paris in 1757, Bosc d'Antic was offered the opportunity to establish a manufacture of flint glass in the village of Rouelles (in Haute-Marne). The associates were the naturalist Etienne-Clément de Marivetz, who owned a mirror manufacture in Lyon, François Véron de Forobonnais, Inspecteur general des monnais, and the glassmaker Antoine Allut, later author of the articles on glass in the *Encyclopédie Méthodique*. The manufacture was not successful but Bosc d'Antic's reputation did not diminish. During this period he wrote several scientific memoirs devoted to glassmaking and in 1760 his memoir on the means for improving glassmaking in France was awarded a prize by the Académie des sciences (Bosc d'Antic, 1761). Encouraged by this recognition, Bosc d'Antic's scientific ambitions grew. In 1775 he published a comprehensive *Mémoire sur les manufactures à feu* in which he pointed out that the main weaknesses of the French glass industry was the lack of chemical knowledge and he urged the government to employ experts ('personnes éclairées') in order to instruct manufacturers to perfect their technical know-how. Another measure recommended by Bosc d'Antic was the creation of technical schools devoted to the metallurgical arts. Bosc d'Antic's recommendations were ignored and only during the Revolution, with the creation of the Bureau de consultation des arts et métiers, the question of public technical education became an issue again. Bosc d'Antic's plea for a centralized control of French metallurgical education was still in line with Colbert's guidelines and explicitly ignored the more flexible and successful model set forth in England (Clow and Clow, 1952, pp. 269–292; Taylor, 1957, pp. 69–80).

The reason for the French preference for the centralized system was not merely economic; the prestige and authority of the Académie des sciences was so high that it was difficult to think of a successful alternative. Bosc d'Antic's subsequent actions underlines this dependence upon the academic world. After his success as a technical expert, he outlined a theory of matter based on the experience he gained in Saint-Gobain. In 1777 he submitted to the Académie des sciences a memoir entitled *Observations sur les matières à convertir en verre, sur la nature du verre et sur le principe vitrifiant* (Bosc d'Antic, 1780, vol. 1, pp. 241–250) in which he claimed that by submitting the fondants and the silica to calcination separately before combining them, they became infusible and lost their property of producing glass. This led Bosc to conclude that both the fondants and silica, as well as most substances, contained a vitrifiable principle and that this was volatile. The alchemical background of glassmaking was resurgent with new arguments and through the authoritative voice of a respected expert. Glass was the key of the understanding of chemical transmutations, a theory which had already been set forth by Buffon in the *Théorie de la terre* (1749).

At the Académie the memoir was examined in January 1778 by Lavoisier and Macquer and in their severe report the two chemists pointed out that the sustainability of a theory had to be supported by a more thorough knowledge of the recent development of theoretical chemistry and by a greater deal of experimental evidence than the one set forth by Bosc-d'Antic (Lavoisier and Macquer, 1778). Glassmaking was becoming a too important part of chemistry to allow bizarre theories to be legitimized. On the other hand, Bosc-d'Antic's reputation as a glassmaker was so well established that a more thorough response was necessary. It is certainly not a coincidence that following his report, Lavoisier multiplied his efforts to make glassmaking an integral part of the new chemistry. In addition to the new prize for the production of flint-glass and glassmaking that he promoted in 1788 (Lavoisier, 1788), Lavoisier resumed the work on the publication of the *Descriptions des arts et métiers* and appointed a young, promising chemist, Pierre Loysel, to write a new systematic work on the *art de la verrerie* and its relation to physics and chemistry. Apart from being pone of the few naturalists who put his life in danger in order to safe Lavoisier from the guillotine, little is known of Loysel's scientific education. With a background as an engineer, a long experience at Saint-Gobain, Loysel was one of the few non academic chemists to be admitted in Lavoisier's laboratory for the preparation of a comprehensive work on chemistry. During this period, Loysel was able to produce a first draft of a treatise on glass which was finally embodied into the realm of modern chemistry. This quality, among others, was highlighted in a long report drafted by a commission of the Académie des sciences composed of three of Lavoisier's distinguished collaborators: Jean Darcet, Claude Louis Berthollet and Antoine François Fourcroy. At the beginning of their report the three chemists remarked:

> La verrerie est peut-être de tous les arts, celui qu'on peut soumettre le plus rigoureusement à des principes déterminés par la physique, celui par conséquent qui peut parvenir à la plus grande précision; mais il demandoit un observateur qui fût également familiarisé avec tous ses procédés et instruit en physique. M. Loysel possède ces deux qualités (Darcet et al., 1791, p. 113).

During the preparation of his work, Loysel wrote to Lavoisier at the end of December 1787 and justified his delay with the need of updating his knowledge on the recent discoveries made in pneumatic chemistry:

> Il faudroit qu'un verrier fût bon chimiste et les verriers ne veulent pas entendre parler de chimie. L'explication de la plupart des phénomènes de la vitrification dépend de la science des gaz. (Lavoisier, 1993, p. 103).

Furthermore the precision achieved in this branch of chemistry suggested that the application of the quantitative methods would be fruitful in glassmaking. Indeed Loysel's *Essai* was centered on quantification and, on specific topics, even on the mathematization of glassmaking. Loysel pointed out in the synthetic historical sketch introducing the *Essai,* that Neri's *Arte vetraria* contained nothing useful on the principles of the art and that it was only thanks to the works of recent chemists that more significant progress had been made (Loysel, 1799–1800, pp. xiii–xvi). The exact and systematic determination of the specific gravities of the ingredients contained in glass was one important key to the understanding of the chemical properties of this material. All different kinds of glass were qualified by Loysel on the sole basis of the specific gravities of their ingredients, an analytical method which had been advocated by Lavoisier and several other modern chemists (Loysel, 1799–1800, pp. 200–203). Moreover, the emphasis on the central importance of precise measurements of heat had to rely on Laplace's and Lavoisier's researches on caloric rather than on the empirical assessment acquired by experience.

Unlike his predecessors, Loysel paid no tribute to the history of glassmaking and claimed that his work represented the opening a new era which finally emancipated the art from the secrecy and the ambiguities of alchemy. Loysel ultimately proved to be right and glassmaking became an integral part of applied and industrial chemistry. However, the innovative methods proposed in the *Essai* were not equally successful in guiding glassmakers in the production of artworks as perfect as those made during the Antiquity and the Renaissance.

Bibliography

Beretta, M., *The Alchemy of Glass: Counterfeit, Imitation, and Transmutation in Ancient Glassmaking,* Sagamore Beach: Science History Publications/USA, 2009.

Bosc d'Antic, P., *Mémoire qui a remporté le prix extraordinairement proposé par l'Académie royale des sciences, pour l'année 1760. Quels sont les moyens les plus propres à porter la perfection & l'économie dans les verreries de France?,* Paris: A Paris, Chez Durand, 1761.

Bosc d'Antic, P., *Oeuvres de M. Bosc d'Antic ; contenant plusieurs mémoires sur l'art de la verrerie, sur la faïencerie, la poterie, l'art des forges, la minéralogie, l'électricité, & sur la médicine,* Paris: Rue & Hôtel Serpente, 1780, 2 vols.

Buffon, G. L. L. de, "Nouvelle invention des miroirs ardens", in: *Mémoires de mathématique et de physique, tirés des registres de l'Académie Royale des Sciences* (Année 1748).

Clow, A. and Clow N. L., *The Chemical Revolution. A Contribution to Social Technology,* London: The Batchworth Press, 1952.

Darcet J., Berthollet, C. L., Fourcroy, A. F., "Rapport su un ouvrage de M. Loysel, qui a pour titre *Essai sur les principes de l'art de la verrerie*", in: *Annales de chimie, 9* (1791), pp. 113–137.

Frémy, E., *La manufacture royale de glaces en France au XVIIe et au XVIIIe siècles,* Paris: Plon, 1909.

Gillispie, C. C., "The Discovery of the Leblanc Process", in: *Isis, 48* (1957), pp. 152–170.

Gillispie, C. C., *Science and Polity in France: The end of the Old Regime,* Princeton: Princeton University Press, 1980.

Goupil, M., "La *Description des arts et métiers*", in: Lavoisier, *Correspondance,* vol. 5, Paris: Académie des Sciences, 1993, pp. 277–286.

Guerlac H., "Some French Antecedents of the Chemical Revolution", in: *Chymia, 5* (1959), pp. 73–112.

Hamon M., Mathieu C., *Saint-Gobain 1665-1937. Une entreprise devant l'histoire*, Paris: Fayard, 2007.

Lavoisier, A.-L., "Prix extraordinaire proposé par ordre du Roi par l'Académie des sciences pour l'année 1788 (précis de ce qui a été fait pour perfectionner le flint-glass et le verre en général)," in Lavoisier, *Oeuvres*, vol. 6, Paris: Imprimerie Nationale, 1893, pp. 20–30.

Lavoisier, A.-L., *Correspondance*, vol. 5, Paris: Académie des Sciences, 1993.

Lavoisier, A.-L., Macquer, P. J., "Rapport sur un mémoire relatif à la nature du verre," (1778) in: Lavoisier, vol. 4, *Oeuvres*, Paris : Imprimerie Impériale, 1868, pp. 290–292.

Lespinasse, R. de, *Les métiers et corporations de la ville de Paris*, Paris: Imprimerie Nationale, 1892, vol. 2.

Loysel, P., *Essai sur les principes de l'art de la verrerie*, Paris: Desenne, 1799–1800.

Mahoney, M. S., "Organizing Expertise: Engineering and Public Works under Jean-Baptiste Colbert, 1662-83", in: *Osiris*, *25* (2010), pp. 149–170.

Réaumur, R. F. de, *Réflexions sur l'utilité dont l'Académie des sciences pourroit être au Royaume, si le Royaume luy donnoit les Secours dont elle a besoin* (ca. 1716–1727), in : Ernest Maindron, *L'Académie des sciences*, Paris: Alcan, 1888, pp. 103–110.

Saint-Léon, E. M., *Histoire des corporations de metiers depuis leurs origins jusqu'à leur suppression en 1791*, Paris: Alcan, 1922.

Scoville, W. C., "State Policy and the French Glass Industry, 1640-1789", in: *The Quarterly Journal of Economics*, *56*, No. 3 (May 1942), pp. 430–455.

Smeaton, W. A., "Some large Burning Lenses and their Use by Eighteenth-Century French and British Chemists", in: *Annals of Science, 44* (1987), pp. 265–276.

Stahl, G. E., *Physica subterranea profundum subterraneorum genesin, e principiis hucusque ignotis ostendens: opus sine pari … et Specimen Beccherianum, fundamentorum documentorum, experimentorum*, 2nd edition, Lipsiae: ex officina Weidmanniana, 1738.

Taylor, F. S., *A History of Industrial Chemistry*, London: Heinemann, 1957

Véron, P., *Verre d'optique et lunettes astronomiques*, Florence: Fondazione Giorgio Ronchi, 2009.

Chapter 21
Balloons, Hydraulic Machines and Steam Engines at War and Peace: Jean-Pierre Campmas, a Visionary or an Inefficient Inventor?

Patrice Bret

Charles Gillispie's outstanding contribution to "science and polity in France" carefully analyses the links between science and the modern state in the last decades of the eighteenth century and the first of the nineteenth. Beside major scientists and administrators, he considers several second rank scientists, engineers or inventors who also contributed to modernity, as well as charlatans and others who challenged scientific institutions. In his thorough investigation through the national, scientific and military archives in Paris and Vincennes, he might have come across Jean-Pierre Campmas. That inventor was not directly relevant for Gillispie's investigations. He was nevertheless involved in many of the same areas as Gillispie's protagonists, including ballooning, artillery and other matters that he submitted to French scientific and administrative institutions from the beginning of Louis XVI's reign to the early Napoleonic period.

With few exceptions, historians have paid little heed to Campmas.[1] Despite the abundance of primary sources (printed and manuscript[2]), it is difficult to

---

[1] Mainly regarding his 1784 balloons (Julien Turgan, *Les ballons*, Paris, Plon, 1851, p. 178; Michael R. Lynn, *Popular Science & public opinion in eighteenth-century France*, Manchester, Manchester University Press, 2006, p. 133; Marie Thébaud-Sorger, *L'Aérostation au temps des Lumières*, Rennes, Presses universitaires de Rennes, 2009, p. 89) and their later military development (P. Bret, "Napoléon et les technologies militaires nouvelles: essai d'analyse à partir des exemples de l'aérostation et de la fusée de guerre", *Revue de l'Institut Napoléon*, n° 148 (1987), 46–60: 49, 54, 58, 59). Also: other military inventions during the Revolution (André Duvignac, *Histoire de l'armée motorisée*, Paris, Imprimerie nationale, 1947, pp. 17–19; P. Bret, *L'État, l'armée, la science. L'invention de la recherche publique en France, 1763–1830*, Rennes, Presses universitaires de Rennes (coll. Carnot), 2002, 124, 126, 127). Strangely enough, the most cited proposal is his unique joint monumental building project in 1801 (Jean Davallon, *Claquemurer, pour ainsi dire, tout l'univers: la mise en exposition*, Paris, Centre Georges Pompidou, 1996, p. 64; Dominique Poulot, *Surveiller et s'instruire: la Révolution française et l'intelligence de l'héritage historique*, Oxford, Voltaire Foundation, 1996, p. 26, and *Une histoire du patrimoine en Occident, XVIIIᵉ-XXᵉ siècle*, Paris, Presses universitaires de France, 2006, pp. 35–36).

[2] Besides the numerous manuscript sources referred hereafter in the footnotes, we must also cite a file about Campmas which is among those of inventors at the National Archives (AN),

P. Bret (✉)
Centre Alexandre Koyré – Centre de recherches en histoire des sciences et des techniques, Paris, France
e-mail: patrice.bret@yahoo.fr

J.Z. Buchwald (ed.), *A Master of Science History*, Archimedes 30,
DOI 10.1007/978-94-007-2627-7_21, © Springer Science+Business Media B.V. 2012

construct the career and life of such a person, for he is not worth extensive research on his own accord, and my aim here is not to put stress on a second or third rank individual. Nevertheless, Campmas' work and practice are significant for, as an inventor, he is representative of hundreds of others in his time, ones who were prompt to seize fashionable topics and eager for money and glory, usually with little success. The boundaries between fakers and inventors of true genius are not always impermeable. The worst among them might have visionary views scattered amidst their dreams or buried in the most trivial technical details, even when they do not share the rational and science-based approach of experts who assessed their work.

The known biographical data on Jean-Pierre Campmas are less numerous and reliable than the sources regarding his work.[3] Probably born in the mid-1740s, the inventor may have died in 1804, since no further information can be found in later sources. He has often been confused with other personages in the province of Guyenne, in Southwestern France, where that name was rather common in various fields.[4] It would be good to know whether he was a relative of Pierre Jean-Louis Campmas (1756–1821), an attorney and a member of the National Convention (*Convention nationale*).[5] All the same, Jean-Pierre Campmas provides a good example of the various ways in which inventors could ply their trade, both at the end of the Old Regime and during the Revolutionary and early Napoleonic times. In *Science and Polity*, Gillispie pointed to some of them.[6] Like many others, Campmas used the full range of means at inventors' disposal to achieve recognition at the Royal Academy of Sciences or elsewhere,

---

which contains material about his "mobile suitable for a great number of hydraulic machines" (AN, F^{14} 3187) and another about his inventions of 1796 to 1802 (F^{12} 2325). Unfortunately, I have not been able to see it.

[3] Even his name has been wrongly read and written, then printed, as "Campenas", "Campinas", "Campmar" and even "Champmas" or "Canas", and his first name turned into Jean-René.

[4] A fountain maker (*fontainier*) at Revel, Pierre Campmas, helped Pierre-Paul Riquet to provide the *Canal du Midi* with water from the Black Mountain (*Montagne noire*) in the 1670s. The physician of the Countess of Artois, Jean-François Campmas, born in Monestiès (Tarn), formerly demonstrator of physics at Montauban, and author of reflexions on childbirth, was sent to the *États généraux* in 1789, but *La France Littéraire* (since 1784, p. 199) wrongly attributed to him a "Machine for rising water with a rope" invented by J.-P. Campmas.

[5] P.J.L. Campmas – also known for his translation of Scipione Breislak's *Institutions géologiques*, from Italian, in 1819 – was born at Blaye (Tarn) and Jean-Pierre was the name of both grandfather, a notary at Carmaux – let us notice that the inventor's signature was followed by a complex flourish very similar to the notary's ones – and his uncle, a priest (E. Appolis, "Un conventionnel régicide" *Revue du Tarn*, 1943–1944, p. 142).

[6] Especially vol. I, Chapter VI, Industry and Invention, pp. 388–478, and vol. II, Chap. VI, Scientists at War, pp. 339–444.

and felt frustrated by lack of credit and funds to develop his devices.[7] He was mainly concerned with hydraulics, but, given the interests of the period, he dealt also with balloons, steam engines and weaponry. Though tedious it may be, a detailed account is necessary to apprehend the inventor's daily struggle for his mechanical ideas.

21.1 An "Hydraulic Engineer" Seeks Recognition

Campmas' first attempt to gain the Academy's approval occurred on March 12, 1774. He presented three different machines for grinding wheat, sawing stones and pumping water, but the commission (Le Roy, d'Arcy and Vaucanson) reported negatively on his proposals six weeks later.[8] After this first failure, Campmas kept his distance for several years. He worked for noblemen and entrepreneurs in his province of Guyenne, mainly in the castles around Bordeaux and in that colonial and wine seaport itself, settling thereafter in Paris in early 1781.[9] On March 7 of that year, he presented to the Academy a "novel naval carriage" supposed to sail upstream on its own using the stream's own power. Gaspard Monge, named as reporter with Bossut, probably told him that the memoir was to be rejected, and the inventor withdrew it two weeks later.[10] Campmas had no more success with the various inventions he presented later: a capstan on March 2 and 9, 1782 (report by Coulomb and Bossut, May 8, 1782),[11] four hydraulic machines on April 24, same year (report by Le Roy and

---

[7] Liliane Hilaire-Pérez has given an outstanding analyze of the strategies of inventors in her comparative approach with Britain (*Inventions et inventeurs en France et en Angleterre au XVIIIe siècle*, Ph.D., Université Paris I/Panthéon-Sorbonne, 1994, 4 vols.; *L'invention technique au siècle des Lumières*, Paris, Albin Michel, 2000). See also Roger Hahn, *The Anatomy of a scientific Institution: the Paris Academy of Sciences, 1666–1803*. Berkeley: University of California Press, 1971, for the competition with the Academy. Thébaud-Sorger, *L'Aérostation...*, op. cit.; Catherine Lanoë, *La poudre et le fard. Une histoire des cosmétiques de la Renaissance aux Lumières*, Seyssel, Champ Vallon, 2008; Christiane Demeulenaere-Douyère, "Entre obscurité individuelle et gloire collective? Une société d'inventeurs sous la Révolution", in Patrice Bret, Hélène Gispert, Gérard Pajonk eds., *Savants et ingénieurs entre la gloire et l'oubli. Figures du progrès, imaginaires sociaux et construction historique de catégories culturelles*, Paris, Éditions du CTHS, in press.

[8] Archives of the Academy of Sciences, Paris (AAdS), "*pochettes de séance*" (*pochette*) March 12 and Apr. 23, 1774; "Procès-verbaux de l'Académie des sciences", mss. (PVAS) 1774, f° 98 and 137v–138v.

[9] His memoirs and plans presented at the Academy in 1782 (AAdS, *pochette* Dec. 7, 1782) are signed at Paris (April and July 1774 and again from April 1781 onwards), and meanwhile at Bordeaux (Oct. 1776, March 1778) and in various neighbouring places: castles of La Fite-en-Médoc (Apr. 1775), of Granet entre-deux-mers (Oct. 1776), of Saint-George-en-Pui (Oct. 1777), of Citran-en-Médoc (May to Sept. 1779).

[10] Ibid.

[11] AAdS, *pochette*, May 8; PVAS 1782 f° 32, 37v and 82.

Coulomb, March 1st, 1783)[12]; towing machines and other hydraulic devices on February 18, 1784, completing a 5-foot model submitted two weeks earlier (Legendre, Laplace and Le Roy were named reporters).[13] Two years later, the same group was commissioned to assess his "novel kind of pump", but before providing a definitive opinion they required experiments, which were never performed (April 7, 1786).[14]

Campmas on occasion preferred registering sealed envelopes (*plis cachetés*) to be opened only at his request. That permitted him to have his inventions secured and duly dated. The first and heaviest among the envelopes, n° 181 (December 7, 1782), contained "figures or drawings of machine" (see Figs. 21.1 and 21.2). It included the same 14-page memoir that he had withdrawn from

Fig. 21.1 Sealed envelope (*pli cacheté*) #181 (AAdS, *pochette*, Dec. 7, 1782). Up to down: brief description by Campmas (date of closing: Apr. 15, 1782); registration by Condorcet (Dec. 7, 1782); mention of opening by the Academy (March 7, 1977). Courtesy of the Academy of Sciences, Paris

---

[12] AAdS, *pochette*, May 8; PVAS 1782 f° 76 and 1783, f° 70. See Matthias Thomas-Hercent, "Les innovations du cabestan dans la France des Lumières", Master thesis, University François Rabelais, Tours (Pascal Brioist supervisor, Centre des études supérieures de la Renaissance), 1999, 185 p.

[13] AAdS, *pochette*, May 8; PVAS 1782 f° 76 and 1783, f° 70.

[14] AAdS, PVAS 1786, f° 146v–147v.

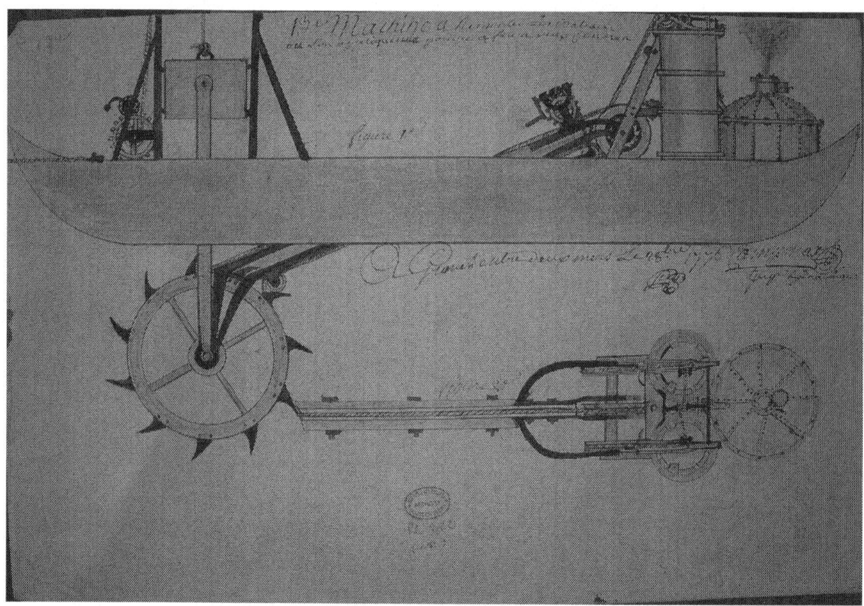

Fig. 21.2 "13th machine for towing boats, where a two-cylinder steam engine is applied", Granet entre deux mers, Oct. 2, 1776 (AAdS, *pochette*, Dec. 7, 1782). Here steam-driven, Campmas' first model of mobile cogwheel could use several sorts of power. Courtesy of the Academy of Sciences, Paris

Monge the previous year, together with sixteen additional pages of new details. Several other sealed envelopes were registered in 1783.[15]

Campmas then decided to participate in the public contest opened this year by the Academy on behalf of the King for restoring or replacing the famous Marly machine, built a century earlier by Rennequin for Louis XIV in order to pump water from the Seine to his chateaux at Versailles and Marly. The proclamation was first issued in 1785, but the contest was then prolonged two more years, with the hope of obtaining better proposals. All the applicants presented anonymously, identified only by a motto or an epigraph. Campmas proposed four different projects, in at least two sets of anonymous memoirs and plans: a 30-page one bearing the motto *Multiplex et una*, registered under n° 31, and a longer one, *Mobilitate firma* (69 p.) under n° 70. The reporters named by the Academy were Borda, Bossut, Coulomb, Perier and Monge, later replaced with Vandermonde – most of them could have recognized that these

---

[15] AAdS, *pochettes*: n° 223 (January 18) was a "parcel" concerning four new hydraulic machines, including a permanent fountain and a pump without piston; n° 224 (January 22) contained a "very small discover, detailed with nine figures and explanations" about "new permanent pens" (i.e. metallic and fountain pens) – Campmas asked for its opening and resealed in on July 24, 1797; and n° 232 (December 13) had "useful discoveries detailed with six figures", dealing with air navigation. There is some doubt about n° 222, which is mixed in n° 181, but another invention by Bonnemain is registered under this number (*Plis cachetés*, "Liste des dépôts faits à la cidevant Académie des Sciences depuis l'année 1776 jusqu'en l'année 1792").

anonymous submissions were his. Although he did not win the competition, which had a monetary award, he did receive one of the six special mentions (*accessits*) among more than 45 applicants.[16] On May 12, 1787, only a few days after the prize was proclaimed at the Easter public meeting of the Academy, the inventor borrowed his own memoirs from the permanent secretary (*secrétaire perpétuel*) Condorcet, promising to bring them back when the secretary asked. Unfortunately, that never happened, and we are therefore unable to know the details of these projects, though elements of them can be guessed from Campmas' previous and subsequent applications.

The Royal Academy was not the only place to have one's inventions approved and to obtain further funding.[17] An alternative was to apply to the ministerial authorities. Thus, in April 1782, Campmas wrote directly to Amelot de Chaillou, State Secretary of the King's Household (*Maison du Roi*), requesting an "exclusive privilege" to exploit machines that enabled boats to sail upstream on their own: he later enclosed essentially the same concepts in his sealed envelope for the Academy, propelled by various kinds of power (man, water, steam) and with several underwater driving mechanisms (wheels, legs) (see Fig. 21.3). Amelot immediately forwarded the request to the mayor (*Prévôt des marchands*) of Paris, Lefèvre de Caumartin, who returned the advice of the Town Board (*Bureau de la Ville*). The Board wisely proposed experiments to see whether the machines would work, since the inventor would not describe his jealously-guarded mechanical secrets.[18] These Campmas had safely enclosed in his first sealed envelope at the Academy.

Another alternative to the Academy's support was aristocratic patronage. In a later (1800) claim for priority, Campmas provided the National Institute (*Institut national*) with a copy of a plan dated 1773. He asserted that the pump which he had presented in 1786 was the same suction and force device that he had designed thirteen years earlier, and on which Borda and Meusnier had reported that same year. Although members of the Academy, these two (both engineers and physicists) were not at that time acting within an academic commission, but as experts on behalf of Lieutenant-General Duke of Harcourt, governor of Normandy (1775–1789) and governor of the Dauphin (1786–1789),[19] whose support Campmas was then seeking. While claiming priority in 1800, the inventor referred neither to the negative assessment that the Academy had produced in 1774, nor

---

[16] See Mathilde Lardit, "Les concours de l'Académie royale des sciences", Master thesis, Université Paris I Panthéon-Sorbonne (Daniel Roche supervisor), 1997, pp. 116–117.

[17] See Hilaire-Perez, L'invention technique, op. cit.

[18] AN, O¹ 1294, #12–14. Quoted by Liliane Hilaire-Pérez, *Inventions et inventeurs en France et en Angleterre au XVIIIᵉ siècle*, Ph.D., Université Paris I/Panthéon-Sorbonne, 1994, vol. I, p. 63.

[19] François-Henri d'Harcourt (1726–1802), count of Lillebonne, 5th duke of Harcourt (1783). Let us notice that the three people involved in the alleged 1786 assessment were no longer able to testify in 1800: Meusnier was killed at the siege of Mayence in 1793, Borda died in 1799 and the Duke of Harcourt was still in emigration, as the representative of Louis XVIII to the British government.

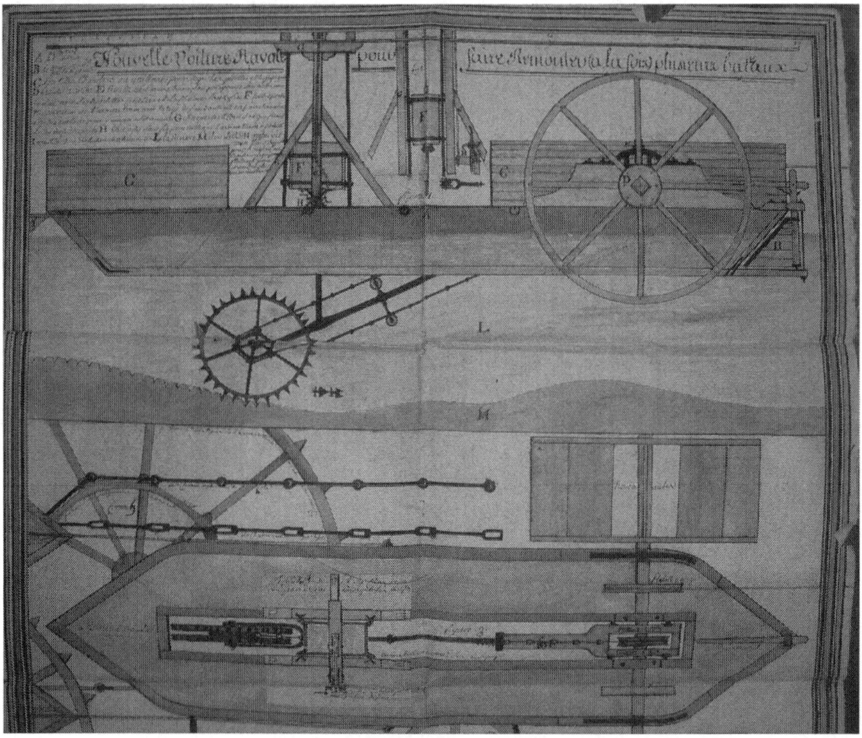

Fig. 21.3 "New naval carriage for towing several boats at the same time", Paris, March 1, 1781 (AAdS, *pochette*, Dec. 7, 1782). Paddle wheeled tugboat here driven by tugboat tracted bequipped with Campmas' second model of mobile cogwheel. Courtesy of the Academy of Sciences, Paris

to the above mentioned memoir that he did submit on the machine, but which Legendre, Laplace and Le Roy did not fully evaluate in 1786.[20]

In yet another effort to make money from his inventions, Campmas claimed he had improved the recently invented Vera's machine that pumped water with an endless rope, as early as November, 1781 in the gardens of Lieutenant-General Marquis de Crussol d'Amboise,[21] rue de Sève [Sèvres], at Paris. In June 1782, the jealous inventor gave a description of his improvement in a letter to the *Journal de Paris*, and announced the publication of a print describing a "Machine for rising water with a rope" in the *Mercure de France*: the engraving

---

[20] *Procès verbaux de l'Académie des sciences*, printed (*PVAS*) II, Germinal 26, Year 8, pp. 142–143.

[21] Anne Emmanuel François Georges de Crussol, Count d'Amboise (1726–1794) had also a military command in Normandy in 1789, when he was elected to the *Etats généraux*. He was beheaded on July 26, 1794 – one day before Thermidor 9.

sold for either 3 £ with explanations, or 1 £ 16 s. (1 *livre* 16 *sols*) without.[22] There was no need do keep his plans secret by that time. Numerous other improvers had made experiments and had provided descriptions of this simple, cheap and therefore fashionable machine, which could soon be found elsewhere, notably in Pilâtre de Rozier's experiments, which were published in the August issue of the *Journal de physique*,[23] or in Marsilio Landriani's *Description d'une machine propre à élever l'eau par la rotation d'une corde verticale* (Geneva, 1782). Campmas' print may have sold well, although the mathematician and physicist Antoine Deparcieux soon proved that the output of the machine that it pictured was only half that of an ordinary pump with a piston.[24]

When Campmas left Bordeaux for Paris in 1781, and soon failed again at the Academy, he came to the conclusion that inventions were better judged publicly, especially for those who, like him, were not able to get any institutional support.[25] Publishing pamphlets and announcements in journals was the best way to get publicity, notoriety, and support. Campmas thereafter frequently published either in the only French daily newspaper, the *Journal de Paris*, in the *Journal général de France*, which appeared every two days, or in a dozen or so weekly, bimonthly and monthly literary journals and other magazines. As in his surviving letters and memoirs, these pieces of information merely show his slant on the events. Fortunately the journals also provided a venue for critics, and here Campmas' story meets history, in particular around the birth of aviation.

Campmas began an offensive strategy. In October 1781, he announced his arrival from Guyenne in the *Journal de Paris*.[26] Two months later, he published a lengthily list of his inventions in two different periodicals: he began with twenty-one inventions for the capital city in the *Journal politique, ou Gazette des gazettes*, followed by twelve more for Paris and fifteen for the provinces in the *Journal encyclopédique*,[27] including self-towing machinery for boats, various devices powered by men, horses, wind, fire, and several others, such as a powder mill and new machinery for Marly. Then, he wrote a letter to Pahin de La Blancherie – who published it in his *Nouvelles de la République des Lettres et des Arts* – in which he referred to his first proposal for pumping water out of ships as early as 1774.[28] Cleverly, he was even able to find unexpected support through the fame of one

---

[22] *Journal de Paris*, June 19, 1782, p. 689. *La Machine à élever l'eau par une corde sans fin, perfectionnée par M. Campmas, Ingénieur-Hydraulique* is the print wrongly attributed to the physician J.-F. Campmas after *La France littéraire*.

[23] See also *Encyclopédie méthodique. Arts et métiers mécaniques*, Paris, Panckoucke; Liège, Plomteux, vol. III, 1784, pp. 688–690.

[24] *Dissertation sur le moyen d'élever l'eau par la rotation d'une corde verticale sans fin*, Amsterdam, 1782 (not seen).

[25] Campmas lived at the Hôtel St-Louis, rue Gît le Cœur, at least from December 1781 up to May 1787.

[26] Oct. 2, 1781, vol. 2, p. 1109.

[27] Respectively Dec. (second fortnight), pp. 78–81 and vol. 8, part 3, Dec. 15, pp. 516–519.

[28] 1782, p. 4.

competitor, when he claimed in several journals that he had been the first to take advantage of Vera's machine.[29] And he took advantage of every opportunity to have his name and address published: regarding the recent terrible earthquakes in Calabria, while confessing his lack of scientific knowledge, especially in chemistry and mineralogy, he nevertheless offered hypotheses about volcanoes and earthquakes in the *Journal de literature, des sciences et des arts*.[30] But the *Mercure de France* and its attached second part (the *Journal politique de Bruxelles*) became his favored media for advertising.[31]

As early as 1779, Campmas had also published a small pamphlet on hydraulic work under water.[32] In 1782, as we already noted, he published the engraving of his "Machine for rising water with a rope". It seems that neither the pamphlet nor the engraving exists today, unlike two later ones that appeared after the climax of the French Revolution: *Plan général des finances, nouvelles fabriques monétaires, moulins nationaux et greniers d'abondance d'un nouveau genre...* at the very beginning of the Directoire regime (October 1795),[33] and *Lettre du Cit. Campmas, ... sur son projet de deux grands établissements pour la ville de Paris, divisés en six objets d'utilité majeure* at the eve of Bonaparte's government (November 1799).[34] This last was an offprint of a letter sent to the *Moniteur universel* calling for funds – through a notary and a police superintendent – to support his project to convert the *Salpêtrière Hospital* in the eastern districts of Paris and either the *École militaire* or the island of Swans (*île des Cygnes*), in the Seine River, in the western part, into arms factories, powered with a machine similar to the one he had built in Amboise a dozen years earlier.

21.2 From Industrial Plants to Priority Disputes

From 1789 onwards, Campmas was able to base new proposals on his own experience in Amboise on the Loire River, east of Tours, where he got his best opportunities for developing in his engineering at the very end of the Ancien

---

[29] Letter to the *Journal de Paris*, June 19, 1782, p. 689, *Journal politique, ou Gazette des gazettes*, Aug. (first fortnight), 1782, p. 90.

[30] "Réflexions de M. Campmas, Ingénieur-hydraulique, sur les désastres de la Calabre & de Messine", 1783, vol. 3, pp. 7–12.

[31] For example, Dec. 28, 1782, n° 52, p. 184; March 22, 1783, p. 238.

[32] *La France littéraire*, 1779, p. 589.

[33] *Plan général des finances, nouvelles fabriques monétaires, moulins nationaux et greniers d'abondance d'un nouveau genre, présenté au Conseil législatif des Cinq Cens, par le citoyen Jean-Pierre Campmas... 4 brumaire an IV* [Oct. 26, 1795], Paris, n.d., 4p., in 4° (Bibliothèque nationale de France, Paris (BnF), Rp 13446). The proposal to the newly elected legislative assembly was made the last day of the National Convention.

[34] *Lettre du Cit. Campmas, ingénieur en hydraulique, sur son projet de deux grands Etablissemens pour la ville de Paris, divisés en six objets d'utilité majeure; au rédacteur du Moniteur universel. Paris, le 20 nivôse an VIII* [Jan. 10, 1800]. Paris, impr. de H. Agasse, n.d., in 8°, 13 p. (BnF, Vp 13079).

Regime. His aristocratic patronage, the positive appreciation by the academy of
his project at the Marly Prize, and his own advertising strategy at last yielded a
major mechanical building operation indeed. After borrowing his memoirs on
the Marly machine from Condorcet in May 1787, he moved to Amboise. It
seems that he was called there – possibly via Crussol d'Amboise's influence – by
the entrepreneur Sanche (since 1782 manager in Amboise of the ironworks) in
order to build hydraulic machines to drive the hammers of his "Royal manufacture of fine steels". In 1789, this large industrial plant had six working
furnaces (of the twelve that had been built), forty hammers and eighty steel
forges.[35] Even though the local production was not good enough to challenge
British steel in quality, official certificates attest that Campmas' hydraulic
machinery itself was quite successful. In 1793, the Consultative Board for arts
and crafts (*Bureau de Consultation des arts et métiers*) therefore gave him its
maximum "national award" (*récompense nationale*) of 6,000£ for "having
conceived and executed successfully the mechanical part of the steel plant
at Amboise", i.e. "the mills to hammer the steel and to blow the forges".[36]
Campmas asserted that his equipment had even withstood the breaking up of
the frozen river which had carried away the bridge in January 1789. From this
point on he could rely on strong references.

Yet things at the Amboise steel plant were not quite so positive. A report by a
local administrator (*subdélégué de l'intendant*), in March 1789, states that the
mills on the river were not powerful enough, and that the dyke built to supply
them with water was itself charged by the inhabitants with the recent disaster
that had destroyed the bridge when the river unfroze.[37] But the construction
of his equipment nevertheless continued while the engineer rented a flat at a
local lawyer's residence (Gitton, later vice-president of the district).[38] There
Campmas also designed one of the various projects for rebuilding the bridge,[39]

---

[35] Hilaire-Pérez, *L'invention technique…*, op. cit., pp. 281–282, 295; Denis Woronoff,
L'industrie sidérurgique en France pendant la Révolution et l'Empire. Paris: EHESS, 1984,
p. 352; Charles Ballot, *L'introduction du machinisme dans l'industrie française* (Paris, 1923).
Reprint, Geneva: Slatkine, 1978.

[36] Archives du Musée national des arts et métiers, Paris, 10–544 #105. The experts named by
the *Bureau de consultation* on May 9, 1792 were not members of the Academy, but representative of inventors societies (Lucotte, Trouville and Dumas). On August 25, Campmas only
received 950£ after the legal tax of 5%, since he had already obtained 4,750£ on July 21 (AN,
F⁴ 1316). See Charles Ballot, "Procès-verbaux du Bureau de Consultation des Arts et
Métiers", *Bulletin d'histoire économique de la Révolution*, 1913, pp. 95, 98, 140.

[37] AD Indre-et-Loire, C 143, quoted by Georges Rosenberger, "Production et usage de l'acier
en France au XVIIIe siècle. Tentative de bilan", in Philippe Dillmann, Liliane Pérez et
Catherine Verna (eds.), *L'acier en Europe avant Bessemer*, Toulouse, CNRS/FRAMESPA,
Editions Méridiennes, 2011, pp. 339–357 (p. 355).

[38] Service historique de la défense/ Département armée de terre, Vincennes (SHD/DAT), 6 W
66, dos. 6033(7), p. 3.

[39] AD Indre et Loire, C. 738.

and he made numerous tests on the devices that he had invented, continuing at least until the summer of 1791.[40]

After he moved back to Paris, Campmas joined the *Société du Point central*, one among a number of newly-founded ones for inventors which challenged the old Academy.[41] In Year II (1793–1794), he settled in the cloister of the former cathedral church Notre-Dame, soon to be renamed the Temple of the Reason, on the île de la Cité, where many instrument-makers and artisans were housed.[42] On the 18th of the first month of Year II (October 9, 1793), his new project for restoring the Marly machine was entrusted to Coulomb, Trouville and Le Roy at the Consultative Board for Arts and Crafts. This time the Board granted him the medium award of the first class (5,000£ in *assignats*) on April 3, 1795.[43] Moreover, since a new contest on Marly had been proposed by the Committee of Patrimony and Alienation (*Comité des domaines et aliénation*) of the National Convention, his proposal was also included in the final report for that contest by Prony and Molard, thereby providing information on Campmas' project, along with those of his competitors. This was likely similar to the one that he had proposed six years earlier, which involved a unique suction and force pump using only one wheel that was designed to bring water from the river to the aqueduct in a single flow. Though they congratulated Campmas for his hard work and clever ideas, the committee still thought that the flow should be divided in two or three parts, as it had been in the 1680s by Rennequin's device.[44]

When the new Constitution of Year III restored the ministries and created the National Institute – the first class of which replaced the Academy of Sciences – Campmas first neglected the new academic power and preferred applying to the political power. But the latter usually forwarded to the former. Thus, in September 1796, his memoir to the Minister of the Interior on three

---

[40] SHD/DAT, 6 W 66, dos. 6033(7), p. 3. In 1793, the steel plant was converted into blades manufacturing when Amboise was besieged by the royalist rebellion (*Vendéens*), but Campmas' mills were destroyed in the Revolutionary years.

[41] He was registered after Oct. 9, 1792 (Archives du Musée national des arts et métiers, 10–932).

[42] Enclos de la Raison (alias Cloître Notre-Dame), #46 until 1801 at least. Previously : Hôtel de Provence, rue St-André des arts (May 1792), Caffé de Conty, descente du Pont neuf (October 1792).

[43] Archives du Musée national des arts et métiers 10–544 #290. At that time (Ventose 20, Year 3/March 10, 1795), Campmas asked the Provisional Commission of Arts to seek in Condorcet's papers so as to find out three of the memoirs he had presented to the formerAcademy (Louis Tuetey, Procès-verbaux de la Commission temporaire des arts. Vol. II, Paris, 1917, p. 167).

[44] Convention nationale, *Rapport de Prony et Molard sur les projets présentés au comité des domaines et aliénation, pour remplacer la machine de Marly. Imprimé par ordre de la Convention nationale...*, Paris, Impr. nationale, Du 15 vendémiaire, l'an III de la République [Oct. 6, 1794], pp. 12–14 (description) and 17–19 (comments).

naval machines came to the National Institute, where Prony and Bory decided to postpone the assessment until the Consultative Board had been replaced.[45] The Minister of Finances did the same with a more original project for stamping money, including a hydraulic machine in the Mint, and an underground water supply for it. That project had been first alluded to one year earlier by Campmas in his *Plan général des finances*: it was intended to be less expensive and "at least one hundred time faster" than existing machines in order to keep only one Mint in the Republic.[46] This was now assessed by Prony and Darcet, later by Perier. They were clearly annoyed with the inventor's wish to keep his plans secret and by his criticisms of the judges themselves: he challenged Perier's objectivity as being both judge and judged (Perier controlled the water supply in Paris), and Prony, whose previous report on the Marly machine had brought "a fatal strike to his reputation". They asked for more detailed explanations, and waited for "a sufficient knowledge about his projects and plans".[47] Then, in February 1797, both ministers forwarded several new letters and memoirs from Campmas, in particular one concerning a steam engine for the Mint, which were once more entrusted to Perier and Prony.[48]

Despite these disputes, Campmas later applied directly to the National Institute. At first, in Summer 1797, he did not request an assessment, but only borrowed his sealed envelope on metallic pens that had been registered in 1783; he registered it again, and soon registered as well two new sealed envelopes on "a discovery which could become useful", probably on the same object.[49] Twice, he asked only for a dated certificate: one for the new, stronger and longer metallic pipes to supply water and the model of a hydraulic machine that he presented next spring, the other for a memoir on "new vaporous pneumato-chemical baths and aromatic fumigations" in December 1798.[50] But in and after the previous spring, he had also asked for regular applications. On March 21, with Bonaparte chairing the meeting, Campmas read a memoir on the construction of a hydraulic machine; he read another on October 2 and a third, on the Marly machine again, was forwarded by the Minister of the Interior on November 6. The same reporters (Charles, Brisson and Le Roy) were named the three times, but no report was yielded and one year later, on October 28, 1799, Campmas presented his umpteenth hydraulic machine.

---

[45] *PVAS* I, pp. 96 (Fructidor 26, Year 4/Sept. 12, 1796) and 106 (Vendémiaire 1st, Year 5/Sept. 22, 1796).

[46] AN, F^{17} 1240 B. *Plan général des finances...*, op. cit., p. 3. From April 1796, Campmas was replacing Dumas as hydraulic engineer in charge of setting up a copper rolling mill on a floating factory on the Seine River (AAdS, pochette, Vendémiaire 11, Year 5/Oct. 2, 1796).

[47] *PVAS* I, pp. 113 (Vendémiaire 11, Year 5/Oct. 2, 1796), 136 (Frimaire 6/Nov. 26), 150 (Nivôse 6 and 11/Dec. 26 and 31).

[48] Ibid., pp. 171 (Pluviôse 21, Year 5/Feb. 9, 1797), 172 (Pluviôse 22/Feb. 10).

[49] Ibid., pp. 241–242 (Thermidor 6 and 11, Year 5/July 24 and 29, 1797), 258 (Fructidor 1/Aug. 18).

[50] Ibid., pp. 362 (Ventôse 26, Year 6/March 16, 1798), 502 (Frimaire 21, Year 7/Dec. 11, 1798).

That very day, two other inventors, Vincent Bidot and Charles Thilorier, also presented pumps: the reporters in charge for Campmas were Le Roy, Brisson and Charles; for Bidot there were Perier and Prony, later joined by two naval engineers, Forfait – also Minister of the Navy – and Sané, while Thilorier had Perier and Forfait.[51] In February 1800, as already noted, Campmas asserted his priority, claiming that he had invented a pump similar to Bidot's as early as 1773 and proposed an improved version in 1786.[52] Bidot's reporters mentioned these claims, but declared that they were far from thinking that he was not able to invent his own device: "We think very probable that he has only met the idea of another, very common accident among those who busy themselves with applied mechanics".[53] They also added that "checking Citizen Campmas' original titles are not included in the objects the [first] Class [of the Institute] has to know."[54] The day the report was read, on March 27, Bonaparte – now the First Consul – was chairing, and Campmas presented a model of a machine that used rolling and pitching to drive onboard pumps. At the next meeting, five days later, the illustrious chairman put Carnot and Bory in charge of this invention.[55] Campmas then argued against the resulting report, and asked that a due record of his counter-claims be made. He sent it to the Ministry of the Interior, Napoleon's brother Lucien Bonaparte, to complain about the behavior of the Institute towards him, with copy to the accused, and he wrote again to the Institute, reclaiming against the order of the day, which neglected his letter.[56] By the way, he also asked for the final report concerning his memoir of September 1796 on naval machines – Monge had joined his colleagues Prony and Bory for that evaluation, but the commission had not completed its work.[57] The contentious was no longer between the inventor and his competitor, but now with his National Institute itself. It now reached its climax, and Campmas never applied any more.

Moreover, Bidot was not the only inventor attacked by Campmas. At the same time, the bitter *ingénieur en hydraulique* published his above mentioned letter to the *Moniteur universel*, regarding the conversion of the *Salpêtrière Hospital* and the *École militaire* into arms factories: his aim in doing so was to assert priority against similar projects previously exposed in a letter from another engineer, Brullée, to this journal: "For a long time, I have let know my views on that matter to many people, and indeed to Citizen Brullée's associates; they are not less my property, whatever distorted shape he might

---

[51] *PVAS* II, pp. 20–21 (Brumaire 6, Year 8/Oct. 28, 1799).

[52] Ibid., p. 111 (Ventôse 6, Year 8/Feb. 25, 1800).

[53] Ibid., p. 142 (Germinal 26, Year 8/Apr. 16, 1800).

[54] Ibid., p. 143 (Germinal 26, Year 8/Apr. 16, 1800).

[55] Ibid., pp. 128–129 (Germinal 6 and 11, Year 8)

[56] Ibid., pp. 146 (Floréal 6, Year 8/Apr. 26 1800), 156 (Floréal 16/May 5), 173 (Prairial 11/May 31), 175 (Prairial 21/June 10).

[57] Ibid., p. 146 (Floréal 6, Year 8/Apr. 26, 1800).

have presented them to the government".[58] More and more, he felt robbed of both credit and possible funds by his competitors.

In Fall 1801, the Society for the Encouragement to National Industry (*Société d'encouragement pour l'industrie nationale*) was created in order to join administrators, scientists, bankers, entrepreneurs and inventors in a common endeavor. Chaptal, industrial chemist and minister of the Interior, was the president.[59] Campmas applied to the new institution as soon as January 25, 1802. The "intolerable conditions" attached to his proposal for 24 machines – these conditions are not known, but seem to be financial, either to sell models or to get funding to develop his devices – were rejected by the Committee of Mechanical Arts one month later, as being contrary to the by-laws of the society.[60] He also participated twice in the 1801 "Exhibition of the products of the national industry" at the Louvre, with a "machine for preventing shipwreck", and in the following year's exhibition as well, but the *Jury des arts* in charge of distribute premiums and medals did not grant him any award. In fact, that machine – a force pump – had already been sent to the Minister of the Interior, tested at the Swimming School (*École de natation*) on the Seine River, and even awarded 1,200 francs after a report by Bralle and Montgolfier for the new Consultative Board for arts and manufactures.[61] But the inventor's last alternative, which had little chance of success, was to find private support, and through the *Moniteur universel* he invited people to visit his cabinet of machines at the cloister Notre-Dame in October 1801.[62]

Campmas did not merely work on hydraulic matters. Certainly, he had presented himself as a "hydraulic engineer" (*ingénieur hydraulique*) at the end of the Old Regime – and occasionally as an "engineer privileged by the King" (*ingénieur privilégié du Roi*), despite the fact that he had obtained nothing more than a privilege to sell his prints on hydraulic machines – then as an "engineer in hydraulics" (*ingénieur en hydraulique*) during the Revolution. But he also called himself "*physicien*". This equivocal term, which could be used as well for physicists and natural philosophers, was mainly prized by the first aeronauts – Pilâtre de Rozier was a true demonstrator of physics indeed – and other balloon demonstrators, most often amateurs of various professions with a more or less light tincture of chemistry and physics. Campmas became deeply involved in ballooning from 1783 onwards.

---

[58] *Lettre du Cit. Campmas… sur son projet de deux grands Etablissemens…*, op. cit., pp. 1–2. Brullée's letter had been published on January 1, 1800.

[59] Serge Benoît, Gérard Emptoz, Denis Woronoff (eds.), *Encourager l'innovation en Europe. La Société d'encouragement pour l'industrie nationale*, Paris, Éd. du CTHS, 2006.

[60] Costaz, Baillet, Molard, Bardel and Conté belonged to the Committee. Archives of the Société d'encouragement pour l'industrie nationale, Procès-verbaux, Pluviôse 5 and Ventôse 5, Year 10, and Campmas' file (CME 1/1).

[61] AN, F[12] 2422. I am grateful to Marie Thébaud-Sorger, who drew my attention on this file.

[62] Brumaire 8, Year 10 (Oct. 30, 1801).

21.3 Ballooning in War and Peace

After the Montgolfier brothers and "the invention of aviation", as Gillispie termed it, what had been a spectacular social phenomenon were followed by large numbers of aerial projects throughout France. Campmas could not stand aside from this fashion.[63] Moved by his passion for inventing passion and his hope to earn money, like many other "*physiciens*" Campmas tried to turn to aerial demonstration and navigation. Again like so many, he did not succeed, for he failed to solve the major problem of steering balloons.

As he had for hydraulics, Campmas undertook both public announcements in the journals and secret applications to the Academy. As early as September 1783, he published his first papers on that problem, announcing forthcoming trials both in the *Gazette d'agriculture* (n° 76) and the *Mercure de France* (n° 40). On September 28, he produced a design for a hot air balloon, with an alembic as an improved heater to catch the resulting "aqueous vapors" (see Fig. 21.4). Three months later, he enclosed the design and registered it under n° 232 with the Academy, as "useful discoveries detailed with six figures". In a note following this two-and-a-half-page memoir, he remarked: "If sphericity were not the most advantageous shape, I shall give the Globe that of an egg, a rounded-off [h]exagon, in order to bring back more quickly the vapor condensate around the active firebox".[64] Meanwhile, in a letter to Étienne de Montgolfier dated November 28, 1783, he claimed that he had built several hot air balloons in Paris and in an unspecified province, but that he had mistakenly confided in "an Englishman" (who remains unknown), and that man had published Campmas' letter in the *Mercure de France* and the *Journal de Paris*.[65] Like Jean-Pierre Blanchard and other famous aeronauts, and even Lavoisier and Meusnier at the Academy of Sciences, Campmas was confident in the potential of aerial navigation.[66] He soon projected an "aerial diligence" (*diligence aérienne*) and pretended that he was able to move it as he pleased, with man-powered steers and rows, and other secret means. He sent entrance tickets to a public

---

[63] Charles C. Gillispie, *The Montgolfier Brothers and the Invention of Aviation, 1783–1784. With a Word on the Importance of Ballooning for the Science of Heat and the Art of Building Railroads.* Princeton: Princeton University Press, 1983. On the fashion, see James M. Hunn, *Balloon Craze in France: A Study of Popular Culture, 1783–1799*, PhD Vanderbildt University, microfilm, Nashville, 1982; Lynn, *Popular Science…*, op. cit.; Thébaud-Sorger, *L'Aérostation…*, op. cit.

[64] AAdS, *pochette* Dec. 13, 1783.

[65] Musée de l'Air et de l'Espace, Le Bourget, fonds Montgolfier, Box XXI, #24.

[66] Van Marum reported a discussion on that point, after the July 20 meeting at the Academy, between Lavoisier, who was confident in Meusnier's work, and Laplace, who considered steering a balloon impossible (Martinus Van Marum, *Journal physique de mon séjour à Paris 1785*, in *Martinus Van Marum. Life and Work*, ed. R. J. Forbes, Haarlem, 1970, vol. II, pp. 220–239). Regarding the academic assessment, see "Lavoisier et les deux commissions académiques successives pour l'étude des aérostats", Œuvres de *Lavoisier – Correspondance*. Vol. IV, Michelle Goupil, ed., Paris, Belin, 1986, pp. 293–297.

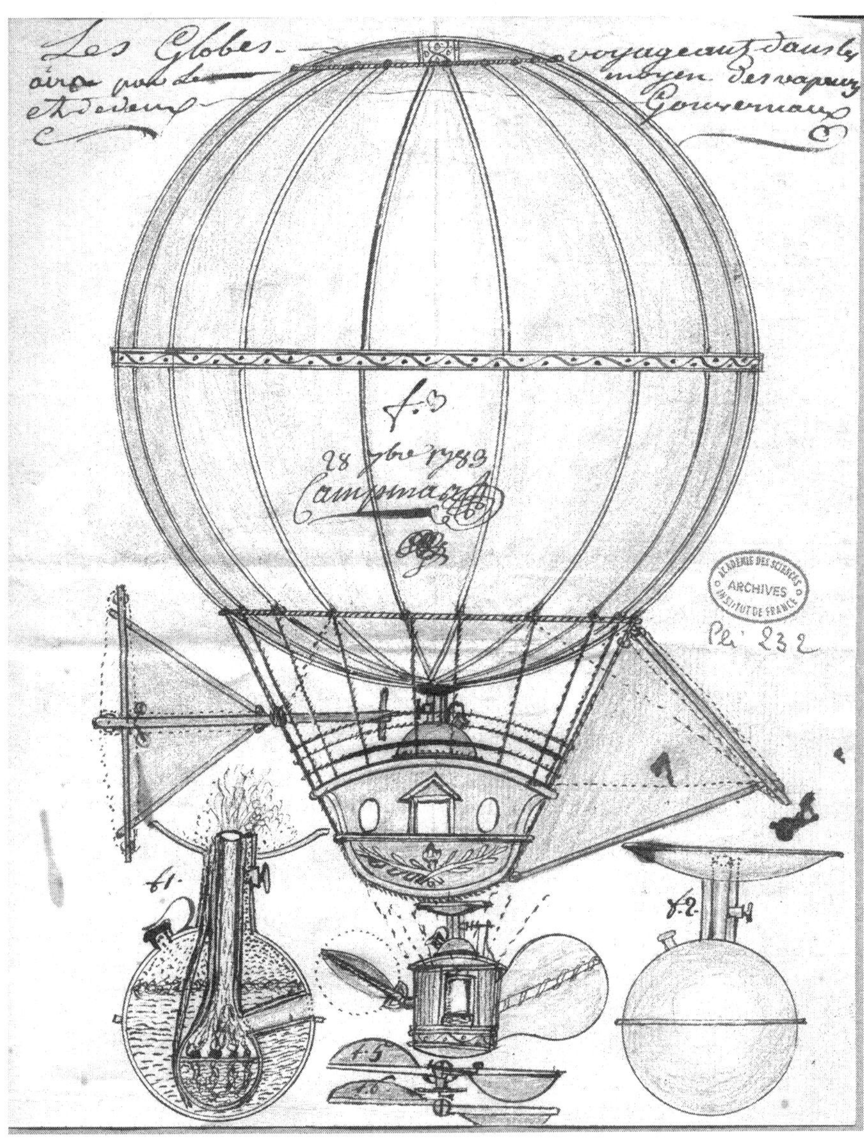

Fig. 21.4 "The Globes travelling in the air by means of vapors and two rudders", Paris, Sept. 28, 1783 (AAdS, *pli cacheté* #232, *pochette*, Dec. 13, 1783). Courtesy of the Academy of Sciences, Paris

demonstration of this aerial steering machinery that he claimed to have built to Montgolfier,[67] and sold them through public advertising. He arranged for three successive campaign announcements in various journals, each corresponding to one of his planned aeronautic exhibitions.

Campmas wrote one letter about his forthcoming exhibitions dated February 20, 1784 that was published four days later in *Annonces, affiches et avis divers, ou Journal général de France*, which was soon followed by his "observations". His proposed balloon was no longer spherical, but had its larger circumference of 152 feet and the narrower 82.[68] On March 9, he gave the date of his test, "next Friday"; two days later, he indicated the place and time: the Spiritual Concert Hall at the Tuileries, on March 12, at noon.[69] Nothing more is known about it: a model was probably on display, a means commonly used to obtain funds in order to construct a full-scale device.[70] Three months later, on Saturday June 12, Condorcet read at the Academy a letter from Campmas asking the members to visit his machine in the Queen's Gallery at the Tuileries after their meeting. The inventor urged them to come, because the Queen herself would soon arrive at her apartments there. Brisson and Meusnier were named by the Academy to do it, but they left no written account of their visit.[71] This exhibition of his machine – now "a rather high box fixed on castors", with moving strings or rows – had been announced in the *Journal politique* and the *Journal encyclopédique*.[72] After these preliminary exhibitions of the machine, Campmas announced his ascent around October 20.

On the 23rd of this month, he wrote again to the *Journal de Paris* to explain he had to postpone it, and gave a detailed account of the daily occupations of the forty workers involved in the preparation. "Nothing more entertaining than this description, which bears all the pomposity of a gasconade" the author of the *Mémoires secrets* reported.[73] The *Correspondance secrète*

---

[67] Musée de l'Air et de l'Espace, Le Bourget, fonds Montgolfier, Box VI, #58. The tickets bore an engraving of the steering machine, without the balloon: "Diligence aérienne de Mr. Campmas, ingénieur et physicien./Billet pour le départ de Mr Campmas et compagnie de voyage./Se trouve à Paris chez l'auteur rue Gît-le-Cœur quay des Augustins à l'Hôtel St-Louis. 1784" (BnF, Estampes, coll. Hennin, 10018).

[68] Feb. 24, 1784, pp. 112–113; Feb. 26, 1784, pp. 117–118.

[69] March 9, p. 142; March 11, p. 149.

[70] Bienvenu and Launay did the same with their heavier than air model. See P. Bret, "Un bateleur de la science : le 'machiniste-physicien' François Bienvenu et la diffusion de Franklin et Lavoisier", in Jean-Luc Chappey (ed.), "La vulgarisation des savoirs et des techniques sous la Révolution", *Annales historiques de la Révolution française*, n° 338 (Oct.–Dec. 2004), pp. 95–127.

[71] AAdS, PVAS 1784, f° 148v.

[72] Respectively June (first fortnight), 1784, pp. 39–40 and June 1784, pp. 497–498.

[73] The French term contains an ethnic connotation: "*gasconnade*" refers to inhabitants of Gascogne, a part of Guyenne – where Campmas came from – the Gascons, whose boasting was legendary. *Mémoires secrets pour servir à l'histoire de la République des lettres en France...*, London, John Adamson, 1786, t. 26, .pp. 258–259.

mocked Campmas again on December 1, 1784: Pilâtre's next aerial travel was now expected with much impatience and his machine was on display at the Tuileries, the very place where Campmas got hissed for his failed experiments. The inventor has now changed the shape of his device: "It is no more a tower, it is a horizontal cylinder that will be able to move in every direction. (...) But prudently, he never fixes the day of departure, & one does not hasten to subscribe, because the Abbot Miolan's promises are still reminded."[74] After so many failures of pseudo-aeronauts, people had confidence only in those who had already proved their ability to deliver, like Pilâtre: the first to have flown, he now planned to cross the Channel, which proved fatal to him and his companion the following month.

Thanks to this activity, and despite his failure, Campmas soon entered the annals of flying, for instance in David Bourgeois' 1784 *Recherches sur l'art de voler*, which mentioned Campmas' rowing wheel and "aqueous vapors", adding: "He has forgotten to explain how he could prevent the condensation of these vapors. He will be showered with praises, if he succeeds removing this obstacle".[75] The procedure for handling the "aqueous vapors" had been designed the previous year, as already mentioned. The rowing wheel was a human-driven wheel device that he had drawn on February 1 that same year for towing boats upstream, which he claimed to apply also to aerial navigation later (see Fig. 21.5). Referring to his previous sealed envelope, he then added: "Two months or so ago, I registered at the Royal Academy of Sciences, a memoir through which I propose to apply a similar mechanism in order to move horizontally Monsieur de Montgolfier's aerostatic machien. We will see the method in due time."[76] This was supposedly to be done, but the fact is that Campmas' aeronautical projects always remained fanciful.

When Campmas proposed his *Plan général des finances*, in 1800, a member of the Legislative Corps raised against it Campmas' previous failure in aeronautics. In reply the inventor added a special note following the main prospectus. He explained that he had lost all his wealth in the aerial enterprise before he had been called to "very major operations" – at Amboise – and had therefore not been able to carry on when he came back, since all his materials and machinery, kept in barracks, had been "devastated and carried away in the first troubles of the Revolution". Since he had no more devices to test nor money to build new ones, Campmas undertook a 400-page "Treatise of Balloonning" ("Traité d'aérostation") and presented it to the *Lycée des Arts* in

---

[74] *Correspondance secrète, politique et littéraire*, London, John Adamson, 1789, t. 17, 177–178. On July 11, Miollan and Janinet's balloon could not ascend and was eventually destroyed by furious – paying – spectators.

[75] David Bourgeois, *Recherches sur l'art de voler, depuis la plus haute Antiquité jusqu'à ce jour*, Paris, Cuchet, 1784, p. 88.

[76] AAdS, *pochette*, Feb. 18, 1784 (signed by Condorcet that day).

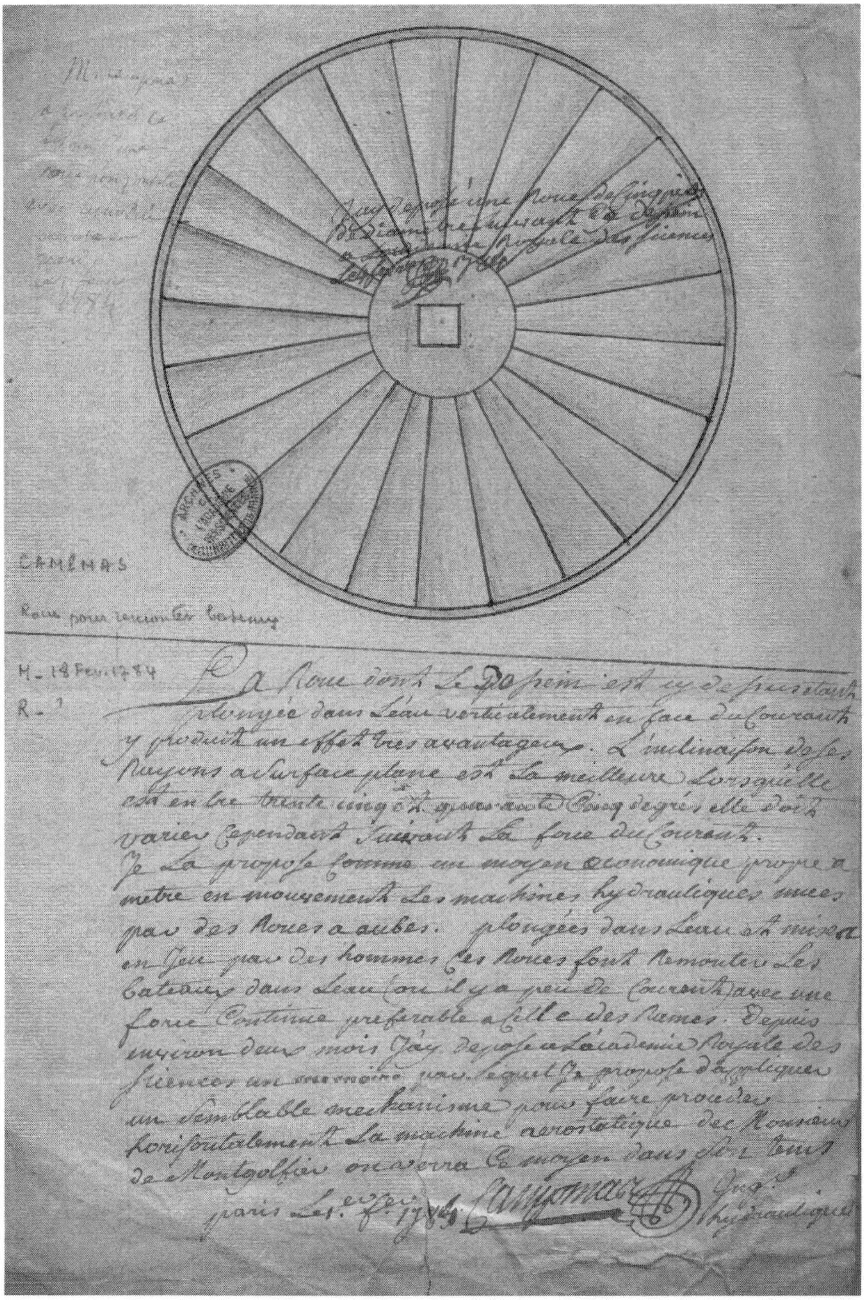

Fig. 21.5 Naval and aerial towing wheel, Feb. 4, 1784 (AAdS, *pochette*, Feb. 18, 1784). Top left: registration by Condorcet, same day, when Campmas presented a 5-feet model of the wheel at the Academy. Courtesy of the Academy of Sciences, Paris

1794. He asserted it was unanimously approved and awarded a medal and the most honorable mention.[77] The judges concluded that Campmas needed collaborators to build his machines, the smallest of which could carry 230 men, and added: "Considering attentively the enormous expenses that maritime navigation absorbs, we can easily think of spending three or four millions, if necessary, so as to realize an attempt, the success of which will bring rapture and happiness to all the peoples of the earth".[78] The treatise was never published. Probably it was not worth printing and nothing is known about the contents or the possible location of the manuscript, had it even survived. But Campmas did not drop his project for aerial navigation, and eventually adapted it for war.

On January 28, 1798, he sent to the Minister of War a "project for raid (*descente*) on England with a crew of 200 men, by a novel aerial vessel, intended to make France rich and bring happiness to all countries" – revolutionary rhetorics! – "an aerial vessel of a new kind, able to carry a crew of more than 200 men, who will be able to fly as they please and to cast lightning onto the British cabinet in order to make it capitulate".[79] On February 6, the minister, Scherer, named "Buonaparte", Fourcroy, Borda and Le Roy – of the National Institute – as reporters for the ministry. The general was busy inspecting the northern coastal forces, but the other members received the inventor at the Institute after their meeting on the 14th. Borda was a major theoretical and applied scientist, a pioneer of experimental hydraulics, the inventor of the geodetic "repeating circle" used for the new measurement of the Paris Meridian, and both a naval and military engineer, the reformer of the French naval building in the 1780s and currently a member of the commission for artillery trials at Meudon.[80] Rather skeptical about Campmas' proposal, he compared naval and aerial manœuvres and asserted that the power of aerial rowers would be nearly null – amounting on average to a fifth of their effort.[81] Campmas cited these doubtful remarks and dared to remark that Borda was a beginner in the field of aeronautics; in his next letter to the minister, he even asserted that the scientist had not read a line of either Campmas' memoir or the report of the other commissioners.[82] Indeed, a rather strange report was presented, apparently to Scherer, just before he was dismissed on February 21. The eleven-page minutes of the report concluded that the machine should be tried using governmental funding and asserted that the proposed means to move and steer the

---

[77] *Plan général des finances...*, op. cit., p. 4; SHD/DAT, 1 M 1161, #26 (Ventôse 14, Year 6/March 4, 1798); 6 W 66, dos. 6033(8), p. 3.

[78] *Plan général des finances...*, op. cit., p. 4.

[79] SHD/DAT, 1 M 1161, #23 (Pluviôse 27, Year 6/Feb. 15, 1798), #26 (Ventôse 14, Year 6/March 4, 1798).

[80] Jean Mascart, *La vie et les travaux du chevalier Jean-Charles de Borda (1733–1799). Épisodes de la vie scientifique au XVIIIe siècle* (1919), 2nd ed. Paris, Presses de l'université de Paris-Sorbonne (Bibliothèque de la revue d'histoire maritime), 2000.

[81] SHD/DAT, 1 M 1161, #23 (Pluviôse 27, Year 6/Feb. 15, 1798).

[82] *Ibid.* and #26 (Ventôse 14, Year 6/March 4, 1798).

balloon were "suitable for fulfilling their object".[83] After Borda's comments, it is doubtful that a positive report would have received his agreement: this could explain why Campmas claimed that Borda had not read the report. Bonaparte and Borda now being off the committee, and Fourcroy's expertise being mainly chemical, Le Roy was probably the reporter – he who had first supported Marat's works before retracting his recommendation.[84] On March 4, Campmas wrote to Milet-Mureau, the new Minister of War. He argued that the members of the Academy, who had all, excepting only Meusnier, rejected aerial navigation, would adopt his balloon only once it had actually returned from London.[85] Milet-Mureau responded that he had put the same group in charge of discussing Campmas' means of carrying out his plans with the inventor.[86] The project then spent two years moving from one minister to another, and from one commission to another.

Bonaparte's coup d'état changed the context. On January 25, 1800, General Berthier, the new Minister of War, put the Central Committee of Fortifications in charge of assessing the project, and the committee named General Chasseloup and Prieur reporters.[87] Four weeks later, Campmas was heard by the committee about his "aerial vessel intended to fight the fleet of England". As usual he refused to give any information, "saying that he was reserving the knowledge of the art processes as his own particular property",[88] and before any definitive decision, he asked for a provisional indemnity, proportionate to the sacrifices he had made for sixteen years to improve his machine, costs that he estimated at 25,000£. He further claimed that the various means he proposed for piloting balloons would be tried and that the best one would then be subjected to further tests. Therefore considering that the documents gave too vague and incomplete descriptions, the committee understandably refused on May 26 support the proposed experiments without further technical information. Since the inventor's works had not been demanded by the War department and had been of no use to it, the reporters judged that the requested indemnity depended on the Minister of the Interior, as "objects of arts and industry".[89]

---

[83] Public auction, Collection Chavaillon, #324, December 2–3, 2005, Bordeaux. I had seen only the first and last pages of this document.

[84] In addition to Gillispie, *Science and Polity in France at the End of the Old Regime.* Princeton, Princeton University Press, 1980, pp. 305–306, see Olivier Coquard, *Jean-Paul Marat.* Paris, Fayard, 1993; Jean Bernard, Jean-François Lemaire, Jean-Pierre Poirier eds., *Marat homme de science?*, Paris, Les Empêcheurs de tourner en rond, 1993, that includes an « Intervention du professeur Charles C. Gillispie », pp. 151–157.

[85] SHD/DAT, 1 M 1161, #26 (Ventôse 14, Year 6/March 4, 1798).

[86] SHD/DAT, 1 M 1161, #24 (Ventôse 19, Year 6/March 9, 1798).

[87] SHD/DAT, Génie, register Comité central des fortifications (CCF), Pluviôse 5, Year 8/Jan. 25, 1800, f° 133.

[88] SHD/DAT, Génie, CCF, Ventôse 2, Year 8/Feb. 21, 1800, f° 165.

[89] SHD/DAT, Génie, CCF, Prairial 6, Year 8/May 26, 1800, f° 290–291.

It is worth noting that at the same time that these last two proposals were made, another curious person, the attorney Charles Thilorier – whom we encountered above as one of Campmas' hydraulic competitors – proposed similar projects for attacking England. Campmas clearly belonged to a milieu of inventors, but his work on military matters reveals an unusual, concealed aspect to his particular inventiveness.

21.4 Steam Engine and Artillery in the Revolutionary Wars

The first mention of Campmas' interest in the military field was a gunpowder mill he claimed he had designed in the *Journal encyclopédique* in December 1781, followed by a steam boat and a steam carriage, both equipped with steam artillery, whose designs he presented to the Amboise authorities in August 1791. In these cases, his military inventions were listed among civilian ones.[90]

The context was quite different on August 15, 1792, when he offered ten military inventions to the Legislative Assembly. Accused of double-dealing, the monarchy had effectively collapsed, having fled the Tuileries palace to escape the furious crowd five days previously, France had been at war for nearly four months, and the Duke of Brunswick was close to invading the country with the formidable Prussian army and its Austrian ally. Since his return from Amboise, Campmas had shown himself to be in accord with revolutionary aims, and he received honors at the Assembly's meetings in July for a projected monument entitled the "Tree of Liberty".[91] He now proposed a large number of inventions for military manufacturing (steam-driven gun casting and boring, firearms factory), as well as new weapons (recoilless firearms, canister guns, muskets and pistols), mobile carriages (steam cart, steam oven and mills), and fortifications (mobile redoubts). The most ambitious was a "New portable arms factory with a novel steam engine (*pompe à feu*) for main power, the fuel of which would cost very little, because it would be used at the same time to heat, forge, anneal, pierce, file down, turn and polish pieces, thanks to secondary machines driven by the first one".[92]

The day following Campmas' proposal, the Assembly forwarded his memoir to the Arms Committee (*Commission des armes*).[93] However, having received no answer after several months, the engineer withdrew his proposals on May 21, 1793, which was by then in the hands of the Arms Section (*Section des armes*) of the Committee of Public Safety of the new National Convention, which had

---

[90] *Journal encyclopédique*, Dec. 15, 1781, pp. 516–519; SHD/DAT, 6 W 66, dos. 6033(8), pp 3–5.

[91] James Guillaume, *Procès-verbaux du Comité d'Instruction publique de l'Assemblée législative*. Paris, Imprimerie nationale, 1889, p. 281.

[92] SHD/DAT, 6 W 66, dos. 6033(7), p. 2, item 10.

[93] For the organization of assessing inventions in the military field, see Bret, *L'Etat, l'armée, la science...*, op. cit.

proclaimed the French Republic the previous September. Campmas in any case confessed three years later that he had at the time possessed "no other knowledge in artillery than that inspired by the natural language of reason and the will to be useful to my motherland".[94] He then proposed to the same committee to convert a steam engine he was building on his own for "another public utility use" into a piercing machine, which would be able to produce 48,000 musket guns per year. However, the delays of the bureaucratic treatment of the affair were so long that the Paris Manufacture of portable arms ceased its own activity before his proposal was set up.[95]

In addition to steam engineering, Campmas continued work in a field for which he did have practical knowledge, namely hydraulics. In May, 1794, the war effort was at its zenith. Referring to his success at the Amboise manufacture and his award from the Consultative Board, he proposed to the Convention to convert the Marly machine into an arms factory, and after the peace into "the most brilliant factory in the whole world, for metals suitable for all the arts". Lazare Carnot, who was chairing the assembly, applauded and offered Campmas the meeting's honors, while the project was forwarded to the Committee of Alienations, then to the Provisional Commission of Arts (*Commission temporaire des arts*), where it was entrusted to the mechanical section of *Ponts et chaussées*. In July, Campmas sought an answer concerning the military part of his proposal. As we have seen above, Prony and Molard had considered only the mechanical part of the Marly machine in their report, having ignored the arms factory project, the one to which the inventor referred six years later when he published similar projects for the *Salpêtrière* and the *Ecole militaire*.[96]

New artillery carriages Campmas designed met a better reception. Two months earlier, on March 21, the revolutionary authorities of his district (*section de la Cité*) sent a deputation to the National Convention, offering to it thirteen models that the engineer had designed and built, including nine artillery carriages (for land and naval service) – in addition to two light ambulances and two pikes, the latter symbolically representing the *sans-culottes'* valor. These were forwarded to the War Committee, to which the inventor hastened to write.[97] Then, ten days after a positive report was issued by the *Jury des armes et inventions de guerre*, signed by the mining engineer Alexandre Miché as secretary, the military engineer and principal in charge of arms production, Prieur de la Côte-d'Or, proposed, and the Committee of Public Safety accepted on June 14, that Campmas should test his "flying artillery" two-wheel carriage

---

[94] SHD/DAT, 6 W 66, dos. 6033(8), p. 1.

[95] *Ibid.*, p. 3.

[96] *Réimpression de l'ancien Moniteur*, vol. 20, Paris, Bureau central, 1841, pp. 525–526 (Floréal 30, Year II/May 19, 1794); Louis Tuetey, *Procès-verbaux de la Commission temporaire des arts*. Vol. I, Paris, 1912, pp. 211–212 (Prairial 15/June 3), 275 (Messidor 20/July 8).

[97] SHD/DAT, 6 W 66, dos. 6033(7).

for a 4-pounder gun, which was the lightest caliber among the French ord-nance.[98] This was in fact done at Vincennes on July 25. Generals Favereau and Drouas, director and deputy director of the Paris Arsenal, favorably wrote one week later of "the genius and the intelligence that had directed the fabrica-tion of [Campmas'] field carriage", asking to have it tested at least on the field of the *École de Mars*, in the Paris suburbs, until the end of the campaign the following Fall.[99]

Artillery officers, military engineers and several inventors from the Consulta-tive Board, the *Jury des armes*, and the *Lycée des arts* also attended those trials at Vincennes, notably Targe, president of the *Lycée*, Detrouville, Dumas, and Desaudray, members of both the Board and the *Lycée*. An improved version of the full size carriage, now including a fountain for watering the cannon, was subsequently exhibited in the centre of the main hall of the *Lycée des arts*, in the gardens of the Palais Royal. At its general assembly of November 30, in view of the far cheaper price of Campmas' carriage (500£ vs. 2 to 3,000 for a regular field carriage) and the great advantages in its construction and mobility, the *Lycée des arts* awarded Campmas honorable mention and a medal bearing a crown. It also decided to send copies of the report by the above mentioned members to the committees of the National Convention....[100]

Despite these positive evaluations and recommendations, the inventor faced a bureaucracy which resulted in his proposals being "endlessly forwarded from one office to another by second rank agents". Two years later, on October 3, 1796, Campmas once again took up his the pen and wrote to the Minister of War, claiming that he had spent the equivalent of more than 50,000£ in gold on the military inventions alone, and had now to sell them at a cheap price "in order to be able to survive". He also sent abstracts of his previous proposals to the Legislative Assembly and National Convention and copies of the reports that had been issued concerning them. Finally, on November 17 the minister Petiet asked the Central Committee of the Artillery to send a report about these military proposals, which Generals Aboville, Fabre de La Martillière and Drouas soon wrote.

The worst military crisis of the 1790s was now past, and the treasury was empty, while the innovative actions taken in the Revolutionary Year II were regressing: the Consultative Board had closed in spring, as had the research center for secret weapons created at Meudon in Fall 1793, while the twin center on military ballooning was now attacked, both at the legislative assemblies, like the *École polytechnique*, and by the Generals, especially since the balloon and aeronautic gear of the Rhine Army had been taken by the Austrians at Würzburg in September, when Jourdan retreated.[101] Given all of this,

---

[98] *Ibid.*, Titre 3, p. 5 (Prairial 16, Year II/June 4, 1794); AN, AF II 220, #2 (Prairial 26/June 14).

[99] *Ibid.*, Titre 4, pp. 5–7. The *École de Mars* was a revolutionary training school for soldiers.

[100] *Ibid.*, Titre 5, pp. 8–11.

[101] Emmanuel Grison, "Les premières attaques contre l'Ecole polytechnique (1796–1799). La défense de l'École par Prieur de la Côte-d'Or et Guyton de Morveau", *Bulletin de la Société des amis de la bibliothèque de l'Ecole polytechnique (SABIX)*, 8 (Dec. 1991), 1–24; P. Bret,

Campmas' praised field carriage was neither adopted nor further tested, but rejected by the Committee. However, the inventor was still encouraged by the opinion of the "three Generals learned in the science of machines" concerning his steam carriage and gun: "New fire-powered carts to facilitate the hauling of artillery gear and war munitions. It can also shoot bullets without powder".[102]

To Campmas' new memoir, on November 23, 1796, a table titled "The propagation of fire, or the Twelve Sisters" was added. It contained drawings and descriptions of various steam engines, together with the minutes of the visit of his collection of models at Amboise by fourteen administrators of the district and town, which had taken place as early as August 18, 1791:[103]

The first, called *La Division*, the model of which M. Campmas showed us, is intended to saw (reffendre) big pieces in the forests by the action of fire so as to make beams, ribs and planks, and to carry them out of the woods.

The second, called *La Générale*, the model of which we have also seen, is intended to replace by the action of fire all the machines run by streams of water.

The third, called *L'Orageuse* (The Thundery), the model of which we have also seen, is intended to run by fire all the machines necessary in metals manufacturing.

The fourth called *L'Amazone* is a steam carriage intended to run without the assistance of horses.

The fifth, called *La Baleine* (The Whale) is intended to sail ships by the action of fire and without the assistance of wind; to provide them with drinking water; to pump out water if necessary; to renew the air; and to shoot bullets without gunpowder.

The sixth, called *L'Abeille* (The Bee), is intended to have boats sailing upstream, and to pilot some aerial vessels that this engineer promises to let soon be known.[104]

The seventh called *Le Bon Patriote* (The Good Patriot) is a novel, fire-powered hammer that could be said universal, because it could be set up in all the countries around the world next to an oven, and the fuel burnt [by that oven] will be sufficient to run it.

The eighth, called *La Révolution*, is a novel, cheap means to run by continuous fire of the ironworks all the machines necessary to their exploitation, and to bore guns without any water stream or special fuel.

The ninth called *L'Abondance* (the Abundance) is a new, fire-powered cart intended for plowing and for every kind of haulage, even for shooting bullets without gunpowder, thanks to secondary means.

The tenth called *La Française* (the French) is a new and simple fire pump, without any apparent pendulum or injector. M. Campmas showed us a model conformable to the drawing.

The eleventh called *L'Aube* (the Dawn), is a steam engine (*machine à feu*) without any piston. The author dedicates it to the memory of the famous Papin, who was the first to think of pumping water by means of fire. The author proposes to use it for the

---

"Recherche scientifique, innovation technique et conception tactique d'une arme nouvelle: l'aérostation militaire (1793–1799)", in Jean-Paul Charnay ed., *Lazare Carnot ou le Savant-citoyen*, Paris, Presses de l'Université de Paris-Sorbonne, 1990, pp. 429–451.

[102] SHD/DAT, 6 W 66, dos. 6033(8), p. 6. The steam cart was Item 8 in August 1792 (dos. 6033(7), p. 2).

[103] SHD/DAT, 6 W 66, dos. 6033(8), p. 5.

[104] This item was developed in the above mentioned "Treatise on Ballooning" (cf. footnote 76).

maintenance of new seaports and ship canals that he has invented. We have also seen the arrangement of a model he had used.

Finally, the twelfth, named *La Balance* (the Scale) is a novel, double-effect steam engine (*pompe à feu*), simpler and less expensive than English machines. Its purpose is to raise water to any required level.

Did these engines actually work? Gitton, the vice-president of the district of Amboise, and his son Sylvain testified that Campmas had brought a model of *L'Orageuse* to Paris in May 1789, in order to have it made in full size, at a time when France was mainly concerned with the opening of the *États généraux*. That model was eventually made more powerful and faster in operation later, thanks to an alembic, what had not been used at first. All the local authorities who visited him at Amboise in August 1791 were granted a demonstration:

Soon after the fire came into action, M. Campmas simply turned the key of a small tap on the top of the machine. Immediately, the latter began a very fast rotary movement that made the hammer stamp. We are not able to calculate its speed. But the author was able to moderate the action as he pleased by the means of the key of the same tap, by turning it to a lesser or greater extent. We noticed that the machine kept working as long as the fire was active.[105]

In November 1796, Campmas proposed to build for military haulage a full size steam carriage based on his ninth device, called *L'Abondance* because it was originally designed for plowing, this being the one first proposed to the Legislative Assembly four years earlier. He required 12,000£ to build it, for which he offered all his present and future belongings as collateral and asserted that, although four months would suffice to build it, the carriage should be delivered only after a two hundred-league experimental journey on roads. He promised further advantages: a crew of only two men (one for fuel, one for steering), obviously no need to feed horses even when they are not working, very low costs, etc. He asked as well support in the amount of 6,000£ to build full size models of his field carriages for 4, 8 and 12-pounder guns in order to field-test them.[106]

Although Generals Aboville, Fabre de La Martillière and Drouas again rejected both of Campmas' proposals three weeks later, they did propose to grant him 1,340£, i.e. the cost of a regular 24-pounder carriage, seen as a fair indemnity for his construction of the "flying artillery" carriage,

1° because his patriotism has led him to turn his inventive genius to objects that he thought the most useful to the Republic

2° because he has built his carriage at his own expense, and did that only following an order of the Committee of Public Safety on Prairial 26, Year 2nd [June 14, 1794].[107]

---

[105] Ibid., pp. 3–4.

[106] Ibid., p. 6. In February 1798, Campmas gave a note on his field carriages to the National Institute, that was read and sealed on March 1 (*PVAS* I, p. 353).

[107] SHD/DAT, 6 W 66, dos. 6033(9), Frimaire 26, Year 5 (Dec 16, 1796). See also 6 W 65 and 6 W 75.

Even though the context was not propitious for developing inventions, the above reporters remained open to innovations. General Aboville himself, the president of the Central Committee of the Artillery, often proved his continuing interest in such things.[108] At Campmas' workshop, they had visited the collection of military machines which he had invented, including carriages, guns, and a caisson ambulance, and remarked that all of them should be purchased by the government, since many contained in their smallest details ideas that could be of profit to artillerists in charge of improving ordnance. They were especially impressed by Campmas' new proposal: "a steam engine (*machine à feu*), the power and the robustness of which will be such that it will be easy to carry on the main roads ten thousand pounds with a speed of two leagues per hour; its daily consumption of coal will not exceed 600 pounds. Therefore, the cost will be only 20 francs for carrying 10 thousand on 48 leagues, what is less than one penny (*denier*) per pound". More generally, the experts of the Artillery appreciated his other major inventions in the field:

> It seems to us, that [Campmas' inventions] must lead to most useful things. There are applications of the steam engine (*machine à feu*), which has been too neglected as a means for replacing everywhere the raising forces that Nature does not everywhere nor every time supply. The assistance of fire is very powerful and less expensive than that of horses. The extreme complication of those first machines has probably delayed the progress of this precious invention. But it is only by repeating and multiplying attempts, that we shall reach that greater simplicity which most machines acquire only over time. Citizen Campmas seemed to us have made a few steps forward for those machines of which we speak. He showed us the designs of a great number of the applications which he had envisioned, including one for replacing horses for hauling. We know that this test has already been done.[109] Even should Citizen Campmas' efforts prove fruitless, he nevertheless deserves the gratefulness of the government: he will have smoothed the path by which another will succeed. Therefore, we think that Citizen Campmas' works for improving steam engines and their various applications are worth being encouraged.[110]

Three days later, General Milet-Mureau, on behalf of the minister (Petiet) made an offer to the inventor. Agreeing to sell his artillery models, Campmas suggested that he should add their plans too, which had been at the Consultative Board, then the National Institute for the past fifteen months.[111] He also asked for an advance of 900£ to execute a full size model of his three-horse ambulance which could carry some twenty people, with couchettes for the most wounded.

---

[108] He defended Fabre's shells (*boulets creux*), and was later himself an inventor. Charles Gillispie, "Science and secret weapons development in Revolutionary France, 1792–1804: A documentary history", *Historical Studies in Physical Sciences*, 23:1 (1992), 35–152; Bret, *L'Etat, l'armée, la science...*, op. cit., pp. 310–312, 346.

[109] Reference to Cugnot's experiments around 1770.

[110] SHD/DAT, 6 W 66, dos. 6033(9).

[111] Lastly, on February 19, he presented a memoir on a new construction for field carriages and a new shape of gun to the president of the first class of the National Institute, that was read at the next meeting (Feb. 24), then sealed on March 1. *PVAS* I, 353.

Finally, he said he would sell the military carriages only after receiving a patent (*brevet d'invention*).[112]

On January 5, 1797, Petiet urged the Central Committee of the Artillery to speak with Campmas about his steam-driven hauler. Encouraging private research and funding its development are not the same, particularly in a time of budgetary restrictions. The negotiations failed, since Campmas wanted "to keep, as his secret and property, the means he uses", as duly recorded by the still innovation-minded members of the committee (Aboville, Pommereul, Lariboisière, Villantroys), who delivered further, severe conclusions on February 1. The 12,000 £ asked by Campmas to build his carriages and to prove the feasibility of his proposals were judged to be highly exaggerated. Moreover, steam carriages had never proved their efficiency, and even though they might succeed on the roads, they could not be used in artillery movements on the field, and could therefore not replace horse carriages. Consequently the inventor was invited to "to turn his aims and applications toward commercial haulage, which will always present him with more assured awards than those the government can offer."[113]

Fig. 21.6 "12th machine for towing boats, 1st means for towing boats with a steam engine", Paris, Apr. 23, 1774 (AAdS, *pochette*, Dec. 7, 1782). Fanciful punting with mechanical legs, which preceded Campmas' mobile cogwheel could use several sorts of power. Courtesy of the Academy of Sciences, Paris

---

[112] SHD/DAT, 6 W 66, dos. 6033(10) (Nivôse 4, Year 5/Dec. 24, 1796).

[113] SHD/DAT, 6 W 66, dos. 6033(11) (Pluviôse 13, Year 5/Feb. 1, 1797).

Several inventors and entrepreneurs dealt with the steam engine in France, including major ones like the Perier brothers or the Spanish engineer Betancourt, who did have some industrial success, or Cugnot, Jouffroy d'Abbans, Robert Fulton and Charles Dallery who succeeded with machines for mobility. But none proposed such a huge range of applications as Campmas. For decades he conceived many applications of the steam engine, both as fixed devices in arms factories and at the Mint, and in mobile devices, from his very first project of a legged steam tugboat in 1774 (see Fig. 21.6) to his carriages for hauling, which were definitively rejected in 1797. There was even the last work that he presented to the National Institute on December 5, 1803, for which he asked only a dated certificate: "New naval and continental artillery, proved by twenty-three drawn figures and accompanied by manuscript details", which probably included steam-powered carriages and boats and steam guns, as in Campmas' previous inventions.[114] For him, who was never able to build a full size version of his designs, the steam engine was a kind of universal power source adaptable to any and every mechanical purpose.

21.5 Conclusion

In November 1796, Campmas decried what he saw as inconstant governmental decisions concerning the inventions that he had proposed and even constructed over "twenty three years (...) in Paris and in the departments, with money, sometimes from companies and sometimes from ordinary persons who trusted me". He bitterly constructed a story of bitter rejection, the story of an inventor harassed by the experts of administration and scientific institutions:[115]

Twenty six years of my live have been spent in useless proposals that I made to the government. I say useless, because, people more skilled than me in dealing with the government have been able to profit from the proposals I started, and from the far too long delays of the reports required by my applications; they have been able, I say, to bury my work, while they had their own, useless projects adopted, which only caused the ruin of the Nation. As evidence of my assertion, I can cite the enormous amount of quite expensive machines set up for portable firearms, machines that were eventually destroyed.

Complicated, repetitive and tedious though Campmas' story maybe, it deserves to be known in detail, as a way to look inside the black box of invention and assessment, to visualize and measure the tremendous work required from ordinary inventors to gain credit, as well as their need to work within existing and to forge a new social network, in both the previous aristocratic as well as the new revolutionary contexts – as nicely exemplified by the societies of inventors. Campmas deployed about twenty different direct and indirect ways to have his

---

[114] *PVAS* III, pp. 29–30 (Frimaire 13, Year 12).

[115] SHD/DAT, 6 W 66, dos. 6033(8), p. 6.

works approved and to generate support from many different kinds of institutions – a dozen members of the Royal Academy being the better known public ones, followed by the more than thirty civilian and military experts who assessed his inventions. There were also private sources to be sought, sources that he hoped to activate by means of announcements in journals and through the sale of his engravings. The only source that he never used was the new system of patents that the societies of inventors obtained in January 1791,[116] for it appeared to him to be too expensive and insufficiently protective, unlike the sealed envelopes of the Academy.

This case study illuminates, on the one hand, the increasing bitterness of the rejected inventors towards the Academy and the bureaucracy; and, on the other hand, the increasing irritation of the experts, who could not carry out their tasks without full information concerning the proposed inventions, information that the inventors were reluctant to yield for fear of losing control. This was the major stumbling block between those who were in charge of assessing invention and the numerous frustrated inventors, many of whom – like Campmas – eventually joined the societies of inventors during the Revolution. The gap grew ever wider between inventors to one side and the savants and even the military experts to the other. The latter two groups argued for an open science and technology, while many inventors who thought secrecy to be their best protection. Nevertheless, Campmas, like many inventors, did benefit to some extent from the Revolution, thanks to new institutions and the growing presence for the first time of non-academic experts. The academicians of the Consultative Board of arts and crafts – half of the members – and the *Lycée des arts* remained uninvolved when the institutions to which they belonged strongly supported Campmas: the reporters for the Amboise machinery at the Board and for his artillery models at the *Lycée*, for example, were all members of the *Société du Point Central*, like Campmas himself.

His too short or too long but still vague memoirs often palmed off old projects with small corrections. His desperate applications, his bitter claims and plaintive logorrhea are similar to those of many other inventors, who jealously guarded their inventions, and who were quick to challenge their judges, to claim priority and to charge their competitors with theft. Their bothersome behavior repulsed the experts, who could not correctly assess proposals on the basis of insufficient data. Nuisance to the experts their behavior may have been, it often saved the inventor from failure. Most likely they would often have failed since they lacked appropriate knowledge for the elaborate devices that they had in mind. But these failures themselves might have led

---

[116] See Gabriel Galvez-Behar. *La République des inventeurs: Propriété et organisation de l'innovation en France (1791–1922)*. Rennes, Presses Universitaires de Rennes, 2008; Christiane Demeulenaere-Douyère, "Les pétitions et le vote des lois protectrices de l'invention en 1791", in *L'individu face au pouvoir: les pétitions aux assemblées parlementaires, Revue administrative*, special issue 2008, pp. 61–69; id., "Inventeurs en Révolution: la Société des inventions et découvertes", *Documents pour l'histoire des techniques*, 17 (2009), pp. 19–56.

to further improvements and cumulative technological progress. Campmas clearly lacked scientific knowledge, and so he was not able to compete successfully with increasingly better-trained engineers. Still, though he nearly always failed to have his inventions developed, there was the exception of his participation in a major hydraulic operation for an importantly-innovative industrial project supported by the government, namely the one at Amboise. Whatever the actual feasibility and reliability of his devices may have been, it is worth noting the visionary spirit of his proposals, especially regarding steam engines. His drafts of these sorts of inventions deserve to be known, not because he was a precursor of later developments – that would be nonsense – but to emphasize the particular quality of his inventiveness in an unusually inventive context, that was encouraged and intensified by the government in wartime. Campmas' vision of one and the same device for both plowing and cannoneering was in a sense realized in the twentieth century through the famous T-34 Soviet tank, which was designed at the Stalingrad tractor plant.[117]

Braggadocio and reality mix together in the proposals of these fertile minded but paranoiac inventors, and it is often difficult to sort truth from half-lies and outright falsehood. Yet Campmas was not a faker, and he did have some expertise, enough to achieve the occasional, if rare, success, at the two opposite sides of the process of technological invention, namely in his first general idea of a novel device for a new application, which could seem fanciful to others, and in his micro-innovation of details, i.e. in the design of minor, but sometimes crucial, improvements. Unlike Conté with his balloons at Meudon and other achievements in Egypt, and unlike many other inventors who also belonged to the scientific milieu, Campmas never accepted that he ought to work in an open world. Unlike Captain Fabre and his secret weapons at Meudon, carefully studied by Charles Gillispie,[118] Campmas never joined alliance between the military and scientific worlds of the Year II. His story bears witness to the difficulties faced by an ordinary inventor in France at the end of the Old Regime and in the Revolutionary and early Napoleonic years.

Acknowledgement I am grateful to Jed Buchwald for his careful editing and polishing my text, and to Florence Greffe, Claudine Pouret, Marie Thebaud-Sorger and Christiane Demeulenaere-Douyère for their kind help.

---

[117] Yves Cohen, « Technique et politique : une histoire réciproque (France et Union soviétique entre les deux guerres), in *Artisans, industrie. Nouvelles révolutions du Moyen Âge à nos jours*, eds. Natacha Coquery, Liliane Hilaire-Pérez, Line Sallmann and Catherine Verna, Lyon, ENS Editions – SFHST, 2004, 227–236 (p. 235). The anti-Bolshevik writer Jean M. Rivière prophesized that "the inoffensive tractors that serve as agricultural machines at peace can easily be transformed into gun tractors or tanks at war" (*L'URSS dans le monde: l'expansion soviétique de 1918 à 1935*, Paris, Payot, 1935).

[118] Gillispie, "Science and secret weapons development...", op. cit.

Chapter 22
Cauchy's Theory of Dispersion Anticipated by Fresnel

Jed Z. Buchwald

In 1836 Augustin-Louis Cauchy (1789–1857), having left Paris and settled in Prague following the July Revolution, published a memoir on the dispersion of light under the auspices of Prague's Royal Society of Sciences.[1] In it he produced an equation that is even today known as Cauchy's formula for dispersion. It works reasonably well for normally dispersive bodies and was only replaced towards the end of the 19th century following the discovery of anomalous dispersion in Denmark by C. Christiansen in 1870 and consequent changes in theory by Wolfgang Sellmeier and Hermann von Helmholtz (1821–1894) in Germany.[2] In his publication Cauchy nowhere referred for inspiration to Augustin-Jean Fresnel (1788–1827), the originator in France of wave optics. Instead, he wrote that Gustave-Gaspard Coriolis (1792–1843), having read Cauchy's earlier work on the equations of motion that govern a system of material points, suggested that terms which Cauchy had there neglected might account for dispersion – assuming that the medium, or ether, that was presumed to carry optical radiation is itself so constituted.[3]

Cauchy's effort in optics was preceded by his major innovations in elasticity theory, which he was stimulated to investigate when he read a paper by Claude-Louis Navier (1785–1836) on the theory of elastic plates. Navier had submitted the paper to the *Académie des Sciences* in Paris on August 14, 1820 and had given lithographic copies to a number of academicians, including Cauchy. An abstract was printed but only in 1823. However, in 1821 Navier developed a full theory of elasticity based on a consideration of forces between particles, which

---

[1] Cauchy, A.-L., 1836; Cauchy, A.-L., 1836 (1895). The memoir had originally appeared the previous year, published as a separate installment of Cauchy's ongoing series, the *Exercises de Mathématiques* (Cauchy, A.-L., 1835).

[2] On which see Buchwald, J. Z. (1985, pp. 233–37).

[3] Cauchy, A.-L., (1836, p. 1 1836 (1895), p. 196).

J.Z. Buchwald (✉)
Division of Humanities and Social Sciences, California Institute of Technology, Pasadena, CA 91125, USA
e-mail: buchwald@its.caltech.edu

J.Z. Buchwald (ed.), *A Master of Science History*, Archimedes 30,
DOI 10.1007/978-94-007-2627-7_22, © Springer Science+Business Media B.V. 2012

"

he read to the *Académie* on May 14 but which, like his paper of 1820, remained out of print and under evaluation. This major work on elasticity was not published until 1827.[4]

Cauchy informed the *Académie* of his own results concerning elasticity on September 30, 1822, though he neither read his paper at a weekly meeting of the organization nor did he deposit a manuscript with it, as was customary. That paper nevertheless contained his path-breaking introduction of the concept of stress, there introduced without any consideration of the physical structure of the elastic body, which he treated as continuous. Cauchy contributed a summary account of his results in the *Journal de la Société Philomatique* four months later in its January issue, wherein he did note that he had undertaken his research "on the occasion of a memoir published [publié] by Navier on August 14, 1820."[5] At that time Navier's paper had of course not been printed in a journal, just lithographically copied, though Cauchy's "publié" no doubt simply meant "made known." The abstract of Navier's 1820 paper on elastic plates in fact appeared in the same journal pages after Cauchy's.

In the summary account, Cauchy recalled having spoken with Fresnel. He had just developed his own approach to elasticity when "… M. Fresnel came to talk with me about some investigations on light that, as yet, he had only presented in part to the Institut. I learned that he had obtained a theorem analogous to my own, his result being based on certain laws according to which elasticity emanating from a single given point varies in different directions."[6] The "theorem" to which Cauchy referred derived ultimately from Fresnel's efforts to deduce what he termed an optical "surface of elasticity," whose properties he succeeded in developing shortly before March 1822. The problem that had given rise to his search for such a thing concerned the behavior of light in crystals. To solve this required a generalized form of the wave front, and that, he knew, would be a complicated matter to deduce.[7] Fresnel accordingly needed a reasonably straightforward way to reach it, hence the much simpler elasticity surface, which has the form of an ellipsoid with three unequal axes. It has special sectioning properties from which the intricate wave surface could ultimately be found (though even here Fresnel had to take a shortcut), and he justified these properties by invoking the physical characteristics of the ether.[8] As in contemporary French understanding of material elastica, Fresnel's

---

[4] Navier, C.-L. (1827).

[5] Cauchy, A.-L., 1823 (1958).

[6] Translated in Belhoste, B., 1991, p. 94 from Cauchy, A.-L., 1823 (1958), p. 301. Belhoste details the events surrounding Navier's and then Cauchy's presentations.

[7] The general wave surface is indeed complicated, satisfying as it does the following equation for a surface of two sheets: $\frac{a^2 x^2}{r^2-a^2} + \frac{b^2 y^2}{r^2-b^2} + \frac{c^2 z^2}{r^2-c^2} = 0$ where a,b,c are constants pertaining to the specific biaxial crystal. If two of the constants are equal to one another, say b,c then this reduces to the wave surface in a uniaxial crystal (viz a sphere and an ellipsoid), which was the only type known until the 1810s.

[8] Buchwald, J. Z. (1989b, pp. 260–90).

optical ether was thought to consist of interacting particles. Navier had in fact used just that model in his 1820 paper for material bodies, as well as in the later paper of 1821, though in 1820 he had not developed it in detail – with the result that a major controversy with Siméon-Denis Poisson (1781–1840) was to erupt over the subject in later years.

After considerable efforts beset by initially deceptive paths, Fresnel in due course succeeded in constructing an empirically workable "surface of elasticity," one which has the necessary property for his purposes that a displacement along any one of its semiaxes gives rise to a parallel restoring force. He was able to deduce the surface from the proposition that the force generated in reaction to a displacement is a linear function with constant coefficients of the displacement's components along three mutually-orthogonal directions. That is, the relation between reaction and its generating displacement involves what in modern terms is a linear transformation with constant coefficients, whose symmetry Fresnel demonstrated on the basis of a balance of moments. Fresnel did not however by his own admission provide a physically-acceptable foundation for that critical proposition since he had found it necessary to make the obviously unphysical assumption that the reaction to an ether particle's displacement could be calculated by shifting it alone, leaving all the others in situ.[9]

That linear transformation was the specific result to which Cauchy referred in his 1823 remark, though his form of it for material bodies emerged, unlike Fresnel's for the optical ether, from general considerations of symmetry. His version specified that the directed force on a given plane subject to elastic deformation can be found from a linear transformation applied to the plane's normal. In effect, Fresnel's ether displacement stood in the same relation to the force that is associated with it as Cauchy's normal to a given plane within a deformed elastic body stood to the corresponding force on that plane. Cauchy accordingly recognized that his transformation was the same in essence as Fresnel's, and that it led to what has since been termed the stress quadric – which is also Fresnel's "surface of elasticity."[10] These results eventually enabled Cauchy to develop general equations of motion for elasticity without relying on any particular physical model, e.g. of material points.

We do not know precisely what Fresnel and Cauchy discussed in their meeting, which must have taken place between late 1821 and the early fall of 1822. He left vague just how far their colloquy had extended, for he mentioned only that Fresnel had obtained a "theorem analogous to my own," viz the linear transformation in question, as well as the associated surface. He did not write anything about Fresnel's foundation of the "theorem" in particle interactions (faulty though that foundation was even to Fresnel). However, it seems quite

---

[9] The sequence of Fresnel's investigations is complex: see Buchwald, J. Z. (1989b, pp. 260–90) for details. Fresnel's final surface of elasticity has the equation $r^4 = a^2x^2 + b^2y^2 + c^2z^2$.

[10] He remarked in a note that from the 'theorem' in question there resulted a surface with properties that "agree with the final researches of Fresnel."

likely that, until then, Cauchy had not considered the ways in which his new conception of stress could be derived from, or at least connected to, the kind of model that Fresnel had deployed, and with which Navier had also worked in the paper that he had read to the *Académie* in 1821. Still, Cauchy did note that they had discussed Fresnel's "investigations on light," and it is entirely possible – indeed, we shall see quite likely – that their discussions ranged over more than the issues raised by the transformation, the stress quadric, and its consequences for developing the several surfaces needed for Fresnel's final theory of birefringence. In any case, it seems reasonably certain that Cauchy's further developments were at least stimulated by his discussion with Fresnel.

The detailed presentation of Cauchy's new theory of stress appeared in print five years after his meeting with Fresnel, in the second volume of his *Exercises de Mathématiques* (1827). An addendum to the piece presented a deduction of the symmetric transformation that is implied by the three general theorems which Cauchy developed in the body of the article. That deduction relied for the first time on the basic elements of the particle-based model that Fresnel (and Navier) had deployed. Here Cauchy noted that the transformation and "many propositions which can be deduced from it and which are analogous to the theorems I, II, III" of his own presentation were "due to Fresnel." He continued to leave open the question of whether he, Cauchy, had also learned anything from Fresnel with respect to the underlying deductive structure whose elements he outlined.[11]

Years before, in fact not many months after their meeting in 1822, a tense situation had developed between Cauchy and Fresnel when Cauchy preempted Navier by publishing his 1823 summary in the *Philomatique*. This had generated an angry reaction on Navier's part that eventually drew in Fresnel. Neither of Navier's memoirs had been printed or even formally evaluated by the *Académie*, and Cauchy's summary only gave backhanded acknowledgment to Navier's less-general paper on plates as marking the moment at which his own research had begun. In fact, the entire first paragraph of the summary was devoted to dismissing Navier's geometrical distinction between forces of flexure and forces of dilatation, replacing both with a unified concept of stress. Cauchy, recall, had told the *Académie* about his results the previous September 30. That alone had been enough to distress Navier, who wrote the *Académie* on October 6 asking to have his papers rapidly evaluated, remarking without comment that Cauchy was working a similar vein. Matters rapidly turned worse with the *Philomatique* publication, and in March Navier spoke of his own papers before that society, with an "extract" of his research on elastic plates and a second concerning his general theory appearing in the society's journal. Support turned to Navier, and Fresnel jumped in with a strong, public note criticizing Cauchy.[12]

---

[11] See equation 12, Cauchy, A.-L., 1827 (1889), p. 81.

[12] Belhoste, B. (1991, pp. 97–8): *cf* Fresnel, A. (1823) and Navier, C.-L. (1823a); Navier, C.-L. (1823b).

Fresnel's critique primarily concerned Navier's second memoir (his general theory of elasticity). Cauchy's summary seemed to be quite similar to that work, Fresnel remarked, work that had been talked about before the members of the *Philomatique*. "It is important," Fresnel went on, "that the date of this paper [Navier's second, general one] be recalled and certified."[13] The tone of Fresnel's remarks clearly shows, as Belhoste notes, that he believed Cauchy had likely taken results from Navier. At the time Fresnel thought that Cauchy was one of the examiners assigned by the *Académie* to evaluate Navier's memoir, though in this he was mistaken since the official examiners were Gaspard de Prony (1755–1839), Poisson, and Jean-Baptiste Fourier (1768–1830). Fourier himself entered the fray on April 24 when, in reporting on the *Académie*'s events from 1822, he noted that Cauchy had presented a paper on September 30 in which he cited Navier and Fresnel as having "already treated questions of the same kind." He went on explicitly to note that Navier had given two papers, and that Fresnel's optical work had led Fresnel to examine "the properties of vibratory motions that occur within the interior of elastic bodies."[14]

Fresnel's sharp note critiquing Cauchy may have reflected his own experiences at the time with respect to publication. In the early summer of 1821, a confrontation over memoirs by Fresnel had taken place at the *Académie*. François Arago (1786–1853) had written a long-delayed report supporting Fresnel's work on chromatic polarization (in which colors appear when white light is passed through thin crystal slices). There he directly attacked the earlier, lengthy theories of the phenomenon by Jean Baptiste Biot (1774–1862), who continued to think that light consisted of independent rays that mark the paths of optical particles. Stung by the critique from someone with whom he had long had conflicts, Biot accused Arago of having intentionally delayed the report. Fresnel wrote his brother about the events on June 13.

It's entirely clear from the two original papers that Fresnel had written, and on which Arago reported (neither of which was printed in their original form during Fresnel's lifetime), that Fresnel himself had not targeted Biot, likely preferring to keep clear of potential difficulties. These events would no doubt have made him particularly alert to the appearance of remarks about papers still under examination. Moreover the tone of his letters, both then and earlier, clearly indicates that Fresnel was sensitive to issues of priority in discovery, so that Cauchy's having pushed Navier to the side and having barely mentioned Fresnel in his summary report likely rankled.[15] That in turn suggests Fresnel may have discussed a good deal more than Cauchy mentioned either in 1823 or in 1827 (in the latter case, after Fresnel's death).

It may be that Cauchy decided not to fully publish his investigations on elasticity at the time, as Belhoste has suggested, in part because of this

---

[13] Cited and translated in Belhoste, B. (1991, p. 98).

[14] Fourier, J.-B. (1823, p. 258).

[15] On these events see Buchwald, J. Z. (1989a, 1989b, pp. 237–51).

contretemps, waiting until Navier's own paper reached print in 1827.[16] That would have the further advantage of allowing him time to develop a physical foundation for an elasticity theory based on particle interactions, thereby completing his marginalizing of Navier. In fact, Belhoste points out, Cauchy claimed in a published note in the *Philomatique* that even his first memoir, on elastic plates, had been based on particle interactions, though it certainly had not been.[17] Then, on October 1, 1827, Poisson announced that he was working on a "far-reaching study" of elasticity, and this led Cauchy the same day to deposit his own memoir as a *pli cacheté* with the *Académie* to establish priority.[18] Further arguments with Navier and Poisson followed which we need not consider here, because Fresnel had died on July 14 in Ville-d'Avary "in the arms of his mother."[19]

On June 7 and 14, 1830, three years after Fresnel's death, Cauchy presented a comparatively short (given his customary standards) memoir on light before the *Académie* in Paris, which appeared thereafter in the *Bulletin de Férussac*; he also had it printed separately by de Bure Frères – the latter being publishers to the king, among others.[20] He then left the city in the first week of September, probably not at first intending to go into exile but rather to rest after the, to him, dispiriting events of the July revolution and following his exhaustingly extensive record of publication and presentation of memoirs.[21] His absence turned into a true exile, taking Cauchy at first to Fribourg, then Turin, and eventually to Prague in 1833 as tutor in the sciences to the exiled monarch's notably recalcitrant son, the Duke of Bordeaux. Then, in 1835, Cauchy published there an extensive memoir on light that was based on equations he had developed for a system of interacting particles. This memoir contained the expression for optical dispersion that was reprinted the next year under the auspices of Prague's Academy and that continues to appear under his name to this day.

Two years before leaving Paris, Cauchy had written three papers on these particle equations in an attempt to provide a physical basis for his general theory of elasticity. They had appeared in his on-going *Éxercises de Mathématique*. The following year he published a paper in the *Bulletin de Férussac* that

---

[16] Belhoste, B. (1991, pp. 98–99), who remarks that during the intervening years Cauchy modified his continuum theory to incorporate two elasticity constants, thereby making Navier's a special case since its reliance on particle interactions produced a single constant. See Darrigol, O. (2005, pp. 109–25); Grattan-Guinness, I. (1990, pp. 968–1045) for accounts of developments in elasticity theory at the time.

[17] Belhoste, B. (1991, p. 99).

[18] Belhoste, B. (1991, pp. 99–100).

[19] Verdet, E. (1866, p. xcviii).

[20] Cauchy, A.-L. (1830a, 1830b, 1830 (1958)-a). The *Bulletin* was founded in 1823 and continued through 1831 in part with the aim of ensuring rapid publication by young scholars who may not have been held in high regard by the leaders of their fields, as well as by known experts (Taton, R., 1947). Cauchy published papers there in 1828 and 1829.

[21] As suggested in Belhoste, B. (1991, pp. 145–6).

for the first time briefly described applying the results to optics.[22] Then, after the appearance in 1830 of the memoir on light that he had presented on June 14 of that year before the *Académie*, Cauchy published one further paper on light (in *Férussac*, on which more shortly), and nothing more until 1835.[23] We do however know that at the time of his June 14 presentation he also announced to the *Académie* that he "had the formulas relative to the *dispersion of light* that he had read at the last session." The *Procès Verbaux* for the meeting accordingly noted that Cauchy had presented a memoir "on the subject."[24]

Perhaps Cauchy waited until after Fresnel died to move ahead in public with a theory of elasticity based on particles because of the latter's angry note in 1823. That would certainly have avoided any reactions from the departed Fresnel. The question we must now pursue is whether the key new result of that work for optics – Cauchy's formula for dispersion, which reached print for the first time in 1835 – was wholly original in concept and form with him, for there exists an unpublished manuscript by Fresnel dated July, 1822 in which he sketched a theory of, and produced a formula for, dispersion near the very time that we know Fresnel met with Cauchy to show him some results about light.

Cauchy's first published remarks on dispersion had appeared in his second 1830 paper on light, which concerned reflection and refraction.[25] That paper referred back to the particle equations that he had developed over the previous two years and that had been printed in 1828 and 1829. In those papers Cauchy had reached the general equation (22.1) below for the ether (or for that matter for any material body similarly constituted of interacting particles). At that time he had continued by expanding the differences in the particle displacements ($\delta^i \mathbf{u}$) in Taylor series about the particles' equilibrium loci. He had then dropped terms beyond second order and imposed isotropy on the system, thereby obtaining the following equation of motion, which reads in modern notation[26]:

$$m\frac{\partial^2 \mathbf{u}}{\partial t^2} = (R + G)\nabla^2 \mathbf{u} + 2R\nabla(\nabla \bullet \mathbf{u})$$

Here R and G compact the constants of the isotropic system. Cauchy noted that if he had retained the expansion through the fourth order, then terms in $\nabla^4 \mathbf{u}$ would appear, and that these would produce dispersive effects, i.e. that the wave speed would then depend upon the wavelength. This much was entirely obvious,

---

[22] Cauchy, A.-L. 1828 (1890)-a; Cauchy, A.-L., 1828 (1890)-b; Cauchy, A.-L., 1828 (1891); Cauchy, A.-L., 1829 (1958).

[23] Cauchy, A.-L., 1830 (1958)-b.

[24] Anonymous, 1830.

[25] Cauchy, A.-L., 1830 (1958)-b, pp. 155–57.

[26] In the case of Cauchy's work in this area, transforming his equations into vector form does make it much simpler to grasp their structure, but it also traduces to a certain extent the difficulties he faced in forging the system out of a morass of algebraic relations with often perplexing geometric connections.

and not only to Cauchy. The question was how to develop the system's equations to yield a proper formula. At the end of his paper Cauchy remarked that he had described how to do just that in his lectures on June 19 and 22 at the Collège de France, and that he would explain it in "more detail in a new article." That article seems never to have appeared, most likely because of the chaotic events surrounding his departure from Paris. In any case, he may have had in hand many of the results that first reached print in 1835.

The essentials of the structure that Cauchy had begun to develop after Fresnel's death and that he used to produce a dispersion formula are deceptively simple. Imagine an arrangement of point like particles each of which acts on all of the others with a repulsive force. We do not initially make any assumptions about the arrangements of the particles, in particular what symmetries the system might obey, and neither do we specify the form of the force other than to assume that it falls off with distance. Through their interactions these particles establish a pattern that results in wave propagation, and Cauchy aimed rigorously to analyze the system in order to generate optical equations.

Each particle in the system acts to produce an acceleration $f(r)$ on every other one that is directed along the line joining each pair, that depends on their masses and on their mutual distance r, and that, like gravity, satisfies Newton's third law. The system has an equilibrium configuration in which the net force on every particle vanishes, thereby providing Cauchy with a first condition that the constants of the system must satisfy, namely that the following relation for the force on any given particle of mass m must hold:

$$m \sum_i m_i f(r^i) \mathbf{e}_{r^i} = 0$$

Here the r^i represent the distances in equilibrium between the given and the ith particles, $\mathbf{e}_{r^i}$ is a unit vector along $\mathbf{r}^i$, and m, m_i are their respective masses.

If, next, our given particle m is displaced by an amount $\mathbf{u}$, then it will experience a net force. The other elements of the system are also assumed to be shifted from their equilibrium loci (this was the admitted defect in Fresnel's deduction of his optical "surface of elasticity"), as a result of which the distance $\mathbf{r}^i$ changes to $\mathbf{r}^i + \delta^i\mathbf{u}$, wherein $\delta^i\mathbf{u}$ accordingly represents the directed difference between the displacements from equilibrium of m, m^i. The fundamental equation of motion is then:

$$\frac{\partial^2 \mathbf{u}}{\partial t^2} = \sum_i m^i f(\mathbf{r}^i + \delta^i\mathbf{u}) \mathbf{e}_{r^i + \delta^i\mathbf{u}}$$

In consistency with the equilibrium condition, $\mathbf{e}_{r^i + \delta^i\mathbf{u}}$ is a unit vector along $\mathbf{r}^i + \delta^i\mathbf{u}$. The acceleration $f(\mathbf{r}^i + \delta^i\mathbf{u})$ can be expressed in terms of $\mathbf{r}^i$ and $\delta^i\mathbf{u}$ together with the equilibrium values $f(r^i)$ and the latter's derivatives with respect to the r^i by series expansion. If, further, the displacements of the particles are all "small" (meaning through distances much less than the presumed average distance between neighboring particles at any time), then, Cauchy assumed, quadratic terms in $\delta^i\mathbf{u}$ may be neglected in the expansion. Taking into account the equilibrium condition proper, the fundamental equation thereby becomes[27]:

$$\frac{\partial^2 \mathbf{u}}{\partial t^2} = \sum_i m^i \left[\frac{f(r^i)}{r^i} \delta^i\mathbf{u} + \left(r^i \frac{\partial f(r^i)}{\partial r^i} - f(r^i) \right) \left(\frac{\mathbf{e}_{r^i} \bullet \delta^i\mathbf{u}}{r^i} \right) \mathbf{e}_{r^i} \right] \qquad (22.1)$$

Cauchy over time rang several changes on the results that he could draw from (22.1), altering for example his interpretation of how to express the $\delta^i\mathbf{u}$ as functions of the equilibrium distances r^i. In all cases he recurred to what are, in modern terms, eigenvalue techniques in order to obtain expressions for propagation.[28] He was in this way able to obtain a reasonably close approximation of Fresnel's wave surface for biaxial crystals, the crowning glory of the latter's optical theory.[29] And Cauchy obtained as well what he implicitly claimed to be his own expression for dispersion, one that soon generated considerable discussion, particularly in England.

In order to reach tractable equations that would lead to a dispersion formula, Cauchy first imposed central symmetry on his system of particles, followed by complete isotropy.[30] In an extraordinarily intricate analysis running to many pages and equations, he thereby demonstrated that Ω^2, the (squared) rate of propagation for a given wavelength in the medium, can be expressed as an infinite series in the reciprocals of the wavelength's even powers:

$$\Omega^2 = b_o + \sum_{p=1} \frac{b_p}{\lambda^{2p}} \qquad (22.2)$$

---

[27] This equation occurs first in Cauchy, A.-L., 1828 (1890)-b, pp. 227–31 and then in the major dispersion memoirs (Cauchy, A.-L., 1836, pp. 1–5, 1836 (1895), pp. 195–200). It is briefly discussed in Buchwald, J. Z. (1979, p. 251) as well as in Dalmedico, A. D. (1992, pp. 347–50) but is misprinted in both places. Cauchy likely had most of these in hand by 1827 since he had produced formulae for the force on a surface in such a system that (this being Cauchy's main point at the time) are equivalent in form to his symmetric matrix for the continuum case, formulae that contain sums similar to those for the terms in equation (22.3) below (see Cauchy, A.-L., 1827 (1889), p. 81).

[28] For details on this and reactions in England to Cauchy's dispersion theory see Buchwald, J. Z. (1979, pp. 252–56).

[29] On which see Dalmedico, A. D. (1992, pp. 351–76). Along the way Cauchy felt it necessary to change his views concerning the relationship between the displacement vector and the optical plane of polarization.

[30] In central symmetry, if a particle lies on an arbitrary line through any given particle, then a corresponding particle must also lie on the other side of the line at an equal distance from the given particle.

The particle distances in equilibrium as well as the forces involved are packed into the values of the b constants, and so we must next investigate how these constants emerge from the fundamental physics of the situation. That, in turn, will enable us to probe any connections between Fresnel's work on dispersion and this theory of Cauchy's.

We begin with the manner in which Cauchy solved his fundamental equation (22.1). He assumed first of all that its solutions $\mathbf{u}$ could be represented as $\sum_l \mathbf{c}_l e^{i\mathbf{k}\cdot\mathbf{r}}$ wherein the time dependence is assigned to the numbers $\mathbf{c}_l$, taken as functions of position $\mathbf{r}$ and time, and with the $\mathbf{k}$ being real. The latter points in the direction of a disturbance's propagation and is orthogonal to the corresponding (plane) front. Cauchy tacitly took the position vector $\mathbf{r}$ to be effectively the same as the equilibrium loci $\mathbf{r}^i$ of his particles. Taking a single term in the solution series, he then manipulated it into a form containing $\mathbf{u}$, $\mathbf{k}$, and $\mathbf{r}$ which he then inserted into (22.1). Assuming central symmetry, Cauchy could then rewrite (22.1) in the following way:

$$\frac{\partial^2 \mathbf{u}}{\partial t^2} = - \begin{pmatrix} L_x & P_{xy} & P_{xz} \\ P_{xy} & L_y & P_{yz} \\ P_{xz} & P_{yz} & L_z \end{pmatrix} \mathbf{u} \tag{22.3}$$

with the matrix elements having the following forms. In these expressions corresponding subscripts are taken by position in each of the six L,P:

$$L_{x,y,z} \equiv 2\sum_i \left\{ m_i \left[\frac{f(r^i)}{r^i} + \frac{\left(r^i \frac{\partial f(r^i)}{\partial r^i} - f(r^i)\right)}{r^i} \left(\frac{\left|r^i_{x,y,z}\right|^2}{(r^i)^2}\right) \right] \sin^2\left(\frac{\mathbf{k}\cdot\mathbf{r}^i}{2}\right) \right\}$$

$$P_{xy,xz,yz} \equiv 2\sum_i \left\{ m_i \left[\frac{\left(r^i \frac{\partial f(r^i)}{\partial r^i} - f(r^i)\right)}{r^i} \left(\frac{\left(\left|r^i_{y,z,x}\right|\left|r^i_{z,x,y}\right|\right)^2}{(r^i)^2}\right) \right] \sin^2\left(\frac{\mathbf{k}\cdot\mathbf{r}^i}{2}\right) \right\}$$

These could in turn be made more compact by introducing two functions U and V:

$$L_{x,y,z} = U + \frac{\partial^2 V}{\partial x^2, y^2, z^2}$$

$$P_{xy,xz,yz} = \frac{\partial^2 V}{\partial yz, zx, xy}$$

in which

$$U \equiv \sum_i m_i \left\{ \frac{f(r^i)}{r^i} \left[1 - \cos(\mathbf{k} \bullet \mathbf{r}^i) \right] \right\}$$

$$V \equiv \sum_i m_i \left\{ \frac{\left(r^i \frac{\partial f(r^i)}{\partial r^i} - f(r^i) \right)}{r^i} \left[\frac{1}{2} \left(\frac{\mathbf{k} \bullet \mathbf{r}^i}{r^i} \right)^2 + \frac{\cos(\mathbf{k} \bullet \mathbf{r}^i)}{(r^i)^2} \right] \right\} \tag{22.4}$$

Again, recall that the equilibrium loci $\mathbf{r}^i$ are taken to cover the position vector $\mathbf{r}$.

In general, Cauchy's equations always yield three distinct speeds of propagation: two for mutually orthogonal displacements in the surface of a plane wave, and one for a displacement normal to the plane. The latter, he argued, is invisible to the eye, while the former two speeds, which correspond to optical waves, must reduce to one in an isotropic medium. Cauchy had considerable difficulty reducing the system to isotropy and spent a great deal of effort demonstrating that the conditions at which he arrived satisfy the appropriate requirements. He eventually found that under isotropy the equation of motion (22.3) expressed in terms of U, V becomes:

$$\frac{\partial^2 \mathbf{u}}{\partial t^2} = - \left(U + \frac{1}{k} \frac{\partial V}{\partial k} \right) \mathbf{u} - (\mathbf{k} \bullet \mathbf{u}) \frac{\partial \left(\frac{1}{k} \frac{\partial V}{\partial k} \right)}{\partial k} \mathbf{k}$$

If the invisible third wave is ignored, so that the disturbance may be supposed to occur entirely in the plane of the wavefront, then $\mathbf{k} \bullet \mathbf{u}$ vanishes, and the equation transforms into one that has the same form as that of an harmonic oscillator[31]:

$$\frac{\partial^2 \mathbf{u}}{\partial t^2} = -s^2 \mathbf{u} \text{ where } s^2 = U + \frac{1}{k} \frac{\partial V}{\partial k} \tag{22.5}$$

The two equations (22.4) and (22.5) entail a dispersion formula because the wave speed is just s/k, and k, which is inversely proportional to the wavelength, appears in s via U,V. Note that the speed of propagation apparently incorporates the angle d between the vector $\mathbf{k}$ that is normal to the front and the vector directed to the point $\mathbf{r}$ at which the disturbance is evaluated. That angle continues to appear in Cauchy's final dispersion formula, but it merely represents the *apparent* speed that would be measured when looking in any direction other than the one in which the wave is propagating, as in the diagram below. In his final expression for U, V, $\mathbf{k} \bullet \mathbf{r}$ is written $rk\cos(d)$, with $k\cos(d)$ accordingly representing the projection of the propagation vector $\mathbf{k}$ in the direction $\mathbf{r}$. Since

---

[31] Propagation, and not simply oscillation, occurs because s^2, the analog of the harmonic coefficient, is a function of $\mathbf{k}$ and $\mathbf{r}$. Cauchy developed the solution in detail.

dispersion relations – and, in particular, the measured values for refractive indices as a function of wavelength – concern the speed in the direction normal to the front, $\cos(d)$ may be set to one for comparison with experiment. Cauchy retained the angle for generality, but this does not matter in the end, as we will now see, because he replaced all of the lattice-dependent terms with constants.

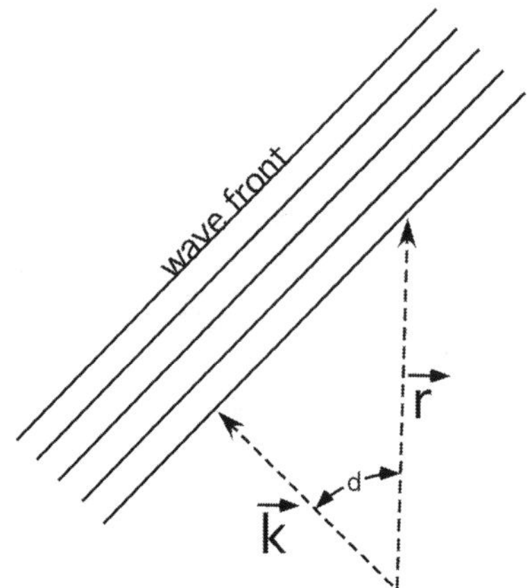

To obtain a dispersion relation, Cauchy, in a crucial step, turned his expressions for U and $\partial V/\partial k$ into series by expanding the term $\cos(\mathbf{k} \bullet \mathbf{r}^i)$ in powers of k. In virtue of (22.5) the following series for Ω^2, the square of the wave speed, results:

$$\Omega^2 = \frac{s^2}{k^2} = \frac{\left[U + \frac{1}{k}\frac{\partial V}{\partial k}\right]}{k^2} = \sum_{j=1} \left[k^{(2j-2)}\right] a_j$$

where $\qquad\qquad\qquad\qquad\qquad\qquad\qquad\qquad$ (22.6)

$$a_j \equiv \left\{ \left(\frac{(-1)^{(j-1)}}{(2j)!}\right) \left[\sum_i m_i (r^i)^{(2j-1)} (\cos(d))^{2j} \left(f(r^i) + \left(\frac{(\cos(d))^2}{2j+1}\right)\left[r^i\frac{\partial f(r^i)}{\partial r^i} - f(r^i)\right]\right)\right]\right\}$$

Cauchy simply took the a_j to be constants that pertain to a particular medium, which accordingly gave him an expression for the wave speed as

$$\Omega = \sqrt{a_1 + a_2 k^2 + a_3 k^4 + \ldots} \qquad\qquad\qquad (22.7)$$

Since k is just $2\pi/\lambda$, where λ is the wavelength, Cauchy's equation is precisely the same in form as the one that Fresnel had obtained for the wave speed in his unpublished manuscript of July 1822, namely $\sqrt{n}\sqrt{P - \frac{Q}{\lambda^2} + \frac{R}{\lambda^4} - \&c.}$ (note that Fresnel included the alternating signs that Cauchy incorporated into his a_j constants). Cauchy's analysis is certainly vastly more intricate and mathematically meticulous, as was his wont, than Fresnel's few pages from 1822.

Nevertheless, Fresnel had reasoned his way to the very same dispersion formula that Cauchy published in 1835 as a rigorous consequence of his elaborate theory.

We know that Cauchy had met with Fresnel to discuss matters of optics around the time that Fresnel wrote down his dispersion formula. And it is highly probable that he had not begun to work on equations for a system of particles until after that meeting. Indeed, his first work on a particle model did not appear until the addendum of 1827 in which he mentioned that Fresnel had arrived at similar results. Turn now to Fresnel's brief, unpublished deduction of July, 1822 in order to expose his principal conceptions and thereby to compare them with the ones that underpin Cauchy's intricate analysis.

What is most interesting about Fresnel's route to dispersion is that he began directly with a propagating wave and considered the forces involved in its motion. In his figure (see below), a wave of form $\sin 2\pi(x/\lambda)$ displaces the ether's particles. Fresnel then examined the effect on a particular "slice" of the wave-bearing medium that is exerted by two neighboring ones. His lines mp, MP, and $m'p'$ each represents a region of the ether displaced by the wave, with the regions spaced a distance h apart. He identified three factors that determine the forces which the neighboring slices exert on MP: first, the difference between their displacements, second their distance apart, and third, the "energy of the elasticity." These first two factors immediately yield expressions for the "action" (i.e. force) on MP by taking the differences $MP\text{-}m'p'$ and $MP\text{-}mp$.

If P is located at x, then p' is at $x+h$, and so the difference in displacements is proportional to $\sin(2\pi x/\lambda) - \sin(2\pi(x + h)/\lambda)$. A similar expression holds for the effect of slice mp, replacing h with $-h$. Fresnel next expanded the differences in the inverse powers of the wavelength to obtain two series which he then added to obtain the combined "action" on the slice MP.[32] At this point we already have, in effect, an appropriate dispersion series in the form $Ca\left[1 - \dfrac{h^2}{3\cdot 4}\dfrac{\varepsilon^2}{\lambda^2} + \dfrac{h^4}{3\cdot 4\cdot 5\cdot 6}\dfrac{\varepsilon^4}{\lambda^4} - \&c.\right]\dfrac{h^2}{2}\dfrac{\varepsilon^2}{\lambda^2}\sin\dfrac{\varepsilon x}{\lambda}$. Since only h and a proportionality constant change for the other slices, Fresnel concluded that the series retains the same form. The factor that multiplies the sine term in his expression, Fresnel asserted, expresses the "energy" of the force for a unit displacement. And, he continued, the period of the oscillation is inversely proportional to the square root of this factor – obviously considering the behavior of a particle through which the wave passes to obey the same rules as an harmonic oscillator whose coefficient is given by Fresnel's factor. The rest follows quite directly, and Fresnel concluded with an admittedly failed attempt to derive a general result that could carry a dispersion relation from one medium to another.

---

[32] He evidently did so by first rewriting $\sin(2\pi(x + h)/\lambda)$ as $\sin(2\pi x/\lambda)\cos(2\pi h/\lambda) - \cos(2\pi x/\lambda)\sin(2\pi h/\lambda)$ and then expanding the terms containing h/λ into series. He did the same for the effect of slice mp.

Fresnel's excursion is vastly less rigorous and detailed than Cauchy's, and yet it yields effectively the same result and, moreover, shares with it an interesting effort to reduce the system to the case of an harmonic oscillator – as Cauchy, after lengthy deductions, did in reaching his equation (22.5) above. Also like Fresnel, Cauchy expanded the factors in his equations that determine the forces involved into a series in the inverse powers of the wavelength. Further, Cauchy's entire structure depended directly on correcting the major *lacuna* in Fresnel's deduction of his surface of elasticity, namely Fresnel's assumption that only the particle in question could be displaced, holding all the others fixed in situ. That assumption led to the linear transformation from which the elasticity surface emerged. Cauchy fixed the *lacuna*.

All of this is not likely to have been coincidental. It seems probable that Fresnel showed Cauchy his notes in the summer of 1822 or thereabouts, and that Cauchy took from those notes the essential idea to express the actions of the particles in terms of a series in the wavelengths, and to do so by generating a related equation of motion in harmonic form, one in which the coefficient contained the requisite wavelength series. Yet Cauchy never mentioned, in print at least, anything about Fresnel's work on dispersion. If the eponymous title of a formula should accrue to its first producer, or at least to the one who first developed the elements, subsequently elaborated, of a foundation for it, then perhaps "Cauchy's dispersion formula" should be reassigned to Fresnel, not least because Cauchy may have seen Fresnel's work. [*Bibliothèque de l'Institut*, MS 3411, pp. 64–7. July, 1822. Unpublished manuscript by Fresnel on dispersion]

Essais théoriques sur la dispersion

La courbe des déplacemens moleculaires est toujours sinusoidale: elle le sera donc à tous les instans dans une [?]mération résultant des détour d'une onde sur elle-même. Soit $y = \sin(x/\lambda)\varepsilon$ L'équation de cette courbe à une certain instant;

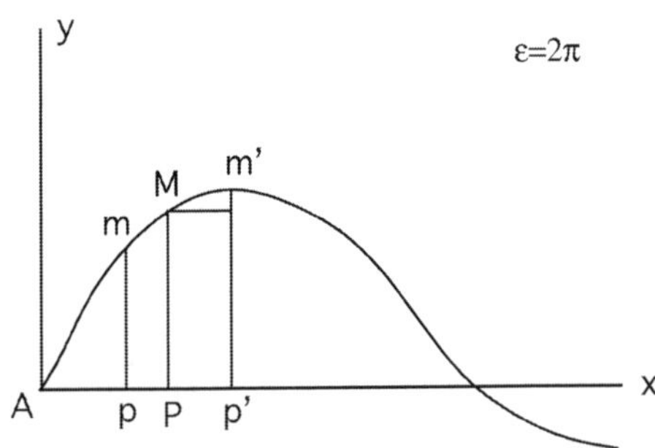

Il s'agit de détermine l'action exercée sur la tranche du milieu vibrant correspondant à l'ordonée MP par deux tranches équidistantes *mp* et *m'p'*. Je représente Pp et Pp' par h. L'action exercée sur la tranche en MP par la tranche en $m'p'$ est toutes choses egales d'ailleurs proportionelle au déplacement relatif $m'p' - MP$: elle dépend en outre de la distance h et de l'énergie de l'élasticité; elle est donc égale à un coëfficient constant dépendant de ces quantités, multiplié par $m'p' - MP$.

$$C(MP - m'p') = Ca\left[-h\frac{\varepsilon}{\lambda}\cos\frac{\varepsilon x}{\lambda} + \frac{h^2}{2}\frac{\varepsilon^2}{\lambda^2}\sin\frac{\varepsilon x}{\lambda} + \frac{h^3}{2\cdot3}\frac{\varepsilon^3}{\lambda^3}\cos\frac{\varepsilon x}{\lambda} - \frac{h^4}{2\cdot3\cdot4}\frac{\varepsilon^4}{\lambda^4}\sin\frac{\varepsilon x}{\lambda} - \&c.\right]$$

$$C(MP - mp) = Ca\left[h\frac{\varepsilon}{\lambda}\cos\frac{\varepsilon x}{\lambda} + \frac{h^2}{2}\frac{\varepsilon^2}{\lambda^2}\sin\frac{\varepsilon x}{\lambda} - \frac{h^3}{2\cdot3}\frac{\varepsilon^3}{\lambda^3}\cos\frac{\varepsilon x}{\lambda} - \frac{h^4}{2\cdot3\cdot4}\frac{\varepsilon^4}{\lambda^4}\sin\frac{\varepsilon x}{\lambda} + \&c.\right]$$

Adjoutant ces deux actions des deux tranches équidistantes:

$$Ca\left[\frac{h^2}{2}\frac{\varepsilon^2}{\lambda^2}\sin\frac{\varepsilon x}{\lambda} - \frac{h^4}{2\cdot3\cdot4}\frac{\varepsilon^4}{\lambda^4}\sin\frac{\varepsilon x}{\lambda} + \frac{h^6}{3\cdot4\cdot5\cdot6}\frac{\varepsilon^6}{\lambda^6}\sin\frac{\varepsilon x}{\lambda} - \&c.\right]$$

$$ou,$$

$$Ca\left[1 - \frac{h^2}{3\cdot4}\frac{\varepsilon^2}{\lambda^2} + \frac{h^4}{3\cdot4\cdot5\cdot6}\frac{\varepsilon^4}{\lambda^4} - \&c.\right]\frac{h^2}{2}\frac{\varepsilon^2}{\lambda^2}\sin\frac{\varepsilon x}{\lambda}$$

On voit que cette force accélératrice, pour les mêmes valeurs de h et de C est toujours proportionelle à $a\sin\frac{\varepsilon x}{\lambda}$, c'est à dire à l'espace à parcourir par la molécule M pour arriver à l'axe AX; ainsi toutes les molécules y arrivent en même tems, et à chaqueinstant de son oscillation la courbe se trouve toujours sinusoïdales, lors même que l'action moléculaire s'étend à des distances sensibles relativement à λ. Notre calcul suppose seulement que la série est convergente, c'est à dire que $\frac{\varepsilon h}{\lambda}$ est plus petit que l'unité, ou h moindre que le tiers de λ.

L'expression de la force exercée par tous les autres couples de tranches équidistantes aurait la même forme; il n'y aurait que h et C qui changerait de valeur:

$$Ch^2\left[1 - \frac{h^2}{3\cdot4}\frac{\varepsilon^2}{\lambda^2} + \frac{h^4}{3\cdot4\cdot5\cdot6}\frac{\varepsilon^4}{\lambda^4} - \&c.\right]a\frac{\varepsilon^2}{\lambda^2}\sin\frac{\varepsilon x}{\lambda};$$

$$C'h'^2\left[1 - \frac{h'^2}{3\cdot4}\frac{\varepsilon^2}{\lambda^2} + \frac{h'^4}{3\cdot4\cdot5\cdot6}\frac{\varepsilon^4}{\lambda^4} - \&c.\right]a\frac{\varepsilon^2}{\lambda^2}\sin\frac{\varepsilon x}{\lambda}$$

Faisant la somme de toutes ces actions, on a:

$$\left[Ch^2 + C'h'^2 + \&c - \left(Ch^4 + C'h'^4 + \&c\right)\frac{\varepsilon^2}{3\cdot4\cdot\lambda^2}\right.$$
$$\left. + \left(Ch^6 + C'h'^6 + \&c\right)\frac{\varepsilon^4}{3\cdot4\cdot5\cdot6\cdot\lambda^4} - \&c\right]a\frac{\varepsilon^2}{\lambda^2}\sin\frac{\varepsilon x}{\lambda};$$

l'écartement $MP = a\sin\frac{\varepsilon x}{\lambda}$.

Par conséquent le facteur constant qui exprime l'énergie de la force pour un écartement égal à 1 est l'expression ci-dessus dans laquelle on aurait supprimé le facteur $a \sin \frac{\varepsilon x}{\lambda}$. Mais la durée de l'oscillation est en raison inverse de la racine carrée de ce coefficient et par conséquent en raison inverse de

$$\frac{\varepsilon}{\lambda}\sqrt{Ch^2+C'h'^2+\&c-(Ch^4+C'h'^4+\&c)\frac{1}{3\cdot4}\frac{\varepsilon^2}{\lambda^2}+(Ch^6+C'h'^6+\&c)\frac{1}{3\cdot4\cdot5\cdot6}\frac{\varepsilon^4}{\lambda^4}-\&c}$$

Pour une onde d'une longeur égale à λ', on aurait:

$$\frac{\varepsilon}{\lambda'}\sqrt{Ch^2+C'h'^2+\&c-(Ch^4+C'h'^4+\&c)\frac{1}{3\cdot4}\frac{\varepsilon^2}{\lambda'^2}+(Ch^6+C'h'^6+\&c)\frac{1}{3\cdot4\cdot5\cdot6}\frac{\varepsilon^4}{\lambda'^4}-\&c}$$

En faisant

$$Ch^2+C'h'^2+\&c = P; \quad \frac{(Ch^4+C'h'^4+\&c)\varepsilon^2}{3\cdot4} = Q; \quad \frac{(Ch^6+C'h'^6+\&c)\{\varepsilon^4\}}{3\cdot4\cdot5\cdot6} = R;$$

Le 1er expression devient, $\dfrac{\varepsilon}{\lambda}\sqrt{P-\dfrac{Q}{\lambda^2}+\dfrac{R}{\lambda^4}-\&c.}$

Et le second $\dfrac{\varepsilon}{\lambda'}\sqrt{P-\dfrac{Q}{\lambda'^2}+\dfrac{R}{\lambda'^4}-\&c.}$

Mais les vitesses de propagation sont en raison inverse des durées d'oscillation; elles sont proportionelles pour le même milieu aux deux expressions,

$$\frac{\varepsilon}{\lambda}\sqrt{P-\frac{Q}{\lambda^2}+\frac{R}{\lambda^4}-\&c.}\,, \quad \text{et,} \quad \frac{\varepsilon}{\lambda'}\sqrt{P-\frac{Q}{\lambda'^2}+\frac{R}{\lambda'^4}-\&c.}$$

Les quantités P,Q,R sont des fonctions des intervalles h, h', $h'''\&\,c.$ des coëfficiens correspondant C, C', $C''\&\,c\ldots$, ou, en d'autres termes, des intégrales dans les diff.^{lles} h est la variable et C une fonction de h qui diminue rapidement à mesure que h augmente, ces intégrales étant prises jusqu'à $h = \infty$.

$$Ch^2 + C'h'^2 + \&c = P = \int_{-\infty}^{+\infty} h^2\varphi(h)dh\ldots;$$

$$Ch^4 + C'h'^4 + \&c = \int_{-\infty}^{+\infty} h^4\varphi(h)dh = \frac{3\cdot4\cdot Q}{\varepsilon^2}$$

$$Ch^{\{6\}} + C'h'^6 + \&c = \int_{-\infty}^{+\infty} h^6\varphi(h)dh = \frac{3\cdot4\cdot5\cdot6}{\varepsilon^4}\cdot R$$

$+\infty$ et $-\infty$ n'indique pas ici des quantités infinies ni mêmes grandes relativement à λ, puisqu'alors nos séries fondamentales n'étant plus convergents deviendrait des expressions illusoires: $+\infty$ et $-\infty$ indiquent seulement des limites de la sphère d'activité sensible de l'action réciproque des tranches du milieu vibrant.

Si sans fixer la loi suivant laquelle cette force décroit avec la distance, on supposait que la loi est semblable dans tous les milieux, c'est à dire que le fonction φ reste la même à facteur près, pour un autre milieux, des quantités P', Q', R' seraient égales à nP, nQ, nR, et les deux vitesses de propagation correspond^{tes} aux long d'ondulat. λ et λ' seraient:

$$\sqrt{nP - \frac{nQ}{\lambda^2} + \frac{nR}{\lambda^4} - \&c.}\;,\;\text{et,}\;\sqrt{nP - \frac{nQ}{\lambda'^2} + \frac{nR}{\lambda'^4} - \&c.}\;;\;\text{ou,}$$

$$\sqrt{n}\sqrt{P - \frac{Q}{\lambda^2} + \frac{R}{\lambda^4} - \&c.}\;,\;\text{et,}\;\sqrt{n}\sqrt{P - \frac{Q}{\lambda'^2} + \frac{R}{\lambda'^4} - \&c.}$$

et par conséquent la dispersion serait la même dans les milieux également réfringens, ce qui est contraire à l'experience. On ne peut donc pas supposer que les coefficiens P', Q', R' soient des coefficiens $P,\ Q,\ R$ multipliées par le même fonction n,

L'expression générale du rapport de réfraction pour le passage des ondes λ de l'air d'un milieu réfringent, est, en appelant v la vitesse de λ dans l'air, $\dfrac{v}{\sqrt{P - \frac{Q}{\lambda^2} + \frac{R}{\lambda^4} - \&c.}}$: λ est la longeur d'ondulat^{n} dans le milieu réfringent

ou, $\dfrac{v}{n\sqrt{1 - \frac{A}{\lambda^2} + \frac{B}{\lambda^4} - \&c.}}$ pour le même milieu $\dfrac{v'}{n\sqrt{1 - \frac{A}{\lambda'^2} + \frac{B}{\lambda'^4} - \&c.}}$, et une

autre longeur d'onde $\dfrac{v''}{n\sqrt{1 - \frac{A}{\lambda''^2} + \frac{B}{\lambda''^4} - \&c.}}$

Bibliography

Anonymous, 1830, "Séance du 14 Juin 1830," *Procès Verbaux de l'Académie des Sciences*: 456–57.

Belhoste, Bruno, 1991, *Augustin-Louis Cauchy. A Biography*. Tr. Frank Ragland, New York, Springer.

Buchwald, Jed Z., 1979, "Optics and the theory of the punctiform ether," *Archive for History of Exact Sciences* 21: 245–78.

——, 1985, *From Maxwell to Microphysics. Aspects of Electromagnetic Theory in the Last Quarter of the Nineteenth Century*, Chicago, The University of Chicago Press.

——, 1989a, "The battle between Arago and Biot over Fresnel," *Journal of Optics/Nouvelle Revue d'Optique* 20: 109–17.

——, 1989b, *The Rise of the Wave Theory of Light*, Chicago, The University of Chicago Press.

Cauchy, Augustin-Louis, 1823 (1958), "Recherches sur l'équilibre et le mouvement intérieur des corps solides ou fluide, élastiques ou non élastiques," in *Oeuvres Complètes, series II*, Vol. 2 (Paris: Gauthier-Villars et Fils).

——, 1827 (1889), "De la pression ou de la tension dans un corp solide," in *Oeuvres Complètes, Series II*, Vol. 7 (Paris: Gauthier-Villars et Fils), pp. 60–81.

——, 1828 (1890a), "De la pression our tension dans un système de points matériels," in *Oeuvres Complètes, Series II*, Vol. 8 (Paris: Gauthier-Villars et Fils), pp. 253–77.

——, 1828 (1890b), "Sur l'équilibre et le mouvement d'un système de points matériels sollicités par des forces d'attraction ou répulsion mutuelle," in *Oeuvres Complètes, Series II*, Vol. 8 (Paris: Gauthier-Villars et Fils), pp. 227–51.

——, 1828 (1891), "Sur les équations différentielles d'équilibre ou de mouvement d'un système de points matériels sollicités par des forces d'attraction ou de répulsion mutuelle," in *Oeuvres Complètes, Series II*, Vol. 9 (Paris: Gauthier-Villars et Fils), pp. 162–73.

——, 1829 (1958), "Mémoire sur le mouvement d'un système de molécules qui s'attirent ou se repoussent à de très petites distances et sur la théorie de la lumière," in *Oeuvres Complètes, series II*, Vol. 2 (Paris: Gauthier-Villars et Fils), pp. 73–74.

——, 1830a, "Mémoire sur la théorie de la lumière," *Bulletin de Férussac* 13: 414–27.

——, 1830b. *Mémoire sur la Théorie de la Lumière*, Paris, Bure Frères.

——, 1830 (1958a), "Mémoire sur la théorie de la lumière," in *Oeuvres Complètes, Series II*, Vol. 2, pp. 119–33.

——, 1830 (1958b), "Sur la réfraction et la réflexion de la lumière," in *Oeuvres Complètes, series II*, Vol. 2 (Paris: Gauthier-Villars et Fils), pp. 151–57.

——, 1835. *Nouveaux Exercises de Mathématiques. Mémoire sur la Dispersion de la Lumière*, Prague, Jean Spurny.

——, 1836. *Mémoire sur la Dispersion de la Lumière*, Prague, J. G. Calve.

——, 1836 (1895), "Mémoire sur la Dispersion de la Lumière," in *Oeuvres Complètes, series II*, Vol. 10 (Paris: Gauthiers-Villars), pp. 191–464.

Dalmedico, Amy Dahan, 1992. *Mathématisations. Augustin-Louis Cauchy et lEcole Française*, Paris, Albert Blanchard.

Darrigol, Olivier, 2005. *Worlds of Flow. A History of Hydrodynamics from the Bernoullis to Prandtl*, Oxford, Oxford University Press.

Fourier, Jean-Baptiste, 1823, "Analyse des travaux de l'Académie des Sciences, pendant l'année 1822. Partie mathématique," *Mémoires de l'Académie Royale des Sciences de l'Institut de France* 5: 231–320.

Fresnel, Augustin, 1823, *Bulletin de la Société Philomatique* 5: 36–37.

Grattan-Guinness, Ivor, 1990. *Convolutions in French Mathematics, 1800–1840*, Basel-Boston-Berlin, Birkhäuser Verlag.

Navier, Claude-Louis, 1823a, "Extrait des recherches sur la flexion des plans élastiques," *Bulletin de la Société Philomathique* 5: 95–102.

——, 1823b, "Sur les lois de l'équilibre et du mouvement des corps solides élastiques. Extrait d'un Mémoire présenté à l'Académie des Sciences, le 14 mai 1821," *Bulletin de la Société Philomathique* 5: 177–81.

——, 1827, "Sur les lois de l'équilibre et du mouvement des corps solides élastiques," *Mémoires de l'Académie Royale des Sciences* 7: 375–96.

Taton, René, 1947, "Les mathématiques dans le Bulletin de Férussac," *Archives internationales d'histoire des sciences* 26: 100–25.

Verdet, Emile, 1866, "Introduction," in *Oeuvres Complètes d'Augustin Fresnel*, edited by Émile Verdet, Henri de Senarmont and Léonor Fresnel, Vol. I (Paris: Imprimerie Impériale), pp. ix–xcix.

Chapter 23
Sadi Carnot on Political Economy. Science, Morals, and Public Policy in Restoration France

Robert Fox

There is good reason to believe that Sadi Carnot took almost as keen an interest in political economy as he did in the theory of the heat engine. His brother Hippolyte implied as much in his *Mémoires sur Carnot par son fils* (1863), where he referred to Sadi as having devoted himself to economics "with remarkable penetration", especially after a visit in 1821 to see his father, Lazare, in exile in Magdeburg.[1] Sadi, it seems, had borne the hopes of his father, who insisted that political economy would never become the rigorous "new science" he wanted it to be until mathematicians turned their minds to the discipline and applied "the experimental method". Hippolyte elaborated the point some years later in the biographical sketch that he appended to the 1878 edition of the *Réflexions sur la puissance motrice du feu*. In the sketch, he reproduced a "fragment sur l'économie politique", which appears either to have been copied from a manuscript of Sadi's now lost or to have been composed by Hippolyte on the basis of his reading of such a manuscript or possibly of one of the few manuscripts in Sadi's hand that have survived.[2] Thereafter, Carnot's ideas on political economy went unnoticed until an important paper by Jacques Grinevald drew attention to them during the conference of 1974 to mark the 150th anniversary of the publication of the *Réflexions*.[3]

---

[1] Hippolyte Carnot, *Mémoires sur Carnot par son fils*, 2 vols. (Paris: Pagnerre, 1863), vol. 2, 616.

[2] Hippolyte Carnot, "Notice biographique sur Sadi Carnot", in Sadi Carnot, *Réflexions sur la puissance motrice du feu et sur les machines propres à développer cette puissance* (Paris: Gauthier-Villars, 1878), 71–87, esp. (for the transcribed text) 84–86.

[3] Jacques Grinevald, "Présentation d'un manuscrit inédit de Sadi Carnot", in *Sadi Carnot et l'essor de la thermodynamique. Paris, Ecole polytechnique 11–13 juin 1974* (Paris: Editions du CNRS, 1976), 383–387, followed by "Manuscrit inédit de Sadi Carnot concernant l'économie politique et les finances publiques", 389–395. The latter text is Grinevald's transcription of a manuscript subsequently published as Appendix B_1 in the edition of Carnot's *Réflexions* cited in the next footnote. The content of the fragment transcribed by Hippolyte (see note 2, above) corresponds very roughly to that of the first seven pages or so of this manuscript, but only a few passages in Hippolyte's fragment appear verbatim in the manuscript.

R. Fox (✉)
Museum of the History of Science, University of Oxford, Oxford, UK
e-mail: robert.fox@history.ox.ac.uk

J.Z. Buchwald (ed.), *A Master of Science History*, Archimedes 30,
DOI 10.1007/978-94-007-2627-7_23, © Springer Science+Business Media B.V. 2012

Any doubts in my own mind about the seriousness of Carnot's interest in economics were removed in the preparation of a critical edition of the *Réflexions* that I published with the Librairie J. Vrin in Paris more than thirty years ago.[4] The manuscripts that appear in Appendix B of that edition, some though not all of them previously published and commented upon by Grinevald, are devoted both to timeless issues in political economy that had been debated since the eighteenth century and to others of more immediate relevance to Carnot's day, in the fifteen years or so after the fall of Napoleon I's Empire in 1815.[5] The uncoordinated character of the manuscripts makes precise identification of Carnot's position difficult. But, as I shall argue, his liberal, humanitarian priorities are unmistakable, as is the independence of his thought, despite signs of a broad affinity with the doctrines of the Genevan-born Sismonde de Sismondi and a corresponding reticence with regard to the *laissez-faire* doctrines of the contemporary classical school, notably of Jean-Baptiste Say and David Ricardo.[6] While my main aim is to set Carnot's ideas in the context of economic debate under the restored Bourbon monarchy, I also explore briefly some of the structural analogies between economic theory in the late eighteenth and early nineteenth centuries and Carnot's theory of the production of motive power.

The chief difficulty in attempting an analysis of Carnot's views on political economy is the fragmentary nature of the papers at our disposal. Since most of his possessions appear to have been destroyed at the time of his death during the cholera epidemic of 1832, the contents of the 16-page notebook and nearly thirty sides of jottings that I reproduce in my edition of the *Réflexions* almost certainly represent a small and arbitrary selection from a larger body of writing on the subject. While the surviving papers may conceivably have been written in preparation for a book that never appeared, they read rather as random notes and reflexions arising from lectures or reading. Nevertheless, it is possible to reconstruct the broad thrust of Carnot's opinions and preoccupations, and although specific references are virtually non-existent, we have enough to show that he was well read in a literature of political economy in which both Lazare and Hippolyte were steeped as well.

The most striking characteristic of the manuscripts is their emphasis on agriculture rather than on manufacturing industry. In this respect, Carnot displays an interest in the favoured problems of François Quesnay, Dupont de Nemours, Turgot, and others who identified more or less closely with the

---

[4] Sadi Carnot, *Réflexions sur la puissance motrice du feu. Edition critique avec introduction et commentaire, augmentée de documents d'archives et divers manuscrits de Carnot par Robert Fox* (Paris: Librairie philosophique J. Vrin, 1978).

[5] Ibid., 273–312. All my subsequent references to Carnot's manuscripts are to the texts on political economy as reproduced in these pages.

[6] On the broad similarity between Carnot's ideas and Sismondi's, I share the tentative opinion of Jacques Grinevald. See Grinevald, "Présentation d'un manuscript", 385.

teachings of the eighteenth-century "économistes" (a new term at the time) or (as they quickly came to be called) Physiocrats. For Quesnay, wealth flowed in a circular fashion between the members of three classes: the "productive" class of agriculturalists, the "propertied" class (consisting primarily of landowners but also of the representatives of the Church and the King, who received respectively tithes and taxes), and the "sterile" class of manufacturers, merchants, and artisans. Quesnay regarded all three classes as essential to the economy and what he called the "natural order". But it was the productive class, those who actually cultivated land, who alone could create new wealth and so invigorate the all-important circular flow. Thanks to the beneficence of nature, the productive class (and only that class) was capable of producing a "net product" (*produit net*), i.e. a disposable surplus of wealth after necessary expenditure on food and the cost of production. By comparison, however Adam Smith and David Ricardo might argue to the contrary, the labour of manufacturers and others in the sterile class was unproductive. Such labour and the commerce that went with it might furnish some of the luxuries and material comforts of life. But they yielded no net product (something that only nature could do), and they therefore contributed nothing to raising the level of economic activity.

Armed with the concept of a net product derived exclusively from agriculture, Quesnay and other Physiocrats developed distinctive principles of taxation, ones that predominantly favoured the members of the propertied class. They allocated to that class the seemingly parasitic but, as they argued, essential function of receiving rent for land which they did not work and which many of them might never even have seen. It was anticipated that, in return, landowners would help to increase the agricultural surplus by reinvesting at least some of their income in their property; self-interest, it was assumed, would be sufficient to ensure this. Quesnay saw such a process as part of the "natural order", as opposed to the artificially created social order. But here lay one of Physiocracy's obvious weaknesses. For it was a matter of bitter experience under the Ancien Régime that absentee landlords felt little incentive to invest in the constructive way the Physiocrats believed they would in the untrammelled, divinely ordained scheme of things.

It was because of the continuing spectacle of France's undercapitalized and flagrantly inefficient system of tenant-farming and dispersed ownership of small plots of land that agricultural taxation retained its importance for the great majority of economic theorists from the heyday of the Physiocrats in the 1760s to the time when Carnot was writing, probably in the last decade of his life. The works of Ricardo, Say, and Sismondi, to name just the best-known of Carnot's contemporaries in political economy, demonstrate the enduring nature of the exchanges; all three wrote on the economics of agriculture in the years of Carnot's maturity, albeit from different perspectives and with different conclusions, and all three were keenly read, as the number of editions of their works testifies. Carnot, in fact, was working in the context of a wider debate which had not only been in progress for half a century or more but which had assumed a new urgency through the social and economic upheavals of, first, the French Revolution and now, since 1815, the Bourbon Restoration and the renewal of contact with Britain.

In a rapidly industrializing and significantly more democratic society, the Physiocrats' system, based on landed privilege and a belief in private property as the essential fount of economic well-being, seemed no longer to point the obvious way forward. Individual elements of Physiocracy still found favour with a dwindling number of followers; the graduate of the pre-revolutionary Ecole des ponts et chaussées Jean-Michel Dutens, for example, continued to defend key Physiocratic ideas until his death in 1848. But the doctrine had generally lost its cogency. It is not surprising, therefore, that the surviving notes show Carnot breaking with Physiocratic principles by exploring methods of taxation designed to encourage landowners to work their own land. To the classic Physiocrats' problem of how best to increase the net product and thereby invigorate the circulation of wealth he offered two possible solutions: one (so far as I can see) entirely novel, the other (for a tax on income from rent) that has parallels in the work of several precursors and contemporaries.

The novel suggestion was for a tax to be levied on ploughs and other agricultural equipment, perhaps also on draught animals.[7] This tax, essentially a tax on capital investment and the transport of goods, had the advantage over a simple tax on the value of land or on rents of being easily assessed and cheaply collected – virtues that always weighed heavily with economists, as they did with Carnot.[8] Since the tax would fall equally on the owners of both good and poor land, it would have the effect of discouraging the cultivation of the least productive areas, where profits were hardest-won. But that, in Carnot's view, might be no bad thing, always provided (to turn to another of his recurring themes – and Sismondi's) that the population was not allowed to increase indefinitely.[9] In his repeated insistence on the need to control population, Carnot was almost certainly taking up the logical consequences of Ricardo's doctrines, which enjoyed something of a vogue in France following the publication of the French edition of the *Principles of political economy, and taxation* in 1819, two years after its original publication in English. According to Ricardo's theory of rent, the price of corn and hence the rent that landowners could charge tenants were determined by the high cost of cultivating the least

---

[7] Carnot, *Réflexions* (1978 edn.), 276, 281–282, and 309–310.

[8] Ibid., 276 and 309–310. Cf. the similar statement of these virtues in J.-C.-L. Sismonde de Sismondi, *Nouveaux principes d'économie politique, ou de la richesse dans ses rapports avec la population*, 2 vols. (Paris: Delaunay, and Treuttel and Wurtz, 1819), vol. 2, 178–179. My subsequent references to the *Nouveaux principes* are to this edition. For a modern English edition with a helpful commentary, see Sismondi, *New principles of political economy. Of wealth in its relation to population*, trans. and ed. by Richard Hyse (New Brunswick, NJ and London: Transaction Publishers, 1991).

[9] Carnot, *Réflexions* (1978 edn.), 284, 293, 295–296. The dangers of an uncontrolled growth in population are the subject of Book VII of Sismondi's *Nouveaux principes d'économie politique*, vol. 2, 248–366.

fertile and consequently least profitable land.[10] It followed that the interest of landowners lay in an increase of population, since this would heighten pressure to cultivate ever poorer land, which (because of the more marginal profit from such cultivation) would in turn push up the price of corn across the board, while also diminishing any incentive for technological improvement. Why should landowners innovate when the effect of a scarcity of corn was to increase prices and hence their income from rent, whatever the quality of the land they owned?

Ricardo's response to this consequence he had himself articulated was that of a champion of free trade who recognized the potential of human suffering that went with high prices. If prices were to be held in check and consumers were not to suffer, his remedy lay in the import of corn, which would keep prices down by easing the pressure to cultivate poor land.[11] Carnot and Ricardo, therefore, shared a concern to reduce the temptation to exploit land of marginal profitability, while diverging sharply on the means to that end. By his warning about the dangers of population growth, Carnot was seeking to diminish demand and so to tackle the problem at its roots and in the long term. Ricardo's solution was immediate and palliative.

Carnot 's second proposal in the realm of taxation was the more orthodox one for a tax on income from rent (*impôt sur les fermages*).[12] His goal, at odds with the *laissez-faire* tendencies of Physiocratic theory but close to Sismondi's more interventionist objectives, was again the discouragement of absentee landlords, "les propriétaires non-cultivateurs" as Sismondi called them.[13] A landowner, Carnot observed, would be able to avoid the tax on rent by cultivating his land himself. The change, for a landowner, from an income derived from rent to one derived from working the land would have the added advantage of encouraging investment in agricultural improvement. The benefits of such improvement were of a kind for which no tenant-farmer could plan, simply because of the time that was needed for the return on long-term investment to be realized. Improvement, of course, called for landowners with sufficient capacity, in terms of capital and labour, and Carnot recognized that the extent of a large estate might make exploitation of the entirety of an individual owner's land impossible. But that very limitation could be made to work in favour of agricultural efficiency, since a landowner always had the option of selling the excess land to smaller owners who would assume the responsibility of working that part of the land themselves and, through labour and improvements of their

---

[10] David Ricardo, *Des principes de l'économie politique et de l'impôt, traduit de l'anglais par F.-S. Constancio, avec des notes explicatives et critiques par M. Jean-Baptiste Say*, 2 vols. (Paris: J.-P. Aillaud, 1819), vol. 1, 63–106 (chapter 2, "Du fermage ou profit des terres [Rent]").

[11] Ibid., vol. 2, 128–167 (chapter 22, "Des primes d'exportation, et des prohibitions d'importation"), esp. 153–154.

[12] Carnot, *Réflexions* (1978 edn.), 276–278.

[13] The consequences for agriculture of absentee landlords are explored in Book III of Sismondi's *Nouveaux principes d'économie politique*; see, for example, chapter 3 ("De l'exploitation patriarcale"), in vol. 1, 166–177; also chapters 8–10, ibid., 217–251.

own, boost its profitability. Carnot's passing comment that a tax on rent might encourage large landowners to divert some of their capital to industry or trade is a rare and fleeting indication of his concern for manufacturing, which (in the surviving notes at least) remains a secondary one.[14]

Carnot's views on agricultural taxation and the vehemence of his approval of governmental intervention in the control of imports and exports amply demonstrate his opposition to the classical school of those he disparagingly called the "économistes modernes", probably Ricardo and Say (although he mentioned no names).[15] Against these advocates of *laissez-faire* and minimal taxation, Carnot saw a positive virtue in a government's capacity to use taxes not merely to raise essential revenue but more importantly to promote useful ends by directing the course of production and trade.

One such end was the curbing of "prodigality", meaning any expenditure that resulted in what Carnot described as "pure loss" or "dissipation".[16] In this, Carnot's position resembled that of virtually all the economic theorists of his day. It had long been a commonplace, in fact, that in any economy the wasteful transfer of wealth from one hand to another had to be minimized; Say's chapter on "unproductive consumption" (*consommation improductive*), with no end in view other than "the mere satisfaction of a want, or the enjoyment of some pleasurable sensation", was a typical expression of a concern that had been voiced in different ways since the days of the Physiocrats.[17] The point clearly weighed heavily with Carnot. Time and again, he insisted that the thrust of a government's taxation policy should be to discourage the consumption of luxury goods, especially those imported from abroad.[18] An import duty on such magnets for prodigal spending as tobacco or silk might not reduce the sums that the wealthy spent on them, any more than the tax levied on imported wine diminished the consumption of wine among the wealthier classes in Britain.[19] And it could be that a limited measure of prodigality was not only unavoidable but also, in the right circumstances, beneficial. At least the revenue from an import duty would enter the coffers of the state and, if properly

---

[14] Carnot, *Réflexions* (1978 edn.), 277.

[15] Ibid., 275.

[16] Ibid., 285.

[17] Jean-Baptiste Say, *Traité d'économie politique, ou simple exposition de la manière dont se forment, se distribuent, et se consomment les richesses*, 4th edn., 2 vols. (Paris: Deterville, 1819), vol. 1, 232–241 (Book III, chapter 4, "Des effets de la consommation improductive en général"), esp. 232. I quote the passage as translated in Say, *A treatise on political economy* [from the American edition of 1836], with a new introduction by Munir Quddus and Salim Rashid (New Brunswick, NJ and London: Transaction Publishers, 2001), 396–401 (396). All subsequent references, however, are to the fourth French edition, the one most probably used by Carnot. For a more diffuse discussion in the same spirit, see Book II ("Formation et progrès de la richesse") of Sismondi, *Nouveaux principes d'économie politique*, vol. 1, 60–149.

[18] Carnot, *Réflexions* (1978 edn.), 285–295.

[19] Ibid., 289

directed, help not only to relieve the burden of taxation on essential goods but also to protect the poor against the temptations of such "pernicious habits" as excessive drinking and smoking.[20]

While Carnot's economic ideal of a system fashioned to maximize efficiency in the circulation of wealth was unremarkable, his secular moral sense lent distinctiveness to the route he advocated to that end. The surviving notes return time and again to the guiding principle that governments should actively work for the welfare of the people as a whole.[21] Governments, in Carnot's view, were emphatically there for the benefit of the governed, and it was their duty to adopt principles of taxation that would promote the welfare of citizens, in the manner of a father acting in the interests of his children.[22] In accordance with that principle, one aim of taxation should be to promote a redistribution of wealth, albeit a redistribution that fell short of complete equalization.[23] A moderate inequality of incomes, in fact, could have advantages, not least in allowing some beneficial activities to be undertaken on a larger scale than would be possible in circumstances of complete equality. The qualification was typical of Carnot's realism. Equality of incomes, as he argued, was like equality between the physical or intellectual gifts of individuals. It was unattainable and hence not worth striving for. What mattered far more was that "men least favoured by fortune should be free from excessive toil and assured of the essentials of life".[24]

The merits of any mechanism for encouraging the circulation of wealth, therefore, had to be judged by the overriding criterion of the benefits that the mechanism yielded for society at large. Circulation, however efficient, counted for nothing in Carnot's eyes if it failed to lessen inequalities of status and fortune and to curb the ills of personal vanity and arbitrary rule. What is the proper function of social institutions?, he asked at one point. To that question, his answer was one of which his father would have been proud. Their function was "to contribute to the happiness of the men for whom they are created. They must guarantee every man a return for his labour, protect the weak against the strong, and so establish, in so far as they are able, the freedom and independence of each member of society".[25] Where liberty and the rule of law held sway (Switzerland, the United States, and Holland were Carnot's examples), prosperity was the norm.[26] Despotism, by contrast, was a curse that brutalized the oppressed, corrupted behaviour, and put culture to flight, in the sciences as in

---

[20] Ibid., 285–289, 292–293, and 294.

[21] The theme dominates the notes in section B$_2$, in Carnot, *Réflexions* (1978 edn.), esp. 297–308. It reappears in the manuscripts in section B$_3$ in Hippolyte's hand (but presumably copied from manuscripts of Sadi's), esp. 311.

[22] For the metaphor of a father's responsibility for his children, see ibid., 285 and 287.

[23] Ibid., 294–295.

[24] Ibid., 295.

[25] Ibid., 300.

[26] The point is made several times in the notes in section B$_2$; ibid., esp. 300–306.

the arts.[27] A contrast between the nobility of ancient Greece, with its wealth of heroes and great men, and the cupidity and ignoble manners of the Greeks now under Turkish rule drove home the point.[28]

I now turn to the far more elusive subject of the analogies between the long-running debates on political economy that I have described and Carnot's ideas on power technology. In doing so, I do not want to advance political economy as a source comparable in importance with contemporary exchanges among engineers and physicists on the theory of the heat engine[29] or with Lazare Carnot's studies of the efficiency of hydraulic machinery, both of which left their mark on the *Réflexions* in ways that Charles Gillispie in particular has explored.[30] Nevertheless, analogies between ways of thinking in economics and power technology did exist, to the point that Carnot must have been conscious that he was reasoning in a similar manner in the two domains.

Take his fundamental notion of the production of power by the circulation of the indestructible entity that he variously called caloric, heat, or fire. In his ideal engine, motive power was yielded in a cycle of isothermal and adiabatic operations in which caloric passed from a high temperature to a low temperature and finally returned to the initial high temperature. In this cyclic process, the heat flowed and returned to its initial state; but, in contrast with thermodynamics as Rudolf Clausius and William Thomson were to formulate it about the mid-century, nothing was consumed. In a strikingly similar way, virtually all economic theorists from the high point of mercantilism in the seventeenth century, through the heyday of the Physiocrats in the eighteenth century, and on into Carnot's day used a model based on the circulation of wealth between the various classes of society or between individuals. Among the relevant classic texts with which Carnot would have been familiar, Quesnay's description of the circulation in the *Tableau économique* (1758) is the best-known,[31] but it was

---

[27] Ibid., 302.

[28] Ibid., 302–303.

[29] On which, see my introduction to Carnot, *Réflexions* (1978 edn.), 16–36 and Fox, "The challenge of a new technology: theorists and the high-pressure steam engine before 1824", in *Sadi Carnot et l'essor de la thermodynamique*, 149–167.

[30] Charles Coulston Gillispie, *Lazare Carnot savant. A monograph treating Carnot's scientific work, with facsimile reproductions of his unpublished writings on mechanics and on the calculus, and an essay concerning the latter by A. P. Youschkevitch* (Princeton, NJ: Princeton University Press, 1971), 90–100 and "The scientific work of Lazare Carnot, and its influence on that of his son", in *Sadi Carnot et l'essor de la thermodynamique*, 23–32. Gillispie identifies parallels between Sadi's conditions for achieving maximum efficiency in a heat engine and Lazare's criteria for maximizing power from water-wheels and other hydraulic machinery. In a water-wheel, for example, Lazare set it as a condition that the water in a millstream should be travelling with the same velocity as the vanes of the wheel, so as to avoid wasteful percussion. See below on Sadi's analogous discussion of the need to avoid a wasteful re-establishment of the equilibrium of caloric.

[31] Although Quesnay's manuscript sketch dates from 1758, the Table was not printed until 1759, when the 12-page "Explication du tableau économique" was also printed with it in a third edition of the *Tableau économique*. For the *Tableau* and its publishing history, see

by no means the only one. As Carnot would have been aware, economists had disagreed ever since Quesnay about the best way of invigorating the cyclic flow. But the model of such a flow was not at issue. The description of national wealth following a "circular motion" in Sismondi's *Nouveaux principes d'économie politique* demonstrates the continued currency of the model in Carnot's day.[32]

An even more specific analogy between political economy and power technology is evident in Carnot's discussions of waste. In his description of the ideal heat engine Carnot warned against the wasteful flow of caloric that occurred whenever heat passed from one temperature to another without being accompanied by the change in the volume of the working substance that yielded power. The example he gave in the *Réflexions* was that of con-duction between a hot and a cooler part of the engine; in such a process, the equilibrium of caloric (to use his language) was re-established, but no mechani-cal benefit ensued.[33] The analogous consideration in Carnot's political econ-omy turned on the waste that followed whenever money changed hands unproductively, as when prodigal landowners frittered away their income from rent on luxuries and unnecessary imported goods or (to cite another of Carnot's examples) when misdemeanours were punished by the costly and unproductive expedient of imprisonment.[34]

In all this, I am doing no more than point to intellectual resources that were available to Carnot. Identifying the recurring analogies of circular-flow and the quest for minimum waste in the domains of political economy and power technology is no substitute for a full explanation of the content of Carnot's theories. With regard to economic theory, for example, I still have to ask the question, to whom specifically did Carnot look as his authority? When I began studying the notes, I expected that there could be only one answer: Jean-Baptiste Say, the disciple of Smith and, from 1819, professor of industrial economy at the Conservatoire des Arts et Métiers in Paris.[35] Carnot is known to have attended lectures at the Conservatoire and to have established such a close friendship with the professor of industrial chemistry there, Nicolas Clément, that it is almost inconceivable that he did not know Say as well.[36] Certainly there are points of similarity between Carnot's ideas and Say's. At a very general level, Say's aspiration for economics to become a science squared with Carnot's (or at least his father's) view of the subject. More specifically, the important

---

François Quesnay et le physiocratie, 2 vols. continuously paginated (Paris: Institut national d'études démographiques, 1958), vol. 2, 667–668 and 675–682, and the plates between 672 and 673.

[32] Sismondi, *Nouveaux principes d'économie politique*, vol. 1, 112.

[33] Ibid., 81–82 and 84, corresponding to *Réflexions* (1824 edn.), 23–24 and 26.

[34] Carnot, *Réflexions* (1978 edn.), 275–276.

[35] Say was appointed as the first holder of the chair in 1819. The designation "industrial economy" appears to have been chosen in preference to "political economy" in an attempt to divert criticism by the reactionary regime of the Bourbon Restoration.

[36] Carnot, *Réflexions* (1978 edn.), 32–36.

distinction between undesirable "unproductive consumption" and profitable "reproductive consumption" in Say's *Traité d'économie politique* is reminiscent of Carnot's discussions of waste in economics (as also in his theory of the heat engine).[37] Even more strikingly, Say's castigation of prodigality as an economic ill that steered profit from agriculture towards expenditure on silks and other luxuries, when it should be spent on the necessities of life or reinvested in production, could almost have been written by Carnot.[38]

But all this can be found in Sismondi too, accompanied by a pervasive humanitarianism akin to Carnot's and a realism far removed from Say's rather detached, academic style of analysis. Sismondi's declared suspicion of "abstractions" and "absolute" solutions conveyed the tenor of his writing, as it did of Carnot's.[39] So too did his eulogy of the benefits of association under a government committed to protecting the weak against the strong.[40] To the question whether governments should exist for the benefit of the mass of citizens or only of some of them, Carnot for his part answered unequivocally that it was for the "governed" and, as a corollary, that at least the more educated in society should have a role in choosing their legislators, military officers, judges, and other senior officials.[41] His ideal entailed a measure of delegation of authority and executive power to those with the necessary experience and skills. But delegation was properly a matter for the people working in an institutionalized form of association in which personal interest would be subservient to the common good. Only in that way could the welfare of the oppressed (and he clearly had the victims of slavery particularly in mind) be assured and advanced. Sismondi would have concurred to the letter.

In conclusion, I have tried not to claim more than my fragmentary sources can bear. While it is possible to identify some of the key intellectual resources on which Carnot may have drawn and to see analogies between contemporary ways of thinking about political economy and Carnot's theory of the heat engine, any more concrete conclusions have to be tempered by the limitations of the evidence. That said, I remain convinced of the broad affinity between Carnot's general approach to political economy and Sismondi's. Admittedly, the case rests more on tone and emphasis than on hard textual evidence, still less on correspondence or known personal contacts. But Carnot and Sismondi were

---

[37] Say, *Traité d'économie politique* (1819 edn.), vol. 1, 226–241, where "consommation improductive" and "consommation reproductive" appear in the titles of chapters 3 and 4 of Book III. The same titles had been used in the first edition (of 1803) and continued to be used in all later editions.

[38] Ibid., vol. 2, 242–263 (in Book III, chapter 5).

[39] Sismondi, *Nouveaux principes d'économie politique*, vol. 1, 288.

[40] See, for example, Sismondi's identification of the overriding goal of government as the advancement of "jouissances nationales" ("national enjoyments", as Richard Hyse's translation has it), in *Nouveaux principes d'économie politique*, vol. 1, 53–55. Cf. also the opening words of the book: "The science of government takes, or should take, as its aim the happiness of men in society" (p. 1).

[41] Carnot, *Réflexions* (1978 edn.), 280, 297–299 and 306–312.

as one in their overriding emphasis on human well-being and in their interest in a controlled distribution of wealth to that end. And both (in contrast with economists working in the manner of the classical school) saw benign government and enlightened interventionist taxation as forces for good against the consequences of the unbridled interplay of private interests.

With regard to the parallels between Carnot's analyses of economic theory and power technology, perhaps only one thing can be said with confidence. This is that as Carnot moved from one domain to the other, he would have recognized, in each of the two realms, the common objective of maximizing output and, as a means to that end, the overriding priority of eliminating waste. In other words, his thinking in one domain constituted a resource on which he may (or may not) have drawn in the other. I doubt whether we can hope to take discussion significantly further. Hippolyte's statement about his brother's engagement with political economy at the same time as he was reflecting on power technology was probably correct. But, the surviving fragmentary manuscripts apart, we have no knowledge of the direction this engagement was taking. If we had, we might well find that Carnot was more interested in the consequences of industry for the French economy of the Restoration than the manuscripts suggest he was. We might even find him exploring the nascent doctrines of Saint-Simon and the Saint-Simonians whom Hippolyte frequented and with whom Carnot himself very probably had some contact. Certainly early passages in the *Réflexions*, where Carnot wrote eloquently of the impact of steam-power in Britain, convey an emphasis on the industrial economy characteristic of leading Saint-Simonians of his generation. But we simply do not have the sources that would allow us to say more. Charles Gillispie, as I have observed, has led the way in identifying sources in power technology. But as far as political economy is concerned, we remain on shaky ground. We remain, in fact, with the Carnot we can reconstruct, however imperfectly, from the limited sources that have come down to us.

Acknowledgements   I am pleased to record my debt to Donald Winch for helpful discussions of Carnot's manuscripts on political economy during our stay at the Institute for Advanced Study, Princeton, where we were both members in 1974–1975. I also have grateful memories of a discussion and subsequent correspondence with Keith Hutchison. Both Donald and Keith will almost certainly have forgotten these exchanges. But this article bears the marks of their comments. Correspondence with Jacques Grinevald too has been important, and I express my equally warm thanks to him.

Index

J.Z. Buchwald (ed.), *A Master of Science History*, Archimedes 30,
DOI 10.1007/978-94-007-2627-7, © Springer Science+Business Media B.V. 2012

Printed by Publishers' Graphics LLC USA
SO20120313-044
2012